Digital Systems Design with FPGAs and CPLDs

Digital Systems Design with FPGAs and CPLDs

Ian Grout

AMSTERDAM • BOSTON • HEIDELBERG • LONDON
NEW YORK • OXFORD • PARIS • SAN DIEGO
SAN FRANCISCO • SINGAPORE • SYDNEY • TOKYO

Newnes is an imprint of Elsevier

Newnes is an imprint of Elsevier
30 Corporate Drive, Suite 400, Burlington, MA 01803, USA
Linacre House, Jordan Hill, Oxford OX2 8DP, UK

♾ Recognizing the importance of preserving what has been written, Elsevier prints its books on acid-free paper whenever possible.

Library of Congress Cataloging-in-Publication Data
Grout, Ian.
  Digital systems design with FPGAs and CPLDs / Ian Grout.
    p. cm.
  Includes bibliographical references and index.
  ISBN-13: 978-0-7506-8397-5 (alk. paper)   1. Digital electronics.   2. Digital circuits — Design
and construction.   3. Field programmable gate arrays.   4. Programmable logic devices.   I. Title.
  TK7868.D5.G76 2008
  621.381—dc22

                                                                                                        2007044907

British Library Cataloguing-in-Publication Data
A catalogue record for this book is available from the British Library.

For information on all Newnes publications
visit our Web site at www.books.elsevier.com

Printed and bound by CPI Group (UK) Ltd, Croydon, CR0 4YY

Transferred to Digital Print 2011

To my family, but especially to my parents and to Jane.

Table of Contents

system

• **noun 1** A set of things working together as parts of a mechanism or an interconnecting network.

Oxford Dictionary of English

Preface

In days gone by, life for the electronic circuit designer seems to have been easier. Designs were smaller, ran at a slower speed, and could easily fit onto a single small printed circuit board. An individual designer could work on a problem and designs could be specified and developed using paper and pen only. The circuit schematic diagrams that were required could be rapidly drawn on the back of an envelope.

Struck by the success of the early circuit designs, customers started to ask for smaller, faster, and more complex circuits—and at a lower cost. The designers started to work on solving such problems, which has led to the rapidly expanding electronics industry that we have today. Driven by the demand from the customer, new materials and fabrication processes have been developed, new circuit design methodologies and design architectures have taken over many of the early traditional design approaches, and new markets for the circuits have evolved.

So how is the design problem tackled today? This is not an easy question to answer, and there is more than one way to develop an electronic circuit solution to any given problem. However, the design process is no longer the activity of a single individual. Rather, a team of engineers is involved in the key engineering activities of design, fabrication (manufacture), and test. All activities now involve the extensive use of computing resources, requiring the efficient use of software tools to aid design (electronic design automation, EDA and computer aided design, CAD), fabrication (Computer Aided Manufacture, CAM), and test (Computer Aided Test, CAT). The circuit is no longer a unique and isolated entity. Rather, it is part of a larger system. Increasingly, much of the design work is undertaken at the system level … at a suitably high level of design abstraction required to reduce design time and increase the designer efficiency. However, when it comes to the design detail, the correctly specified system must also work at the basic electric voltage and current level. How to go from an effective system-level specification to an efficient and working circuit implementation requires the skills of good designers who are aided by good design tools.

For the electronic circuit designer at an early stage in the design process, whether to implement the required circuit functionality using analogue circuit techniques or digital circuit techniques must be decided. However, sometimes the choice will have already been made, and increasingly a digital solution is the preferred choice. The wide use of digital signal processing (DSP) techniques facilitates complex operations that can provide superior performance to an analogue circuit equivalent; indeed some cannot be performed in analogue. Traditionally, DSP functions have been implemented using software programs written to operate on a target processor. The microprocessor (μP), microcontroller (μC), and digital signal processor provide the necessary digital circuits, in integrated circuit (IC) form, to implement the required functions. In fact, these processors are to be found in many everyday embedded electronics that we take for granted. This book could not have been written without the aid of an electronic system incorporating a microprocessor running a software operating system that in turn runs the word processor software.

Increasingly, the functions that have been traditionally implemented in software running on a processor-based digital system in the DSP world and many control applications are being evaluated in terms of performance that can be achieved in software. In many cases, the software solution will be slower than is desired, and the basic nature of the software programmed system means that this speed limitation cannot be overcome. The way to overcome the speed limitation is to perform the required operations in hardware designed for a particular application. However, custom hardware solutions will be expensive to acquire.

If there were a way to obtain the power of programmability with the power of hardware speed, then this would be provide a significant way forward.

Fortunately, programmable logic provides the power of programmability with the power of hardware speed by providing an IC with built-in digital electronic circuitry that is configured by the user for a particular application. Many devices can be reconfigured for different applications. Today, two main types of programmable logic ICs are commonly used: the field programmable gate array (FPGA) and complex programmable logic device (CPLD).

Therefore, it is possible to implement a complex digital system that can be developed and the functionality changed or enhanced using either a processor running a software program or programmable logic with a specific hardware configuration.

For an end-user, the functionality of both types of system will be the same—the design details are irrelevant to the end-user as long as the functionality of the unit is correct. In this book, to provide consistency and to differentiate between the processor and programmable logic, the following terminology will be used:

- A *processor* (microprocessor, μP; microcontroller, μC; or digital signal processor, DSP) will be *programmed* for a particular application using a *software programming language* (SPL).

- *Programmable logic* (field programmable gate array, FPGA; simple programmable logic device, SPLD; or complex programmable logic device, CPLD) will be *configured* using a *hardware description language* (HDL).

The aim of this book is to provide a reference source with worked examples in the area of electronic circuit design using programmable logic. In particular, field programmable gate arrays and complex programmable logic devices will be presented and examples of such devices provided.

The choice whether to use a software-programmed processor or hardware-configured programmable logic device is not a simple one, and many decisions figure into evaluating the pros and cons of a particular implementation prior to making a final decision. This book will provide an insight into the design capabilities and aspects relating to the design decisions for programmable logic so that an informed decision can be made.

The book is structured as follows:

Chapter 1 will introduce the types of programmable logic device that are available today, their differing architectures, and their use within electronic system design. Additionally, the terminology used in this area will be presented with the aim of demystifying the jargon that has evolved.

Chapter 2 will provide a background into the area of electronic systems design, the types of solutions that may be developed, and the decisions that will need to be made in order to identify the right technology choice for the design implementation. Typical design flows will be introduced and discussed for the different technologies.

Chapter 3 will introduce the design of printed circuit boards (PCBs). These provide the mechanical and electrical base onto which the electronic components will be mounted. The correct design of the PCB is essential to ensure that the electronic circuit can be realized (implemented) to operate to the correct specification (power supply voltage, thermal [heat] dissipation, digital clock frequency, analogue and digital circuit elements, etc.) and to ensure that the different electronic circuit components interact with each other correctly and do not provide unwanted effects. A correctly designed PCB will allow the circuit to perform as intended. A badly designed PCB will prevent the circuit from working altogether.

Chapter 4 will discuss the different programming languages that are used to develop digital designs for implementation in either a processor (software-programmed microprocessor, microcontroller, or digital signal processor) or in programmable logic (hardware-configured FPGA or CPLD). The main languages used will be introduced and examples provided. For programmable logic, the main hardware description languages used are Verilog®-HDL and VHDL (VHSIC Hardware Description Language). These are IEEE (Institute of Electrical and Electronics Engineers) standards, universally used in both education and industry.

Chapter 5 will introduce digital logic design principles. A basic understanding of the principles of digital circuit design, such as Boolean Logic, Karnaugh maps, and counter/state machine design will be expected. However, a review of these principles will be provided for designs in schematic diagram form and presented such that the design functionality may be mapped over a VHDL description in Chapter 6.

Chapter 6 will introduce VHDL as one of the IEEE standard hardware description languages available to describe digital circuit and system designs in an ASCII text-based format. This description can be simulated and synthesized. (Simulation will validate the design operation, and synthesis will translate the text-based description into a circuit design in terms of logic gates and the interconnections between the basic logic gates. The gates and gate connections are commonly referred to as the netlist.) The design examples provided in schematic diagram form in Chapter 5 will be revisited and modeled in VHDL.

Chapter 7 will introduce the development of digital signal processing algorithms in VHDL and the synthesis of the VHDL descriptions to target programmable logic (both FPGA and CPLD). Such algorithms include digital filters (low-pass, high-pass, and band-pass), digital PID (proportional plus integral plus derivative) control algorithms, and the FFT (fast Fourier transform, an efficient implementation of the discrete Fourier transform, DFT).

Chapter 8 will discuss the interfacing of programmable logic to what is commonly referred to as the real world. This is the analogue world that we live in, and such interfacing requires both the acquisition (capture) and the generation of analogue signals. To enable this, the digital programmable logic device will require an interface to the analogue world. For analogue signals to be captured and analyzed in digital, an analogue-to-digital converter (ADC) will be required. For analogue signals to be generated from the digital, a digital-to-analogue converter (DAC) will be required.

In this book, the convention used for the word *analogue* will use the -ue at the end of the word, unless a particular name already in use is referred to spelled as *analog*.

Chapter 9 will introduce the testing of the electronic system. In this, failure mechanisms in hardware and software will be introduced, and the need for efficient and cost-effective test programs from the prototyping phase of the design through high-volume manufacture and in-system testing.

Chapter 10 will introduce the increasing need to develop programmable logic–based designs at a high level of abstraction using behavioral descriptions of the system functionality, and the increasing requirements to enable the synthesis of these high-level designs into logic. With reference to a design flow taking a digital design developed in MATLAB® or Simulink® through a VHDL code equivalent for implementation in FPGA or CPLD technology, the synthesis of digital control system algorithms modeled and simulated in Simulink® will be translated into VHDL for implementation in programmable logic.

Throughout the book, the HDL examples provided and evaluated can be implemented within programmable logic–based circuits that may be designed by the user in addition to the PCB design examples that are provided. These examples have been developed to form the basis of laboratory experiments that can be used to accompany the text.

With the broad range of subject material and examples, a feature of the book is its potential for use in a range of learning and teaching scenarios. For example:

1. As an introduction to design of electronic circuits and systems using programmable logic. This would allow for the design approaches, programmable logic architectures, simulation, synthesis, and the final configuration of an FPGA or CPLD to be undertaken. It would also allow for investigation into the most appropriate HDL coding styles and device implementation constraints to be undertaken.

2. As an introduction to hardware description languages, in particular VHDL, allowing for case study designs to be developed and implemented within programmable logic. This would allow for VHDL code developers to see the

code working on real devices and to enable additional testing of the electronic circuit with such equipment as oscilloscopes and spectrum analyzers.

3. As an introduction to the design of printed circuit boards, in particular mixed-signal designs (mixed analogue and digital). This would allow issues relating to the design of the printed circuit board to be investigated and designs developed, fabricated, and tested.

4. As an introduction to digital signal processing algorithm development. This would allow the basics of DSP algorithms and their implementation in hardware on FPGAs and CPLDs to be investigated through the medium of VHDL code development, simulation, and synthesis.

The VHDL examples can be downloaded and run on the hardware prototyping arrangement that can be built by the reader using the designs provided in the book. This hardware arrangement is centered on a Xilinx® Coolrunner™-II CPLD on which to prototype the digital logic ideas, along with a set of input/output (I/O) boards. The full set of boards is shown in the figure below.

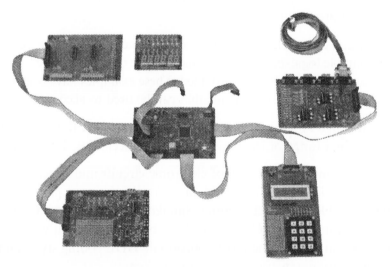

This arrangement consists of five main system boards and an optional seven-segment display board. The appendices and design schematics are available at the author's Web site for this book (refer to http://books.elsevier.com/companions/ 9780750683975 and follow the hyperlink to the author's site).

Abbreviations

A

AC	alternating current
ADC	analogue-to-digital converter
ALU	arithmetic and logic unit
AM	amplitude modulation
AMD	advanced micro devices
AMS	analogue and mixed-signal
AND	logical AND operation on two or more digital signals
ANSI	American National Standards Institute
AOI	automatic optical inspection
ASCII	American Standard Code for Information Interchange
ASIC	application-specific integrated circuit
ASP	analogue signal processor
ASSP	application-specific standard product
ATA	AT attachment
ATE	AT equipment
ATPG	AT program generation
AWG	arbitrary waveform generator
	American wire gauge
AXI	automatic X-ray inspection

B

BASIC	Beginner's All-purpose Symbolic Instruction Code
BCD	binary coded decimal
BGA	ball grid array
BiCMOS	bipolar and CMOS
BIST	built-in self-test

bit	*b*inary dig*it*
BJT	bipolar junction transistor
BNC	bayonet Neill-Concelman connector
BPF	band-pass filter
BSDL	boundary scan description language
BS(I)	British Standards (Institution)
BST	boundary scan test

C

CAD	computer-aided design
CAE	computer-aided engineering
CAM	computer-aided manufacture
CAT	computer-aided test
CBGA	ceramic BGA
CD	compact disk
CE	chip enable
CERDIP	ceramic DIP
CERQUAD	ceramic quadruple side
CIC	cascaded integrator comb
CISC	complex instruction set computer
CLB	configurable logic block
CLCC	ceramic leadless chip carrier
	ceramic leaded chip carrier
CMOS	complementary metal oxide semiconductor
COTS	commercial off-the-shelf
CPGA	ceramic PGA
CPLD	complex PLD
CPU	central processing unit
CQFP	ceramic quad flat pack
CS	chip select
CSOIC	ceramic SOIC
CSP	chip scale packaging
CSSP	customer specific standard product
CTFT	continuous-time Fourier transform
CTS	clear to send
CUT	circuit under test

D

DAC	digital-to-analogue converter
DAE	differential and algebraic equation
DAQ	data acquisition
dB	decibel
DBM	digital boundary module
DC	direct current
DCD	data carrier detected
DCE	data communication equipment
DCI	digitally controlled impedance
DCPSS	DC power supply sensitivity
DDC	direct digital control
DDR	double data rate
DDS	direct digital synthesis
DfA	design for assembly
DfD	design for debug
DFF	D-type flip-flop
DfM	design for manufacturability
DfR	design for reliability
DfT	design for testability
DFT	discrete Fourier transform
DfX	design for X
DfY	design for yield
DIB	device interface board
DIL	dual in-line
DIMM	dual in-line memory module
DIP	dual in-line package
DL	defect level
DMM	digital multimeter
DNL	differential nonlinearity
DoD	U.S. Department of Defense
DPLL	digital PLL
dpm	defects per million
DR	data register
DRAM	dynamic RAM
DRC	design rules checking

DRDRAM	direct Rambus DRAM
DSM	deep submicron
DSP	digital signal processing
	digital signal processor
DSR	data set ready
DTE	data terminal equipment
DTFT	discrete-time Fourier transform
DTR	data terminal ready
DUT	device under test
DVD	digital versatile disk

E

EC	European Commission
ECL	emitter coupled logic
ECU	electronic control unit
EDA	electronic design automation
EDIF	electronic design interchange format
EHF	extremely high frequency
EIAJ	Electronic Industries Association of Japan
ELF	extremely low frequency
EMC	electromagnetic compatibility
EMI	electromagnetic interference
ENB	effective number of bits
EOC	end of conversion
EOS	electrical overstress
EEPROM	electrically erasable PROM
E^2EPROM	electrically erasable PROM
EPROM	erasable PROM
ERC	electrical rules checking
ESD	electrostatic discharge
ESIA	European Semiconductor Industry Association
ESL	electronic system level
ESS	environmental stress screening
EU	European Union
EX-NOR	NOT-EXCLUSIVE-OR
EX-OR	logical EXCLUSIVE-OR operation on two or more digital signals

F

F	Farad
FA	failure analysis
FBGA (FPBGA)	fine pitch ball grid array
FCC	Federal Communications Commission (USA)
FET	field effect transistor
FFT	fast Fourier transform
FIFO	first-in, first-out
FIR	finite impulse response
FM	frequency modulation
FPAA	field programmable analogue array
FPGA	field programmable gate array
FPT	flying probe tester
FR-4	flame retardant with approximate dielectric constant of 4
FRAM	ferromagnetic RAM
FSM	finite state machine
FT	functional tester

G

GaAs	gallium arsenide
GAL	generic array of logic
GDSII	Graphic Data System II stream file format
GND	ground
GPIB	general purpose interface bus
GTL	Gunning transceiver logic
GTO	gate turn-off thyristor
GUI	graphical user interface

H

HBM	human body model
HBT	heterojunction bipolar transistor
HDIP	hermetic DIP
HDL	hardware description language
HF	high frequency
HPF	high-pass filter
HSTL	high-speed transceiver logic
HTML	hyphertext markup language

HVI	human visual inspection
HW	hardware
Hz	Hertz
I	
I_B	base current
I_{BM}	base peak current
I_C	collector current
I_{CC}	power supply current (into V_{CC} pin for bipolar circuits)
I_{CM}	collector peak current
I_{DD}	power supply current (into V_{DD} pin for CMOS circuits)
I_{DDQ}	quiescent power supply current (I_{DD})
I_{EE}	power supply current (out of V_{EE} pin for bipolar circuits)
I_{FS}	full-scale current
I_{GND}	ground current per supply pin
I_{IH}	high-level input current
I_{IL}	low-level input current
I_{LSB}	minimum output current change
I_O	output current
I_{OH}	high-level output current (logic 1 output)
I_{OL}	low-level output current (logic 0 output)
I_{OS}	offset current
I_{OUT}	output current
I_{REF}	reference current
I_{SS}	power supply current (out of V_{SS} pin for CMOS circuits)
I_{SSQ}	quiescent power supply current (I_{SS})
IC	integrated circuit
I^2C (IIC)	inter-integrated circuit (inter-IC) bus
I^2S	inter-IC sound bus
ICT	in-circuit test
	in-circuit tester
IDC	insulation displacement connector
IDE	integrated design environment
	integrated drive electronics
IEC	International Electrotechnical Commission
IEE	Institution of Electrical Engineers
IEEE	Institute of Electrical and Electronics Engineers

IET	Institution of Engineering and Technology
IIR	infinite impulse response
IMAPS	International Microelectronics and Packaging Society
INL	integral nonlinearity
I/O	input/output
IP	intellectual property
IR	instruction register
	infrared
ISO	International Organization for Standardization
ISP	in-system programmable
ISR	in-system reprogrammable
IT	information technology
ITRS	International Technology Roadmap for Semiconductors
I-V	current-to-voltage

J

JDK	JAVA^{TM} Development Kit
JEDEC	Joint Electron Device Engineering Council
JEITA	Japan Electronics and Information Technology Industries Association
JETAG	Joint European Test Action Group
JETTA	Journal of Electronic Testing, Theory, and Applications
JFET	junction FET
JLCC	J-leaded chip carrier
JTAG	Joint Test Action Group

K

KGD	known good die
KSIA	Korean Semiconductor Industry Association

L

LAN	local area network
LC	logic cell
LC^{2}MOS	linear compatible CMOS
LCC	leaded chip carrier
	leadless chip carrier
LCCMOS	leadless chip carrier metal oxide semiconductor (also LC^{2}MOS)
LCD	liquid crystal display
LED	light-emitting diode

LF	low frequency
LFSR	linear feedback shift register
LIFO	last-in, first-out
Linux®	Linux is not Unix
LPF	low-pass filter
LSB	least significant bit
LSI	large-scale integration
LUT	look-up table
LVCMOS	low-voltage CMOS
LVDS	low-voltage differential signaling
LVS	layout versus schematic
LVTTL	low-voltage TTL

M

µBGA	micro ball grid array
µC	microcontroller
µP	microprocessor
MATLAB®	Matrix Laboratory (from The Mathworks, Inc.)
MAX	maximum
MCM	multichip module
MCU	microcontroller unit
MEMs	micro electro-mechanical systems
MF	medium frequency
MIL	military
MIN	minimum
MISR	multiple-input signature register
MM	machine model
MOS	metal oxide semiconductor
MOSFET	metal oxide semiconductor field effect transistor
MPGA	mask programmable gate array
MS	Microsoft®
MSAF	multiple stuck-at-fault
MSB	most significant bit
MSI	medium-scale integration
MSOP	mini-small outline package
MUX	multiplexer
MVI	manual visual inspection (i.e., HVI)

N

NAND	NOT-AND
NDI	normal data input
NDO	normal data output
NDT	nondestructive test
NM_H	noise margin for high levels
NM_L	noise margin for low levels
nMOS	n-channel MOS
NOR	NOT-OR
NOT	logical NOT operation on a single digital signal
NRE	nonrecurring engineering
NVM	nonvolatile memory
NVRAM	nonvolatile RAM

O

OE	output enable
OEM	original equipment manufacturer
ONO	oxide-nitride-oxide
OOP	object-oriented programming
op-amp	operational amplifier
OR	logical OR operation on two or more digital signals
OS	operating system
OSR	oversampling ratio
OTP	one-time programmable
OVI	Open Verilog International

P

P_{tot}	total dissipation
PAL®	programmable array of logic
PBGA	plastic BGA
PC	personal computer
	program counter
PCB	printed circuit board
PCBA	printed circuit board assembly
PCI	PC interface
PDA	personal digital assistant

PDF	portable document format
PDIL	plastic DIL
PDIP	plastic DIP
PERL	practical extraction and report language
PGA	pin grid array
PI	primary input
	proportional plus integral
PID	proportional plus integral plus derivative
PIPO	parallel in, parallel out
PLA	programmable logic array
PLCC	plastic leadless chip carrier
	plastic leaded chip carrier
PLD	programmable logic device
PLL	phase-locked loop
PM	phase modulation
pMOS	p-channel MOS
PMU	precision measurement unit
PO	primary output
PoC	proof of concept
PoP	package on package
POR	power-on reset
PPGA	plastic PGA
ppm	parts per million
PQFP	plastic QFP
PROM	programmable ROM
PRPG	pseudorandom pattern generator
PSOP	plastic SOP
PWB	printed wiring board
PWM	pulse width modulation
	pulse width modulated
PXI	PC extensions for instrument bus

Q

QFJ	quad flat pack (J-lead)
QFP	quad flat pack
QSOP	quarter-size SOP
QTAG	Quality Test Action Group

R

®	trademark (registered; ™ for unregistered)
RAM	random access memory
RC	resistor-capacitor
RD	read
	received data
RF	radio frequency
RI	ring indicator
RISC	reduced instruction set computer
RMS	root mean squared
RoHS	return of hazardous substances
ROM	read-only memory
RTL	register transfer level
RTOS	real-time operating system
RTS	ready to send
RWM	read-write memory (also referred to as RAM)
Rx	receiver

S

ΣΔ	sigma-delta
SA0	stuck-at-0
SA1	stuck-at-1
SAF	stuck-at-fault
SAR	successive approximation register
SCR	silicon-controlled rectifier
SCSI	small computer system interface
SDRAM	synchronous DRAM
SDI	scan data input
SDO	scan data out
SE	scan enable
SFDR	spurious free dynamic range
SG	signal ground
SHF	super high frequency
SI	signal integrity
SIA	Semiconductor Industries Association

SiGe	silicon germanium
SIM	subscriber identity module
SINAD	signal to noise plus distortion (SNR + THD)
SiP	system in a package
SIP	single in-line package
SIPO	serial in, parallel out
SISO	Serial in, serial out
	Single input, single output
SISR	serial input signature register
SLDRAM	synchronous-link DRAM
SMT	surface mount technology
SNR	signal-to-noise ratio
S/(N + THD)	signal to noise plus total harmonic distortion
SOAR	safe operating region
SoB	system on board
SoC	system on a chip
SOC	start of conversion
SOI	silicon on insulator
SOIC	small outline IC
SOJ	small outline J-lead package
SOP	small outline package
SPGA	staggered PGA
SPI	serial peripheral interface
SPICE	simulation program with integrated circuit emphasis
SPL	software programming language
SPLD	simple PLD
SQFP	shrink quad flat pack
SRAM	static RAM
SRBP	synthetic resin-bonded paper
SSAF	single stuck-at-fault
SSI	small-scale integration
SSOP	small shrink outline package
SSTL	stub series terminated logic
STC	Semiconductor Test Consortium
STD	standard
STIL	standard test interface language
SW	software

T

T_L	lead temperature
T_{stg}	storage temperature
TAB	tape automated bonding
TAP	test access port
TCE	thermal coefficient of expansion
TCK	test clock
Tcl	tool command language
TD	transmitted data
TDI	test data input
TDO	test data output
THD	total harmonic distortion
™	trademark (unregistered, ® for registered)
TMS	test mode select
TO	transistor outline package (single transistor)
TPG	test program generation
TQFP	thin QFP
TRST	test reset
TSIA	Taiwan Semiconductor Industry Association
TSMC	Taiwan Semiconductor Manufacturing Company
TSOP	thin SOP
TSSOP	thin shrink SOP
TVSOP	thin very SOP
TTL	transistor-transistor logic
TTM	time to market
TYP	typical
Tx	transmitter

U

UART	universal asynchronous receiver/transmitter
UHF	ultra high frequency
UJT	unijunction transistor
ULSI	ultra large-scale integration
UML	unified modeling language
UNIX™	Uniplexed Information and Computing System (originally Unics, later renamed Unix)
USB	universal serial bus

UTP	unit test period
UUT	unit under test
UV	ultraviolet

V

V_{CB}	collector-base voltage
V_{CC}	power supply voltage (positive, for bipolar circuits)
V_{CE0}	collector-emitter voltage ($I_E = 0$)
V_{CEV}	collector-emitter voltage ($V_{BE} = -1.5$)
V_{DD}	power supply voltage (positive, for CMOS circuits)
V_{EB}	emitter-base voltage
V_{EE}	power supply voltage (negative, for bipolar circuits)
V_{FS}	full-scale voltage
V_{FSR}	full-scale range of voltage
V_I	input voltage
V_{IH}	minimum input voltage that can be interpreted as a logic 1
V_{IL}	maximum input voltage that can be interpreted as a logic 0
V_{LSB}	minimum output voltage change
V_O	output voltage
V_{OH}	minimum output voltage when the output is a logic 1
V_{OL}	maximum output voltage when the output is a logic 0
V_{OS}	offset voltage
V_{OUT}	output voltage
V_{REF}	reference voltage
V_{SS}	power supply voltage (negative, for CMOS circuits)
VASG	VHDL Analysis and Standardization Group
VB	Visual Basic™
VBA	Visual Basic™ for Applications
VCO	voltage-controlled oscillator
VDSM	very deep submicron
VDU	visual display unit
VF	voice frequency
VHDL	VHSIC hardware description language
VHF	very high frequency
VHSIC	very high-speed integrated circuit
VLF	very low frequency

VLSI	very large-scale integration
VQFP	very thin quad flat pack

W

WE	write enable
WEEE	waste electrical and electronic equipment
WR	write
WSI	wafer-scale integration

X

XNF	Xilinx Netlist format

Z

ZIF	zero insertion force socket
ZIP	zig-zag in-line package

CHAPTER 1

Introduction to Programmable Logic

1.1   Introduction to the Book

Increasingly, electronic circuits and systems are being designed using technologies that offer rapid prototyping, programmability, and re-use (reprogrammability and component recycling) capabilities to allow a system product to be developed in a minimal time, to allow in-service reconfiguration (for normal product upgrading to improve performance, to provide design debugging capabilities, and for the inevitable requirement for design bug removal), or even to recycle the electronic components for another application. These aspects are required by the reduced time-to-market and increased complexities for applications—from mobile phones through computer and control, instrumentation, and test applications. So, how can this be achieved using the range of electronic circuit technologies available today? Several avenues are open. The main focus of developing electronics with the above capabilities has been in the digital domain because the design techniques and nature of the digital signals are well suited to reconfiguration.

In the digital domain, the choice of implementation technology is essentially whether to use dedicated (and fixed) functionality digital logic, to use a software-programmed, processor-based system (designed based on a microprocessor, µP; microcontroller, µC; or digital signal processor, DSP), or to use a hardware-configured programmable logic device (PLD), whether simple (SPLD), complex (CPLD), or the field programmable gate array (FPGA). Memory used for the storage of data and program code is integral to many digital circuits and systems. The choices are shown in Figure 1.1.

In Figure 1.1, the electronic components used are integrated circuits (ICs). These are electronic circuits packaged within a suitable housing that contain complete circuits ranging from a few dozen transistors to hundreds of millions of transistors, the

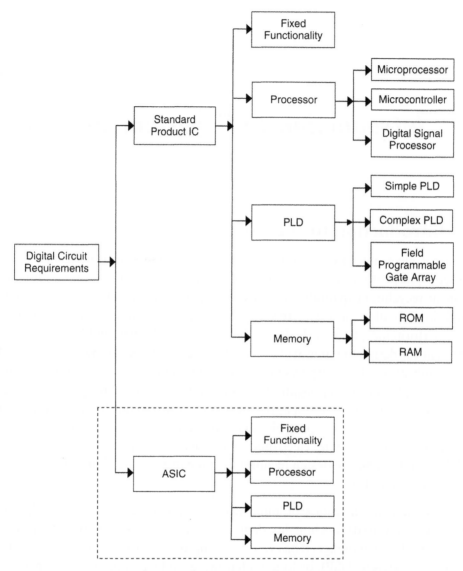

Figure 1.1: Technology choices for digital circuit design

complexity of the circuit depending on the designed functionality. Examples of packaged ICs are shown in Figure 1.2.

In many circuits, the underlying technology will be based on IC, and a complete electronic circuit will consist of a number of ICs, together with other circuit

Figure 1.2: Examples of IC packages with the tops removed and the silicon dies exposed

components such as resistors and capacitors. In this book, the generic word *technology* will be used throughout. The *Oxford Dictionary of English* defines *technology* as "the application of scientific knowledge for practical purposes, especially in industry" [1].

For us, this applies to the underlying electronic hardware and software that can be used to design a circuit for a given requirement. For the arrangement identified in Figure 1.1, a given set of *digital circuit requirements* are developed, and the role of the designer is to come up with a solution that meets ideally all of the requirements. Typical requirements include:

- **Cost restraints**: The design process, the cost of components, the manufacturing costs, and the maintenance and future development costs must be within specific limits.

- **Design time**: The design must be generated within a certain time limit.

- **Component supply**: The designer might have a free hand in choosing the components to use, or restrictions may be set by the company or project management requirements.

- **Prior experience**: The designer may have prior experience in using a particular technology, which might or might not be suitable to the current design.

- **Training**: The designer might require specific training to utilize a specific technology if he or she does not have the necessary prior experience.

- **Contract arrangements**: If the design is to be created for a specific customer, the customer would typically provide a set of constraints that would be set down in the design contract.

- **Size/volume constraints**: the design would need to be manufactured to fit into a specific size/volume,

- **Weight constraints**: the design would need to be manufactured to be within specific weight restrictions (e.g. for portable applications such as mobile phones),

- **Power source**: the electronic product would be either fixed (in a single location so allowing for the use of a fixed power source) or portable (to be carried to multiple places requiring a portable power source (such as battery or solar cell),

- **Power consumption constraints**: The power consumption should be as low as possible in order to (i) minimise the power source requirements, (ii) be operable for a specific time on a limited power source, and (iii) be compatible with best practice in the development of electronic products that are conscious of environmental issues.

The initial choice for implementing the digital circuit is between a standard product IC and an ASIC (application-specific integrated circuit) [2]:

- **Standard product IC**: This is an off-the-shelf electronic component that has been designed and manufactured by a company for a given purpose, or range of use, and that is commercially available for others to use. These would be purchased either from a component supplier or directly from the designer or manufacturer.

- **ASIC**: This is an IC that has been specifically designed for an application. Rather than purchasing an off-the-shelf IC, the ASIC can be designed and manufactured to fulfil the design requirements.

For many applications, developing an electronic system based on standard product ICs would be the approach taken as the time and costs associated with ASIC design, manufacture, and test can be substantial and outside the budget of a particular design project. Undertaking an ASIC design project also requires access to IC design experience and IC CAD tools, along with access to a suitable manufacturing and test capability. Whether a standard product IC or ASIC design approach is taken, the type of IC used or developed will be one of four types:

1. **Fixed Functionality**: These ICs have been designed to implement a specific functionality and cannot be changed. The designer would use a set of these ICs to implement a given overall circuit functionality. Changes to the circuit would require a complete redesign of the circuit and the use of different fixed functionality ICs.

2. **Processor**: The processor would be more familiar to the majority of people as it is in everyday use (the heart of the PC is a microprocessor). This component runs a software program to implement the required functionality. By changing the software program, the processor will operate a different function. The choice of processor will depend on the microprocessor (μP), the microcontroller (μC), or the digital signal processor (DSP).

3. **Memory**: Memory will be used to store, provide access to, and allow modification of data and program code for use within a processor-based electronic circuit or system. The two basic types of memory are ROM (read-only memory) and RAM (random access memory). ROM is used for holding program code that must be retained when the memory power is removed. It is considered to provide *nonvolatile storage*. The code can either be fixed when the memory is fabricated (mask programmable ROM) or electrically programmed once (PROM, Programmable ROM) or multiple times. Multiple programming capacity requires the ability to erase prior programming, which is available with EPROM (electrically programmable ROM, erased using ultraviolet [UV] light), EEPROM or E^{2}PROM (electrically erasable PROM), or flash (also electrically erased). PROM is sometimes considered to be in the same category of circuit as programmable logic, although in this text, PROM is considered in the memory category only. RAM is used for holding data and program code that require fast access and the ability to modify the contents during normal operation. RAM differs from read-only memory (ROM) in that it can be both read from and written

to in the normal circuit application. However, flash memory can also be referred to as nonvolatile RAM (NVRAM). RAM is considered to provide a *volatile storage*, because unlike ROM, the contents of RAM will be lost when the power is removed. There are two main types of RAM: static RAM (SRAM) and dynamic RAM (DRAM).

4. **PLD**: The programmable logic device is the main focus of this book; these are ICs that contain digital logic cells and programmable interconnect [3, 4]. The basic idea with these devices is to enable the designer to configure the logic cells and interconnect to form a digital electronic circuit within a single packaged IC. In this, the hardware resources will be configured to implement a required functionality. By changing the hardware configuration, the PLD will operate a different function. Three types of PLD are available: the simple programmable logic device (SPLD), the complex programmable logic device (CPLD), or the field programmable gate array (FPGA). Figure 1.3 shows sample packaged CPLD and FPGA devices.

Figure 1.3: Sample FPGA and CPLD packages

Both the processor and PLD enable the designer to implement and change the functionality of the IC by changing either the software program or the hardware configuration. Because these two different approaches are easily confused, in this book the following terms will be used to differentiate the PLD from the processor:

- The PLD will be configured using a hardware configuration.

- The processor will be programmed using a software program.

An ASIC can be designed to create any one of the four standard product IC forms (fixed functionality, processor, memory, or PLD). An ASIC would be designed in the same manner as a standard product IC, so anyone who has access to an ASIC design, fabrication, and test facility can create an equivalent to a standard product IC (given that patent and general legal issues around IP [intellectual property] considerations for existing designs and devices are taken into account). In addition, an ASIC might also incorporate a programmable logic fabric alongside the fixed logic hardware.

Figure 1.1 shows what can be done with ASIC solution, but not how the ASIC would achieve this. Figure 1.4 shows the (i) four different forms of IC (i.e., what the IC does) that can be developed to emulate a standard product IC equivalent, and (ii) the three different design and implementation approaches.

In a *full-custom* approach, the designer would be in control of every aspect of ASIC design and layout—the way in which the electronic circuit is laid out on the die, which is the piece of rectangular or square material (usually silicon) onto

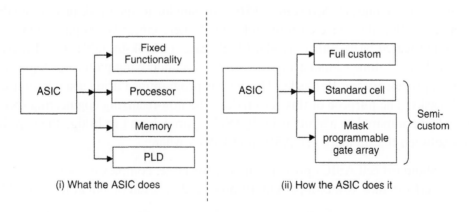

Figure 1.4: ASICs, what and how

which the circuit components are manufactured. This would give the best circuit performance, but would be time consuming and expensive to undertake. Full-custom design is predominantly for analogue circuits and the creation of libraries of components for use in a semi-custom, standard cell design approach. An alternative to the full-custom approach uses a semi-custom approach. This is subdivided into a *standard cell* approach or *mask programmable gate array* (MPGA) approach. The standard cell approach uses a library of predesigned basic circuit components (typically digital logic cells) that are connected within the IC to form the overall circuit. In a simplistic view, this would be similar to creating a design by connecting fixed functionality ICs together, but instead of using multiple ICs, a single IC is created. This approach is faster and lower cost than a *full-custom* approach but would not necessarily provide the best circuit performance. Because only the circuits required within the design would be manufactured (fabricated), there would be an immediate trade-off between circuit performance, design time, and design cost (a trade-off that is encountered on a daily basis by the designer). The MPGA approach is similar to a standard cell approach in that a library of components is available and connected, but the layout on the (silicon) die is different. An array of logic gates is predetermined, and the circuit is created by creating metal interconnect tracks between the logic gates. In the MPGA approach, not necessarily all of the logic gates fabricated on the die would be used. This would use a larger die than in a standard cell approach, with the inclusion of unused gates, but it has the advantage of being faster to fabricate than a standard cell approach.

A complement to the ASIC is the *structured ASIC* [16, 17]. The *structured ASIC* is seen to offer a promising alternative to standard cell ASICs and FPGAs for the mid and high volume market. *Structured ASICs* are similar to the mask programmable gate array in that they have customisable metal interconnect layers patterned on top of a prefabricated base. Either standard logic gates or look-up tables (LUTs) are fabricated in a 2-dimensional array that forms the underlying pattern of logic gates, memory, processors and IP blocks. This base is programmed using a small number of metal masks. The purpose of this is to reduce the non-recurring engineering (NRE) costs when compared to a standard cell ASIC approach and to bridge the gap that exists between the standard cell ASIC and FPGA where:

1. Standard cell ASICs provide support for large, complex designs with high performance, low cost per unit (if produced in volume), but at the cost of long

development times, high NRE costs and long fabrication times when implementing design modifications,

2. FPGAs provide for short development times, low NRE costs and short times to implement design modifications, but at the cost of limited design complexities, performance limitations and high cost per unit.

NRE cost reductions using *Structured ASICs* are considered with a reduction in manufacturing costs and reducing the design tasks. They can also offer mixed-signal circuit capability, a potential advantage when compared to digital only FPGAs.

Hardware configured devices (i.e., PLDs) are becoming increasingly popular because of their potential benefits in terms of logic replacement potential (obsolescence), rapid prototyping capabilities, and design speed benefits in which PLD-based hardware can implement the same functions as a software-programmed, processor-based system, but in less time. This is particularly important for computationally expensive mathematical operations such as the fast Fourier transform (FFT) [5].

The aim of this book is to provide a reference text for students and practicing engineers involved in digital electronic circuit and systems design using PLDs. The PLD is digital in nature and this type of device will be the focus of the book. However, it should also be noted that mixed-signal programmable devices have also been developed and are available for use within mixed-signal circuits that require programmable analogue circuit (e.g. programmable analogue amplifier) components. Whilst this technology is not covered in this book, the reader is recommended to undertake their own research activities to (i) identify the programmable mixed-signal devices currently available (such as the Lattice® Semiconductor Corporation ispPAC and Anadigm™ FPAA (Field Programmable Analog Array)), and also (ii) the history of programmable mixed-signal and devices that have been available in the past but no longer available. The text will introduce the basic concepts of programmable logic, along with case study designs in a range of electronic systems that target signal generation and data acquisition systems for a variety of applications from control and instrumentation through test equipment systems. To achieve this, a range of FPGA and CPLD device types will be considered. The text will also act as a reference from which the sources of additional information can be acquired.

1.2   Electronic Circuits: Analogue and Digital

1.2.1   Introduction

Before looking into detail of what the PLD is and how to use it, it is important to identify that the PLD is digital in nature, and digital circuits and signals are different from analogue circuits and signals. This section will provide an overview of the main characteristics and differences between the continuous- and discrete-time, and the analogue and digital, worlds.

1.2.2   Continuous Time versus Discrete Time

Electronic circuits will receive electrical signals (voltages and/or currents) and modify these to produce a response, which will be a voltage and/or current that is a modified version of the input signal (see Figure 1.5). The signal will be electrical in nature and will convey information concerning the behavior of the related system. The *input* to the *system* will typically be created by a variation of a measurable quantity by the use of a suitable sensor. The *response* will be a modified version of the input that is in a form that can be used. In Figure 1.5, an electronic system receives an input, *x*, and produces a response (output), *y*. The system implements a certain function that is designed to undertake an operation that is of a particular use within the context of the overall system.

Here, the system receives a single input and produces a single response. The term *system* is another generic term which is defined in the *Oxford Dictionary of English* as "a set of things working together as parts of a mechanism or an interconnecting network" [1].

For us, this applies to the overall set of electronic components and software programs that work together to perform the particular set of requirements. In general, there may be one or more inputs and one or more outputs. The system is shown as a *black box* in that the details of its internal operation are hidden and only the *input-output relationship* is known. This black box creates a *signal processor*, and the designer is tasked with

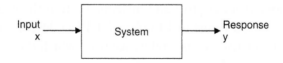

Figure 1.5: Electronic system block diagram

creating the internal details using a suitable electronic circuit technology. The input-output relationship will normally be modeled by a suitable mathematical algorithm. The type of signal [6, 7] that the signal processor accepts and responds to will vary in time but will be classified as either a continuous-time or a discrete-time signal.

A *continuous-time signal* can be represented mathematically as a function of a continuous-time variable. The signal varies in time but is also continuous in time. Figure 1.6 provides four examples of continuous-time signals: (i) a constant value, (ii) a sine wave, (iii) a square wave, and (iv) an arbitrary waveform. Waveforms (i), (ii), and (iv) are continuous in both time and amplitude; (iii) is continuous in time but discontinuous in amplitude. All signals are classified as continuous-time signals.

A *discrete-time signal* is defined only by values at set points in time, referred to as the *sampling instants*. It is normal to set the time spacing between the sampling instants to a fixed value, T, referred to as the *sampling interval*. The *sampling frequency* is $f_S = 1/T$, where T is seconds and f_S is Hertz (Hz). When a signal is sampled at a fixed rate, this is referred to as *periodic sampling*. Figure 1.7 provides examples of discrete-time signals that are sampled values of the continuous time signals shown in Figure 1.6.

When a discrete-time signal is expressed, it will normally be expressed by the sample number (n) where $n = 0$ denotes the first sample, $n = p$ denotes the pth sample, and n increments in steps of 1. For a signal x, then, the samples will be x[0], x[1], x[2], x[3], ..., x[p]. A discrete-time signal would represent a sampled analogue signal. Hence, an electronic circuit would have continuous-time or discrete-time inputs and continuous-time or discrete-time outputs as represented in Table 1.1.

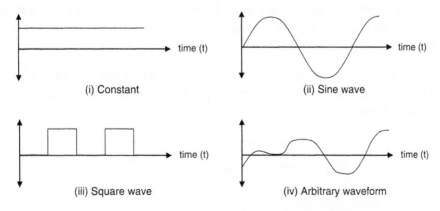

Figure 1.6: Examples of continuous-time signals

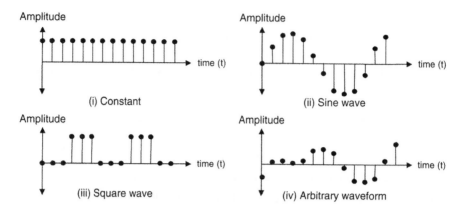

Figure 1.7: Examples of discrete-time signals

Table 1.1: Signal types (continuous- and discrete-time)

Input signal type		Response signal type
Continuous-time	$\longrightarrow$	Continuous-time
Continuous-time	$\longrightarrow$	Discrete-time
Discrete-time	$\longrightarrow$	Discrete-time
Discrete-time	$\longrightarrow$	Continuous-time

1.2.3 Analogue versus Digital

The electronic system as shown in Figure 1.8 will perform its operations on signals that are either analogue or digital in nature, using either analogue or digital electronic circuits. Hence, a signal may be of one of two types, analogue or digital.

An *analogue signal* is a continuous- or discrete-time signal whose amplitude is continuous in value between a lower and upper limit, but may be either a continuous time or discrete time.

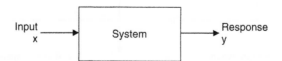

Figure 1.8: Electronic system block diagram

Table 1.2: Signal types (analogue and digital)

Input signal type		Response signal type
Analogue	→	Analogue
Analogue	→	Digital
Digital	→	Digital
Digital	→	Analogue

A *digital signal* is a continuous or discrete-time signal with discrete values between a lower and upper limit. These discrete values will be represented by numerical values and be in a form suitable for *digital signal processing*. If the discrete-time signal has been derived from a continuous-time signal by sampling, then the sampled signal is converted into a digital signal by *quantization*, which produces a finite number of values from a continuous amplitude signal. It is common to use the binary number (i.e., two values, 0 or 1) system to represent a number in a digital representation.

An electronic circuit would have analogue or digital inputs and analogue or digital outputs as represented in Table 1.2. When an analogue signal is sampled and converted to digital, this is undertaken using an analogue-to-digital converter (ADC) [8]. When a digital signal is converted back to analogue, this is undertaken using a digital-to-analogue converter (DAC).

An example of both analogue and digital signals and circuits is shown in Figure 1.9. This electronic temperature controller, as might be used in a home

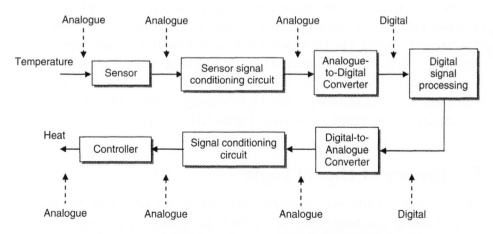

Figure 1.9: Heating control system block diagram

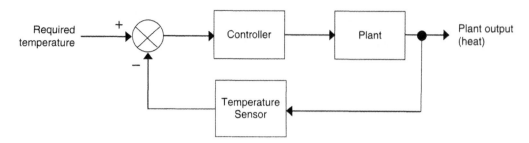

Figure 1.10: General control system

heating system, uses digital signal processing. The system is shown as a *block diagram* in which each block represents a major operation. In a design each block would be represented by its own *block diagram*, going into evermore detail until the underlying circuit hardware (and software) details are identified. The block diagram provides a convenient way to represent the major system operations called a *top-down* design approach, starting at a high level of design abstraction (initially independent of the final design implementation details) and working down to the final design implementation details.

Here, the room temperature is sensed as an analogue signal, but must be processed by a digital signal processing circuit, so it must be sensed and converted to an analogue voltage or current. This is then applied to a sensor signal conditioning circuit that is used to connect the sensor to the ADC. The ADC samples the analogue signal at a chosen sampling frequency. Once a temperature sample has been obtained by the digital signal processing circuit, it is then processed using a particular algorithm, and the result is applied to a DAC. The DAC output is a voltage or current that is used to drive a controller (heat source). The DAC is normally connected to the controller via a signal conditioning circuit. This circuit acts to interface the DAC to the controller in order for the controller to receive the correct voltage and current levels. This particular system is also an example of a closed-loop control system using an electronic controller. The control system is generalized as shown in Figure 1.10 [9, 10].

1.3   History of Digital Logic

Early electronic circuits were analogue, and before the advent of digital logic, signal processing was undertaken using analogue electronic circuits. The invention of the semiconductor transistor in 1947 at Bell Laboratories [11] and the improvements in transistor characteristics and fabrication during the 1950s led to the introduction of linear (analogue) ICs and the first transistor-transistor logic (TTL) digital logic IC in the early 1960s, closely followed by complementary metal oxide semiconductor (CMOS) ICs. The early devices incorporated a small number of logic gates. However, rapid growth in the ability to fabricate an increasing number of logic gates in a single IC led to the microprocessor in the early 1970s. This, with the ability to create memory ICs with ever increasing capacities, laid the foundation for the rapid expansion in the computer industry and the types of complex digital systems based on the computer architecture that we have available today. The last fifty years have seen a revolution in the electronics industry.

Fundamentally, a digital circuit will be categorized into one of three general types, each of which is created and fabricated within an integrated circuit:

- **Combinational logic**, in which the response of the circuit is based on a Boolean logic expression of the input only and the circuit responds immediately to a change in the input.

- **Sequential logic**, in which the response of the circuit is based on the current state of the circuit and the sometimes the current input. This may be *asynchronous* or *synchronous*. In *synchronous sequential logic*, the logic changes state whenever an external *clock* control signal is applied. In *asynchronous sequential logic*, the logic changes state on changes of the input data (the circuit does not utilize a *clock* control signal).

- **Memory**, in which digital values can be stored and retrieved some time later. For a user, memory can be either *read-only* (ROM) or *random-access* (RAM). In ROM, the data stored in the memory are initially placed in the memory and can only be read by the user. Data cannot normally be altered in the circuit application. In RAM (also referred to as *read-write memory*, RWM), the user can write data to the memory and read the data back from the memory.

The digital IC consists of a number of logic gates, which are combinational or sequential circuit elements. The logic gates may be implemented using different fabrication processes and different circuit architectures:

- **TTL**, transistor-transistor logic (bipolar)

- **ECL**, emitter-coupled logic (bipolar)

- **CMOS**, complementary metal oxide semiconductor

- **BiCMOS**, bipolar and CMOS

The material predominantly used to fabricate the digital logic circuits is silicon. However, silicon-based circuits are complemented with the digital logic capabilities of circuits fabricated using gallium arsenide (GaAs) and silicon germanium (SiGe) technologies. Today, silicon-based CMOS is by far the dominant process used for digital logic.

The digital logic gate is actually an abstraction of what is happening within the underlying circuit. All digital logic gates are made up of transistors. The logic gates may take one of a number of different circuit architectures (the way in which the transistors are interconnected) at the transistor level:

- static CMOS

- dynamic CMOS

- pass transistor logic CMOS

Today, static CMOS logic is by far the dominant logic cell design structure used. The number of logic gates within a digital logic IC will range from a few to hundreds of thousands and ultimately millions for the more complex processors and PLDs. In previous times, when the potential for higher levels of integration was far less than is now possible, the digital IC was classified by the level of integration—that is, the number of logic gate equivalents per IC (see Table 1.3). With increasing levels of integration, the following levels were identified as follow-on descriptions from VLSI, but these are not in common usage:

- **ULSI**, ultra-large-scale integration

- **WSI**, wafer scale integration

Table 1.3: Levels of integration

Level of integration	Acronym	Number of gate equivalents per IC
Small-scale integration	SSI	<10
Medium-scale integration	MSI	10–100
Large-scale integration	LSI	100–10,000
Very large-scale integration	VLSI	>10,000

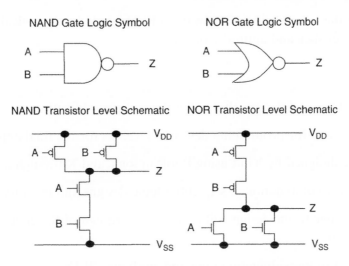

Figure 1.11: Two-input NAND and NOR gates

The equivalent logic gate consists of four transistors. In static CMOS logic, the 2-input NAND and 2-input NOR are four transistor logic gate structures (2 nMOS + 2 pMOS transistors). Figure 1.11 shows the 2-input NAND and NOR gate in static CMOS with both the digital logic gate symbol and the underlying transistor level circuit. At the transistor level, the circuit is connected to a power supply (V_{DD} = positive power supply voltage and V_{SS} = negative power supply voltage). The nMOS transistors are connected toward V_{SS} and the pMOS transistors toward V_{DD}.

1.4   Programmable Logic versus Discrete Logic

When designing a digital circuit or system, there will be the need to develop digital logic designs. One of the initial decisions will be whether to use discrete logic devices (the fixed functionality ICs previously identified) or to use a PLD. This choice will depend on the particular design requirements as detailed in the design specification. In some applications, the choice might be obvious; for other applications, the choice would require careful consideration. For example, if a digital circuit only needs a few logic gates, then a discrete logic implementation would be more probable. However, if a complex digital circuit such as a digital filter design is to be developed, then with the complexity of the resulting logic hardware, a PLD would be the logical choice. These are the characteristics and aptitudes of each:

Discrete logic:

- Suited for small designs that will not require modification

- Can be used for prototyping designs as well as for the final application

- Can be designed by hand using Boolean logic and Karnaugh map techniques

- Suited for combinational, sequential logic designs and memory

- Any change to the design will require the redesign of the circuit hardware and wiring

- No need to know how to design and configure PLDs

- For a particular family of devices, the I/O standard is fixed

- The logic gates may be implemented using different fabrication processes and different circuit architectures: TTL, ECL, CMOS, and BiCMOS.

Table 1.4 identifies selected TTL device family variants in use, Table 1.5 identifies selected CMOS device family variants in use, and Table 1.6 identifies selected low-voltage CMOS device family variants in use.

Programmable logic:

- Suited for all designs from small to large

- Can be used for prototyping designs as well as for the final application

- Suited for designs that might require modification

- Easy to change designs without changing the circuit hardware and wiring that the PLD is connected to by altering the internal PLD circuit configuration

- Can be designed by hand using Boolean logic and Karnaugh map techniques, along with hardware description languages (HDLs) such as VHDL and Verilog®-HDL

- Suited for combinational, sequential logic designs and memory

- The need to know how to design and configure PLDs

Table 1.4: Selected TTL family variants

TTL family variant	Description
74	Standard TTL
74AS	Advanced Schottky
74ALS	Advanced low-power Schottky
74F	Fast
74H	High-speed
74L	Low-power
74LS	Low-power Schottky
74S	Schottky
LVTTL	Low-voltage

Table 1.5: Selected CMOS family variants

CMOS family variant	Description
4000	True CMOS (non-TTL levels)
74C	CMOS with pin compatibility to TTL with same number
74HC	Same as 74C but with improved switching speed
74HCT	As with 74HC but can be connected directly to TTL
74AC	Advanced CMOS
74ACT	As with 74AC but can be connected directly to TTL
74AHC	Advanced high-speed CMOS
74AHCT	As with 74AHC but can be connected directly to TTL
74FCT	Fast CMOS TTL inputs
LVCMOS	Low-voltage CMOS

Table 1.6: Selected low-voltage (LV) CMOS family variants

Low-voltage CMOS variant	Description	
74LV	Low-voltage CMOS	Low-speed operation, 1.0–3.6 V power supply (some functions up to 5.5 V power supply)
74LVC	Low-voltage CMOS	Medium-speed operation, 1.2–3.6 V power supply (5 V tolerant I/O)
74ALVC	Advanced low-voltage CMOS	High-speed operation, 1.2–3.6 V power supply (5 V tolerant I/O on bus hold types)
74AVC	Advanced very low-voltage CMOS	Very high-speed operation, 1.2–3.6 V power supply (3.6 V tolerant I/O)

Table 1.7: Example I/O standards supported by the Xilinx® PLDs

Standard	Standard description
LVTTL	Low-voltage transistor-transistor logic (3.3 V level)
LVCMOS33	Low-voltage CMOS (3.3 V level)
LVCMOS25	Low-voltage CMOS (2.5 V level)
LVCMOS18	Low-voltage CMOS (1.8 V level)
1.5 V I/O (1.5 V levels)	1.5V level logic (1.5 V level)
HSTL-1	High-speed transceiver logic
SSTL2-1	Stub series terminated logic (2.5 V level)
SSTL3-1	Stub series terminated logic (3.3 V level)

- Many PLDs will provide a capability for the designer to set the particular I/O standard to use from those standards supported by the device

- Many PLD vendors provide IP circuit blocks that can be used by the designer within the vendor's PLD, whether free or through royalty payments depending on the licensing arrangement.

Table 1.7 shows example I/O standards that are supported by the Xilinx® [12]. PLDs are configured by the designer. With such programmable I/O capability before the device has been configured with the appropriate standard, the device will default to one of the standards. It is important for the designer to identify the default standard and the implications of using a particular standard on the overall circuit operation.

Early uses of the PLD were for the replacement of standard product discrete logic ICs with a single PLD (see Figure 1.12), allowing for a digital logic circuit to be

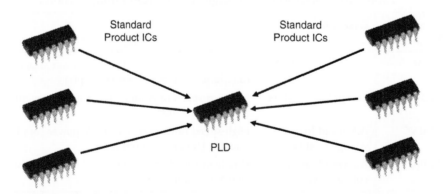

Figure 1.12: Using a PLD to reduce the number of digital logic ICs

implemented in a smaller physical size and therefore reducing the size and cost of the printed circuit board (PCB) on which the logic ICs were to be mounted.

This then led to the use of PLDs for prototyping digital ASIC designs, allowing for real hardware emulation of the ASIC prior to fabricating the ASIC itself. This was useful for design verification and design debugging purposes, but with the early PLDs, the limited speed of operation and size limitations meant that the PLD-based hardware emulation of the ASIC was physically large and slower than the resulting ASIC. Hence, it was not always possible to test the operation of the ASIC hardware emulator at the intended speed of operation of the ASIC.

However, with the high speed and ability to perform complex digital signal processing operations within a single PLD, the PLD itself is becoming in many cases the choice for design prototyping and for use in the final application.

1.5   Programmable Logic versus Processors

The processor is more familiar to the majority of people because it is in everyday use (the heart of the PC is a microprocessor). This component runs a software program to implement required functionality. By changing the software program, the processor will operate a different function. The choice of processor to use will be based on

1. **Microprocessor (μP)**, an integrated circuit that is programmable by the use of a software program. This will be based on an instruction set that the software program uses to perform a set of required tasks. The processor with be based on one of two types of instruction set: a *CISC* (complex instruction set computer) or a *RISC* (reduced instruction set computer). The microprocessor is a general purpose processor in that it is designed to undertake a wide range of tasks. Its architecture would be developed for this purpose and would not necessarily be optimized for specific tasks. The central part of the microprocessor is the central processing unit (CPU) to which external circuits such as memory and I/O interfaces must be added. The CPU has the task of fetching the instructions to be performed from the memory, interpreting the instructions, acting on the instructions, and generating the necessary control signals to fetch, interpret, and act on the instructions. The instructions will be based on arithmetic, logic, and data transfer operations.

2. **Microcontroller (μC)**, a type of microprocessor that contains additional circuitry such as memory and communications ports (such as a UART, universal asynchronous receiver transmitter, for RS-232 communications) along with the CPU, and is aimed at embedded system applications. It would not have the flexibility of the general purpose microprocessor, but instead is aimed at being a self-contained "computer on a chip" with low cost one of the important considerations. The integration of functions that would be in a chip-set mounted on a PCB reduces the design and size requirements on the PCB. The microcontroller is also sometimes referred to as a *microcontroller unit* (MCU).

3. **Digital signal processor (DSP)**, a specialized form of microprocessor aimed at real-time digital signal processing operations such as digital filtering [13] and fast Fourier transforms (FFTs). Although such operations can be performed on a microprocessor, the DSP has an architecture that is optimized for fast computations typically undertaken. For example, a DSP would include a fast hardware multiplier cell that is accessed from the software program that the DSP is running. This allows multiplications to be undertaken on digital data using the fast hardware that would not be possible on a general purpose microprocessor without a hardware multiplier. (A general purpose microprocessor would perform a multiplication in software using shift operations and additions using looping operations that would be slow to undertake.)

The choice of a particular processor to use is based on a number of considerations including:

- final application requirements

- capabilities of the processor

- limitations of the processor

- knowledge and prior experience of the designer

- availability of tools for designing and debugging software applications for the processor

Example processor vendors and products are shown in Table 1.8. This provides a snapshot of the main current companies involved in the processor area. Further information on the range of processors can be obtained from the company web sites.

Table 1.8: Main processor vendors

Company	Example product	Homepage URL
Intel®	Intel Core™ 2 Duo	http://www.intel.com/
Advanced Micro Devices (AMD)	AMD Athon™ 64 FX	http://www.amd.com
Zilog®	Z80180	http://www.zilog.com/
Motorola®	MPC7457	http://www.motorola.com
ARM®	ARM Cortex-A8	http://www.arm.com/
Microchip	PIC 24F MCU	http://www.microchip.com
Texas Instruments, Inc.	TMS320™	http://www.microchip.com
IBM®	PowerPC®	http://www.ibm.com
MIPS Technologies, Inc.	MIPS32® 74K™	http://www.mips.com
Analog Devices, Inc.	ADSP-21262	http://www.analog.com
Freescale Semiconductor, Inc.	MCF5373 ColdFire®	http://www.freescale.com/
Atmel®	AT572D740	http://www.atmel.com

For designers of processor-based systems, the one concern is the possibility of processor *obsolescence*. Here, if a vendor decides to discontinue a processor product or family of products, this would have a major impact on the designer of electronic systems using the particular processor. The designer (and organization that the designer is working in) would potentially have invested a great deal of time and resources in learning and using the processor, associated EDA tools, and design flows—all of which would require reinvestment. A PLD, however, could be used as an alternative to a processor IC purchased from a vendor. With the PLD, it would be possible to implement a processor within the PLD itself. The processor design would be obtained as either a schematic or, more probably, as an HDL description. This HDL description would then be synthesised to map onto the PLD; the PLD would be configured with the same operations as the original processor. This description would not change and would be available for as long as the designer would require it. With this, the processor would be a core (i.e., a block of logic that would be placed within the PLD) and would be provided to the designer as either *hard core* or *soft core*. The *hard core* would be provided as logic gates and interconnect for a particular PLD. A *soft core* would be provided as HDL code describing the processor in terms of functionality, rather than logic gates and interconnect, and would then be synthesised to the required PLD.

An alternative to the predesigned processor architecture is to design the architecture for a specific requirement. This would enable the designer to develop the best architecture for the particular application and not be potentially limited in performance by the availability of an existing processor. Hence, with PLDs, the ability to develop application-specific processors is realistic. This would enable the designer to develop PLD-based systems that can utilize both a processor (running a software application) and dedicated, optimized hardware (for maximum speed of operation) within a single device.

Although there are many potential advantages to using PLDs rather than processors, the design paradigms are different and the need to consider the benefits versus the costs, and the need to learn new design techniques (predominantly hardware rather than software), cannot be underestimated. However, the ability for the designer to choose a solution that provides him or her with the maximum benefit for the particular application is something that cannot be overlooked. It is common to consider the PROM as an SPLD, alongside the PLA, PAL® and GAL (see below), although in this text, only the PLA, PAL® and GAL are only considered in detail.

1.6   Types of Programmable Logic

1.6.1   Simple Programmable Logic Device (SPLD)

The SPLD was introduced before the CPLD and FPGA. The three main types of SPLD architecture—programmable logic array (PLA), programmable array of logic (PAL), and generic array of logic (GAL)—are described below.

The PLA

The PLA consists of two programmable planes AND and OR (see Figure 1.13). The AND plane consists of programmable interconnect along with AND gates. The OR plane consists of programmable interconnect along with OR gates.

In this view, there are four inputs to the PLA and four outputs from the PLA. Each of the inputs can be connected to an AND gate with any of the other inputs by connecting the crossover point of the vertical and horizontal interconnect lines in the AND gate programmable interconnect. Initially, the crossover points are not electrically connected, but configuring the PLA will connect particular crossover points together. In this view, the AND gate is seen with a single line to the input. This view is by convention, but this also means that any of the inputs (vertical lines) can be

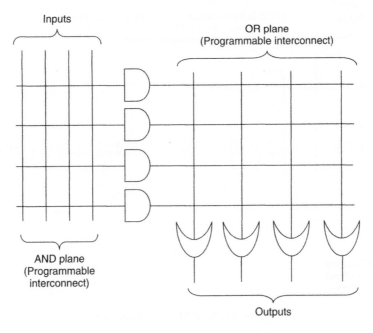

Figure 1.13: PLA architecture

connected. Hence, for four PLA inputs, the AND gate also has four inputs. The single output from each of the AND gates is applied to an OR gate programmable interconnect. Again, the crossover points are initially not electrically connected, but configuring the PLA will connect particular crossover points together. In this view, the OR gate is seen with a single line to the input. This view is by convention, but this also means that any of AND gate outputs can be connected to the OR gate inputs. Hence, for four AND gates, the OR gate also has four inputs.

The PAL®

The PAL® is similar to the PLA architecture, but now there is only one programmable plane, the AND plane, and the AND gate programmable plane is retained (see Figure 1.14). This architecture is simpler than the PLA and removes the time delays associated with the programmable OR gate plane interconnect, hence producing a faster design. However, this comes at a cost of flexibility—the PAL® is less flexible in the ways in which a digital logic design can be implemented than the PLA.

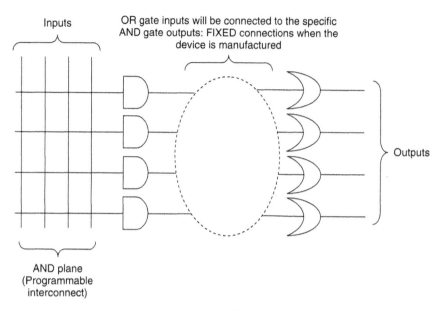

Inputs

OR gate inputs will be connected to the specific AND gate outputs: FIXED connections when the device is manufactured

Outputs

AND plane
(Programmable
interconnect)

Figure 1.14: PAL® architecture

The PLA and PAL® architectures as shown allow combinational logic designs to be implemented. If the design provides for feedback of the outputs to the inputs, then it is possible to implement latches and bistables, thereby also allowing sequential logic circuits to be implemented. This is possible on some commercially available PAL devices. Additionally, some PAL devices also provide the output to be made available from the OR gate output or via an additional bistable connected to the OR gate output. Hence, the types of sequential logic circuits that can be implemented increase and therefore the usefulness of the particular PAL® device increases.

The GAL

PAL and PLA devices are one-time programmable (OTP) based on PROM, so the PAL or PLA configuration cannot be changed after it has been configured. This limitation means that the configured device would have to be discarded and a new device configured. The GAL, although similar to the PAL® architecture, uses EEPROM and can be reconfigured.

1.6.2 Complex Programmable Logic Device (CPLD)

The CPLD is a step up in complexity from the SPLD; it builds on SPLD architecture and creates a much larger design. Consequently, the SPLD can be used to integrate the functions of a number of discrete digital ICs into a single device and the CPLD can be used to integrate the functions of a number of SPLDs into a single device. The CPLD architecture is based on a small number of logic blocks and a global programmable interconnect. A generic CPLD architecture is shown in Figure 1.15.

The CPLD consists of a number of *logic blocks* (sometimes referred to as *functional blocks*), each of which contains a *macrocell* and either a PLA or PAL® circuit arrangement. In this view, eight logic blocks are shown. The macrocell provides additional circuitry to accommodate registered or nonregistered outputs, along with signal polarity control. Polarity control provides an output that is a true signal or a complement of the true signal. The actual number of logic blocks within a CPLD varies; the more logic blocks available, the larger the design that can be configured. In the center of the design is a *global programmable interconnect*. This interconnect allows connections to the logic block macrocells and the *I/O cell*

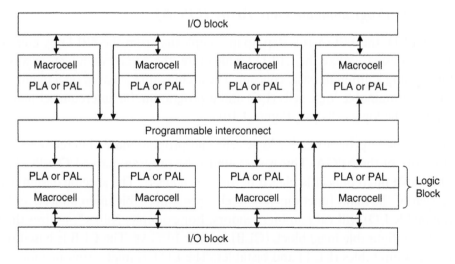

Figure 1.15: Generic CPLD architecture

arrays (the digital I/O cells of the CPLD connecting to the pins of the CPLD package).

The programmable interconnect is usually based on either array-based interconnect or multiplexer-based interconnect:

- *Array-based interconnect* allows any signal within the programmable interconnect to connect to any logic block within the CPLD. This is achieved by allowing horizontal and vertical routing within the programmable interconnect and allowing the crossover points to be connected or unconnected (the same idea as with the PLA and PAL®), depending on the CPLD configuration.

- *Multiplexer-based interconnect* uses digital multiplexers connected to each of the macrocell inputs within the logic blocks. Specific signals within the programmable interconnect are connected to specific inputs of the multiplexers. It would not be practical to connect all internal signals within the programmable interconnect to the inputs of all multiplexers due to size and speed of operation considerations.

1.6.3   *Field Programmable Gate Array (FPGA)*

Like the CPLD, the FPGA is a step up in complexity from the SPLD by creating a much larger design; unlike the CPLD architecture, the FPGA architecture was developed using a different basic concept. The architecture is based on a regular array of basic programmable logic cells (LC) and a programmable interconnect matrix surrounding the logic cells (see Figure 1.16).

The array of basic programmable logic cells and programmable interconnect matrix form the core of the FPGA. This is surrounded by programmable I/O cells. The programmable interconnect is placed in *routing channels*. The specific design details within each of the main functions (logic cells, programmable interconnect, and programmable I/O) will vary among vendors. For example, Xilinx®. utilizes the *logic block* as a configurable logic block (CLB) in their FPGAs. The CLB is based on one or more look-up tables (LUT) and bistables. The LUT is made from memory cells (SRAM cells).

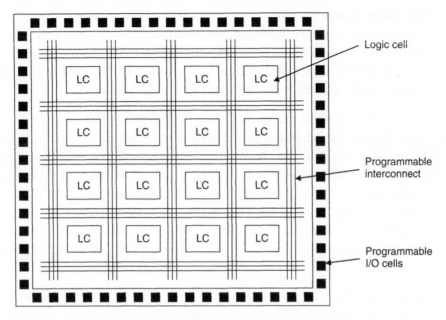

Figure 1.16: Generic FPGA architecture

1.7 PLD Configuration Technologies

The PLD is configured by downloading a particular circuit configuration as a sequence of binary logic values (sequence of 0s and 1s). The configuration will be held in a *configuration file* on the PC or workstation that the design was created on using the required EDA tools. A downloader software application will read the configuration file and download the contents to the PLD. These values are stored in memory within the device, where the memory may be volatile or nonvolatile:

- **Volatile memory**: When data is stored within the memory, the data is retained in the memory as long as the memory is connected to a power supply. Once the power supply has been removed, then the contents of the memory (the data) is lost. The majority of FPGAs utilize volatile SRAM-based memory. Hence, whenever the power supply is removed from the FPGA, then the FPGA configuration is lost and when the power supply is reapplied, then the configuration must be reloaded into the SRAM.

- **Nonvolatile memory**: When data is stored within the memory, the data is retained in the memory even when the power supply has been removed. Some FPGAs utilize *antifuse* technology to store the FPGA configuration; new generation FPGAs will also utilize flash memory. CPLDs utilize nonvolatile memory such as EPROM, EEPROM, and flash memory.

SRAM-based configuration is based on the use of multiple 1-bit memory cells (see Figure 1.17). The cell has write and read modes. In write mode, a data bit (0 or 1) to store in the memory is applied to the *bit line*. The switch transistor is closed (by applying a logic 1 to the transistor *gate*) on the *word line*. When the switch is closed, the logic value on the *bit line* is applied to the input of the top inverter. The inverted output is applied to the input of the bottom inverter, and the output of this inverter is the same logic value as applied on the bit line. When the switch transistor is opened, the inverter arrangement retains the logic value due to the feedback arrangement of the two inverters.

When the value is to be read from the memory cell, the switch transistor is again closed (by applying a logic 1 to the transistor gate) on the word line. The logic value output from the bottom inverter is then applied to the bit line. Each of the inverters contains two transistors (in static CMOS, one nMOS and one pMOS transistor). Hence, the memory cell contains five transistors overall, compared to six transistors in the memory cell of an SRAM memory IC; a second switch transistor is used at the output of the top inverter and creates an output that is the inverse of the bit line value.

Antifuse based configuration uses a two terminal device that is electrically programmed to change from an electrical open circuit to an electrical short circuit. The operation is

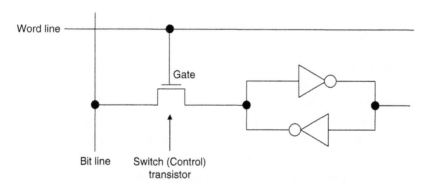

Figure 1.17: SRAM cell based on five transistors

the inverse to that of the fuse. Initially, there is no connection between the two terminals (there is a high resistance). When programmed (blown), a connection (low resistance) is made between the two terminals. This is a *one-time process* (i.e., permanent) and once blown, cannot be undone. The antifuse will be one of two types, amorphous-silicon antifuse or oxide-nitride-oxide (ONO) antifuse.

Figure 1.18 shows the principle of operation. The antifuse material is placed in a via between two metal layers in the circuit (vertical layers). Initially (i), the no connection exists between the two metal layers. Once programmed, a low-resistance link (ii) exists between the metal layers and connects them together.

Configurations based on EPROM, EEPROM, and flash memory use a floating gate transistor. Figure 1.19 shows the basic arrangement for a 1-bit EPROM memory. The transistor acts as a switch. In EEPROM and Flash memories, a second transistor is also used. A more comprehensive description of these memory cells can be found in references [2] and [3].

The switch is closed by the application of a logic 1 on the word line to the control gate of the transistor. However, by applying high voltage during configuration to the control gate of the transistor, a charge is injected into the floating gate and stored on the gate capacitance. When the high voltage is removed, the charge is stored. The effect of this charge is to make the transistor permanently switched off even when the word line signal is applied. (The effect of the stored charge is to increase the threshold voltage of the transistor so that the transistor can never switch on.)

Antifuse-based configuration is a one-time process. That is, once the antifuse has been blown to form the circuit configuration, this cannot be undone. If the design is wrong or requires modification, then the device has to be thrown away and a new device

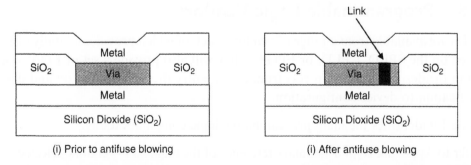

Figure 1.18: Antifuse cell-based configuration (amorphous-silicon antifuse structure)

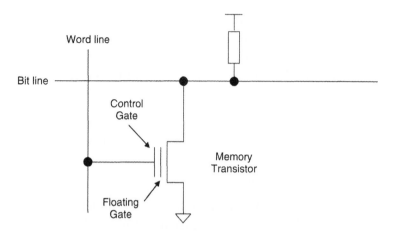

Figure 1.19: EPROM-based configuration

loaded with the new configuration. SRAM-, EPROM-, EEPROM-, and flash-based configurations, however, allow the device to be reconfigured many times.

Electrically programmable (configurable) and erasable PLD configuration allows for the potential for in-system programming (ISP). This means that the PLD can be physically located on its final circuit board (i.e., within a socket or soldered into place onto the board) and via a programming port on the PLD, the configuration data can be loaded into the PLD. The JTAG (Joint Test Action Group) standard is typically used for this purpose. Additionally, for those PLDs that can be reconfigured, the device allows for in-system reprogramming (ISR), meaning that the PLD configuration can be changed while the PLD is located on its final circuit board.

1.8   Programmable Logic Vendors

PLDs are available from a range of vendors, each of which provides a family of PLDs based on the SPLD, CPLD, or FPGA. They will also provide a set of EDA tools to aid in the design creation process from design entry through simulation and design verification to device configuration.

Table 1.9 identifies the main programmable logic companies today.

Refer to Appendix B for a summary reference of the main PLD vendors, selected electronic design companies, electronic component vendors, test equipment vendors, and

Table 1.9: Main programmable logic vendors

Company	Homepage URL
Achronix Semiconductor Corporation	http://www.achronix.com
Actel® Corporation	http://www.actel.com
Altera Corporation	http://www.altera.com
Atmel® Corporation	http://www.atmel.com
Cypress Semiconductor	http://www.cypress.com
Lattice® Semiconductor Corporation	http://www.latticesemi.com
Quicklogic® Corporation	http://www.quicklogic.com
Xilinx®	http://www.xilinx.com

EDA companies. Details on each PLD can be found on the vendor's Internet home page; other useful information usually provided includes:

- device data sheets

- application notes (on how to use the devices)

- white papers (on applications that have been developed with the PLDs)

- audiovisual aids such as tutorial videos and web casts

- vendor EDA tool user guides and tutorials and software download areas

1.9 Programmable Logic Design Methods and Tools

1.9.1 Introduction

To design with a particular PLD, the appropriate design tools are required. In general, free versions of the tools with limited capabilities are available, as well as full versions for purchase. Table 1.10 identifies the tools for each of the main vendors.

Although each software design tool differs in appearance and the manner in which the designer interacts with it, all have a common set of basic features required to create and implement designs within a particular tool. These features are:

- **Project management**: the ability to set up design projects and to manage the design data in a user-friendly manner

Table 1.10: PLD design tool by vendor

Company	Design tool
Actel® Corporation	Libero® IDE
Altera Corporation	Quartus® II
Altium™	Altium Designer
Atmel® Corporation	Integrated Development System (IDS)
Cypress Semiconductor	Warp
Lattice® Semiconductor Corporation	ispLEVER®
Mentor Graphics®	FPGA Advantage®
Quicklogic® Corporation	QuickWorks®
Synplicity®	Synplify Pro®
Xilinx®	ISE™

- **Design entry**: entering the design into the tools using a combination of schematic capture, HDL design entry, state machine flow diagrams

- **Design simulation**: Once the design has been entered, the design can be simulated to check that it performs as required.

- **Design synthesis**: For HDL design entry, typically at the register transfer level (RTL), the HDL description is to be synthesized to produce the digital logic circuit in terms of logic gates and interconnect (netlist).

- **Place and route**: taking the design that has been entered and/or synthesized, and mapping it to the hardware resources on the PLD. This defines which parts of the PLD will contain which functions in the design and how the different parts of the PLD are interconnected.

- **Post-layout delay extraction**: takes the information on the placed and routed design, and extracts timing delays due to the logic gates and interconnect used

- **Post-layout simulation**: Using the layout timing delays, the design is resimulated with these delays included to determine whether the design still functions correctly.

- **Configuration file generation**: creates the PLD configuration data

- **PLD configuration**: downloads the configuration data to the PLD and enables the configuration on the PLD to be verified for correctness

- **Interfacing to external tools**: allows for third-party tools such as simulation and synthesis tools to be interfaced to the main design tools

1.9.2   Typical PLD Design Flow

Whether a CPLD or FPGA is to be used, the designer follows a common design flow for the major stages in the design entry, verification, and device configuration. However, there will be differences in the fine detail between the CPLD and FPGA. Figure 1.20 shows a typical PLD design flow.

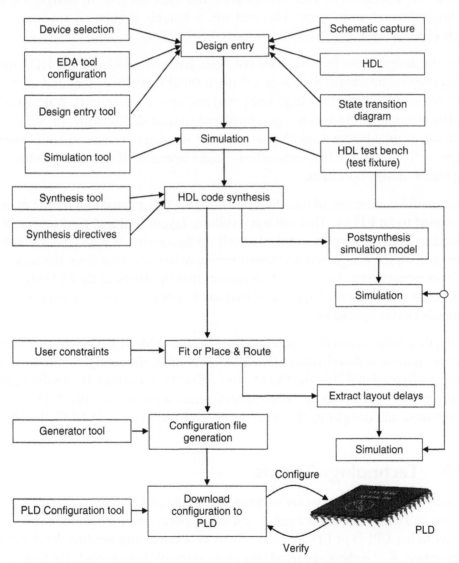

Figure 1.20: Typical PLD design flow

The first step is to enter the design into the appropriate EDA tool, typically using a combination of schematic capture, HDL descriptions, and state transition diagrams (for state machine design). The designs will be added to a design project, and within this project, the target PLD will also be identified, although the target PLD can be changed at a later date. When the designs have been entered, the operation of each design part and then the overall design will be validated through simulation. This will use a suitable simulation tool and test bench (test fixture).

When the design, prior to HDL code synthesis, has been validated, the HDL designs are synthesized into logic. Synthesis will use a suitable synthesis tool and user-generated synthesis directives (e.g., size [area] and power constraints). A postsynthesis simulation model of the design is generated and simulated. Normally, the same test bench as used before would be used and the simulation results on both designs compared to ensure that the postsynthesis design operation is equivalent to the presynthesis design operation.

On successful completion of this stage, the design is either fitted to a CPLD or placed and routed to an FPGA. This will use a suitable layout tool and user-generated constraints (e.g., device pins and the I/O cell configuration). A post-layout simulation is then run on the design and additional timing delays resulting from the logic gates and interconnect used. This simulation ensures that the design at the PLD layout level will operate at the required speed and that the layout delays are not large enough to impede circuit operation.

Finally, the configuration file is generated as a bitstream file or JEDEC format file, then the configuration is downloaded to the PLD. Normally, the configuration tool allows for the configuration within the PLD to be verified by comparing the configuration actually within the PLD to the required configuration (by reading the PLD configuration and comparing this with the original bitstream or JEDEC file [14]).

1.10   Technology Trends

The early SPLDs were, by today's standards, simple and contained few logic gates. They are still used for small designs. For many applications, though, the choice now is whether to use CPLD or FPGA, so the focus of research and product development is on those two. Key technology trends for programmable logic include the following as identified in Table 1.11.

Table 1.11: Technology trends

More functionality per IC	The end-user demands for more functionality within the PLD to enable increased digital signal processing capabilities, as required, for example, in communication system applications.
Emphasis on electronic system level (ESL) design	The majority of design work using HDLs involves writing HDL code at a level referred to as register transfer level (RTL). This level describes the movement and storage of data around the digital system, and synthesis tools have been developed to synthesize RTL-level HDL code into logic gates and interconnect (netlists). As design complexities increase, there is a need for the designer to describe at more abstract levels of description—to describe the behavior of the system and to let the synthesis tool take care of the details. ESL design refers to the design and verification methodologies at higher levels of abstraction from traditional RTL.
Inclusion of hardware macros with programmable logic	Many FPGAs today include dedicated hardware macros such as RAM, hardware multipliers, and processor cores that are seen as resources alongside the programmable logic. When a design is synthesized to a particular PLD, the synthesis tool would know about the available macros and use them appropriately. In addition, the move toward including mixed-signal macros such as ADCs and DACs increases the usefulness of the PLD.
High-level behavioral synthesis	The description of system behavior and the ability to synthesize behavioral descriptions to logic and interconnect, as described above in ESL.
Seamless codesign of hardware-software systems	The ability to design and develop designs based on software operations and hardware operations within a single design environment that seamlessly allows the overall design to be undertaken in a single step.
Increased need for design debug tools	As the types of digital systems being developed increase in complexity, the potential for errors (bugs) increases. The ability to debug PLD designs once configured enables the designer to identify the cause of errors and to remove them—in a similar manner to the debugging arrangements within processor-based designs. The need for more comprehensive design debug tools is increasing.
Higher operating frequencies	As the complexities of the types of digital signal processing algorithms increases, there is a need to perform the algorithm calculations more quickly. This requires faster logic gates, so the PLD can work at higher operating frequencies to enable real-time digital signal processing.

(continued)

Table 1.11 (*Continued*)

Finer fabrication process geometries	To provide for more circuitry within a single device, the size of each of the logic gates and of the interconnect within the device must be reduced. This is achieved by utilizing finer geometry processes. Each process is defined by a technology node that defines the geometries of a particular fabrication process. This is defined by the International Technology Roadmap for Semiconductors (ITRS) [15].
Lower power supply voltages	When the speed of operation of a CMOS design increases, the power consumption increases, and the temperature in turn increases. To reduce power consumption, excessively high operating temperatures, and allow for portable, battery-operated electronics, the power supply voltage is reduced. This reduction in power supply voltage is also required for reliability reasons when using the finer fabrication process geometries.
Newer and faster device test methods	Whenever a PLD is fabricated, the PLD must be tested to ensure that the device was fabricated correctly and without circuit faults. With an ever-increasing device design complexity, the test problem increases. Effective tests are needed to set quality levels at the lowest possible cost.
Lower costs	Driven by the end-user requirements for devices with more functionality but at a lower cost.

References

[1]   *Oxford Dictionary of English*, Second Edition, Revised, eds. C. Soanes and A. Stevenson, Oxford University Press, 2005, ISBN 0-19-861057-2.

[2]   Smith, M., *Application Specific Integrated Circuits*, Addison-Wesley, 1999, ISBN 0-201-50022-1.

[3]   Skahill, K., *VHDL for Programmable Logic*, Addison-Wesley, 1996, ISBN 0-201-89573-0.

[4]   Maxfield, C., *The Design Warrior's Guide to FPGAs*, Elsevier, 2004, ISBN 0-7506-7604-3.

[5]   Cooley, J. W., and Tukey, J. W., "An Algorithm for the Machine Computation of the Complex Fourier Series," *Mathematics of Computation*, Vol. 19, April 1965, pp. 297–301.

[6]   Meade, M., and Dillon, C., *Signals and Systems, Models and Behaviour*, Second Edition, Chapman and Hall, 1991, ISBN 0-412-40110-x.

[7] Parhi, K., *VLSI Digital Signal Processing Systems, Design and Implementation*, John Wiley & Sons, Inc., 1999, ISBN 0-471-24186-5.

[8] Jespers, P., *Integrated Converters D to A and A to D Architectures, Analysis and Simulation.* Oxford University Press, 2001, ISBN 0-19-856446-5.

[9] Astrom, K., and Wittenmark, B., *Computer-Controlled Systems, Theory and Design*, Second Edition, Prentice-Hall International Editions, 1990, ISBN 0-13-172784-2.

[10] Golden, J., and Verwer, A., *Control System Design and Simulation*, McGraw-Hill, 1991, ISBN 0-07-707412-2.

[11] Bell Laboratories (Bell Labs), http:www.bell-labs.com/

[12] Xilinx Inc., USA, http://www.xilinx.com

[13] Ifeachor, E., and Jervis, B., *Digital Signal Processing, A Practical Approach*, Second Edition, Prentice Hall, 2002, ISBN 0-201-69619-9.

[14] Joint Electronic Device Engineering Council (JEDEC), http://www.jedec.org/

[15] International Technology Roadmap for Semiconductors, 2006 Edition.

[16] Zahiri, B., *Structured ASICs: Opportunities and Challenges*, Proceedings of the 21st International Conference on Computer Design, Oct. 2003, pp. 404–409.

[17] Ran, Y., and Marek-Sadowska, M., *Designing Via Configurable Logic Blocks for Regular Fabric*, IEEE Transactions on Very Large Scale Integration (VLSI) Systems, Jan. 2006, pp. 1–14.

Student Exercises

The following exercises will involve the use of suitable reference text books and Internet resources in order to answer.

1.1 The 74LS family of digital logic ICs provides a set of fixed functionality logic gates. For the following logic gates:

- 2-input NAND gate
- 2-input AND gate
- 2-input NOR gate
- 2-input NAND gate
- buffer
- inverter

identify the following characteristics:

- the power supply voltage requirements
- the power supply current requirements
- the number of pins dedicated to the power supply or supplies
- the package type(s) that the IC is available in
- the number of logic gates that a designer has access to use
- the number of I/Os that a designer has access to use

1.2 The 74HC family of digital logic ICs provides a set of fixed functionality logic gates. What are the main differences between 74LS and 74HC logic gates?

1.3 Repeat Question 1.1 using 74HC logic.

1.4 What is an application-specific standard product (ASSP)?

1.5 The majority of integrated circuits are fabricated using silicon-based technology. A particular IC fabrication process will be based on a particular technology node. What is meant by the term *technology node*?

1.6 For the following PLDs:

- Xilinx® Spartan™-3 XC3S1000
- Xilinx® Coolrunner™-II XC2C256-144
- Lattice® Semiconductor MACH4A5-64/32
- Lattice® Semiconductor ispLSI2064E

identify the following from the device datasheets:

- whether the device is a CPLD or FPGA

- the power supply voltage requirements
- the power supply current requirements
- the number of pins dedicated to the power supply or supplies
- the maximum digital clocking frequency
- the package type(s) that the IC is available in
- the number of I/Os that a designer has access to use
- the I/O standards that the designer can set for the I/Os
- the cost of each PLD
- the CAD tools used in the design of circuits and systems with each PLD
- the role of each of the CAD tools used in the design of circuits and systems with each type of PLD

1.7   What is the main difference between a PAL- and a GAL-based SPLD?

1.8   What processors are commonly used in the following:

- desktop PCs
- laptop PCs
- personal digital assistants (PDAs)

Which companies provide these processors?

1.9   Considering the Xilinx® Coolrunner™-II CPLD family, from the datasheet, identify the CPLD architecture used. What is the functional block and what does it do? How does the specific architecture compare to or differ from the generic CPLD architecture identified in this chapter?

1.10   Considering the FPGA, for each of the main PLD vendors who provide FPGA devices, choose one small FPGA and identify:

- the architecture of the particular FPGA
- the particular configuration technology (technologies) used with this device
- the time required to load the configuration into the FPGA

1.11   What are the advantages of using programmable logic over discrete digital logic ICs? Give two examples of where it would be more beneficial to use a PLD.

1.12   Give two examples of where it would not necessarily be beneficial to use a PLD over discrete digital logic ICs.

1.13   What is a structured ASIC? How does this compare and differ from the traditional ASIC and the PLD?

CHAPTER 2

Electronic Systems Design

2.1 Introduction

In this chapter, the design of electronic systems will be introduced by looking at the different parts (subsystems) that are brought together to form the overall system. However, before considering any design three points should always be noted:

1. **Always use common sense**. If something does not seem right, then it probably isn't.

2. **Never leave anything to chance**. What can go wrong will go wrong.

3. **There is almost always more than one way to solve a problem**. The choice for the designer is to determine the most appropriate solution. The first solution developed might not necessarily be the best.

Within the context of this book, the interest lies in the ability to design electronic circuits and systems that can have a wide range of required functions, be practical and useful, and will ultimately use analogue, digital, or mixed-signal circuits. The advantage of each type of circuitry is:

- *Analogue circuits* manipulate electrical signals (voltages and/or currents) that will vary continuously in amplitude between lower and upper limits. Theoretically, the analogue signal is capable of changing by infinitesimally small amounts. Examples of analogue circuits include operational amplifiers, (voltage, current, audio, and power), and analogue filters (low-pass, high-pass, band-pass, band-reject).

- *Digital circuits* manipulate signals that are quantized—that is, using signals that will vary at discrete values between lower and upper limits. Binary (two-level logic,

0 and 1) is most commonly used and is the basis of the majority of computing applications today. Examples of digital circuits include microprocessors, microcontrollers, digital signal processors, digital filters, and programmable logic.

- *Mixed-signal circuits* manipulate both analogue and digital signals and are typically used to interface digital circuits to analogue input and output. Examples of mixed-signal circuits include analogue to digital converters (ADC), digital to analogue converters (DAC), digital processors with on-chip (on-board) ADCs and DACs, comparators, and programmable analogue arrays.

The terms electronic *circuit* and electronic *system* are commonly used and are used throughout this text. The *Oxford Dictionary of English* [1] defines *circuit* as "a complete and closed path around which a circulating electric current can flow: a system of electrical conductors and components forming an electrical circuit," and defines *system* as "a set of things working together as parts of a mechanism or an interconnecting network."

In electronics, there is no clear point at which a circuit becomes a system; a number of different criteria could be found and would make for interesting debate. However, in the context of this book, the distinction is this: an electronic system will be designed to perform a complex function or range of functions and will consist of one or more electronic circuits.

For example, consider the desktop PC in everyday use, as shown in Figure 2.1. This would be considered an electronic system consisting of a number of subsystems, each

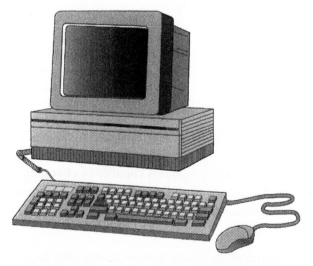

Figure 2.1: Image of a desktop PC

in turn consisting of a number of individual electronic circuits. At the initial visual appearance, the PC consists of a small number of larger units, including:

- case containing the computer electronics
- the visual display unit (VDU)
- the keyboard
- the mouse

The case contains the electronics, which include the following basic subsystems:

- motherboard
- power supply
- hard disk
- floppy disk
- CD-ROM reader
- CD-ROM writer
- DVD reader
- DVD writer
- Input/output (I/O) ports: parallel port (Centronics), serial port (RS-232C), universal serial bus (USB), firewire, local area network (LAN), modem

These are designed to perform specific functions for the manipulation of data and for efficient user interaction. PCs will be available from a number of different manufacturers, with each manufacturer offering their own set of advantages over the competitors (cost, ease of use, etc.). Company and product branding in this highly competitive market is extremely important.

Although the appearance of each PC might vary, the internal arrangement within every PC is basically the same; that is, the architecture of the computer is based on a common architecture. With the side cover taken off the PC, then these internal subsystems will be exposed. Figure 2.2 shows the internal arrangement for an example PC. Here, the PC motherboard is housed vertically and secured to one side of the PC case. Connectors are mounted on the PC motherboard to allow for other subsystems to be connected, for example, the power supply (bottom right)

Figure 2.2: Inside a desktop PC

and disk drives. The disk drives here are placed in slots at the bottom left of the case (empty in this image).

The *motherboard* is of interest here as it is a *printed circuit board* (PCB) that houses the main electronic components, including:

- microprocessor
- memory: ROM and RAM
- clocks, counters, and timers
- miscellaneous logic
- I/O circuitry

The main circuitry is in the form of an integrated circuit (IC). This is shown in Figure 2.3.

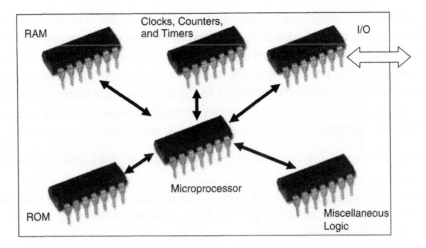

Figure 2.3: PC motherboard electronics (simplified view)

The microprocessor runs a software program that will enable the microprocessor to undertake a number of actions (operations). Read-only memory (ROM) will be used to hold program code. Random access memory (RAM) will be used for temporary storage of data (both program code and variable data). Clocks are used to provide the necessary timing to control the operation of the sequential logic parts of the circuits. Counters and timers are used to provide specific timing signals. The I/O circuitry provides the interfacing between the electronics and the rest of the electronic system. The miscellaneous logic provides specific hardware interfacing between ICs within the overall electronic system.

The software code that the microprocessor runs will be based on the internal instruction set of the microprocessor. This defines what operations the microprocessor can undertake. When a program is written to run on a microprocessor, the programmer uses one of two approaches:

1. **High-level languages** (such as C or Java) are suited for general-purpose programming tasks for which the programmer does not need to understand the details of the target computing hardware. This is an efficient use of the programmer's time but may not produce the most efficient code (in terms of the size of the program code and the time required to execute commands). The high-level program is then compiled into the machine-code form that the microprocessor then uses.

2. **Machine-code** is low-level code that works at the computing hardware level. The programmer must have a good understanding of the internal structure of

the microprocessor and its fundamental instruction set. This is time consuming but can produce efficient code (in terms of the size of the program code and the time required to execute commands). When a program is written in machine-code form, the program is firstly written in the form of standard instruction **mnemonics** that are then converted to the machine-code form. The process of converting the instruction mnemonics to machine-code is referred to as *assembly*. Software programs that undertake this task are referred to as *assemblers*.

Today, most programming is undertaken using a suitable high-level language.

Aside: An interesting read on how the global computer industry developed from the early days in Silicon Valley during the 1970s is the book *Accidental Empires* by Robert Cringley (Harper and Brothers, 1996).

The previous PC example is only one example of how an electronic system utilizes a processor. Increasingly, many other systems utilize programmable logic at the center of the electronics. All designs of this size and complexity need to consider a large number of issues relating to the design, manufacture, and test of the electronic system [2]. The chosen design approach will ultimately be a trade-off in resolving often conflicting requirements, such as performance versus cost. The choices will include:

- **Generating the initial idea**: What must be designed? What functions are to be included? Why? How are ideas to be generated and captured (documented)?

- **Market requirements**: Successful products fulfill a set of market requirements. Identifying what the market requirements are and what the steps are required to develop a product that will be a commercial success are essential.

- **Cost to design, manufacture, and test**: What is the cost to design, manufacture, and test the design?

- **Sales price**: What can the sale price be?

- **Converting the idea into a specification, or family of specifications**: How will the design requirements be captured into a formal document so that the designers and the end users will have a common set of documentation relating to the system? Typically one or more specification documents are created, depending on the type of system to be created and the need for particular types of specification documents (for example, documents to be generated and made available for specific contract requirements).

- **Following a design process**: How will the design be created from the initial idea through to production level manufacture? (Sequential and concurrent design processes are discussed in the next section.)

- **The need for teamwork**: The creation of any system of design complexity requires skills from a number of people who will be formed into teams, each responsible for a specific design task.

- **Choosing the right implementation technology**: Most designs can be implemented in a number of different ways. The choices available can initially be overwhelming, but by suitable care and thought about what exactly is required and how these requirements can be realized in electronic hardware and software, a small number of appropriate choices emerges. There might not necessarily be a right or wrong choice, rather a better or worse choice for the particular design scenario.

- **Incorporating testing and design for testability (DfT)**: During the design and production manufacture of a system, testing ensures that the design itself is correct and that the manufacture of the design has not created defects that result in a faulty operation. To demonstrate the importance of testing and the discovering of faults in an electronic circuit or system after fabrication and before use is referred to as the *Rule of Ten*: the cost multiplies by a factor of ten every time an undetected fault is used to form a large electronic circuit or system (Figure 2.4). Here, if the cost of detecting a faulty device (component) when it is produced is one unit; the cost to detect that faulty device when used at the board level (PCB) is 10 units; and the cost to detect that faulty board when inserted in its system is 100 units, and so on.

- **Setting up and using quality control mechanisms**: Determine the level of quality required of the final system, then adopt the appropriate approach to each stage in design, manufacture, and testing to ensure that the right level of product quality is achieved and maintained. Quality control mechanisms are outside the scope of this text book and so are not considered further.

- **Product branding**: Does the company producing the system and/or the product have a specific and identifiable brand? Does the potential customer associate the company and/or product with price, quality, and reliability?

- **Time to market (TTM)**: How long will it take to get the product into the market so that sales income can be generated?

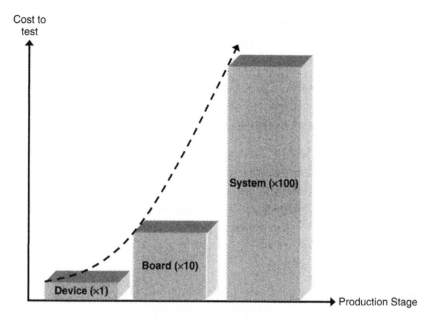

Figure 2.4: Rule of Ten

- **Design simulation**: During the design process and prior to building the prototype, the operation of the design will be simulated. At this stage, many of the bugs in the design can be removed, although care must always be taken because the results of a simulation study are only as good as the simulation set-up (the test stimulus to apply) and the analysis of the simulation results.

- **Design prototyping**: What steps are required to take the initial design idea to a prototyping stage in order (i) to identify the correct operation and that it meets the required specifications, and (ii) where the design does not work correctly, to identify the problem and the correction, whether in the design itself or in the manufacturing. Design prototyping will be undertaken on a physical system that has been built.

- **Design debug**: Debugging is undertaken during design simulation and design prototyping to remove bugs in the design that prevent correct design operation.

- **Production level manufacture**: Once the design prototyping stage has been successfully completed and the design is correct, then the full-scale manufacture of the design can be undertaken. The design is then assumed to be correct.

- **Production level testing**: Testing is undertaken on the systems that have been manufactured to determine that the system has been manufactured without defects that cause faults in the system operation.

- **Future-proofing the design**: Developing a design that is capable of being modified and its operation enhanced in the future according to the market requirements.

- **Aesthetics**: What concerns must be given to the appealing appearance of the product? For example, if the system is to be embedded within a motor car and will not be seen by the user of the motor car (or others), then the appearance is not necessarily of concern. However, if the product is to be used in the home and will be on display, then the aesthetics will be of great concern.

- **Ergonomics**: How will the product be used? Will there need to be a great amount of interaction with the user and so how will the product be designed to make the system both intuitive and easy to use?

The design process itself will not be an isolated activity. It must consider also the need to manufacture the design and the need for testing the design. In recent years, significant emphasis has been placed on the interaction between design and test, leading to the concept of design for testability (DfT). However, DfT is just one example of DfX (design for X). In general the following are also considered and approaches developed:

- **DfA**, design for assembly
- **DfD**, design for debug
- **DfM**, design for manufacturability
- **DfR**, design for reliability
- **DfT**, design for testability
- **DfY**, design for yield

The differentiation between a circuit and a system is further complicated by the increased demands and ability to provide electronic components with ever higher levels of integration—that is, more circuitry placed within individual components. This is leading to the situation in which individual ICs, normally used in an electronic circuit, would themselves be a complete electronic system. Such an IC with a high level of circuit integration is commonly referred to as a *system on a chip* (SoC).

Given the complexities in the circuitry that exists in a modern microprocessor, such a device might be referred to as a *System on a Chip*. However, this could be argued as not being the case. The modern microprocessor might be seen as just a complex integrated circuit which still requires external circuitry in a similar way as to older generation microprocessors. Therefore it would not be seen as an SoC as it is not a complete system within a single integrated circuit. The definition of the SoC is therefore something that needs to be considered carefully. This results in different forms of electronic circuits or systems being available:

- **Integrated circuit (IC)**: An electronic circuit fabricated on a die of semiconductor material, usually silicon based. The die is normally housed within a package although individual bare dies are available.

- **Printed circuit board (PCB)**: An insulating material (substrate) with integrated metal interconnect tracks that is used to mechanically secure and electrically connect electronic components.

- **Multichip module (MCM)**: An insulating material (substrate) smaller than a PCB in size, with metal interconnect tracks that mechanically secure and electrically connect individual ICs (either packaged ICs or bare dies). The MCM was originally referred to as a *hybrid circuit*.

- **System on a chip (SoC)**: A large integrated circuit that contains a complete electronic system.

- **System in a package (SiP)**: An extension to the idea of the MCM, but with the capability of higher levels of integration and three-dimensional (3-D) packaging.

2.2　Sequential Product Development Process versus Concurrent Engineering Process

2.2.1　Introduction

The process undertaken to develop a product is the means by which a design can be developed from an initial concept through to realization as a (commercial or noncommercial) product. One of two approaches can be undertaken to realize the product:

- sequential product development process

- concurrent engineering process

Essentially, these will identify the main steps involved in the development and production of a product and how these steps will interact with each other.

2.2.2 Sequential Product Development Process

In a *sequential design process*, each of the steps involved in the design process—from design concept through to production and testing—is completed before the next step begins. This traditional approach is shown in Figure 2.5.

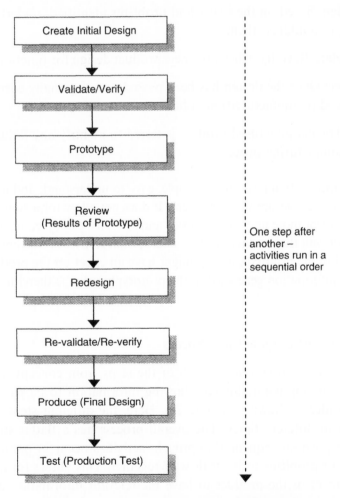

Figure 2.5: Sequential design process [3]

Here, the main steps are:

1. **Design**: Create the initial design.

2. **Validate/Verify**: Check the initial design for functional correctness.

3. **Prototype**: Create a physical prototype of the design and test the functionality of the design.

4. **Review**: Identify whether the design functions as expected and identify any issues raised and/or problems with the design that need to be resolved.

5. **Redesign**: Based on the issues and problems identified, undertake a product redesign to address them.

6. **Revalidate/Reverify**: Check the new product design for functional correctness.

7. **Produce**: Once the design has been passed as functionally correct, then it is produced (manufactured) in volume.

8. **Test**: The manufactured product is tested to identify any failures created by the manufacturing process.

Although this approach appears to be simple, easy to understand, and initially easy to manage, its sequential nature was inefficient. It does not allow for a step to interact with any other step except those immediately prior and after; for example, the prototyping step does not interact with the production step. This in-built restriction can create problems as issues identified in the prototyping step might have an effect on the production step. The important information generated in the prototyping step is therefore lost.

2.2.3 Concurrent Engineering Process

In a *concurrent engineering process*, each of the steps from concept through to production and testing is interlinked, allowing information to be passed among the steps. This idea is shown in Figure 2.6. Here, the different steps in the process appear at different times. The overall process has a flatter structure—in contrast to the previous sequential approach, activities occur in parallel—allowing any issues and/or problems to be dealt with together. This allows for all stakeholders in the development of the product to have the relevant information and assess the impact of design issues and changes on their part of the product development.

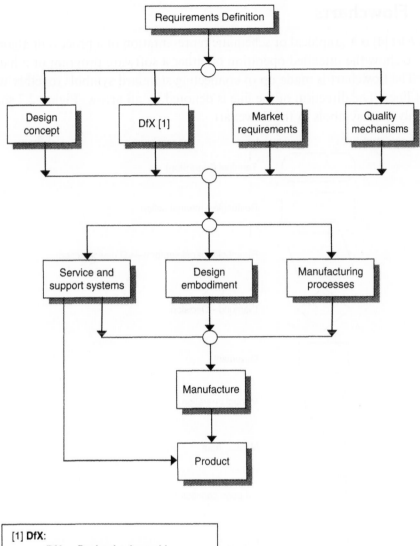

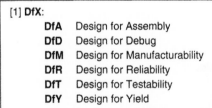

[1] **DfX**:
DfA	Design for Assembly
DfD	Design for Debug
DfM	Design for Manufacturability
DfR	Design for Reliability
DfT	Design for Testability
DfY	Design for Yield

Figure 2.6: Concurrent engineering process (after [3])

2.3 Flowcharts

A *flowchart* [4] is a graphical or schematic representation of a process or algorithm. It is used to show the intended operation of either a software program or a hardware circuit. The flowchart is made up of connecting standard symbols together with straight lines. The direction of the line is denoted by an arrow. Figure 2.7 shows the commonly used symbols in the flowchart.

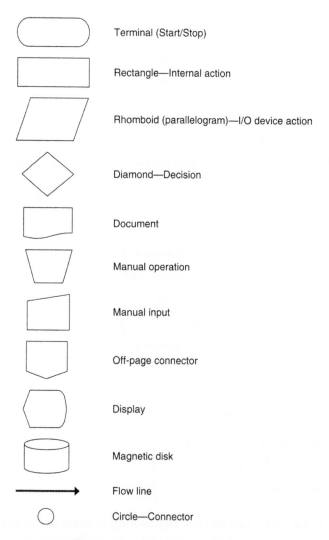

Terminal (Start/Stop)

Rectangle—Internal action

Rhomboid (parallelogram)—I/O device action

Diamond—Decision

Document

Manual operation

Manual input

Off-page connector

Display

Magnetic disk

Flow line

Circle—Connector

Figure 2.7: Flowchart symbols

The *terminal* symbol identifies the start and end of the flowchart. The *rectangle* (*internal action*) symbol identifies an internal action to be undertaken. The *rhomboid* (*I/O device action*) symbol identifies an action to be undertaken by an input or output device. The *diamond* symbol identifies a decision (or branch) to be made. One of two routes out of the diamond symbol will be undertaken depending on the result of the decision. The *document* symbol identifies a document media. The *manual operation* symbol identifies an off-line process to be undertaken by a person at a "human speed." The *manual input* symbol identifies the need for a manual input from a person using a device such as a keyboard or pushbuttons. The *off-page connector* symbol links a flowchart that is drawn on two or more pages. The *display* symbol identifies an output to an online display. The *magnetic disk* symbol identifies an input or output from magnetic disk storage (i.e., data file I/O). The *flow line* identifies the flow of the flowchart based on the actions and decisions. The *circle* symbol identifies a connection of *flow lines*.

An example flowchart is shown in Figure 2.8. Here, a software program detects an input that is a serial bitstream. The pattern to detect is a "101"

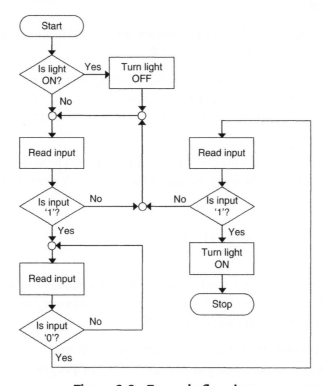

Figure 2.8: Example flowchart

sequence. When this sequence is detected, a light is turned on and the program stops.

2.4   Block Diagrams

A *block diagram* is a circuit or system drawing that identifies major functions and the interconnections between the functions, rather than showing a detailed implementation. Its purpose is to represent graphically a system consisting of subsystems or a subsystem consisting of components. It helps in the creation and interpretation of a design by

- allowing a design concept to be developed in order to identify the required arrangement prior to any detailed design process

- allowing a simplified view of a designed system to be viewed and interpreted

As an example, consider the block diagram for a basic central processing unit (CPU) for a microprocessor as shown in Figure 2.9.

The microprocessor will also contain ROM (holding specific program code for the microprocessor to work), RAM (for temporary storage of data), and a port (for data I/O between the microprocessor and the external electronic system).

The block diagram is a representation of the CPU system. The system itself consists of a number of subsystems. These are modeled by boxes with a text identifier. The identified blocks are:

- **Arithmetic and logic unit (ALU)**: Provides a set of arithmetic and logic functions.

- **Accumulator**: A register used to hold one of the inputs to the ALU and the results of an ALU operation. This is used for temporary storage and is one of the most used registers within the CPU.

- **Program counter (PC)**: This is a counter that increments after each instruction and tracks program execution to ensure that the program executes in the correct sequence.

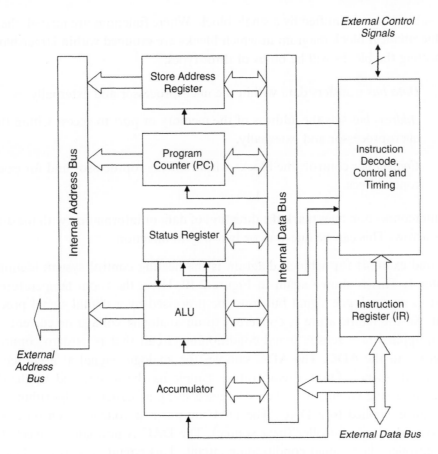

Figure 2.9: Basic CPU block diagram

- **Store address register**: A register that can be loaded with a single address in memory that might be required by the program.

- **Status register**: Also referred to as a *flag register*, it is used to store information relating to the last operation undertaken by the ALU.

- **Instruction decode, control, and timing**: Used for organizing the data flow between the different parts of the CPU.

- **Instruction register (IR)**: Used to store an instruction that the microprocessor is to decode and act upon.

Each subsystem is identified by a single block. Where functions are related, there may be a hierarchical block diagram in which blocks are grouped within larger blocks. Connecting the blocks will be buses of three types:

1. *Data bus* transfers data within the microprocessor and externally.

2. *Address bus* sets the address of the memory or port to access within the microprocessor and externally.

3. *Control bus* controls the blocks within the microprocessor and for external control lines.

The interconnection lines mark the direction of data or information with the direction of the arrow. This can be one-way or two-way in direction.

A second example for a block diagram is the heating control system identified in Chapter 1. This is shown again in Figure 2.10. Here, the room temperature is sensed as an analogue signal but must be processed by a digital signal processing circuit. So the temperature is converted to an analogue voltage or current. This is then applied to a sensor signal conditioning circuit that is used to connect the sensor to the ADC. The ADC samples the analogue signal at a chosen sampling frequency. Once a temperature sample has been obtained by the digital signal processing circuit, it is then processed using a particular algorithm, and the result is applied to a DAC. The DAC output is a voltage or current, which is used to drive a controller (heat source). The DAC is normally connected to the controller via a signal conditioning circuit. This circuit acts to interface the DAC to the controller so the controller can receive the correct voltage and current levels.

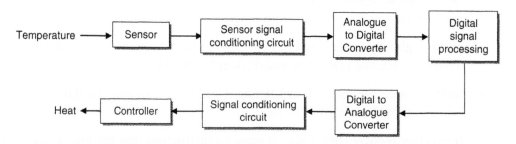

Figure 2.10: Heating control system block diagram

2.5   Gajski-Kuhn Chart

The *Gajski-Kuhn chart* [5, 6] is commonly referred to in the EDA industry [7] in relation to categorizing the different design abstraction levels and design synthesis. As shown in Figure 2.11, the chart takes the form of five concentric circles and three partitions or domains.

The five concentric circles characterize the hierarchical levels of the design process, with increasing abstraction from the inner to the outer circle. Each circle characterizes a model, and the models thus characterized are specific to the three domains.

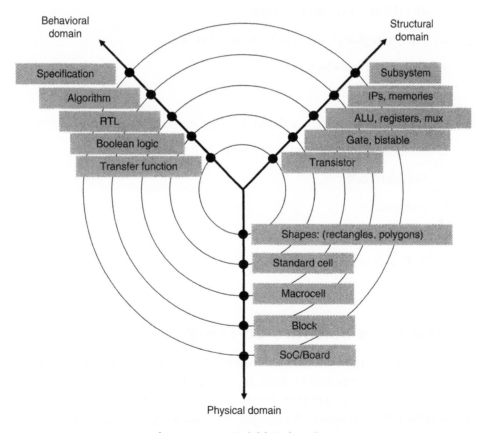

Figure 2.11: Gajski-Kuhn chart

- **Behavior**: describes the functional behavior of the system

    1. Specification

    2. Algorithm

    3. RTL

    4. Boolean logic

    5. Transfer function

- **Structure**: describes the circuits and subsystems that will be connected together to form the required system

    1. Subsystem

    2. IPs (IP blocks) and memories

    3. ALUs, registers, and multiplexers (MUX)

    4. Gates and bistables

    5. Transistors

- **Physical domain**: describes the underlying implementation of the system

    1. Shapes (rectangles and polygons)

    2. Standard cells

    3. Macrocells

    4. Blocks

    5. SoC and board

2.6   Hardware-Software Co-Design

Many digital circuits and systems are based on digital logic hardware only. However, many other digital circuits and systems are based on processors running a software program. These processors will then interface to external hardware circuitry. For such hardware (HW) and software (SW) designs, it is necessary to design the hardware and software parts together to create

- a working design (Designing a software program without knowing the hardware it will run on will ultimately result in a failure.)

- a design that uses the best set of hardware components

- a design that efficiently uses the available hardware

- a design that runs an efficient software program

- a design that is maintainable and can be upgraded

- a design that is cost-effective

Hardware-software co-design [8–10] is an idea that has been around for a long time, being continually refined and updated to adapt to emerging technologies. However, the fundamental basis remains the same: to provide an approach for the cooperative or collaborative design of electronic systems with hardware and software parts.

An approach to hardware-software co-design is shown in Figure 2.12. The design approach initially starts with the *system specification*, which contains a document or set of documents that define what exactly the system is intended to do. The *design choices* are then made to identify which parts are to be undertaken in hardware and which parts are to be undertaken in software. This is followed by the partitioning of the design into the *hardware parts* and *software parts*, along with the parts that provide the *interface* between them. It is at this point that the design implementation typically comes to the hardware and software designers. Given that this initial partitioning of the design has been completed, then the system design is refined to develop the specifications for the hardware and software parts.

When those specifications have been developed and formally agreed on, the design can be undertaken. Specific EDA tools relevant to the electronics or the software programming are used. When hardware and software designers work in close co-operation, EDA tools that support an integrated hardware-software co-design approach can be used. Simulation (validation) and formal verification support the design process. On integration of hardware and software, a *hardware-software co-simulation* might be undertaken that will simulate the operation of the software program on the actual hardware. *Design prototyping* creates a physical prototype of the overall system that allows the operation of the real design to be evaluated. On successful completion of the design prototyping, the final design would be ready for *design production*. Depending on the required application, the number of systems to be produced can range from one to millions.

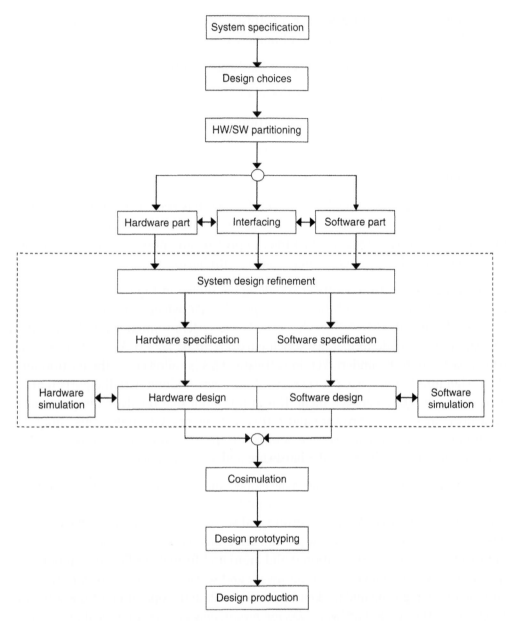

Figure 2.12: Hardware-software co-design

2.7   Formal Verification

Formal verification is essentially concerned with identifying the correctness of hardware [11] and software design operation. Because verification uses formal mathematical proofs, a suitable mathematical model of the design must be created. Today, both verification and validation processes are typically undertaken to analyze a design implementation. Verification differs from validation in that:

- *Validation* seeks to examine the correctness in the operation of the electronic circuit or software program implementation by examining its behavior (e.g., through simulation or prototype evaluation).

- *Verification* seeks to examine the correctness in the operation of the electronic circuit or software program implementation by a mathematical proof.

An example where both verification and validation can be undertaken is during the design of digital circuits and systems using hardware description languages (HDLs). This idea is shown in Figure 2.13. Here, the process starts with an RTL (register

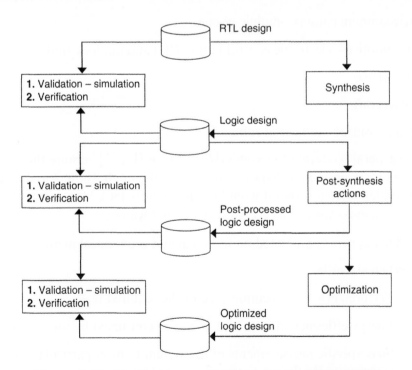

Figure 2.13: Verification and validation of an RTL design

transfer level) description of a digital circuit. This is synthesized using a suitable synthesis tool.

After the design has been synthesized into a netlist, postsynthesis actions are undertaken on the design such as clock tree insertion and testability (typically a scan path test). The design is then optimized to form the final design, then simulated. Validation is undertaken via simulation, and verification is undertaken using a mathematical model of the design.

2.8   Embedded Systems and Real-Time Operating Systems

A *real-time operating system* (RTOS) is a software operating system that is intended for use in real-time applications such as:

- consumer electronics—household appliances, cameras, audio equipment
- telecommunications—mobile phones
- automotive—electronic control unit (ECU) and antilock brakes
- aerospace
- spacecraft
- plant control—industrial robots

These are generally referred to as *embedded systems* [12, 13] because they include computing functions and are dedicated to a particular application. An obvious aspect of an embedded system is that it would not necessarily look like a computer, but instead are enclosed within the everyday items that we use.

An embedded system is evaluated on technical and economical merits:

- **Technical merits:**
  - **Performance**: the execution time of the required tasks
  - **Energy efficiency**: the amount of power consumed by the system
  - **Size**: specific measurements of the system to meet particular size constraints for the application

o **Flexibility**: the ability to reconfigure the system for different applications

o **Deterministic operation**: the system performs tasks within a guaranteed time period.

- **Economical merits**:

  o **Unit cost**: cost to manufacture a unit, excluding nonrecurring engineering (NRE) costs

  o **Nonrecurring engineering (NRE) costs**: costs to design and manufacture the system. For example, if an ASIC is to be part of the system, then there would be NRE costs associated with designing and manufacturing the mask sets required in the lithographic steps in the ASIC wafer fabrication.

  o **Flexibility**: the ability to redesign the system, or parts of the system, without incurring high NRE costs

  o **Time to market (TTM)**: the time required to develop the system so that it is in a state that can be sold to the customer

The operating system running on the embedded system processor is a multitasking operating system in that it is required to execute multiple processes concurrently by multitasking the CPU of the processor used within the embedded system. Tasks would be executed using one of two basic design approaches:

1. **Event-driven**: The CPU switches to a particular task when the task itself requests servicing (via interrupts on the CPU). Tasks are prioritized, and a task with a higher priority will be serviced before a task with a lower priority.

2. **Time-sharing**: The CPU switches to between tasks on a time-sharing basis.

An important aspect of the embedded system would be that its operation is deterministic. This means that, if designed correctly, it can undertake specific tasks within a specific, guaranteed time period. This feature differs from the general purpose computer (such as a desktop or laptop computer), whose operation would not be deterministic.

2.9 Electronic System-Level Design

With the increasing complexities of digital systems to be created today, particularly for applications such as communications, there is a need to enable the designer to work at higher levels of design abstraction and away from the detailed design aspects. Designing at such high levels is referred to as *electronic system-level* (ESL) design [14, 15]. ESL design is an emerging area for the design community and is a response to the emerging needs of the designers (both hardware and software) to support their need to develop more complex systems designs but in a reduced time. This allows the designer to:

- raise the design entry point to a design abstraction level to make the complex design problem manageable

- concentrate on high-level design concept issues rather than low-level design implementation issues

- reduce design time by automating specific time-consuming tasks that are suited to automation

- explore the design space at the abstraction level and explore trade-offs (in size, performance, power consumption) in the design decisions

ESL design is a response to designers working at a behavioral level, as has become more prevalent in recent years, with behavioral-level modeling of designs being developed for synthesis into logic. However, ESL design is required to overcome limitations with working at design behavioral level and considers higher levels of design abstraction and complexity.

To facilitate this design approach, then, the designer requires:

- design entry tools to support ESL design

- design languages (either textural or graphical) that effectively model the wide range of designs to be encountered and the different levels of design abstraction

- design simulation tools to simulate complete systems at different levels of design abstraction

For ESL design, suitable EDA tools are required to enable high-level designs to be automatically translated to HDL code, which can then be synthesized in the normal manner.

2.10   Creating a Design Specification

A *design specification* describes the detailed operation and attributes of a system and is used as the basis of the design concept. With small designs, developing a clear and concise design specification is a relatively straightforward task. However, as designs become more complex, with increased functionality and more customer requirements, then the task of writing a design specification becomes more complex.

In most cases, a specification is a document that can be referred to by all or some of the stakeholders (active participants—the designers and the customers) involved in the design process. Normally two or more specification documents are required for internal use (by the designers only) and for external use (by the designers and the customer). The purposes of the design specification are to:

- involve all stakeholders in the plans for the system development—the specification should be written for the particular audience (technical, nontechnical, management, etc.)

- identify potential problems and risks before they are encountered to save time and money

- be used as the basis for project planning and review

- be used as the basis for the design itself

Whatever the use of the design specification, it follows the same set of requirements:

1. Be clear.

2. Be concise.

3. Avoid general statements and be specific.

4. Avoid statements that are open to multiple interpretations.

5. Be accurate.

6. Be available in a format that is agreed by its users.

7. Adhere to specific requirements and standards adopted by the organizations involved.

8. Be readable.

When considering the creation of a design specification, it is sometimes easier to identify what *not* to do rather than what to do. For example avoid using statements such as "The user interface should be user friendly." After all, what is actually meant by *user friendly*? An interface that appears user friendly to one person may be impossible to use by someone else! For example, a software programmer who works at a UNIX™ or Linux command line and never touches a graphical user interface (GUI) would not necessarily appreciate a highly complex GUI with many unnecessary options. Hence, the requirements of end-user must always be considered.

Aside: A humorous read on how engineers, scientists, and software programmers think is in "The Dilbert Principle" by Scott Adams (Boxtree, 1997). Particularly illuminating is Chapter 14, "Engineers, Scientists, Programmers, and Other Odd People"!

Although a design specification is generally a document, it can also take other forms: diagrams, charts, tables, databases, prototypes, or mock-ups. Mock-ups are different from prototypes in that mock-ups are scaled models to show what the system would look like, whereas the prototype is a fully functional system used for evaluating the system prior to manufacture.

2.11   Unified Modeling Language

UML (unified modeling language) [16] is a standardized specification language used in software engineering for object modeling—specifically, for software specification, visualization, construction, and documentation of the software system and its component parts. UML was conceived with the aims to:

- provide software developers with a visual programming language with which to develop models of the software

- provide a means to extend the core concepts

- be independent of any particular programming language and software development process

- provide a basis on which to formally understand the modeling language

- integrate best practices in software development

- support high-level software development concepts

Although conceived for software engineering, UML is not restricted to modeling software, but also has applications in such areas as systems engineering modeling and process modeling. When a model is developed in UML, the UML model forms the basis to translate the UML model to other languages such as Java™.

Because UML is a visual language, a UML diagram is created to allow developers and customers to view the software system from their different perspectives and at different levels of abstraction. UML diagrams commonly include the following:

- **Use case diagram**. This displays the relationship between actors and use cases. An *actor* is a user of the system who applies a stimulus to the system and cannot be controlled by the system itself. The actor is seen as a role rather than a physical person. *Use cases* are services that the system knows how to perform. Figure 2.14 shows an example case diagram for a user of a bank ATM machine. The actor is drawn as a stick figure, and the use case is drawn as an ellipse. The lines show the interactions.

- **Class diagram**. This display provides a static view of the classes in a model. It also shows the relationships such as containment, inheritance, and associations.

- **Interaction diagram**. The two types of interaction diagram are the sequence diagram and the collaboration diagram:

  o The *sequence diagram* displays the time sequence of the objects participating in a particular interaction. The objects will interact by passing messages among themselves. On the diagram, the vertical direction represents the time, and the horizontal direction represents the different objects.

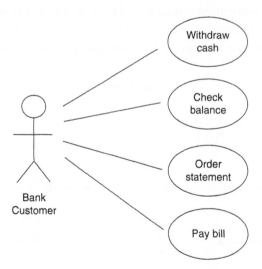

Figure 2.14: Example case diagram for a bank ATM machine

○ The *collaboration diagram* displays the interaction among objects and the links between objects. Numbers are used to show the sequence of messages passed among objects.

• **Activity diagram**. This displays a state diagram that focuses on flows driven by internal object processing. This provides a means to describe *workflow*.

• **Statechart diagram**. This displays the sequences of states that an object will go through during an interaction with a received stimulus and the object's responses and actions. This diagram is closely related to the activity diagram. Statechart diagrams provide a means to describe the behavior of dynamic model elements.

• **Implementation diagram**. The two types of implementation diagram are the component diagram and the deployment diagram:

○ The *component diagram* displays the relationships among the software components in the system.

○ The *deployment diagram* displays the hardware configuration used to implement the system and the links between the hardware components.

2.12   Reading a Component Data Sheet

All components that are available to purchase for use within an electronic circuit or system will have an associated *data sheet*. The data sheet provides the necessary information for the designer of an electronic circuit to determine whether the component is suitable for the particular application. The data sheet (see Figure 2.15)

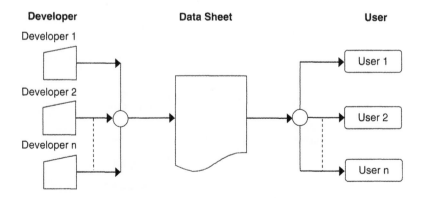

Figure 2.15: Data sheets

should be presented in a style that is quick and easy to read, and allows the designer to evaluate the information to determine component suitability. Reading a component datasheet takes practice and familiarity with the typical style of presentation [17]. Writing a data sheet takes much more practice.

There is no single style to the presentation of information within the datasheet, but the following style for a digital IC is a good generic model:

- Company logo and part number (and name)

- Features: the general electrical and thermal features to be found within the component

- Description: an introduction to the component

- Package types and pinout: the package types that the IC can be obtained in and the pin designation (pinout) for each package type. Appendix C identifies the main IC packages commonly used (see the last paragraph of the Preface for instructions regarding how to access this online content).

- Functional block diagram: a block diagram of the internal architecture of the IC

- Absolute maximum ratings: the absolute maximum ratings give the values of voltage, current, and temperature that, if exceeded, could cause permanent damage to the component. Table 2.1 provides example absolute maximum ratings for an example digital IC.

- ESD warning: a warning logo and description to identify the potential damage to the component from electrostatic discharge (ESD)

- Terminology: identifies the terminology and abbreviations used in the data sheet and their meaning

Table 2.1: Example absolute maximum ratings for a digital IC

Symbol	Parameter	Value	Unit
V_{CC}	D.C. power supply voltage	−0.5 to +7.0	V
V_I	D.C. input voltage	−0.5 to +7.0	V
V_O	D.C. output voltage	−0.5 to +7.0	V
I_O	D.C. output current	±50	mA
I_{CC}	D.C. output current per supply pin	±100	mA
I_{GND}	D.C. ground current per supply pin	±100	mA
T_{stg}	Storage temperature	−65 to +150	°C
T_L	Lead temperature (10 sec)	300	°C

- Thermal information: temperature range and package thermal resistance information

- Operating conditions: The D.C. power supply voltage and D.C. input voltage range (minimum and maximum) expected for normal operation

- Static electrical specifications: voltage and current specifications—minimum (MIN), typical (TYP) and maximum (MAX), or a subset of these—that must be applied to the IC, or that will be guaranteed by the IC, for correct operation. In addition, input and output capacitance of the inputs and outputs would normally be provided.

- Description of operation: a detailed description of the operation of the IC, including how the designer would use the features of the IC in his or her own application

- Function pin definition: identification of the name and description of operation for each pin on the IC in a table format

- Dynamic electrical specifications: timing information for the system timing waveforms

  - System timing waveforms: timing diagrams showing the required digital timing for operation of the IC

  - Example use: also shows how they can be interfaced to other electronic circuits

  - Package dimensions: for the different packages in which the component is available

The parameters for the device will be taken for specific test conditions, such as ambient temperature and power supply voltage. These conditions should be noted with care as the quoted parameters are only valid at these operating conditions.

T_{stg} identifies the storage temperature for the IC. However, the IC will have temperature ratings for different scenarios:

- **Storage**: the range in temperature that the IC can handle without damage during component storage (before power is applied to the IC)

- **Lead**: the absolute maximum temperature (for a given duration) that the IC can handle at the IC lead (pin) without damage during component soldering to a PCB

- **Junction**: the maximum temperature that the die within the IC can reach under any condition without damage

- **Operating**: the range in temperature that the IC can handle without damage during component use. This will depend on the application, and the IC will be one of the following types:

    ○ **Commercial**: 0°C to +70°C

    ○ **Industrial**: –40°C to +85°C

    ○ **Military**: –55°C to +125°C

2.13 Digital Input/Output

2.13.1 Introduction

When preparing to transmit digital data in the electronic system, these questions need to be asked:

- What is a logic level (0 or 1) in terms of the voltage levels in the circuit?

- How is the digital data to be transmitted? What is the communications channel?

- What preprocessing must the data undergo before it can be transmitted, and what postprocessing must the data undergo after it has been received?

- What effect does the communications channel have on the signal?

Data transmission can take a number of forms and serve different purposes; an example of this is shown in Figure 2.16. Here, a number of PCs are locally connected on a LAN and connected to the external world using the telephone line (modem), the Internet (telephone or dedicated lines), and satellite.

Figure 2.16 shows the communications between large electronic systems. Communications will also occur locally within the system itself, whether within individual ICs, between ICs on a PCB, or between subsystems (e.g., between separate PCBs). Whatever the purpose of the communications is, there will be a need to design to particular standards for the correct transmission and receipt of data at various speeds of data transmission. Each digital IC will have pins to be used for creating (transmitting) and capturing (receiving) digital data. The digital inputs to an IC and

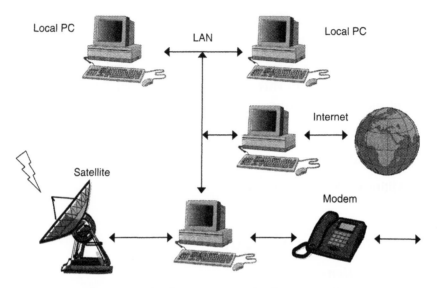

Figure 2.16: Data communication examples

the digital outputs from an IC will adhere to particular standards. A number of the main standards will be identified and discussed in Section 2.14, "Parallel and Serial Interfacing."

The I/O signals will be either single ended or differential depending on the particular standard. A digital IC will adhere to one or a number of standards. For example, the Xilinx® range of field programmable gate arrays (FPGAs) and complex programmable logic devices (CPLDs) can be configured by the user to adhere to one of a number of standards. Table 2.2 shows example I/O standards that are supported by the Xilinx® PLDs and configured by the designer. With such a programmable I/O

Table 2.2: Example I/O standards supported by the Xilinx® PLDs

Standard	Standard Description
LVTTL	Low-voltage transistor-transistor logic (3.3 V level)
LVCMOS33	Low-voltage CMOS (3.3 V level)
LVCMOS25	Low-voltage CMOS (2.5 V level)
LVCMOS18	Low-voltage CMOS (1.8 V level)
1.5 V I/O (1.5 V levels)	1.5 V level logic (1.5 V level)
HSTL-1	High-speed transceiver logic
SSTL2-1	Stub series terminated logic (2.5 V level)
SSTL3-1	Stub series terminated logic (3.3 V level)

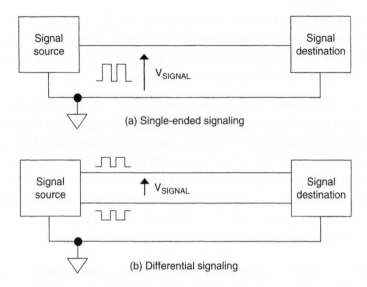

Figure 2.17: Single-ended versus differential signals

capability, before the device has been configured with the appropriate standard, the device will default to one of the standards. It is therefore important for the designer to identify the default standard and the implications of using a particular standard on the overall circuit operation.

A *single-ended signal* is a single signal on a single wire that creates a voltage that is referenced to a common point in the circuit (usually the 0 V common connection). *Differential signals* utilize two wires to carry complementary signals, and the signal is the difference in voltage between the two wires (Figure 2.17). Differential signaling is suitable for use with low-voltage electronics (such as mobile devices that obtain power from batteries) and is robust against noise added during data transmission.

Two important points to note with digital logic ICs are:

1. No input to an IC input is to be left unconnected (referred to as *floating input*). If an input to an IC is not required, then it must be tied to logic level (0 or 1). This is usually achieved by connecting a high-resistance value resistor (typically 10 to 100 kΩ in value) between the unused input and one of the power supply connections (V_{DD} for logic 1, V_{SS} or GND for logic 0). In some ICs, specific inputs might be designed to be used only for specific circumstances and

will have integrated into the IC input pin circuitry a *pull-up* (to logic 1) or *pull-down* (to logic 0) component. Such integrated pull-up or pull-down components alleviate the need for the designer to place resistors on the PCB and so reduce the PCB design requirements.

2. Where a logic gate only produces a logic 0 or 1 output, then no two or more logic gate outputs are to be connected together unless the implementation technology (the circuitry within the logic gate) allows this. Certain logic gate outputs can be put into a high-impedance state, which stops the output from producing a logic output and instead turns the output into a high-impedance electrical load. Circuits with a high-impedance output are used where multiple devices are to be connected to a common set of signals (a bus) such as a microprocessor data bus.

Whenever an FPGA or CPLD is used, there may be situations where not all of the available digital I/O pins are used. In this case, the unused pins are not connected to any circuitry and would be left unconnected on the PCB. However, internally within the FPGA or CPLD, the pin circuitry would be arranged so that it would not be left floating. The designer of a system using FPGAs or CPLDs should check what happens when the pin is not used (i.e, not configured) given the particular arrangement of the device.

In telecommunications systems, the transmission of high-speed digital data is often tested using an *eye diagram* (or *eye pattern*). Essentially, this is an oscilloscope display where the received data is sampled at a fixed rate and applied to the vertical input of the oscilloscope. The data rate is then used to trigger the horizontal sweep of the oscilloscope. The eye diagram is so called because, for several types of signal, the pattern looks like a series of eyes. In Figure 2.18, the top eye diagram is for an undistorted signal, and the bottom eye diagram includes the noise in the signal and signal timing errors.

Analysis of the eye diagram can identify issues such as:

- signals that are poorly synchronized to the system clock

- noise

- overshoot and undershoot

- signal jitter (variance in signal transmission timing)

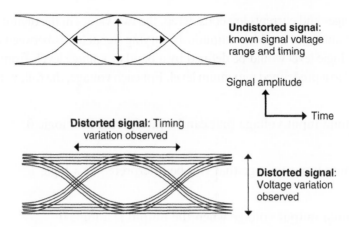

Figure 2.18: Eye diagram: undistorted signal (top) and distorted signal (bottom)

2.13.2 Logic-Level Definitions

When designing with logic gates, the primary concern is to consider the logic levels (logic 0 and logic 1) and ensure that the correct logic levels appear at the required nodes in the circuit at the right time. However, the underlying circuitry within the logic gates is analogue (using transistors), so the voltages and currents in the design must be considered. Shown in Figure 2.19 is a two-input AND gate with voltage signal generators connected to the inputs *A* and *B*, and the resulting voltage is monitored at the output *Z*.

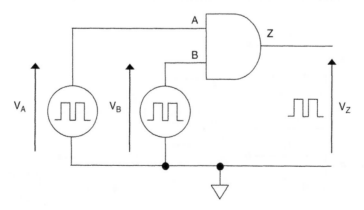

Figure 2.19: Two-input AND gate with voltage sources

When the voltages and currents are considered, the two values in the digital world (0 or 1) become, in the analogue world, continuously varying signal levels between a lower and upper limit. A logic level would be defined by a band of voltage levels from a predefined minimum level to a predefined maximum level. For each voltage, the following are defined:

- V_{IL}
  Maximum input voltage that can be interpreted as a logic 0

- V_{IH}
  Minimum input voltage that can be interpreted as a logic 1

- V_{OL}
  Maximum output voltage when the output is a logic 0

- V_{OH}
  Minimum output voltage when the output is a logic 1

These voltage levels are discussed in the next section.

In addition to the voltages defined above, the logic gate will also have low-level and high-level input and output currents as shown in Figure 2.20:

- I_{IH}
  High-level input current: the current that flows into an input when a high-level voltage (value to be specified) is applied

- I_{IL}
  Low-level input current: the current that flows out of an input when a low-level voltage (value to be specified) is applied

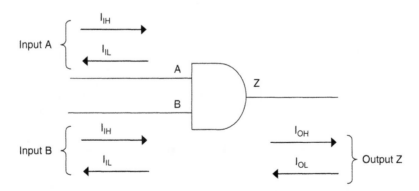

Figure 2.20: Two-input AND gate with current definitions

- I_{OH}

  High-level output current: the current that flows out of an output when a high-level voltage (logic 1 output) is created. The output load conditions will need to be specified.

- I_{OL}

  Low-level output current: the current that flows into an output when a low-level voltage (logic 0 output) is created. The output load conditions will need to be specified.

When designing with digital ICs, these voltage and current figures should be provided in the particular device data sheet.

2.13.3 Noise Margin

In digital logic, two logic levels are defined: logic 0 and logic 1. Each logic level will represent a voltage the analogue circuit level (the transistor operation within the digital logic gate). In the digital logic inverter, the input and output voltages and how they will create the required logic levels can be considered. Consider the static CMOS inverter, which uses one nMOS and one pMOS transistor as shown in Figure 2.21. Here, the logic symbol and the transistor level connections are shown.

The circuit requires a DC power supply voltage (V_{DD}/V_{SS}) to operate. Here, two signal voltages are identified (V_{IN} and V_{OUT}), which represent the input and output voltages. A logic 0 will be considered as an input voltage at the V_{SS} (0 V) level, and a logic 1 will be considered an input voltage at the V_{DD} (+3.3 V) level. For each voltage, the following are defined:

- V_{IL}

  Maximum input voltage (V_{IN}) that can be interpreted as a logic 0

- V_{IH}

  Minimum input voltage (V_{IN}) that can be interpreted as a logic 1

- V_{OL}

  Maximum output voltage (V_{OUT}) when the output is a logic 0

- V_{OH}

  Minimum output voltage (V_{OUT}) when the output is a logic 1

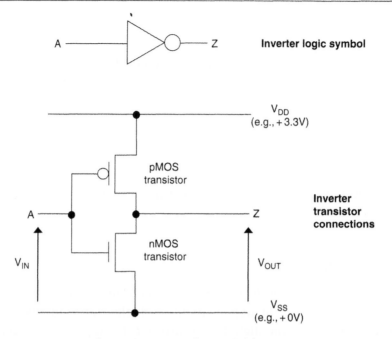

Figure 2.21: Static CMOS inverter

This means that the input and output voltages will not be a single value, but rather the logic level will represent a band of voltage levels from a predefined minimum level to a predefined maximum level. Two values for *noise margin* are then identified:

- **NM_L**
  Noise margin for low levels: $NM_L = V_{IL} - V_{OL}$

- **NM_H**
  Noise margin for high levels: $NM_H = V_{OH} - V_{IH}$

Figure 2.22 graphically displays the noise margin and hence the tolerance of the circuit to variations in voltage level so the logic levels can be viewed. The noise margin for a circuit becomes increasingly important for low-voltage systems (moving down to and below $1.0\,V$ V_{DD}) as the noise margin decreases and the potential for noise to corrupt values can increase (a logic 0 level becomes a logic 1, and vice versa).

Table 2.3 provides the V_{IL}, V_{IH}, V_{OL}, and V_{OH} voltage levels for several TTL and CMOS family variants [18] when V_{DD}/V_{CC} is $+5.0\,V$.

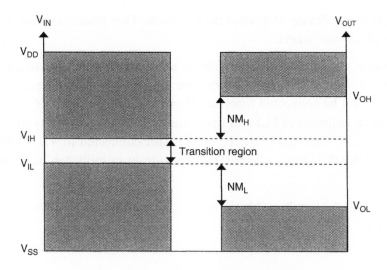

Figure 2.22: Noise margin definitions

Table 2.3: TTL and CMOS family variants

Parameter/Device	V_{IL} (max)	V_{IH} (min)	V_{OL} (max)	V_{OH} (min)
CMOS				
4000B	1.5	3.5	0.05	4.95
74HC	1.0	3.5	0.1	4.9
74HCT	0.8	2.0	0.1	4.9
74AC	1.5	3.5	0.1	4.9
74ACT	0.8	2.0	0.1	4.9
TTL				
74LS	0.8	2.0	0.5	2.7
74AS	0.8	2.0	0.5	2.7

2.13.4 *Interfacing Logic Families*

In an electronic system, ICs must be connected at the PCB level. When using digital logic ICs, the designer may need to interface ICs that are based on different circuit architectures (basically the different variants of TTL and CMOS logic), and that may also operate at different power supply voltage levels. In such situations, the designer will need to ensure that the device providing a signal can meet the voltage and current requirements of the device or devices being driven. Two figures are normally quoted for *fan-in* and *fan-out*, where:

- *Fan-in* is the number of logic outputs that can be connected to a logic gate input. Standard TTL and CMOS logic outputs (providing logic levels 0 and 1) should not be connected together. However, certain digital ICs provide for open-collector (TTL) and open-drain (CMOS) outputs as shown in Figure 2.23. External to the IC is a resistor connected to V_{CC} (TTL) or V_{DD} (CMOS). Open-collector and open-drain outputs can be connected together.

- *Fan-out* is the number of logic inputs that can be driven from a logic gate output.

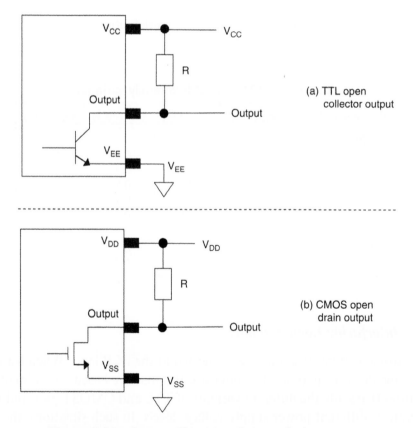

Figure 2.23: Open-collector and open-drain outputs

TTL Logic IC Driving a CMOS Logic IC

Considering both digital logic ICs operating on the same power supply voltage, then with a CMOS logic gate input, the current that would flow into an input would be low and a TTL device would be able to provide the necessary current to drive one or more CMOS logic IC inputs. However, problems will occur when considering the voltage levels required by the different technologies (V_{IL}, V_{IH}, V_{OL}, and V_{OH}). Table 2.3 shows several examples. Some CMOS family variant devices (e.g., 4000B, 74HC, and 74AC series) have V_{IL}, V_{IH}, V_{OL}, and V_{OH} levels different than TTL, whereas other family variant devices (e.g., 74HCT and 74ACT series) have V_{IL}, V_{IH}, V_{OL}, and V_{OH} levels compatible with TTL. A common solution to overcoming the problem for non-TTL level CMOS devices is to use an external pull-up resistor as shown in Figure 2.24. Here, the power supply voltage is $+5.0$ V. A typical value would be 10 kΩ.

When the TTL output is a logic 1, then the pull-up resistor will pull the voltage to approximately $+5.0$ V, which produces a voltage high enough for the CMOS input to receive a logic 1 input.

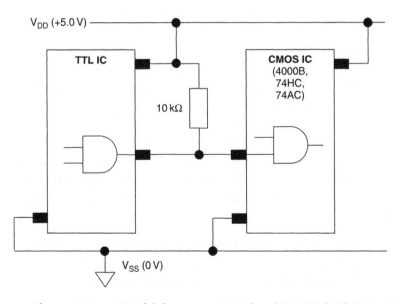

Figure 2.24: TTL driving a non-TTL level CMOS logic IC

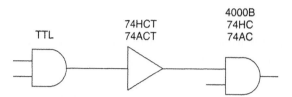

Figure 2.25: TTL to CMOS using an HCT or ACT interface IC

An alternative interfacing method, as shown in Figure 2.25, is to use a 74HCT or 74ACT device as a buffer between the TTL and non-TTL level CMOS devices.

CMOS Logic IC Driving a TTL Logic IC

When a CMOS logic IC is to drive a TTL logic IC (+5.0 V power supply), then:

- A 74HCT or 74ACT IC can be connected directly to a TTL IC.

- A 74HC, 74AC, or 4000B IC can be connected directly to a TTL IC.

Lower Power Supply Voltages

In past times, the +5.0 V DC power supply was commonly used. Now, however, many digital ICs operate at +3.3 V, +2.5 V, or +1.8 V, with some operating as low as +1.0 V. In this case, care is needed when using different power supply voltages, particularly in many microprocessors, FPGAs, and CPLDs that operate on a dual power supply (one power supply for the core of the IC and a second for the I/O circuitry). The I/O power supply tends to be higher than the core power supply to enable connections to other ICs.

In some cases, an IC would operate at a power supply of +3.3 V, with the digital logic levels created by 0 V (logic 0) and +3.3 V (logic 1), but would also be capable of accepting a higher input voltage (+5.0 V tolerant) to enable direct connections to +5.0 V logic devices.

Where mixed power supply voltages are to be used in a circuit, and the ICs working at different power supply voltage levels and signals are to be connected, this is typically achieved by:

1. Direct connection, if the ICs allow for this capability

2. Using a pull-up resistor where a lower-voltage device is to drive a higher-voltage device

3. By using a special level translator IC

4. By configuring the I/O pin to the required standard (if possible)

Techniques 1 to 3 are shown in Figure 2.26 and Figure 2.27 in relating +2.5 V logic to +3.3 V logic. A similar approach would be taken for interfacing +3.3 V logic to +5.0 V logic. Technique 4 would be identified in the particular IC data sheet.

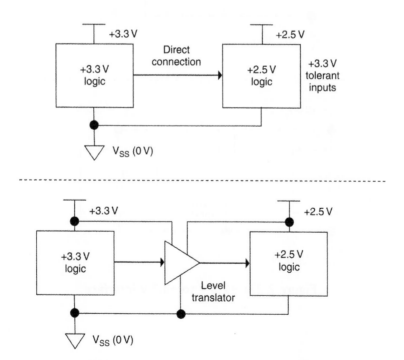

Figure 2.26: +3.3 V to +2.5 V interface

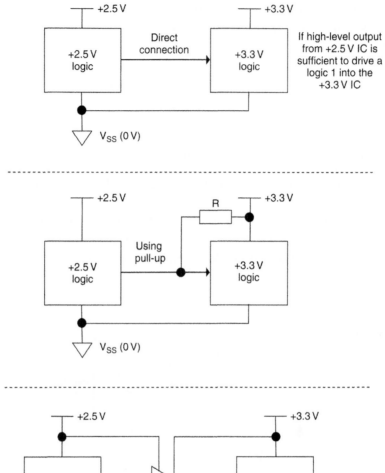

Figure 2.27: +2.5 V to +3.3 V interface

2.14   Parallel and Serial Interfacing

2.14.1   Introduction

Interfacing the electronic system allows the electronic circuit or system to communicate internally and externally. The communications interface allows the transmission of either analogue signals or digital data. A system that transmits data to and receives data from an external source is shown in Figure 2.28.

Each electronic system communicates with other systems by transmitting data via a *transmitter* (Tx) subsystem and receives data via a *receiver* (Rx) subsystem. The medium between the two systems is the *communications channel*. However, when analogue signals or digital data are transmitted through the communications channel, noise might be added to the signal, potentially corrupting the data. A great deal of care must be taken to ensure that the electronic systems do not use corrupted information.

Although information can be sent or received as analogue signals or digital data, digital data transmission is increasingly common and occurs as either *parallel* or *serial* data transmission:

- **Parallel data transmission**. Multiple bits of data are transferred simultaneously, allowing high-speed data transfer.

- **Serial data transmission**. One bit of data is transferred at a time (a serial bitstream). Serial data transmission takes longer, but when the data is transmitted on electrical wires (typically copper wires), fewer wires are required than with the parallel data transmission. Serial data transmission also lends itself to data transmission via optical fibers and wireless methods.

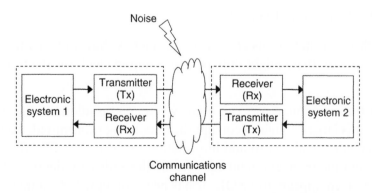

Figure 2.28: Data transmission and receipt

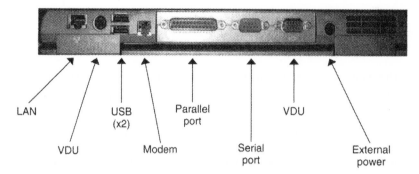

Figure 2.29: Rear view of laptop identifying PC connections

Many systems allow several parallel and serial communications standards. The PC is a good example. Figure 2.29 shows the rear view of a PC, with several connections identified.

When the data are transmitted, they must be received and stored for use. Data transmission will be either *synchronous* or *asynchronous*:

- **Synchronous**, in which a continuously running clock is carried along with the data, and the data are synchronized with the clock. Both of these signals are received by the receiver circuit, and the receiver uses both the clock and the data inputs to capture and store the data for use.

- **Asynchronous**, in which only the data are transmitted. An internal clock within the receiver is used to synchronize the receiver with the data in order to capture and store the data for use.

The basic idea is shown in Figure 2.30.

For the synchronous data transfer, a separate clock is shown for the transmitter and receiver. In practice, there might only be one common clock for the transmitter and receiver.

During data transmission, errors can occur when noise is added to the signal and when the noise is large enough to corrupt the data being transmitted. The transmitter circuit can include the ability to add information to the data before they are transmitted, and the receiver circuit can include the ability to identify whether the data it has received appears to be OK or has been corrupted. A simple method for error checking is to use *parity checking*, in which a bit is added and transmitted with

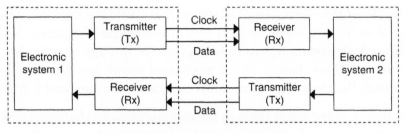

(a) Synchronous data transmission

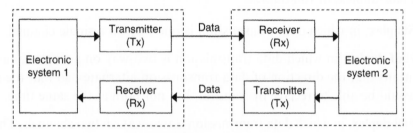

(b) Asynchronous data transmission

Figure 2.30: Synchronous and asynchronous data transfer

the data. Considering a byte of data (8 bits) as an example, parity checking is of two types:

- *Odd parity coding* will set the parity bit to a logic 1 if the number of logic 1s in the byte is even, so that the total number of logic 1s is an odd number. If the receiver receives an odd number of logic 1s, then it will identify that the byte was transmitted correctly.

- *Even parity coding* will set the parity bit to a logic 1 if the number of logic 1s in the byte is odd, so that the total number of logic 1s is an even number. If the receiver receives an even number of logic 1s, then it will identify that the byte was transmitted correctly.

Parity checking is a rudimentary method, and most communications systems include more sophisticated capabilities.

The characteristics of the channel must also be considered, the data may need to be modulated before transmission. Modulation takes either of two forms:

- *Baseband signals* in digital are the 1s and 0s being generated. On a PCB and communicating between ICs on the PCB, baseband signals are used. These signals cover a frequency range from DC to an upper frequency value.

- *Modulated signals* are baseband signals that have been modulated by a carrier signal so that the entire signal is now at some higher frequency. Modulation allows the baseband signals to be transmitted through a particular communications channel. When modulated signals are transmitted and received, the electronic system must include a *modulator* and a *demodulator*.

The transmission of the signal through the communications channel can be either one-way or two-way, so the designer must decide whether the communication is to be simplex, half-duplex, or full-duplex:

- **Simplex**, in which data transmission is one-way on a single channel.

- **Half-duplex**, in which data transmission is two-way on a single channel. This means that the direction of data transmission alternates, so that the system would be able to receive or transmit, but not both at the same time.

- **Full-duplex**, in which data transmission is two-way on two channels. This means that an electronic system would be able to receive or transmit at the same time.

This idea is shown in Figure 2.31.

Finally, the signal will be transmitted through the communications channel via electrical wires, optical fibers, or using wireless methods.

- **Wired**, in which metal wires, typically copper, are used to transmit the electrical signal.

- **Optical fiber**, in which an electrical signal is converted to an optical (light) signal and transmitted along the optical fiber. This allows high transmission rates and low loss, so that signals can be transmitted over long distances, and a low bit error rate. The electrical signal is generated either by a light-emitting diode (LED) creating noncoherent light or by a laser creating coherent light. At the receiver end, the signal is converted back to an electrical signal using a photodiode or phototransistor.

- **Wireless**, in which an electrical signal is modulated and applied to an antenna. The more popular modulation methods are AM (amplitude modulation), FM (frequency modulation), and PM (phase modulation). The signal is transmitted through free space, and at the receiver, another antenna picks up the transmitted signal, demodulates it, and restores it. It must then be amplified before it can be used.

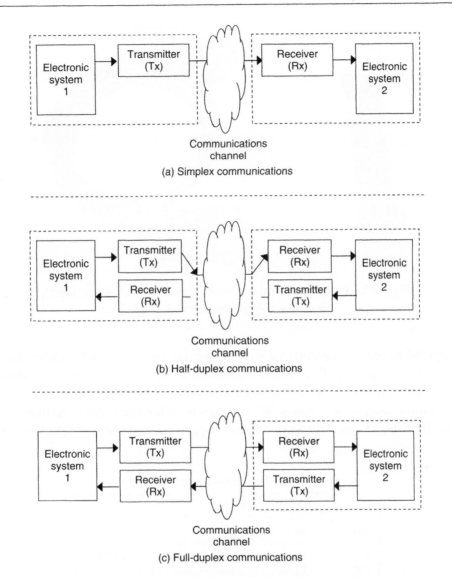

(a) Simplex communications

(b) Half-duplex communications

(c) Full-duplex communications

Figure 2.31: Simplex, half-duplex, and full-duplex communications

For wired communications, two example cable assemblies are shown in Figure 2.32. The cable assembly on the left consists of a ribbon cable with IDC (insulation displacement connector) terminations. The assembly on the right consists of a multicore cable terminated at each end with a nine-way D-type connector (female); this type would be used to connect an external electronic circuit to a PC via the RS-232C standard.

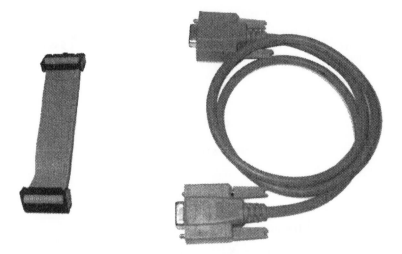

**Figure 2.32: Example cable assemblies: IDC connector (left),
nine-way D-type connectors (right)**

Both optical fiber transmission and wireless use the electromagnetic spectrum in the transmission of signals [19]. Wireless transmission occurs at the lower frequencies, and optical communications use infrared and visible light at the higher frequencies. Wireless transmission frequencies fall into bands within the radio spectrum, from 3 Hz to 300 GHz. Table 2.4 shows the radio spectrum and the corresponding bands.

Table 2.4: Radio spectrum

Frequency		Band
From	To	
3	300	Extremely low frequency (ELF)
300	3 kHz	Voice frequency (VF)
3 kHz	30 kHz	Very low frequency (VLF)
30 kHz	300 kHz	Low frequency (LF)
300 kHz	3 MHz	Medium frequency (MF)
3 MHz	30 MHz	High frequency (HF)
30 MHz	300 MHz	Very high frequency (VHF)
300 MHz	3 GHz	Ultra high frequency (UHF)
3 GHz	30 GHz	Super high frequency (SHF)
30 GHz	300 GHz	Extremely high frequency (EHF)

Figure 2.33: Example antenna (60 kHz)

An example of a low-frequency antenna, consisting of an inductor wound on a ferrite core with a parallel capacitor to form a 60 kHz tuned circuit, is shown in Figure 2.33. This antenna is secured to a PCB.

2.14.2   Parallel I/O

Parallel I/O allows groups of data bits to be transmitted simultaneously. In early versions of the microprocessor, data was grouped into bytes (8 bits). Today, microprocessors work with 8, 16, 32, 64, and 128 bits of data. Access to more memory requires address buses with an increased number of bits and the required control signals. The variety of parallel I/O standards available for use today include:

- Centronics (PC printer port)
- IEEE 488-1975 (also known as GPIB, general purpose instrument bus)
- SCSI (small computer system interface)
- IDE (integrated drive electronics)
- ATA (AT attachment)

PC Parallel Port (Centronics)

The PC parallel port (by Centronics) was until recently the port used primarily to connect the PC [20, 21] to a printer device, as shown in Figure 2.34. Here, each device is fitted with a 36-pin connector, and byte-wide data are sent from the PC to the printer (the peripheral) with *handshaking*—i.e., both the PC and the peripheral communicate with each other to control data transmission to be at a time suitable for both.

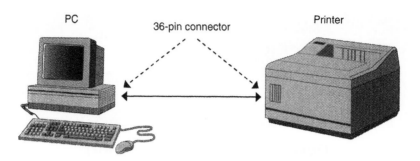

PC

36-pin connector

Printer

Figure 2.34: Connecting a PC to a printer using the parallel port

Table 2.5 identifies the cable connections for the Centronics printer port. Signals are transmitted on a twisted-pair (i.e., two wires twisted together) with its own common connection. Signal directions are shown from the perspective of the PC rather than the peripheral.

Today, the parallel port connection to the printer is usually replaced by a USB interface.

Table 2.5: Centronics (printer) port signals (PC connector)

Name	Pin Number		Direction (PC)	Meaning
	Signal	Common		
STROBE	1	19	OUT	Data strobe
D0	2	20	OUT	Data bit 0 (LSB)
D1	3	21	OUT	Data bit 1
D2	4	22	OUT	Data bit 2
D3	5	23	OUT	Data bit 3
D4	6	24	OUT	Data bit 4
D5	7	25	OUT	Data bit 5
D6	8	26	OUT	Data bit 6
D7	9	27	OUT	Data bit 7 (MSB)
ACKNLG	10	28	IN	Finished with last character
BUSY	11	29	IN	Not ready
PE	12	30	IN	No paper
SLCT	13	–	IN	Pulled high
AUTO FEED XT	14	–	OUT	Auto LF
INIT	31	16	OUT	Initialise printer
ERROR	32	–	IN	Can't print
SLCT IN	36	–	OUT	Deselect protocol
GND	–	33	–	Additional ground
CHASSIS GND	17	–	–	Chassis ground

2.14.3   Serial I/O

To connect an electronic system to an external device such as a PC or instrumentation, serial I/O is often preferred because it reduces the amount of wiring required. This is particularly important when dealing with large data and address buses, as when parallel I/O is used, and the IC and wiring connectors need to have more pins. This leads to larger IC packages and the need to route a large number of tracks on the PCB. Many digital ICs (such as memories) now provide serial I/O rather than parallel I/O to reduce the package requirements. In the circuits within such serial ICs, however, data serial-to-parallel and parallel-to-serial conversion capabilities are needed. Among the serial I/O standards available for use today are:

- RS-232C
- RS-422
- RS-423
- RS-485
- Ethernet
- USB
- I^2S (inter-IC sound bus)
- I^2C (inter-IC bus)
- SPI (serial peripheral interface)
- Firewire (IEEE Std 1394a-2000)
- Serial ATA
- Bluetooth (wireless)
- Wi-Fi (wireless, based on IEEE Std 802.11)
- Zigbee (wireless, IEEE Std 802.15.4).

For serial data transmission, each bit is sent one at a time. The *bit rate* is the number of bits sent per second. For serial data transmission, the *baud rate* is the same as the *bit rate*.

RS-232C

This has been a serial I/O available on PCs until the last couple of years, when it has been replaced by a USB. However, it is an important standard and provides an important introduction to serial communications. Bytes of data are sent as a serial bitstream asynchronously between terminals (such as between a PC and another PC or a modem), as shown in Figure 2.35, first the LSB (least significant bit) then the MSB (most significant bit). Typical baud rates for RS-232C used for data transmission on PCs are:

- 9,600 baud

- 19,200 baud

- 38,400 baud

- 115,200 baud

Serial data is transmitted and received via a circuit called a UART (universal asynchronous receiver transmitter). One example is the CDP6402 CMOS Universal

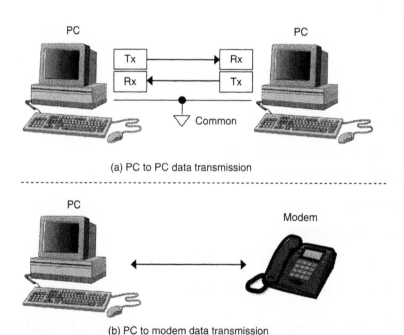

(a) PC to PC data transmission

(b) PC to modem data transmission

Figure 2.35: Uses for RS-232C

Asynchronous Receiver/Transmitter (UART) [22] from Harris Semiconductor. This circuit provides a 40-pin DIP device with internal serial-to-parallel, parallel-to-serial conversion and control logic.

RS-232C provides a means to send bytes of ASCII data between devices. ASCII is the most widely used *alphanumeric code* in use and stands for American Standard Code for Information Interchange. The ASCII code is a seven-bit code, so there are $2^7 = 128$ possible codes. The first 32 are control codes (nonprintable), and the remaining 96 character codes are printable characters. Table 2.6 shows the ASCII *character set*. This contains columns (0–F) and rows (0–7).

This panel is organized as follows: the code is presented in hexadecimal number format with:

- row numbers representing the first digit (0–7), 3 bits
- column numbers representing the second digit (0–F), 4 bits

For example, the letter *A* is ASCII code *41$_{16}$* (*64$_{10}$*).

A byte of data is sent serially in the form shown in Figure 2.36. Here, when data is not being sent, the level is a logic 1. A *start bit* (logic 0) indicates the start of the byte transmission. Eight data bits (or seven data bits and a parity bit) are then sent, beginning with the LSB (*data bit 0*). A logic 1 indicates a *stop bit*, and the signal then remains at a logic 1 until the next *start bit* occurs.

Within an electronic system, the logic levels are generated by a digital IC typically operating on a +5.0 V or +3.3 V power supply. For transmission, these voltage levels must be increased to achieve the voltage limits set by the standard. A logic 0 is a voltage between +3 V and +15 V (also referred to as the *space*), whereas a logic 1 is a voltage between –3 V and –15 V (also referred to as the *mark*). This idea is shown in Figure 2.37, where a digital signal is shown with the voltage levels for signal transmission.

The last bit of data (*data bit 7*), noted as the MSB, can also be used as a parity bit. If the MSB is used as a parity bit, then the data is reduced to 7 bits. As the data is sent asynchronously, the receiver and transmitter must create their own internal clocks. With the UART, this clock is set to be sixteen times that of the baud rate. Table 2.7 shows the UART clock frequencies required for different baud rates.

To translate the voltage levels generated by a digital IC with those required for transmission, a suitable transceiver such as the MAX-232CPE [23] (3.0 V

Table 2.6: ASCII codes

	0	1	2	3	4	5	6	7	8	9	A	B	C	D	E	F
0	NUL	SOH	STX	ETX	EOT	ENQ	ACK	BEL	BS	TAB	LF	VT	FF	CR	SO	SI
1	DLE	DC1	DC2	DC3	DC4	NAK	SYN	ETB	CAN	EM	SUB	ESC	FS	GS	RS	US
2	SP	!	"	#	$	%	&	'	(	)	*	+	,	-	.	/
3	0	1	2	3	4	5	6	7	8	9	:	;	<	=	>	?
4	@	A	B	C	D	E	F	G	H	I	J	K	L	M	N	O
5	P	Q	R	S	T	U	V	W	X	Y	Z	[	\	]	^	_
6	`	a	b	c	d	e	f	g	h	i	j	k	l	m	n	o
7	p	q	r	s	t	u	v	w	x	y	z	{	|	}	~	DEL

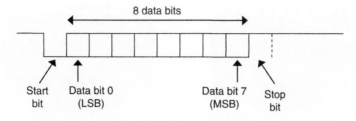

Figure 2.36: RS-232 timing waveform (logic levels)

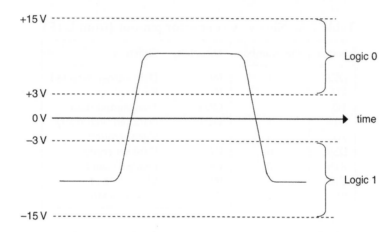

Figure 2.37: RS-232 timing waveform (voltage levels)

Table 2.7: Baud rate and UART clock frequency

Baud Rate	UART Clock Frequency (Hz)
9,600 baud	153.6 k
19,200 baud	307.2 k
38,400 baud	614.4 k
115,200 baud	1.8432 M

power supply version) from Maxim Integrated Products is typically used. This accommodates an IC (in a 16-pin DIL package) with external capacitors, thereby providing the necessary circuitry to connect devices such as FPGAs and CPLDs to transmit and receive RS-232C level signals.

The connector for the RS-232C wiring is either a 25-pin or a 9-pin D-type connector. Both male (plug) and female (socket) connectors are used. Figure 2.38 shows PCB

Figure 2.38: Nine-way PCB mount D-socket (left) and D-plug (right)

Table 2.8: Nine-way connector pin-out (from DTE)

Name	Pin Number	Direction	Function
DCD	1	IN	Data carrier detected
RD	2	IN	Received data
TD	3	OUT	Transmitted data
DTR	4	OUT	Data terminal ready
SG	5	–	Signal ground
DSR	6	IN	Data set ready
RTS	7	OUT	Ready to send
CTS	8	IN	Clear to send
RI	9	IN	Ring indicator

mount 9-pin D-type plug and socket. In the standard, two types of equipment were originally considered, *data terminal equipment* (DTE) and *data communication equipment* (DCE). Care must be taken when connecting equipment together to ensure that the right connections are established.

Table 2.8 identifies the connections for the 9-way connector for *data terminal equipment*. Because signals will be transmitted both ways, care must be taken to ensure that the correct connections are established. In a minimal form, with no handshaking needed, only the TD, RD, and SG connections are needed.

2.15   System Reset

At some point during the operation of a digital circuit or system, there will be the need to reset the circuit into a known state. This is particularly important when the power supply is first switched on to an electronic circuit as the state of the circuit is not then known. Circuits typically include a reset input connection in the pins of their ICs to reset internal connections (on the bistables) within the design. The reset signal will be designed to occur (be asserted) when:

- the power supply is initially switched on

- some time during the normal circuit operation when the circuit must be reset for normal circuit operation.

When the power supply is initially switched on, the power supply voltage at the power supply pins of the ICs in the circuit will take a finite time to increase from 0 V to the normal operating voltage (e.g., +3.3 V). During this power supply voltage rise time, the power supply voltage of the ICs increases to the normal operating voltage of the power supply, which will be sufficient to operate the ICs:

- The power supply voltage is designed to have a typical value (e.g., +3.3 V) with a tolerance (e.g., ±10%). If a tolerance of ±10% is set for a nominal +3.3 V power supply, then the power supply voltage would be in the range +3.0 to +3.6 V.

- The ICs used in the circuit have a typical power supply voltage value (e.g., +3.3 V), but with a tolerance over which the operation of the IC is guaranteed.

The tolerance of the power supply voltage must be such that all components in the circuit will operate correctly over the normal power supply voltage range variance.

When the power supply is initially switched on, the power supply voltage will rise to a level at which the IC will start to operate correctly (the power supply *threshold voltage*), as shown in Figure 2.39. When this threshold voltage has been reached, the circuit will operate correctly. During the device power-up, the device should be held in its reset state (i.e., the reset input is asserted). After the threshold voltage has been reached, the reset should be removed. The top graph of Figure 2.39 identifies the power supply voltage rise (in time), and the bottom graph identifies the reset (/reset as it is active low here) signal being asserted (logic 0) and removed (logic 1).

The reset signal can be generated in one of three ways:

1. by using a discrete RC (resistor-capacitor) network

2. by using a discrete power-on reset (POR) circuit

3. by using an integrated POR circuit

In a discrete RC network, the resistor and capacitor are connected in series across the power supply. Initially the voltage across the capacitor is zero, and when the power supply

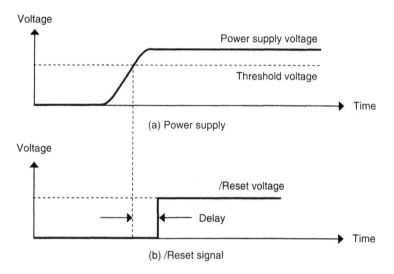

Figure 2.39: Power supply threshold voltage

is switched on, the capacitor starts to charge (an exponential rise in voltage) with a time constant set by (R.C). This is the reset voltage and can be applied directly to the reset pin of the IC. Although this is a simple circuit to implement, it is limited by the rise time of this signal, particularly for high-speed logic. The input to the IC should be a *Schmitt Trigger* input rather than a simple digital input buffer. Figure 2.40(a) shows an addition to this circuit, a push-switch across the capacitor to allow for a manual (user) reset.

In a discrete POR circuit, an external IC acts to create the reset signal for the circuit. An example arrangement with a manual reset switch input is shown in Figure 2.40(b). The choice of which POR circuit to use, discrete or integrated, depends on the threshold required [24]:

1. The power supply voltage has a nominal value with a tolerance.

2. The IC to be reset requires a nominal power supply voltage with a tolerance to operate correctly.

3. The circuit is designed so that it will tolerate short power supply glitches, and the POR does not assert a reset signal if a short power supply glitch occurs but would not affect circuit operation.

Where multiple ICs are to be reset, the order in which the resets are to be asserted and removed is a consideration. Additionally, the circuit may contain ICs operating on different power supply voltages, and so multiple reset signals will be needed.

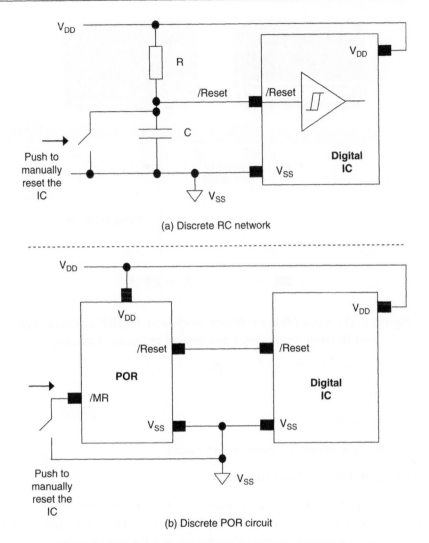

(a) Discrete RC network

(b) Discrete POR circuit

Figure 2.40: Different circuit reset methods

2.16   System Clock

In many electronic circuits and systems, one or more clock signals are required to control the timing of circuit operations. These clock signals are needed to generate the required clock frequencies and to operate at the required power supply voltage levels, and must remain stable (in the generated frequency) over variations in the power supply voltage, over the operating temperature range, and over time.

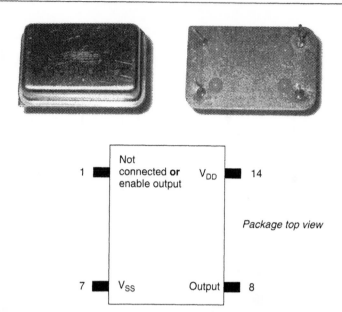

Figure 2.41: Four-MHz oscillator module in 14-DIP package: top and bottom views (top) and typical pin-outs (bottom)

A clock is generated using one of four types of circuit:

1. RC network

2. Quartz crystal

3. Through-hole-mounted oscillator modules

4. Surface-mount oscillator modules

For simple clocks, then an RC network connected to suitable circuitry within the IC is sufficient (a simple example of this would be the 555 timer IC). However, accurate timing can be difficult because of tolerances in the values for the resistor and the capacitor. A quartz crystal (available in either a through-hole or surface-mount package) connected to suitable circuitry within the IC provides a more accurate clock. This two-terminal device is connected to circuitry internal to the IC so that the crystal creates an oscillatory electrical signal. Oscillator modules, which are complete clock signal generators, are available in either through-hole or surface-mount packages. Figure 2.41 shows an example of a through-hole-mounted 4 MHz oscillator module in a metal case. This is in a metal 14-pin DIP package with four pins: two for the power supply, one for the oscillator output, and one which would either be unused (not connected) or used in some modules for a clock enable signal.

2.17 Power Supplies

Whether AC or DC, the power supply provides the necessary power to operate the circuit. It requires an energy source and will modify the energy to provide the necessary voltages and currents required by the circuit, as shown in Figure 2.42. This power supply must guarantee circuit operation within a set range (a nominal value with a tolerance), be stable over the operating temperature range, be stable over time, and provide the necessary voltages and currents required by the electronic circuit or system.

The choice of power supply is concerned with:

- the means by which to obtain the energy input

- the required AC and DC voltage and current outputs

- the size and weight of the power supply

- whether the electronics are static (located in a single location) or portable (mobile)

- the length of time that the power supply is required to operate before it must be recharged or replaced

A *fixed power supply* that is to operate indefinitely without being recharged or replaced will operate from either the domestic or industrial mains power supply or from a generator (such as a wind turbine or solar panel). A *portable power supply* utilizes batteries, whether disposable or rechargeable (from a fixed power supply). In addition, voltage must be converted from AC input to DC output (using a transformer and diode-based rectifier circuit or a switched-mode power supply), or from DC input to AC output (using an inverter, for example, to operate mains powered electronic equipment from a car battery).

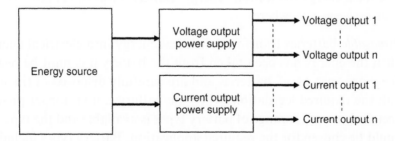

Figure 2.42: Power supply generating multiple voltage and current power supplies

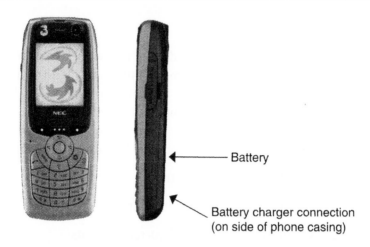

Battery

Battery charger connection
(on side of phone casing)

Figure 2.43: Mobile phone (portable electronics using a battery power supply)
Images courtesy of NEC, © NEC 2001–2004, no longer in stock

Figure 2.43 shows the example of a mobile phone (specifically, the NEC e228). This is a third generation (3G) mobile phone for use with the 3G mobile phone standards and technology. Such devices provide for a wide range of services for individuals to effectively communicate with each other using voice, text and video communications means. The left view shows the front of the phone (the user interface). Because this is a portable device, the phone will operate on a rechargeable battery (3.7 V DC and 1,100 mAh rated lithium-ion) with a charge lifetime in hours. The battery location is shown in the right view, housed in the rear of the mobile phone with the back removed.

The battery is recharged using a battery charger that operates from a domestic electricity connection. A battery consists of one or more electrochemical cells that converts chemical energy to electrical energy. Batteries will be classed as either *disposable* or *rechargeable*, where:

1. *Disposable* batteries transform chemical energy into electrical energy and when the energy has been taken from the battery it cannot be restored. These are "use once" batteries and are carefully disposed of (in accordance with the required legislation) when the battery can no longer provide electrical energy. A range of battery types is available and the type of battery would be chosen for the required application. Battery types include *alkaline* and *silver-oxide*.

2. *Rechargeable* batteries also transform chemical energy into electrical energy, but the energy can be restored by the supply of electrical energy to the battery. These batteries can be recharged and so can be used multiple times. A range of battery types is available and the type of battery would be chosen for the required application. Battery types include *nickel-cadmium (NiCd), nickel-metal hydride (NiMH)* and *Lithium-ion.*

2.18   Power Management

When an electronic circuit or system is operating, it will consume power from either a fixed or portable power supply. The power consumption for some circuits can be large, so any reduction in the power consumption of the circuit is beneficial:

- It will consume less power and so be cheaper to operate.

- It will be suitable for portable, battery-operated systems required to operate for long durations between charges.

- It will require less heat removal (some ICs such as the microprocessor will generate heat, which must be removed so the microprocessor can operate without failure), and so the heat removal system would be smaller and cheaper.

- The power supply would be smaller, lighter, and cheaper.

Power consumption can be considered by looking at all stages in the creation and use of the design, in particular by considering:

1. **Design architecture**. Design circuits using circuit architectures that will consume less power.

2. **Fabrication process**. Within an IC the circuits consist of transistors, resistors, and capacitors. Most ICs are silicon based, and the circuits are bipolar and MOS transistors. CMOS is suited for low-power, low-voltage circuits, and static CMOS circuits provide low-power consumption when the circuit activity is low.

3. **Reduced power supply voltage**. Using electronic components that can operate at low power.

4. **Minimized circuit activity**, keeping signal logic transitions from 0 to 1 and 1 to 0. In static CMOS logic gates, current flows when nodes in a digital logic

design change their logic levels, which happens when the transistor switches move from closed to open and open to closed positions. If this activity is reduced, then less current would be required to flow from the power supply.

5. **Power management features**. Some ICs provide the ability to shut down parts of the circuit when they are not used. (For example, RF transmitters consume considerable power when the RF circuitry is active, but this circuitry might only be required to be operational for short periods of time.) Additionally, some microprocessors allow reduced clock frequency within the microprocessor itself when the required activity of the microprocessor is low.

2.19   Printed Circuit Boards and Multichip Modules

An electronic system consists of a number of subsystems and components that are connected together to form the required overall system. In many cases, the main functions of the system are created using integrated circuits mounted onto a PCB. There are four package levels between a circuit die (within a package) and the PCB [25]:

1. **Die level**—Bare die (predominantly based on silicon).

2. **Single IC level**—Packaged silicon die (considering a single packaged die).

3. **Intermediate level**—Silicon dies (die level) and/or packaged dies (single IC level) are mounted onto a suitable substrate that may or may not be further packaged.

4. **PCB level**—Printed circuit board level.

Combining these four levels creates four types of packaged electronics:

1. **Type 1**—Packaged silicon die mounted onto a PCB.

2. **Type 2**—Packaged silicon die mounted onto an intermediate substrate that is then mounted onto a PCB.

3. **Type 3**—A bare silicon die mounted onto an intermediate substrate that is then mounted onto a PCB.

4. **Type 4**—A bare die mounted directly onto a PCB.

Many semiconductor devices contain a circuit fabricated on a single die (as in the single IC level). However, sometimes multiple dies are housed within the package, such as a device that contains a sensor (e.g., accelerometer) along with sensor signal conditioning circuitry and a communications interface. For either technical or cost reasons, the sensor and circuitry cannot be fabricated on a single die. Where multiple dies will be housed within the package, this device is referred to as a multichip module (MCM, originally referred to as a *hybrid circuit*). The MCM consists of two or more integrated circuits and passive components on a common circuit base (substrate), and interconnected by conductors fabricated within the substrate. The ICs may be either packaged dies or bare dies (an unpackaged known good die, KGD).

The MCM was developed to address a number of issues relating to the need to reduce the size of increasingly complex electronic circuits and to the degradation of signals passing through the packaging and interconnect on a PCB. The MCM can provide advantages in certain electronic applications over a conventional IC on a PCB implementation such as:

- increased system operating speed

- reduced overall physical size

- ability to handle ICs with a large number of I/Os

- increased number of interconnections in a given area (higher levels of interconnect density)

- reduced number of external connections for a given functionality (as the majority of the interconnect is within the MCM itself)

In addition, an MCM may contain dies produced with different fabrication processes within a single packaged solution (e.g., mixing low-power CMOS with high-power bipolar technologies). There are a number of types of MCMs that can be realized:

- **MCM-D**—MCMs whose interconnections are formed by thin film deposition of metals on deposited dielectrics. The dielectrics may be polymers or inorganic dielectrics.

- **MCM-L**—MCMs using advanced forms of PCB technologies, forming copper conductors on laminate-based dielectrics.

- **MCM-C**—MCMs constructed on co-fired ceramic substrates using thick film (screen printing) technologies to form conductor patterns. The term *co-fired* relates to the fact that the ceramic and conductors are heated in the oven at the same time.

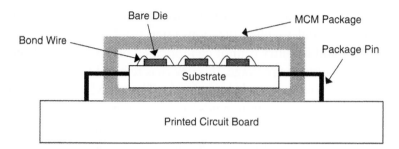

Figure 2.44: Example MCM structure

- **MCM-D/C**—MCMs using a deposited dielectric on co-fired ceramic.

- **MCM-Si**—MCMs using a silicon-based substrate similar to conventional silicon ICs.

The MCM typically uses a similar package as that used for the integrated circuit, so it is not obvious that the package contains multiple dies and sensors unless the structure and operation of the packaged device is known. Figure 2.44 shows the cross-section of a MCM in which the dies are mounted onto a substrate and electrically connected to the substrate using bond wires. This MCM is mounted directly to the PCB. The substrate contains additional interconnect in a similar way to the PCB.

2.20   System on a Chip and System in a Package

An extension to the basic integrated circuit is the system on a chip (SoC) [26]. This is essentially a complex (mainly digital) IC that can be considered as a complete electronic system in a single IC. Modern communications ICs are examples of SoC design. The need to develop such complex ICs has been in response to the end-user requirements, who need:

- increased device functionality (more circuitry per mm^2 of silicon area)

- higher operating frequencies

- reduced physical size (more circuitry in a smaller package)

- lower cost

The ability to integrate complex digital circuits and systems on a single circuit die has led to incorporating the functionality that was once manufactured as a discrete chip-set within the single IC itself. The SoC includes a number of interconnected circuits:

- one or more processor cores

- one or more embedded memory macros (RAM and ROM)

- dedicated graphics hardware

- dedicated arithmetic hardware (e.g., adder, multiplier) for high-speed arithmetic

- bus control circuitry for data, addresses, and control signals between the main circuit blocks

- serial and parallel I/Os

- glue logic—miscellaneous logic for subsystem interfacing purposes

- data converters, ADCs and DACs

- Phase-locked loop (PLL)

An extension to the multichip module is the system in a package (SiP) [27]. The ITRS [28] definition for the SiP is "any combination of semiconductors, passives, and interconnects integrated into a single package." SiP designs extend the concept of the MCM from devices placed horizontally side-by-side and bonded to a substrate to include the ability to vertically stack dies with bonding to the substrate.

2.21   Mechatronic Systems

Mechatronics [3, 29, 30]—*mecha*nical and elec*tronics*—is the combined design of products and processes containing mechanical, electrical or electronic software, and information technology parts. Systems that contain these parts are referred to as *mechatronic systems*. The concept is shown in the Venn diagram in Figure 2.45. The computer science set encompasses software engineering and information technology. The union of the three sets is the mechatronic domain.

Mechatronics provides the focus required to bring together different disciplines and create mixed-technology design. Traditionally, these have been housed in separate departments within an organization, which has blocked effective communications in

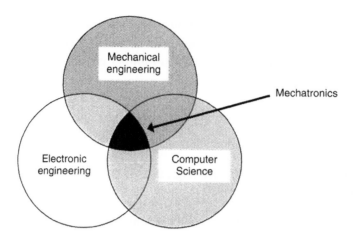

Figure 2.45: Mechatronics, combining the disciplines

the design process, with each discipline providing its own set of terminology and competition instead of collaboration. The combined approach naturally removes barriers and allows effective communications, thereby leading to an improved design process and a higher-quality end product.

Example application areas of mechatronics include automotive, aerospace, space, biomedical, and industrial control. Consider the motor control example shown in Figure 2.46. Here, a DC electric motor is to be controlled by a CPLD, the heart of the electronic controller, which is configured to provide the closed loop control. A number of subsystems are required to implement the overall system design, with each subsystem drawing on the expertise of one or more engineering disciplines, including:

- *Electronic engineer* to design the CPLD configuration (digital logic), power electronics, and sensor interface electronics

- *Communications engineer* to design the communications interface (wired, optical fiber, or wireless)

- *Software engineer* to design the software application to run on the PC required to interface to the controller

- *Control engineer* to design the underlying closed-loop control algorithm to control the electric motor to given design requirements

- *Mechanical engineer* to design the mechanical load

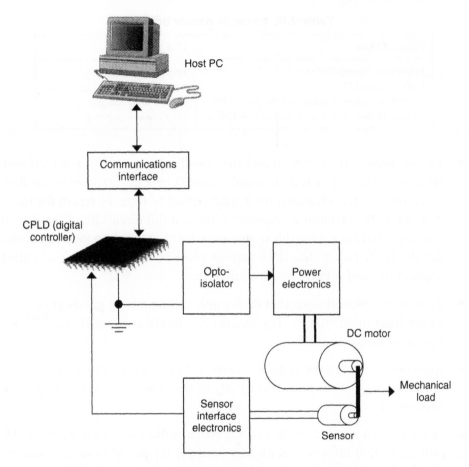

Figure 2.46: CPLD control of a motor in a mechatronic system

2.22 Intellectual Property

Intellectual property (IP) allows people to own things that they have created, similar to owning a physical item, so they can control their use and reap the rewards [31]. There are five types of IP:

- *Copyright* protects material such as literature, art, music, sound recordings, films, and broadcasts. It can also cover software. Copyright allows the right for someone to reproduce their own original work.

- *Design rights* protect the visual (aesthetic) appearance of a product. Design rights may be unregistered or registered.

Table 2.9: Example patent offices

Patent Office	URL
European Patent Office	http://www.epo.org/
Irish Patents Office	http://www.patentsoffice.ie/
United Kingdom Intellectual Property Office	http://www.ipo.gov.uk/
United States Patent and Trademark Office	http://www.uspto.gov/

- *Patents* protect the technical and functional aspects of both products and processes. The patent is a monopoly granted by a government to the first inventor of a new invention for a fixed period of time. In return for this monopoly, the inventor is required to make a full disclosure of the invention. This information is available to anyone who might wish to view the invention details. To be patentable, the invention must be new, be capable of industrial application, and involve an inventive step.

- *Trademarks* protect signs that distinguish a company or goods of one trader from other traders. Trademarks can be either *unregistered* (TM) or *registered* (®).

- *Know-how*, also known as *trade secrets*, refers to secret (or proprietary) information. It is not protected by any of the above means, but only by being kept secret.

Table 2.9 identifies a number of the existing patent offices and their websites. These offices provide further information on how to apply for patents and also search engines for finding existing patents.

2.23   CE and FCC Markings

For electronic circuits and systems to be available for commercial sale, they must meet the requirements of specific legislation. If electronic products meet the requirements, they will have a verifying marking on the outside, usually either CE or FCC. Figure 2.47 shows part of an electronic product (in this case a power supply) with both CE and FCC markings.

The *CE marking* is a declaration by a product manufacturer that the product meets all of the appropriate provisions of the relevant legislation required to implement specific

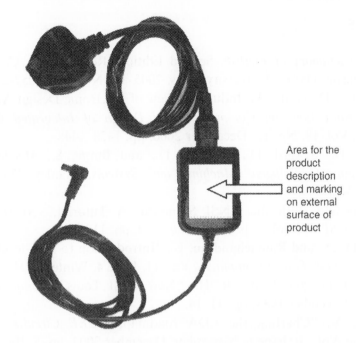

Area for the product description and marking on external surface of product

Figure 2.47: Electronic product with CE and FCC marking

European Directives [32, 33]. *CE* is not an abbreviation for any specific words, nor is it meant to be a mark of product quality.

The *FCC marking* is for commercial electronic devices for sale in the United States that are unintentional radio-frequency radiators intended for operation without an individual broadcast license [34]. It covers devices that use clocks or oscillators, operate above a frequency of 9 kHz, and use digital techniques. The specific requirements are set down in the FCC Rules and Regulations, Title 47 CFR Part 15 Subpart B. Most processor-based systems, for example, fall into this category. This is regulated by the Federal Communications Commission (FCC) and categorizes the parts into one of two classes:

- **Class A**: A device intended for an industrial or business environment and not intended for use in a home or a residential area

- **Class B**: A device intended for use in a home or a residential area

References

[1] *Oxford Dictionary of English*, Second Edition, Revised, eds. C. Soanes and A. Stevenson, Oxford University Press, 2005, ISBN 0-19-861057-2.

[2] MacMillen, D., et al. "An Industrial View of Electronic Design Automation," *IEEE Transactions on Computer Aided Design of Integrated Circuits and Systems*, Vol. 19, No. 12, December 2000, pp. 1428–1448.

[3] Bradley, D., Seward, D., Dawson, D., and Burge, S., *Mechatronics and the Design of Intelligent Machines and Systems*, Stanley Thornes, 2000, ISBN 0-7487-5443-1.

[4] "Flowcharting With the ANSI Standard: A Tutorial," *ACM Computing Surveys (CSUR)*, Vol. 2, Issue 2, June 1970, pp. 119–146.

[5] Gajski, D. D., and Ramachandran, L., "Introduction to high-level synthesis," *IEEE Design & Test of Computers*, Vol. 11, Issue 4, Winter 1994, pp. 44–54.

[6] Gajski, D. D., and Kuhn, R. H., "New VLSI Tools," *Computer*, Vol. 16, Issue 12, December 1983, pp. 11–14.

[7] Hemani, A., "Charting the EDA roadmap," *IEEE Circuits and Devices Magazine*, Vol. 20, Issue 6, November–December 2004, pp. 5–10.

[8] Wolf, W. H., "Hardware-Software co-design of embedded systems," *Proceedings of the IEEE*, Vol. 82, Issue 7, July 1994, pp. 967–989.

[9] Balarin, F., et al. *Hardware-software Co-design of Embedded Systems: The Polis Approach*, Kluwer Academic Publishers, 1997, ISBN 079239936.

[10] Gajski, D. D., and Vahid, F., "Specification and design of embedded hardware-software systems," *IEEE Design & Test of Computers*, Vol. 12, Issue 1, Spring 1995, pp. 53–67.

[11] Kropf, T., *Introduction to Formal Hardware Verification*, Springer, 1999, ISBN 3-540-65445-3.

[12] Marculescu, R., and Eles, P., "Guest Editors' Introduction: Designing Real-Time Embedded Multimedia Systems," *IEEE Design & Test of Computers*, September–October 2004, pp. 354–356.

[13] Edwards, S. A., "The Challenges of Synthesizing Hardware from C-Like Languages," *IEEE Design & Test of Computers*, September–October 2006, pp. 375–386.

[14] Grant, M., Bailey, B., and Piziali, A., *ESL Design and Verification: A Prescription for Electronic System Level Methodology*, Morgan Kaufmann Publishers Inc., 2007, ISBN 0123735513.

[15] Densmore, D., et al. "A Platform-Based Taxonomy for ESL Design," *IEEE Design & Test of Computers*, September–October 2006, pp. 359–374.

[16] Bennett, S., Skelton, J., and Lunn, K., *UML*, McGraw-Hill, 2001, ISBN 0-07-709673-8.

[17] Mancini, R., "How to read a semiconductor datasheet," *EDN*, April 14, 2005, pp. 85–90, http://www.edn.com

[18] Tocci, R. J., Widmer, N. S., and Moss, G. L. K., *Digital Systems*, Ninth Edition, Pearson Education International, USA, 2004, ISBN 0-13-121931-6.

[19] Sears, F., Zemansky, M., and Young, H., *University Physics*, Seventh Edition, Addison-Wesley, 1987, ISBN 0-201-06694-7.

[20] Mueller, S., *Upgrading and Repairing PCs*, Sixteenth Edition, Que Publishing, 2005, ISBN 0-7897-3210-6.

[21] Horowitz, P., and Hill, W., *The Art of Electronics*, Second Edition, Cambridge University Press, 1989, ISBN 0-521-37095-7.

[22] Harris Semiconductor, "CDP6402, CDP6402C CMOS Universal Asynchronous Receiver/Transmitter (UART)," product datasheet, March 1997.

[23] Maxim Integrated Products, "MAX232-CPE RS-232 Transceiver," product datasheet, 2000.

[24] Maxim Integrated Products, "Power-on Reset and Related Supervisory Functions," application note 3227, May 11, 2004.

[25] Doane, D. A., and Franzon, P. D., *Multichip Module Technologies and Alternatives, The Basics*, Van Nostrand Reinhold, New York, 1993, ISBN 0-442-01236-5.

[26] Rajsuman, R., *System-on-a-Chip Design and Test*, Artech House Publishers, USA, 2000, ISBN 1-58053-107-5.

[27] Rickett, P., "Cell Phone Integration: SiP, SoC and PoP," *IEEE Design & Test of Computers*, May–June 2006, pp. 188–195.

[28] International Technology Roadmap for Semiconductors (ITRS), 2003 Edition, "Assembly and Packaging."

[29] Bolton, W., *Mechatronics: Electronic Control Systems in Mechanical Engineering*, Second Edition, Longman, 1999, ISBN 0582357055.

[30] Walters, R. M., Bradley, D. A., and Dorey, A. P., "The High Level Design of Electronic Systems for Mechatronic Applications," IEE Colloquium on Structured Methods for Hardware Systems Design, 1994, pp. 1/1–1/4.

[31] Wilson, C., *Intellectual Property Law*, Second Edition, Sweet & Maxwell, 2005, ISBN 0-421-89150-5.

[32] Department for Trade and Industry (United Kingdom), http://www.dti.gov.
      uk/innovation/strd/cemark/page11646.html

[33] European Commission, *Guide to the Implementation of Directives Based on New
      Approach and Global Approach*, http://ec.europa.eu/enterprise/newapproach/
      legislation/guide/

[34] Federal Communications Commission (United States of America), http://
      www.fcc.gov/

Student Exercises

1.1 Draw a flowchart for the following processes:

   a. Changing a broken light bulb in a home
   b. Changing the tire of a car
   c. Driving correctly through a crossroad with a set of traffic lights
   d. Making a cup of tea

1.2 Consider the following scenario:

   A user of an electronic system enters three different integer numbers from a keypad (possible numbers are 0 to 9). The electronic system determines which number is the highest in value and displays this on a two-line LCD display.

   Draw a flowchart for the operation of this electronic system function.

   Write a design specification for this electronic system.

1.3 Consider the following scenario:

   A software program running on a PC is to open a text file and read the contents of the file character by character until the end of the file is reached. If the character is upper case (A–Z), then it is displayed on the computer VDU.

   Draw a flowchart for the operation of this electronic system function.

   Write a design specification for this software program.

1.4 Modify the operation of the software program in Exercise 1.3 so that it now also writes the uppercase character (A–Z) to a second text file.

   Draw a flowchart for the operation of this electronic system function.

   Write a design specification for this software program.

1.5 Identify the types of batteries available for use. For each type of battery, identify its output voltage level and its ampere-hour rating. How does battery operation vary with temperature?

1.6 Identify the principle of operation of the switched-mode power supply.

CHAPTER 3

PCB Design

3.1 Introduction

Within an electronic system, the printed circuit board (PCB) fulfils an essential role in which to mount the main electronic components, whether by soldering or by the use of fixing aids such as screws, and the means by which the electronic components are electrically connected to form the required electrical circuit, using metal tracks patterned onto the PCB and solder joints.

Figure 3.1 shows a 3-D graphical representation of an example PCB with models for the components placed on the PCB in their intended positions. A number of PCB design tools (for example, the Altium^{TM} Protel PCB design software) provide for a 3-D viewing capability that enables the designer to view the PCB as it would appear in the final fabricated PCB with components inserted prior to PCB fabrication. The main base (commonly referred to as the *substrate*) is the insulating material, and tracks are patterned into it. Here, the electronic components are mounted to the top of the board, although components may also be mounted to both the top and bottom.

In this example, the board is rectangular and 1.6 mm thick; actually this PCB was designed to be *Eurocard size* (160 mm × 100 mm [6.3″ × 3.94″]). However, the actual shape of the PCB can be decided by the designer (restricted only by the manufacturing capabilities and cost to manufacture) to fit into the appropriate housing requirement for the electronics.

To develop a working PCB that operates according to the required functionality, three key steps must be successfully completed:

- **Design**. First develop a suitable design specification for the required circuit [1], then develop the circuit schematic (the components to use and interconnect

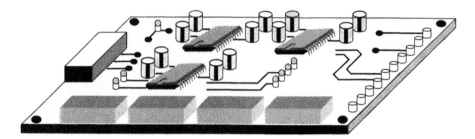

Figure 3.1: Graphical representation of an example PCB (top view)

between the components) to meet the initial design specification, and finally develop the PCB layout (the actual representation of the design that will be manufactured). The designer will work with different design representations (in which to view the design and understand the design functionality) to arrive at a solution that can work.

- **Manufacture**. The manufacture, or fabrication, of the printed circuit board itself must adhere to the design details. The two main steps are manufacturing the PCB base (insulating base with metal interconnect), and electrically and mechanically connecting the electronic components to the PCB base. Connecting the components to the PCB base is commonly referred to as *populating the board.*

- **Test**. The purpose of testing the design and manufactured PCB is to ascertain whether or not the design is working [2, 3]. Testing is undertaken at a number of points during the design and manufacture. Testing includes both simulation testing of a model of the PCB design prior to manufacture to determine the functional correctness of the design and physical testing of the manufactured PCB to take electrical measurements to determine the functional correctness of the manufactured design.

PCB design can take a number of different approaches, which initially arose from the lack of a suitable standard adopted by all PCB designers. More recently, there has been a move to standardize PCB design approaches and terminology used by the design community, in particular the activities of the IPC Designers Council. In this text, the descriptions presented in the next section are used to identify the approaches and terminology commonly used.

3.2   What Is a PCB?

3.2.1   *Definition*

A printed circuit board (PCB) is an electrical component [4, 5] made up of one or
more layers of electrical conductors that are separated by insulating material. Other
electrical components are secured to the top and bottom of the PCB to create a
complete electrical circuit. An example PCB with components soldered to the top
is shown in Figure 3.2.

Here, five connectors are used to connect the board to the remainder of the electronic
system (the board here is only a small part of a larger electronic system). Four D-type
connectors are placed along the bottom edge of the PCB and a single IDC (insulation
displacement connector) is placed on the left edge of the board. Along the right edge
of the board are small terminals to connect test equipment to electrical signals
generated on the board for test and evaluation purposes. The main circuit is in the
center the board, with three integrated circuit (IC) sockets (the ICs themselves are not
yet placed in the sockets) [6], seven light-emitting diodes (LEDs), fifteen capacitors,
seven resistors, and one diode. The patterned metal tracks can be seen as narrow lines
on the top of the board. The thickness of the board is 1.6 mm, and the thickness of the
copper tracks is 35 μm (0.035 mm).

Figure 3.2: Manufactured PCB (top)

This circuit will be discussed in further detail in Section 3.5 (case study design), but four key things can be immediately noted from this board:

1. All components are through-hole mounted; that is, they are placed on the top of the board, their electrical connections (legs) pushed through holes in the PCB, and then soldered from the bottom of the board. The bottom of this board is shown in Figure 3.3. The patterned metal tracks can be seen as narrow lines on the bottom of the board. There are two thicknesses of track: the thin tracks are used for signals requiring little current flow, and the thicker tracks are used for the component power supply (positive and negative) that requires a greater current flow.

2. The tracks on the bottom of the board are connected to the top of the board through metal-plated holes (vias) drilled into the base insulation.

3. The color of the board is green in appearance. This results from the *solder mask* material covering the entire board. The base insulator is made of FR-4 material, which is typically yellow in color.

4. This particular board does not have many components, and they are not densely packed; that is, the few components on the board are not placed close to each other. This eases physical access to the components for probing with test equipment.

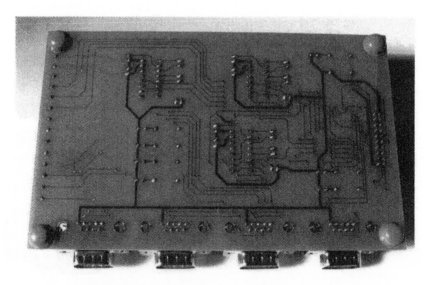

Figure 3.3: Manufactured PCB (bottom)

In some texts, the PCB is referred to as a PWB (printed wiring board) [7], however in this text, the term PCB will be used throughout.

3.2.2 Structure of the PCB

Overview

A PCB consists of an electrically insulating base onto which conducting metal tracks are patterned to form electrical connections for electronic components mounted to the top, and sometimes the bottom, of the insulating based. The PCB has electrical, mechanical, and thermal properties that must be considered when creating a design for a particular application.

The insulating material commonly used is FR-4 (flame retardant with a dielectric constant of approximately 4), also referred to as the *dielectric*. FR-4 is usually preferred over cheaper alternatives such as synthetic resin bonded paper (SRBP) as it can operate at higher electrical frequencies (important for high frequency applications), is mechanically stronger, absorbs less moisture (any moisture from the surrounding ambient conditions), and is highly flame resistant. However, note that the choice of material for the PCB will depend on the final application requirements and cost.

Simple Single-Sided PCB

The simplest type of PCB consists of a square or rectangle of insulating material with patterned metal tracks on one side only. The metal is usually copper. This is suitable for the simplest of circuits but cannot hold a larger number of components because all of the tracks cannot be physically routed on one side of the board. The electronic components are placed on the opposite side of the board, and holes (called vias) drilled through the board allow for the ends of the electronic component legs to be located on the same side of the board as the metal tracks.

When the leg of the component passes through the board, the component is referred to as a *through-hole component*. Where the legs are to be in contact with the tracks, the tracks are shaped to form pads that are normally larger than the tracks and allow the component leg to be soldered to a suitably large amount of metal track material. Figure 3.4 illustrates the placement and soldering of a component. Traditional solder is an alloy of tin and lead (typically 60–40), which melts at a temperature of about 200°C. (Coating a surface with solder is called tinning because of the tin content of solder.)

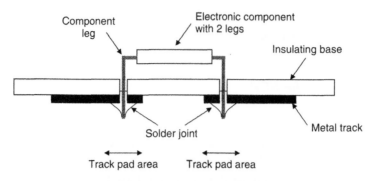

Figure 3.4: Single-sided PCB

However, as lead is poisonous, the solder in use today does not contain lead, and alternative alloys are used. The solder used for electronic circuit manufacture also contains tiny cores of flux. The flux cleans the metal surfaces as the solder melts. Without the flux, most solder joints would probably fail because the metals quickly oxidize, and the solder itself will not flow properly onto a dirty, oxidized, metal surface.

Two-Sided PCB

A more sophisticated and more common PCB has metal tracks on both sides of the board. This allows twice the area to pattern the tracks, and the electrical connections formed by the tracks can move between the top and bottom of the board through holes (vias) drilled in the board. Vias are of two types, plated through-hole and nonplated through-hole. A nonplated through-hole via is simply the hole that was drilled. To make an electrical connection through the hole, a piece of metal wire is soldered top and bottom. The plated through-hole via has a metal plating connecting the top and bottom track pads formed during the PCB base manufacture (Figure 3.5).

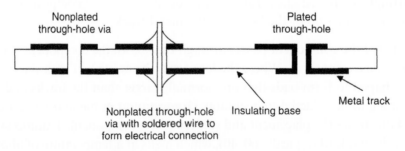

Figure 3.5: Through-hole vias

PCB Pads

The shapes of the pads are typically round, oval, square, rectangular, or octagonal, shown in three different sizes in Figure 3.6. The round center of each pad is sized to the hole that is to be drilled to fit the component leg; different components require legs of different sizes, which should be specified on the data sheet for the component. The outside part of the pad is the metal (track material) to which the solder adheres.

The different shapes signify different pins. For example, in Figure 3.7, a 14-pin DIP (dual in-line package) IC pad placement is shown. The number 1 pin is shown on the

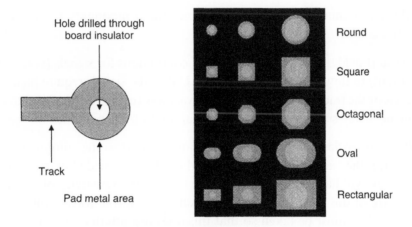

Figure 3.6: Pad shapes and sizes

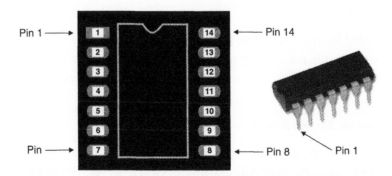

Figure 3.7: 14-pin DIP pad placement (through-hole component) and image of DIP package

top left of the image and is signified by a rectangular shape. The other 13-pin pads are oval in shape. Here, the pad metal is placed on both sides of the board (top and bottom). For a through-hole component, the pad need only be on the bottom side of the board where the leg of the component is to be soldered to the pad. However, if the pad is placed on both sides of the board, and the via is a plated through-hole via, then tracks can connect to the bottom and top sides of the pad. This will provide a benefit for track routing in that both the bottom and top of the board can be used for routing the required power and signal tracks.

Tracks

The metal tracks connecting the pads and hence the components are for two types of electrical use:

1. *Signal* provides the necessary electrical connections for signals (voltage and current) to flow between components. Unless the signals require high current levels or the track carrying the signal is very low resistance, the signal track widths are normally small to allow many signal tracks to be patterned on the PCB.

2. *Power* provides the required voltage and current to the components. In general, they are wider than the signal tracks to provide low-resistance paths. A track will have a certain resistance (due to the resistivity of the metal and the size of the metal), and when currents flow in the track, voltage will drop. Care must be taken so that this does not interfere with or degrade the operation of the circuit.

Components on Two Sides

When a two-sided board is used, tracks can be created on the top and bottom of the board. Components are usually mounted only on the top side of the board, but they can be mounted on both top and bottom, as shown in Figure 3.8.

Through-Hole versus Surface Mount

The earliest components, and those still in many everyday electronic circuits, are through-hole components, as previously discussed. However, the space requirements for the legs and the need to fit the legs through the PCB itself for soldering has created the need for considerable surface area and physically large PCBs. The lengths of the

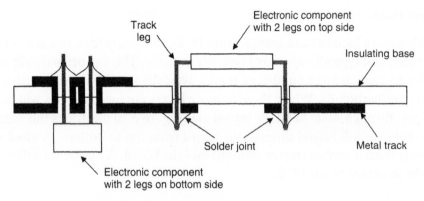

Figure 3.8: Two-sided board with components on top and bottom

leads also add parasitic circuit elements (resistance, capacitance, and inductance) that can seriously affect high-frequency performance. An alternative to through-hole components are *surface mount components*. (The technology associated with surface mount components is generically referred to as *surface mount technology*, SMT.) Rather than having legs that are pushed through the board, the connections for soldering the component to the PCB pads are on the same side of the PCB as the component itself (Figure 3.9), allowing physically smaller components that can be mounted onto smaller PCBs, with superior high-frequency performance when compared to a through-hole equivalent.

Figure 3.9: Surface mount component (eight-pin surface mount MSOP, mini-small outline package)

Multilayer PCBs

Some fabrication facilities can manufacture PCBs with more than two layers of metal interconnect, and typically up to six layers are possible. This can dramatically increase the ability to route a large number of tracks, typically for applications such as computer motherboards. Where three or more layers are used, the vias will be one of three types: through-hole, blind, or buried via. Figure 3.10 illustrates this idea. The through-hole via will extend through the board from top to bottom. A blind via will extend only from a surface (top or bottom) into the board. A buried via will be buried within the structure of the PCB.

Ground and Power Planes

A metal layer within the PCB structure can be used as a ground or a power plane. These are large areas of metal that can span all, or nearly all, of a metal layer to provide a large area for current to flow, accommodating the power supply connections (positive and negative) and the common connection (ground for both analogue and digital circuitry). This creates a low-resistance path for the current and allows for substantially more current than would be possible in a thin track. One or more of the metal layers can be used for a power or ground plane. When one layer is used for a single power or ground plane, this is referred to as a *single plane*. However, a single layer can be used for multiple power or ground planes, where the metal is separated into different areas, one for each connection; this is referred to as a *split plane*.

Protective Coating

A protective coating is normally applied to the surface of the PCB to prevent damage from the environment in which it will be used. This protective coating can be applied

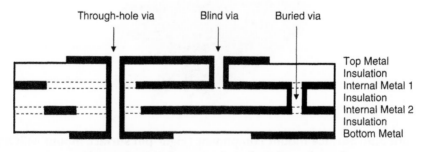

Figure 3.10: Via types in a four-layer board

after manufacture and either before or after the electronic components have been soldered on. The protective coating protects in several different ways:

- The copper commonly used for the tracks will be corroded by exposure to oxygen in the air, and the protective coating (a *passivation layer*) puts a barrier between the oxygen and the metal):

  ○ If the copper must be accessible, either for soldering (on to pads) or for electrical contact (such as edge connectors off the PCB), then the copper is plated with another metal such as tin or nickel. This additional metal forms a passivation layer that protects the copper from oxidation.

  ○ Where the copper need not be accessible, then an electrically insulating protective coating is applied over the metal. This has the additional advantage of preventing dirt and moisture from reducing the insulation resistance between the tracks.

- The insulating material used in the substrate (e.g., FR-4) will readily absorb moisture from the air, thereby reducing the electrical properties of the substrate. The protective coating puts a barrier between the substrate and the moisture in the air.

- The protective coating also controls the flow of solder during the soldering process. This prevents solder from jumping across tracks and causing short circuits.

When a protective coating is applied prior to soldering the components onto the board, it is usually referred to as a *solder mask*. When applied after the components have been soldered, it is usually referred to as a *conformal coating*.

Silk Screen

Screen printing techniques using a silk screen can be used to apply solder paste to the PCB for the attachment of the electronic components when board assembly is automated. Here, the solder paste is applied only to the places on the PCB on which solder is required. Additionally, a silk screen is used to create legends, text or shapes, on top of the protective coating (sometimes referred to as a *top overlay*). Figure 3.11 shows the legends for four capacitors and one IC created in white ink on the top layer of a PCB.

Figure 3.11: Silk screen, top overlay

Track Thickness

The thickness of the copper track is normally specified in ounces per square foot, which refers to the weight if the copper were laid out flat in one square foot of area. Most common is 1 oz copper, although increased metal thicknesses such as 0.5 oz, 2 oz, and 4 oz are possible. Table 3.1 identifies the resulting thicknesses of the common specified values.

Thicker copper PCBs are usually for high-current circuits. Calculations for track width based on a particular track thickness are usually made by considering the maximum current flow and maximum rise in temperature of the board. The IPC provides a detailed method to calculate the required track width for given circuit requirements [8].

Track Resistance

Metal tracks have electrical resistance, determined by both the metal resistivity (ρ) [9] and the track dimensions. Example resistivity values for different metals and alloys are identified in Table 3.2. The units for resistivity are Ω.m (ohm.meter).

Table 3.1: Common copper track thickness values

oz/ft^2	Thickness		
	μm	inches	mils
0.5	17.5	0.0007	0.7
1	35	0.0014	1.4
2	70	0.0028	2.8
3	105	0.0042	4.2

Table 3.2: Metal and alloy resistivity values

Metal	Resistivity (ρ) – Ω.m
Aluminum	2.63×10^{-8}
Copper	1.72×10^{-8}
Iron	1.0×10^{-7}
Gold	2.44×10^{-8}
Lead	2.08×10^{-7}
Platinum	1.1×10^{-7}
Silver	1.47×10^{-8}
Tin	1.15×10^{-7}
Tungsten	5.51×10^{-8}
Alloy	**Resistivity (ρ) – Ω.m**
Brass (an alloy of zinc and copper)	0.8×10^{-7}
Steel (alloy of iron and carbon)	1.0×10^{-7}

When a track is formed on the PCB insulating substrate, it will have a cross-sectional area and length (Figure 3.12). The resistance of the track (Ω) from end to end (A to B) is given by:

$$R = \frac{\rho.L}{A}$$

where:

ρ is the resistivity of the metal (Ω.m)

R is the resistance of the track (Ω)

L is the length of the track (m)

A is the cross-sectional area of the track, width (W) $\times$ thickness (T) (m^2).

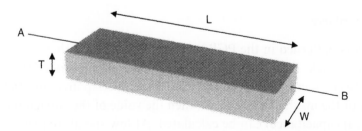

Figure 3.12: Metal track resistance calculation

The thickness of the track material is a fixed value set by the PCB manufacturing process, so the resistance will be set by the designer from the given track length and width. For a given length of track, a wide track will have less resistance than a narrow track.

For example, a track is created using copper, with a length of 100 mm and a width of 0.25 mm. The track thickness is 17.5 μm (i.e., 1 oz copper). What is the resistance of the track?

$$R = \frac{\rho . L}{A}$$

$$R = \frac{(1.72 \times 10^{-8}) \times (0.1)}{(25 \times 10^{-5}) \times (17.5 \times 10^{-6})} = 0.393\,\Omega$$

Electromigration

A phenomenon known as electromigration can occur when a high current level flows in a track. If the current density (amount of electrical current flowing per cross-sectional area, A/m^2) is high, then electromigration is the gradual movement of the ions in a conductor due to the momentum transfer between conducting electrons and diffusing metal atoms. The effect is for the metal to move, causing a reduction in the width of one part of the metal as the metal atoms "flow." Eventually, the track width reduces to a narrow enough cross-section for the metal to "fuse." That is, it becomes an open circuit, in the same manner as a fuse would be designed to intentionally fuse (or "blow") when the current passing through the fuse exceeds a maximum permitted value.

Insulation Capacitance

When a track is patterned in the PCB, and a second track, either above or below, crosses the first track, then the area created by the combination of the tracks and insulation between them creates a capacitor. If the overlap area and the capacitance per unit area of the insulation is known, then the value of the capacitance (a parasitic [i.e., unwanted] capacitance) can be calculated. At low signal frequencies, this capacitance does not necessary affect the operation of the circuit. However, as the

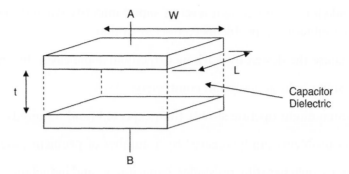

Figure 3.13: Track-track capacitance calculation

signal frequency increases, the effect of the capacitance also increases (as its impedance decreases), which can have a serious effect on circuit operation. Capacitance value (Figure 3.13) is calculated by:

$$C = \frac{\varepsilon_o . \varepsilon_{ins} . A}{D}$$

where:

C is the value of the capacitance of the overlapping area of the two tracks A and B (in Farads, F)

A is the area of overlap of the two tracks, width (W) × length (L) (cm^2)

D is the thickness of the insulator (dielectric) (cm)

ε_o is the relative permittivity of the insulating material (for FR-4, this is approximately 4)

ε_{ins} is the permittivity of free space ($\approx 8.85 \times 10^{-14}$ F/cm).

Signal Integrity

Signal integrity affects the electrical signals as they pass through the tracks in the PCB. Ideally, the signal should not be altered by the electrical properties of the track. However, a real track will alter the shape of the signal and so corrupt its integrity.

If care is not taken to ensure a high level of signal integrity when designing the PCB layout, then manufacturing problems can occur in that:

- It will cause the design to work incorrectly in some cases, but not all cases.

- The design might actually fail completely.

- The design might operate slower than expected (and required).

Signal integrity problems can be created by a number of problems, including:

- The tracks' own parasitic resistance, capacitance, and inductance will be altered.

- Cross-talk between two or more different tracks will occur because of a capacitive coupling between the tracks resulting from the PCB substrate insulation.

- For high-frequency signals, the characteristic impedance of the transmission line that the track creates does not match the signal source and destination.

An example where the track resistance and capacitance can create a parasitic resistor-capacitor (RC) network that is modeled as a single resistor and capacitor is shown in Figure 3.14. Applying a digital clock signal, a square wave voltage waveform, to the RC network causes a change in the observed waveform at the output. The output becomes an exponential waveform with a time constant $\tau =$ R.C. Such an effect can cause circuit failure.

Drawing Units

When designing the PCB layout, considering both the component placement and the interconnect placement, the designer is working with physical dimensions. Component placement and routing depends on a number of

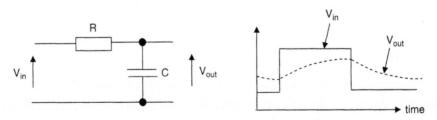

Figure 3.14: RC time constant effect

Table 3.3: Imperial-to-metric conversion

Imperial (mils)	Metric (mm)
1,000	25.4
100	2.54
10	0.254
1	0.0254
Eurocard size PCB	
3937 × 6299 (3.94" × 6.3")	100 × 160 (0.1 m × 0.16 m)

considerations. The particular PCB manufacturer will provide the necessary minimum (and possibly maximum) dimensions that can be used for their manufacturing process.

The dimensions will be provided in either Imperial measurements (using mils) or metric (using mm). There are 1,000 mils and 25.4 mm in 1 inch. Table 3.3 is a conversion chart.

3.2.3   Typical Components

The PCB will be populated with a number of components, using both through-hole and surface mount packages. Component location on the PCB is critical for:

1. Efficiently routing the PCB tracks (signal and power)

2. Accounting for thermal effects when components heat up during normal operation. The temperature rise must not be too large on any single part of the board. Suitable placement of components and the addition of heat sinks (components that absorb heat and allow it to be dissipated away from the component that generated it).

3. Ergonomic considerations where a user may need to access part of the PCB to control components (e.g., switches) or for test and evaluation purposes.

Table 3.4 identifies some of the electronic components more commonly found on typical PCBs.

Table 3.4: Typical components on a PCB

Component	Description	
Resistor	The resistor is a 2-terminal electronic component that resists the flow of current and produces a voltage drop across the component that is proportional to the current flow as given by Ohms Law. The image to the right shows a single resistor and a resistor array (in a14-pin DIL package).	
Variable resistor	The variable resistor (potentiometer or preset) is a 3-terminal device that acts to vary the resistance between two connections as a mechanical screw is rotated. The image to the right shows three different preset packages.	
Capacitor	The capacitor is a 2-terminal device that consists of two metal plates separated by a dielectric material that creates a specific value of capacitance. A range of materials are used as the dielectric. Specific capacitors are used for particular requirements within the circuit, and specific capacitor types are polarized; that is, one connection has a positive potential to the other connection. The image to the right shows four different capacitor types and packages.	

Table 3.4 *(Continued)*

Component	Description	
Inductor	The inductor is a 2-terminal device that consists of a winding of metal that creates a specific value of inductance. The image to the right shows an inductor that is created in a package similar in size and shape to a through-hole resistor package.	
Diode	The diode is a 2-terminal semiconductor device that allows current to flow in one direction through the device but blocks the flow of current in the opposite direction. The image to the right shows a through-hole package diode.	
Transistor	The transistor is a 3-terminal semiconductor device that is either use to amplify a signal (voltage or current) in analogue circuits or acts as an electronic switch in digital circuits. Both bipolar (npn and pnp) and CMOS (nMOS and pMOS) transistors, along with unijunction and JFET transistor structures, can be created. The image to the right shows three of the different package sizes and shapes that are available.	
Integrated circuit	The integrated circuit is a semiconductor device that consists of a packaged circuit die (silicon, silicon germanium, or gallium arsenide semiconductor material) that contains an electronic circuit consisting of transistors, resistors, capacitors, and possibly inductors. The image to the right shows a surface mount package.	

(continued)

Table 3.4 *(Continued)*

Component	Description	
Switch	The switch is a device that mechanically opens or closes metal contacts to connect or disconnect parts of an electrical or electronic or circuit. The image to the right shows a PCB mount toggle switch.	
Connector	The connector provides a mechanism to connect different electronic circuits together using wires. The image to the right shows three of the different package sizes and shapes that are available.	
Transformer	The transformer is a device consisting of two sets of wire coils to form a mutual inductance. The transformer is used to step up (increase) or step down (decrease) an AC voltage. The image to the right shows an example transformer package.	
Light emitting diode (LED) Available colors are red, yellow, green, blue, white	The light emitting diode is a 2-terminal semiconductor device that produces light when a current is passed through it. The image to the right shows two LED, for soldering to a PCB. The LED can be obtained in various shapes and sizes and also as 7-segment displays.	

Table 3.4 *(Continued)*

Component	Description	
Liquid crystal display (LCD)	The liquid crystal display is a device that is used to present either images or text. The image to the right shows a 2-line, 16-character LCD display.	
Test probe point	The test probe point is a 1-terminal device that allows external text and measurement equipment to be connected to a point in an electronic circuit for test and evaluation purposes. The image to the right shows eyelet probes that allow for test equipment probes to be hooked to the test probe point.	
Crystal oscillator	The crystal oscillator is a device that produces an oscillating signal at a particular frequency for the generation of clock signals within a digital circuit. The image to the right shows an example oscillator module that is housed in a 14-pin DIL package.	
Jumper terminal	The jumper terminal is a 2-terminal device that connects two points of a circuit together when a metal header is placed across the terminals, or disconnects two points of a circuit together when the header is physically removed by a user. The image to the right shows a jumper terminal (+ header) mounted to a PCB.	

3.3   Design, Manufacture, and Testing

3.3.1   PCB Design

Overview

PCB design begins with an insulating base and adds metal tracks for electrical interconnect and the placement of suitable electronic components to define and create an electronic circuit that performs a required set of functions. The key steps in developing a working PCB are shown in Figure 3.15 and briefly summarized below:

- Initially, a *design specification* (document) is written that identifies the required functionality of the PCB. From this, the designer creates the circuit design, which is entered into the PCB design tools.

- The *design schematic* is analyzed through simulation using a suitably defined test stimulus, and the operation of the design is verified. If the design does not meet the required specification, then either the design must be modified, or in extreme cases, the design specification must be changed.

- When the design schematic is complete, the *PCB layout* is created, taking into account layout directives (set by the particular design project) and the manufacturing process design rules.

- On successful completion of the layout, it undergoes *analysis* by (i) resimulating the schematic design to account for the track parasitic components (usually the parasitic capacitance is used), and (ii) using specially designed signal integrity tools to confirm that the circuit design on the PCB will function correctly. If not, the design layout, schematic, or specification will require modification.

When all steps to layout have been completed, the design is ready for submission for manufacture.

PCB Design Tools

A range of design tools are available for designing PCBs, running on the main operating systems (Windows®, Linux, UNIX™). The choice of tool depends on the actual design requirements, but must consider:

- **Schematic capture capabilities**: the ability to create and edit schematic documents representing the circuit diagram

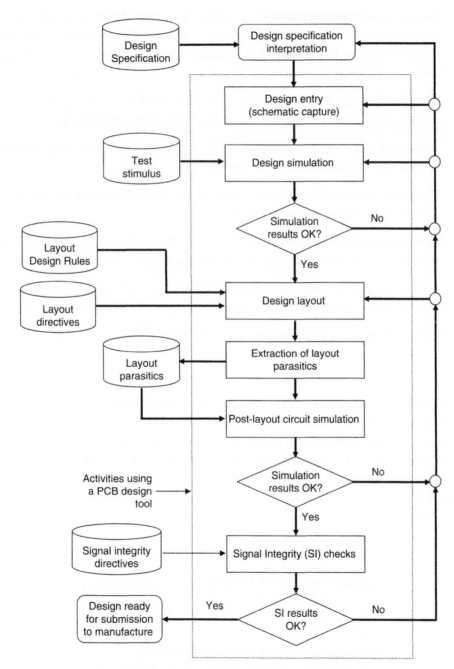

Figure 3.15: Steps to PCB design

- **Layout generation capabilities**: the ability to create the PCB layout either manually or using automatic place and route tools. Some design tools will link the schematic to the layout so that changes in the schematic are reflected as changes in the layout (and vice versa).

- **Circuit simulation capabilities**: the ability to simulate the design functionality using a suitable simulator such as a simulator based on SPICE.

- **Supported operating systems**: What PC or workstation operating systems are needed for the software tool to operate?

- **Company support**: What support is available from the company if problems are encountered using the design tools?

- **Licensing requirements and costs**: What are the licensing arrangements for the software, and is there an annual maintenance fee?

- **Ease of use and training requirements**: How easy is the design tool to use, and what training and/or documentation is available to the user?

Table 3.5 shows the main PCB design tools currently used.

LVS

Layout versus schematic (LVS) checking is a process by which the electronic circuit created in the final PCB layout is compared to the original schematic circuit diagram. This check is undertaken to ensure that the PCB layout is electrically the same as the original schematic, and errors have not been introduced. LVS can take a manual approach, in which the designer manually checks the connections in the layout and compares them to the schematic connections, or it can be automated using an LVS software tool.

Table 3.5: Example PCB design tools

Design Tool Name	Company
Allegro®	Cadence™ Design Systems Inc.
Board System®	Mentor Graphics®
Eagle	CadSoft
Easy-PC	Number One Systems
Orcad®	Cadence™ Design Systems Inc.
Protel	Altium™

DRC

Design rules checking (DRC) is a process by which the PCB layout is checked to confirm that it meets manufacturing requirements. Each manufacturing process has a set of design rules that identifies the limitations of the manufacturing process and ensures a high manufacturing yield. Design rules are rarely violated, and only then if clearance is given by the manufacturer and the designer is aware of and accepts any inherent risks.

Layout Design Rules and Guidelines

To produce a well-designed and working PCB, design guidelines (should be followed but are not mandatory) and rules (must be followed to avoid manufacturing problems) are to be followed. For example:

- Do not violate the minimum track widths, track spacing, and via sizes set by the PCB manufacturer. Table 3.6 provides a set of minimum dimension constraint examples.

- Avoid exposed metal under component packages. Any metal under a package should be covered with solder mask.

- Make the pads for soldering the electronic components to the board as large as possible to aid component soldering.

- Avoid the placement of components and tracks (and ground and power planes) that will require the removal of a great amount of copper from parts of

Table 3.6: Layout design considerations

Layout consideration	Meaning
Internal line width	Minimum the width of a metal track inside the PCB structure.
Internal line spacing	Minimum the distance between two metal tracks inside the PCB structure.
External line width	Minimum the width of a metal track on an outside surface (top or bottom) of the PCB.
External line spacing	Minimum the distance between two metal tracks on an outside surface of the PCB.
Minimum via size	The minimum size allowable for a via.
Hole to hole	The minimum distance between adjacent holes in the PCB insulating material.
Edge to copper	The minimum distance from the edge of the PCB to the copper that is designed for use.

the board, and leaving large amounts of copper in the remainder of the board. Where possible, have an even spread of tracks and gaps between the tracks across the entire board. (The copper layer starts as a sheet of metal covering the entire surface, and an etching process removes the unwanted copper to pattern the tracks.)

- Use ground (and power) planes for the component power supplies. Where possible, dedicate a layer to a particular power level (e.g., 0 V as ground). Use split planes if necessary; these are multiple planes on a layer where a part of the layer is dedicated to a particular power level.

- Use power supply decoupling capacitors for each power pin on each component. Place the decoupling capacitor as close as possible to its component pin. For example, data converter data sheets normally provide information for the PCB designer in relation to the decoupling capacitor requirements.

- Use decoupling capacitors for each DC reference voltage used in the circuit (e.g., reference voltages for data converters). For example, data converter data sheets normally provide information for the PCB designer in relation to the decoupling capacitor requirements.

- Use separate digital and analogue power supply planes and connect at only one point in the circuit. For example, a data converter package normally has separate power (V_{DD} and GND) pins for the analogue and digital circuitry. The device analogue and digital power will be provided by connecting the IC to separate power planes. The GND connection is connected at one point only underneath the IC (see Figure 3.16). Data converter datasheets normally

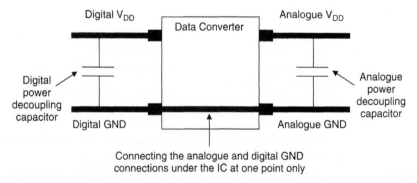

Figure 3.16: Example data converter GND ("common") connection (top down)

provide information for the PCB designer relating to the placement of signal and power connections.

- Minimize the number of vias required.

- Avoid ground loops, which can form when the ground connections on the electronic components are laid out to the common track (or plane) so that loops of metal are formed. They can cause noise problems in analogue signals.

- For the particular PCB, consider which is more important, the placement of the components or the routing of the tracks? Adopt a layout design procedure to reflect this.

- Separate the digital and analogue components and tracks to avoid or reduce the effects of cross-talk between the analogue signals and digital signals.

Ground Planes

Ground (GND) and power planes on the PCB are large areas of metal that are connected to either a power supply potential (e.g., V_{DD}) or the common (0 V) connection (commonly referred to as *ground*). They appear as low-impedance paths for signals and are used to reduce noise in the circuit, particularly for the common signal. In a multilayer PCB, one or more of the layers can be dedicated to a plane. Any given metal layer can have a single plane or multiple planes (split plane), shown in Figure 3.17. Signals will pass through the plane where the metal is etched away at specific points only, signified by the white dots in the illustration.

PCBs for Different Applications

Certain PCB manufacturers will provide a range of different PCB fabrication facilities to support different applications including:

- **High-frequency circuits**: Specific materials will be required for the insulating base and the track metal for the circuit to operate at the required frequencies [10, 11].

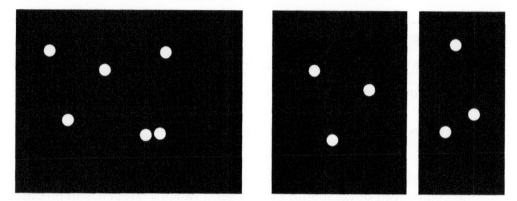

Figure 3.17: Single (left) and split (right) planes

- **Power supplies**: Power supplies may be required to operate at high voltages and high currents to meet performance requirements.

- **Controlled impedance**: This is required in applications in which the interconnecting track acts as a transmission line and must have a known and controlled impedance. Such applications include radio frequency (RF) circuits and high-speed digital switching circuits.

3.3.2   PCB Manufacture

When the design layout has been completed, it is submitted for manufacture. Depending on the manufacturer and design project requirements, either one or several prototype PCBs will be manufactured and populated for design debug and verification purposes prior to a full-scale production run.

The design layout will normally be submitted in electronic format using one of the PCB layout file tools supported by the manufacturer.

Figure 3.18 shows the different layers that are used to make a two-sided PCB with through-hole plated vias and top overlay layers for information text (and graphics). This is the board shown in Figure 3.2.

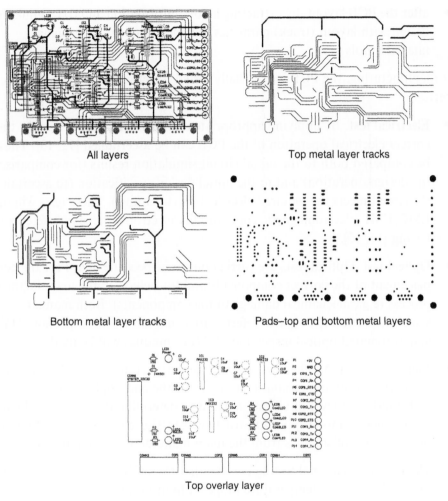

Figure 3.18: PCB layers

3.3.3   PCB Testing

To verify that the circuit design is functionally correct, the design is tested both prior to and after manufacturing. Prior to manufacturing, the design is simulated using an appropriate simulation model of the circuit and a suitable test stimulus. Simulation is undertaken twice:

- prior to creating the PCB layout, to verify the correct electrical functionality of the circuit schematic diagram

- after the PCB layout, by extracting layout information and (i) resimulating the design with layout (track) parasitics included, and (ii) using signal integrity tools

After manufacturing, the PCB is physically tested for electrical and nonelectrical properties:

- **Electrical test**. By applying appropriate analogue and digital signals, the correct electrical operation of the PCB can be ascertained [12, 13]. These will be compared both to the initial circuit simulation results (for comparison of the design operation) and to the initial design specification (to ascertain that the circuit meets the required electrical specifications). These tests will include EMC/EMI (electromagnetic compatibility/electromagnetic interference) testing [14, 15].

- **Optical test**. Optical tests are carried out to inspect the board for the correct placement of the correct components and for defects in the manufacturing process (e.g., mechanical damage to the components). Both manual visual inspection (MVI, also referred to as human visual inspection, HVI) and automated optical inspection (AOI) techniques will be used.

- **Mechanical test**. Mechanical testing is undertaken to ensure that the PCB meets the required mechanical strength for the end application (e.g., it can withstand a set level of vibrations) and to determine the mechanical strength of the solder joints [16]. For destructive tests (those that stress the PCB until it breaks), a set of samples from the main manufacturing run are used.

- **Thermal test**. Thermal testing ensures that the PCB will operate over the required operating temperature range without failure.

- **WEEE and RoHS compliance**. Tests undertaken to ensure that the PCB is compliant with the required legislation (discussed further in the next section).

3.4   Environmental Issues

3.4.1   Introduction

Increasingly, the whole process of design, manufacture, and test is required to consider their impact on the environment. There is a need, guided by legislation, to reduce that environmental impact.

3.4.2 WEEE Directive

The WEEE Directive (waste electrical and electronic equipment) was introduced by the European Union (EU) to increase the electrical and electronic equipment recycling [17]. A key part of this is to make manufacturers and importers (also referred to as *producers*) responsible for meeting the costs of the collection, treatment, and recovery of equipment that has come to the end of its life span. This encourages the designers of such equipment to create products with recycling in mind.

The WEEE Directive covers a number of items, such as:

- small and large household appliances
- Information technology (IT) and telecommunications equipment
- consumer equipment
- lighting
- electrical and electronic tools (except large-scale stationary industrial tools)
- toys, leisure, and sports equipment
- medical devices (with exceptions)
- monitoring and control instruments
- automatic dispensers

3.4.3 RoHS Directive

The RoHS Directive (return of hazardous substances) supports the WEEE directive by covering the use of certain hazardous substances used in electrical and electronic equipment [18]. The European Union directive, effective July 1, 2006, limits the use of certain substances to prescribed maximum concentration levels in electrical and electronic equipment unless the equipment is exempt from the directive. The banned substances are:

- lead
- cadmium

- mercury

- hexavalent chromium

- polybrominated biphenyl ethers

- polybrominated diphenyl ethers

Equipment is categorized as RoHS compliant, not RoHS compliant, or RoHS compliant by exemption. Equipment that is required to be compliant must have a Certificate of RoHS compliance.

3.4.4   Lead-Free Solder

In electronic circuits, traditional (lead) solder was comprised of 60% tin and 40% lead (Sn60/Pb40) by mass to produce a near-eutectic mixture. (A eutectic or eutectic mixture is a mixture of two or more phases at a particular composition of materials that have the lowest melting point, at which temperature the phases will simultaneously crystalize.) For Sn60/Pb40, the lowest melting point is below 190°C.

Since the introduction of the WEEE directive and RoHS, lead has been removed from electrical and electronic equipment. The resulting lead-free solders contain other metals such as tin, copper, and silver [19]. Lead-free solders have higher melting points, which has necessitated re-engineering the electronic components to withstand the higher solder melting points.

3.4.5   Electromagnetic Compatibility

When an electronic circuit is to operate in a particular environment, it will be required to operate:

- without producing any interference to the operation of any other electronic circuit

- without itself being interfered to by any other electronic circuit

Electromagnetic noise is produced when an electronic source produces rapidly changing current and voltage. Nearby electronic circuits that are coupled to the source (by conductive, radiative, capacitive, or inductive coupling) can receive noise through this coupling mechanism, and electromagnetic interference (EMI) will occur. Electromagnetic compatibility (EMC) is the ability of an electronic circuit to function in its operating environment without causing or experiencing performance degradation resulting from unintentional EMI.

Unless circuits are designed to be coupled, circuit designs can be made to reduce the noise effect by any of several means:

- reducing any signal switching frequency to as low as possible to maintain the circuit operation

- physically separating the circuits

- suitably shielding the circuit using shielding material and enclosures

3.5 Case Study PCB Designs

3.5.1 Introduction

The case study designs presented within the book are based on the development of a complete mixed-signal electronic system, as shown in Figure 3.19, using a complex programmable logic device (CPLD) development board with plug-in modules (Eurocard-sized PCBs). As such, the modules can be developed on a need-to-use basis. With this arrangement, the single experimentation arrangement will enable a wide range of designs to be designed, developed, and tested.

Each of the boards can be designed and manufactured as and when required, depending on the type of system to develop and experiments to undertake. The core of the system is the CPLD development board, containing a XC2C256 Coolrunner™-II CPLD, SRAM (static RAM) memory, and connectors for connecting the other boards. Hence, the development board must be designed and manufactured first. The board operates on a single $+3.3\,$V power supply for both the CPLD and SRAM. Additionally, a $+1.8\,$V power supply is derived from the $+3.3\,$V input power to provide the necessary power to the CPLD; this particular CPLD requires a $+1.8\,$V power supply for the core and a $+3.3\,$V periphery interface level to the external circuitry.

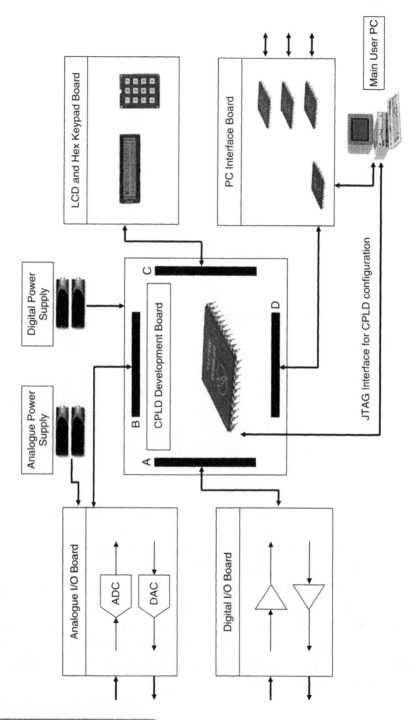

Figure. 3.19: Case study board set-up

Aside: In this section, the PCB board operation and connections are identified, along with the potential uses. The circuit schematics are not provided here, but are available in Appendix F, Case Study PCB Designs (see the last paragraph of the Preface for instructions regarding how to access this online content).

The digital logic uses LVCMOS standard (logic 1 = +3.3 V, logic 0 = 0 V), and the analogue circuits operate on a +/−5.0 V dual rail power supply. The digital logic power supply for all boards is derived from the CPLD development board, but the analogue I/O board requires a separate dual rail power supply for the analogue parts. Hence, the circuit is designed to operate at the lower voltage levels.

If signal voltage levels exceeding the designed levels are required, they must be generated externally. The external circuit levels must be compatible with the designed levels for the system and must not under any circumstances exceed the absolute maximum ratings for each component in the circuit. Absolute maximum ratings for each component are identified in the datasheet for the particular component.

3.5.2 System Overview

Once the CPLD development board has been designed, manufactured, and successfully tested, it can be used for developing digital circuit and systems designs. Those designs are developed based on logic schematic diagrams and/or hardware description language (HDL) and using an appropriate CPLD design tool. (The Xilinx® ISE™ tools available from the company will be required.) It is possible to use both VHDL (VHSIC hardware description language) and Verilog®-HDL to develop the digital designs, and synthesizing the resulting HDL RTL (register transfer level) code into a netlist targeting the CPLD, but in this book, only VHDL will be considered. Attached to the CPLD development board (the motherboard) will be four daughter boards, each with a different function as follows:

1. LCD (liquid crystal display) and hex keypad board

2. PC interface board

3. digital I/O board

4. analogue I/O board

With this arrangement, it is possible to develop a wide range of digital and mixed-signal electronic circuits based on a central digital core, for applications such as:

- general computing

- communications

- digital signal processing (DSP)

- digital control

- security and alarm systems

- instrumentation

- environmental monitoring

- mixed-signal electronic circuit test equipment

- analogue signal generation (using an arbitrary waveform generator, AWG)

- direct digital synthesis (DDS)

The CPLD I/O connections will be configured to adhere to the LVCMOS (3.3 V level) standard so that the I/O will interface to the digital circuitry it is attached to. However, the digital circuitry will be required to adhere to the LVCMOS (3.3 V level) standard for compatibility, unless suitable level shifting circuitry is utilised to interface the CPLD to the digital circuitry.

3.5.3   CPLD Development Board

The CPLD development board is the heart of the electronic system. This contains a XC2C256 Coolrunner™-II CPLD, SRAM memory, and connectors for connecting the other boards. The CPLD development board operates on a single +3.3 V power supply, used to power both the CPLD and SRAM. Additionally, a +1.8 V power supply is derived from the +3.3 V input power to provide the necessary power to the CPLD; this particular CPLD requires a +1.8 V power supply for the core and a +3.3 V periphery interface level to the external circuitry.

The CPLD operates using a 50 MHz crystal oscillator IC and has a power-on reset circuit (with additional manual reset switch).

The CPLD is programmed from a PC using its built-in JTAG (Joint Test Action Group) interface. The ISE^{TM} tool is to be used for CPLD design entry, simulation, layout, and configuration. An introduction to the design tool used is provided in Appendix E, Introduction to the Design Tools (see the last paragraph of the Preface for instructions regarding how to access this online content).

The CPLD I/O connections are configured to adhere to the LVCMOS (3.3 V level) standard. However, the CPLD I/O can be configured to operate with the following digital logic standards:

- LVTTL, Low-voltage transistor-transistor logic (3.3 V level)

- LVCMOS33, Low-voltage CMOS (3.3 V level)

- LVCMOS25, Low-voltage CMOS (2.5 V level)

- LVCMOS18, Low-voltage CMOS (1.8 V level)

- 1.5 V I/O (1.5 V levels), 1.5 V level logic (1.5 V level)

- HSTL-1, High-speed transceiver logic

- SSTL2-1, Stub series terminated logic (2.5 V level)

- SSTL3-1, Stub series terminated logic (3.3 V level)

The I/O standard is set during the design entry within the CPLD design tools and is one of the design constraints that the user will set.

The CPLD development board (see Figure 3.20) is based around the use of the Coolrunner^{TM}-II CPLD (XC2C256-144) device using a 144-pin package (in a TQFP [thin quad flat pack]) package, connected to IDC connectors to interface the CPLD to the daughter boards. The board also houses a Cypress Semiconductor CYC1049CV33 512x8 SRAM IC that can be used for temporary data storage whenever the CPLD is configured to undertake either digital signal processing or data capture operations.

The CPLD is automatically reset whenever the power is applied using a power-on reset circuit. (The configuration is held in nonvolatile memory so that whenever the power is removed from the CPLD, the last configuration is retained.) This reset can also be manually applied using a push-switch at any time by the user. This circuit uses the Maxim MAX811-S voltage monitor IC with manual reset input.

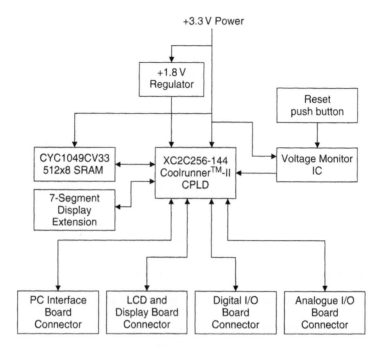

Figure 3.20: CPLD development board

The IDC connector pin allocation for the CPLD development board to connect to the four daughter boards is the same as for each of the daughter boards.

The circuit diagram for this PCB is provided in Appendix F, Case Study PCB Designs (see the last paragraph of the Preface for instructions regarding how to access this online content).

Table 3.7 identifies the component list for the CPLD development board.

3.5.4 LCD and Hex Keypad Board

A 12-key hex keypad is used for data entry into the CPLD (whether at a data entry terminal, security keypad, telephone keypad, for instance), and data is displayed on a MDLS162653V 2-line, 16-digit LCD (see Figure 3.21). The LCD can be used for a range of scenarios such as message boards and prompts for the user to take specific actions. This user I/O mechanism is based on typical portable electronics used today (e.g., the mobile phone). The circuit is designed to operate with a logic 1 value of +3.3 V and a logic 0 value of 0 V, and the LCD display is designed for 3.3 V

Table 3.7: CPLD development board component list

Component no.	Component description	Quantity
1	XC2C256-144 Coolrunner™-II CPLD	1
2	CYC1049CV33 512x8 SRAM	1
3	PCB mount Push-switch	1
4	1N4001 diode	1
5	150 Ω resistor (0.6 W, ±1% tolerance)	1
6	Blue LED (20 mA)	1
7	20-way IDC connector	4
8	2.1 mm power connector	1
9	50 MHz crystal oscillator (8-pin DIP)	1
10	REG1117 +1.8V voltage regulator	1
11	MAX811-S voltage monitor IC	1
12	14-way connector (specific to JTAG programmer cable)	1
13	16-way connector (for LED display board extension)	1
14	100 nF capacitor	13
15	10 μF electrolytic capacitor	1
16	Eyelet test probe point	8

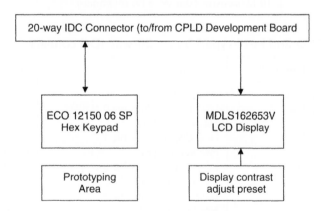

Figure 3.21: LCD and hex keypad board

operation. The data sheet for the MDLS162653V display obtained from the appropriate manufacturer can be referred to for precise logic I/O specifications (power supply operation, logic I/O voltage levels, and speed of operation).

A preset (variable resistor) is connected to the LCD display to allow the user to adjust the display contrast. The free space on the PCB (i.e., the area not used by the components and interconnect track) is filled with a prototyping area consisting of through-hole plated vias spaced at 2.54 mm in a 24 × 12 array. The via spacing is set to that of through-hole DIL (dual in-line) packages. The circuit diagram for this PCB is provided in Appendix F, Case Study PCB Designs (see the last paragraph of the Preface for instructions regarding how to access this online content).

Table 3.8 identifies the component list for the LCD display and hex keypad board.

Table 3.9 identifies the 20-way IDC connector pin allocation for the LCD display and hex keypad board.

Table 3.8: LCD and Hex keypad board component list

Component no.	Component description	Quantity
1	20 way IDC plug (PCB mount)	1
2	1N4001 diode	1
3	150 Ω resistor (0.6 W, ±1% tolerance)	1
4	Blue LED (20 mA)	1
5	10 kΩ preset	1
6	LCD display (16 character, 2 line), MDLS162653V	1
7	12-key hex keypad – ECO 12150 06 SP	1
8	10 kΩ resistor (0.6 W, ±1% tolerance)	7

Table 3.9: LCD and hex keypad board 20-way IDC connector pin allocation

Pin no.	Identifier	Function	Direction
1	VDD	+3.3 V DC	Power supply
2	DB0	LCD display data bit 0 (LSB)	Input
3	D	Hex keypad (D)	Input/output
4	DB1	LCD display data bit 1	Input
5	E	Hex keypad (E)	Input/output
6	DB2	LCD display data bit 2	Input
7	F	Hex keypad (F)	Input/output
8	DB3	LCD display data bit 3	Input
9	G	Hex keypad (G)	Input/output
10	DB4	LCD display data bit 4	Input
11	H	Hex keypad (H)	Input/output
12	DB5	LCD display data bit 5	Input
13	J	Hex keypad (J)	Input/output
14	DB6	LCD display data bit 6	Input
15	K	Hex keypad (K)	Input/output
16	DB7	LCD display data bit 7	Input
17	RS	LCD register select control	Input
18	Enable	LCD enable control	Input
19	R/W	LCD read/write control	Input
20	VSS	0 V DC	Power supply

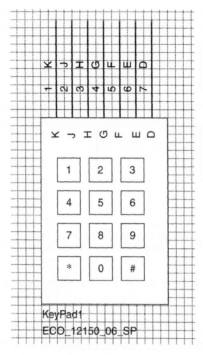

Figure 3.22: Hex keypad pin identification

The CPLD must be configured so that it will suitably access the keypad. This involves providing output logic levels to specific connections on the keypad and reading in from the remaining connections. This is a common technique adopted when using hex keypads of this type. Figure 3.22 shows the keypad pin identification.

3.5.5   PC Interface Board

The PC interface board (Figure 3.23) uses RS-232 communications protocol, which allows digital circuits to communicate with each other using a UART (universal asynchronous receiver transmitter) circuit. The board contains three-level shifting ICs (the MAX3232CPE) to provide four serial links (COM links 1 to 4) that can be connected to the CPLD board. The MAX2323CPE level shifting ICs (IC pin connections) translate the PC UART output levels to +3.3 V/0 V levels compatible

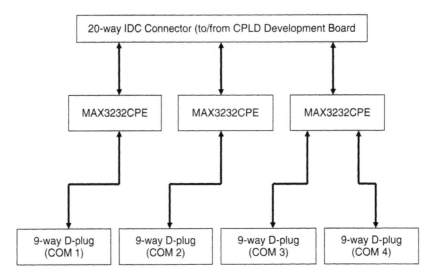

Figure 3.23: PC interface board

with the CPLD board (Figure 3.24). The datasheet for the MAX3232CPE obtained from the appropriate manufacturer provides precise logic I/O specifications (power supply operation, logic I/O voltage levels, and speed of operation).

The circuit diagram for this PCB is provided in Appendix F, Case Study PCB Designs (see the last paragraph of the Preface for instructions regarding how to access this online content).

Table 3.10 identifies the component list for PC interface board.

Table 3.11 identifies the 20-way IDC connector pin allocation for the PC interface board.

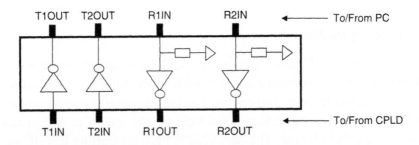

Figure 3.24: MAX3232CPE circuit with pin identifiers

Table 3.10: PC interface board component list

Component no.	Component description	Quantity
1	20 way IDC plug (PCB mount)	1
2	1N4001 diode	1
3	150 Ω resistor (0.6 W, ±1% tolerance)	7
4	Blue LED (20 mA)	1
5	Red LED (20 mA)	2
6	Yellow LED (20 mA)	4
7	MAX3232CPE Level-Shifter IC	3
8	10 μF electrolytic capacitor	15
9	PCB mount D-connector (plug)	4
10	Eyelet test probe point	14

Table 3.11: PC interface board 20-way IDC connector pin allocation

Pin no.	Identifier	Function	Direction
1	VDD	+3.3 V DC	Power supply
2	Tx_LED	Transmitter indicator LED	Input
3	COM1_Rx	COM 1 data receiver Input	Output
4	Rx_LED	Receiver indicator LED	Input
5	COM1_Tx	COM 1 data transmitter output	Input
6	COM4_LED	COM 4 selected indicator LED	Input
7	COM1_CTS	COM 1 clear to send	Output
8	COM3_LED	COM 4 selected indicator LED	Input
9	COM1_RTS	COM 1 ready to send	Input
10	COM2_LED	COM 4 selected indicator LED	Input
11	COM2_Rx	COM 2 data receiver input	Output
12	COM1_LED	COM 4 selected indicator LED	Input
13	COM2_Tx	COM 2 data transmitter output	Input
14	COM4_Tx	COM 4 data transmitter output	Input
15	COM2_CTS	COM 2 clear to send	Output
16	COM4_Rx	COM 4 data receiver input	Output
17	COM2_RTS	COM 2 ready to send	Input
18	COM3_Tx	COM 3 data transmitter output	Input
19	COM3_Rx	COM 3 data receiver input	Output
20	VSS	0 V DC	Power supply

There are four possible COM ports, using the following connections:

- COM 1—Tx, Rx, CTS, and RTS signals
- COM 2—Tx, Rx, CTS, and RTS signals

- COM 3—Tx and Rx signals

- COM 4—Tx and Rx signals

The CPLD can identify which COM port it is currently accessing using the four yellow LEDs on the PC interface board (where a logic 1 output from the CPLD turns ON the LED and a logic 0 output from the CPLD turns OFF the LED). The CPLD can also identify Tx and Rx signal activity by using the two red LEDs on the PC interface board.

3.5.6   Digital I/O Board

The digital I/O board (Figure 3.25) uses octal three-state (tri-state) buffers using 74HC240 logic ICs that provide a digital buffer between the CPLD and external digital logic circuitry. This both allows the CPLD logic outputs to be applied to external circuitry and provides protection; if a fault in the external circuitry causes a situation that can damage the CPLD (e.g., electrical overstress), then the 74HC240 logic ICs will be damaged before the CPLD. The 74HC240 logic ICs (IC pin connections, Figure 3.26) are cheaper to replace and will be mounted in sockets, thereby avoiding the need to unsolder the CPLD surface mount package. The circuit is designed to operate with a logic 1 value of +3.3 V and a logic 0 value of 0 V.

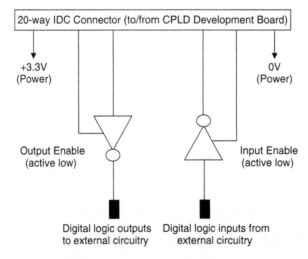

Figure 3.25: Digital I/O board block diagram

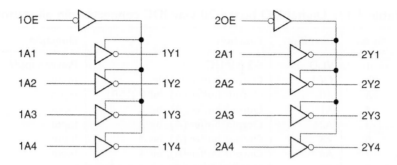

Figure 3.26: 74HC240 octal buffer circuit schematic with pin identifiers

The data sheet for the 74HC240 IC obtained from the appropriate manufacturer provides the precise logic I/O specifications (power supply operation, logic I/O voltage levels, and speed of operation).

The circuit diagram for this PCB is provided in Appendix F, Case Study PCB Designs (see the last paragraph of the Preface for instructions regarding how to access this online content).

Table 3.12 identifies the component list for the digital I/O board.

Table 3.13 identifies the 20-way IDC connector pin allocation for the digital I/O board.

The circuit requires a +3.3 V power supply via the IDC connector from the CPLD development board to provide the necessary power to the buffer ICs. A protection diode (1N4001) is reversed-biased across the power supply so that when the power supply is connected in the correct orientation, the diode does not

Table 3.12: Digital I/O board component list

Component no.	Component description	Quantity
1	20 way IDC plug (PCB mount)	1
2	74HC240 (Octal buffer with 3-state outputs)	2
3	Blue LED (20 mA)	1
4	Yellow LED (20 mA)	2
5	150 Ω resistor (0.6 W, ±1% tolerance)	3
6	1N4001 diode	1
7	9-way terminal block	2
8	100 nF ceramic capacitors	2
9	Eyelet test probe point	18

Table 3.13: Digital I/O board 20-way IDC connector pin allocation

Pin no.	Identifier	Function	Direction
1	VDD	+3.3 V DC	Power supply
2	OE1	Output buffer enable	Input
3	A0	Output buffer (A), bit 0 (LSB)	Input
4	A1	Output buffer (A), bit 1	Input
5	A2	Output buffer (A), bit 2	Input
6	A3	Output buffer (A), bit 3	Input
7	A4	Output buffer (A), bit 4	Input
8	A5	Output buffer (A), bit 5	Input
9	A6	Output buffer (A), bit 6	Input
10	A7	Output buffer (A), bit 7 (MSB)	Input
11	OE2	Input buffer enable	Input
12	B0	Input buffer (B), bit 0 (LSB)	Output
13	B1	Input buffer (B), bit 1	Output
14	B2	Input buffer (B), bit 2	Output
15	B3	Input buffer (B), bit 3	Output
16	B4	Input buffer (B), bit 4	Output
17	B5	Input buffer (B), bit 5	Output
18	B6	Input buffer (B), bit 6	Output
19	B7	Input buffer (B), bit 7 (MSB)	Output
20	VSS	0 V DC	Power supply

have any effect. If, however, the power supply orientation is reversed (i.e., +3.3 V and 0 V are connected the wrong way around), then the diode will conduct for a short time until it is damaged, then the IC V_{DD} will be limited to approximately −0.6 V (because of the forward-biased diode voltage drop), and during this time, the ICs will be protected from damage resulting from electrical overstress.

Three LEDs are also included on the board: one blue to indicate the power supply is connected, and two yellow to indicate that the buffers are enabled (LED is OFF) or disabled (LED is ON).

3.5.7 Analogue I/O Board

The analogue I/O board generates and samples analogue voltages under the control of the CPLD (Figure 3.27).

A stereo DAC (digital-to-analogue converter) provides two analogue output voltages digitally generated by the CPLD. The DAC is a Wolfson® Microelectronics WM8725

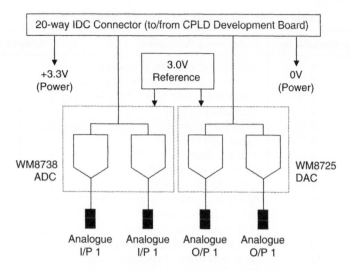

Figure 3.27: Analogue I/O board block diagram

stereo DAC with a serial interface, which requires seven digital signals for control and data, analogue and digital power supplies, and an analogue reference voltage. A stereo ADC (analogue-to-digital converter) is used to sample two analogue input voltages into the CPLD. The ADC is a Wolfson® Microelectronics WM8738 stereo ADC with a serial interface, which requires six digital signals for control and data, analogue and digital power supplies, and an analogue reference voltage.

Table 3.14 identifies the component list for the analogue I/O board.

Table 3.15 identifies the 20-way IDC connector pin allocation for the analogue I/O board.

Both the DAC and the ADC require a reference voltage to operate. This is externally generated using a Reference Voltage IC (REF3230), which provides an accurate +3.0 V voltage to supply the analogue power to both the DAC and ADC, which in turn internally generates the required reference voltage.

Each of the analogue inputs and outputs to and from the board are connected via an op-amp operating as a unity gain buffer to BNC connectors on the board. The output voltage range is set by the output range of the DAC (minimum to maximum output voltage values) and the input range of the ADC (minimum to maximum input voltage values). The outputs and inputs are also unipolar (positive voltages only).

Table 3.14: Analogue I/O board component list

Component no.	Component description	Quantity
1	20-way IDC plug (PCB mount)	1
2	150 W resistor (0.6 W, ±1% tolerance)	5
3	1 MW resistor (0.6 W, ±1% tolerance)	2
4	Blue LED (20 mA)	1
5	Red LED (20 mA)	2
6	Green LED (20 mA)	2
7	1N4001 diode	5
8	WM8725 stereo DAC	1
9	WM8738 stereo ADC	1
10	REF3230 (3.0 V) voltage reference IC	1
11	LM324 quad op-amp	1
12	PCB mount BNC connector	4
13	PCB mount screw terminal (3-way)	1
14	10 µF electrolytic capacitor	5
15	100 nF ceramic capacitor	6
16	Eyelet test probe point	6

Table 3.15: Analogue I/O board 20-way IDC connector pin allocation

Pin no.	Identifier	Function	Direction
1	VDD	+3.3 V DC	Power supply
2	ADC_FMT	WM8738 ADC signal FMT	Input
3	ADC_NOHP	WM8738 ADC signal NOHP	Input
4	ADC_SDATO	WM8738 ADC signal SDATO	Output
5	ADC_LRCLK	WM8738 ADC signal LRCLK	Input
6	ADC_BCLK	WM8738 ADC signal BCLK	Input
7	ADC_MCLK	WM8738 ADC signal MCLK	Input
8	DAC_FORMAT	WM8725 DAC signal FORMAT	Input
9	DAC_SCKI	WM8725 DAC signal SCKI	Input
10	DAC_LRCIN	WM8725 DAC signal LRCIN	Input
11	DAC_DIN	WM8725 DAC signal DIN	Input
12	DAC_BCKIN	WM8725 DAC signal BCKIN	Input
13	DAC_DEEMPH	WM8725 DAC signal DEEMPH	Input
14	DAC_MUTE	WM8725 DAC signal MUTE	Input
15	—	—	—
16	ADC_1_LED	ADC input 1 selected indicator LED	Input
17	ADC_2_LED	ADC input 2 selected indicator LED	Input
18	DAC_1_LED	DAC input 1 selected indicator LED	Input
19	DAC_2_LED	DAC input 2 selected indicator LED	Input
20	VSS	0 V DC	Power supply

Therefore, for bipolar (positive and negative voltages) and for a wider range of I/O voltages, external circuitry is required to appropriately condition the signals.

Four yellow LEDs are also mounted on the PCB so the CPLD can indicate which DAC or ADC is actually selected at any one time.

The circuit requires a +3.3 V digital power supply via the IDC connector from the CPLD development board to provide the necessary power to the buffer ICs. A protection diode (1N4001) is reversed-biased across the power supply so that when the power supply is connected in the correct orientation, the diode does not have any effect. If, however, the power supply orientation is reversed (i.e., +3.3 V and 0 V are connected the wrong way around), then the diode will conduct and for a short time until it is damaged, then the IC V_{DD} will be limited to approximately −0.6 V (because of the forward-biased diode voltage drop), and during this time, the ICs will be protected from damage resulting from electrical overstress.

The analogue power for the op-amps is provided by a separate screw terminal connector. This additional power supply also incorporates protection diodes.

3.6   Technology Trends

The areas of PCB design, manufacture, and test are taking on an increasingly important role in ensuring that a circuit design will operate correctly once manufactured. Among the technological and commercial drivers requiring these improvements are:

- **Cost reduction**: End users requiring more product for less cost

- **Higher quality levels**: The need to continually improve the quality of the manufactured PCB

- **Adherence to legislation directives**: Increased implementation of legislation that requires particular design, manufacture, and test specifications

- **Adherence to standards**: The development of standards by organizations to ensure consistency in the design, manufacture, and testing of PCBs

- **Higher density interconnect**: Reduced interconnect track widths and spacing between the tracks to provide more interconnect on the PCB—particularly important for computer and communications applications

- **Higher density of electronic components**: Reduced spacing between the electronic components to provide more circuit functionality on the PCB—particularly important for computer and communications applications

- **Reduced electronic component package size (the "footprint" on the PCB)**: Reduced packaging dimensions for the electronic components to provide more circuit functionality on the PCB—particularly important for computer and communications applications

- **Increased use of surface mount technology**

- **Less empty space**: A reduction in the amount of PCB surface area left unused to provide more functionality for the PCB and to reduce overall production costs

- **Higher operating frequencies**: Driven by computer and communications applications so the electronic circuit can undertake more operations in a reduced time: for computer applications, the need for high-speed digital data transfer; for communications applications, both high-speed digital data transfer, wired and wireless data transfer (RF)

References

[1] Horowitz, P., and Hill, P., *The Art of Electronics*, Second Edition, Cambridge University Press, 1989, ISBN 0-521-37095-7.

[2] O'Connor, P., *Test Engineering, A Concise Guide to Cost-effective Design, Development and Manufacture*, John Wiley & Sons, Ltd., 2001, ISBN 0-471-49882-3.

[3] Bushnell, M., and Agrawal, V., "Essentials of Electronic Testing for Digital, Memory & Mixed-Signal VLSI Circuits," Kluwer Academic Publishers, 2000, ISBN 0-7923-7991-8.

[4] Hughes, E., *Electrical and Electronic Technology*, Ninth Edition, Pearson Education, 2005, ISBN 0-13-114397-2.

[5] Floyd, T., *Electronics Fundamentals, Circuits, Devices, and Applications*, Fifth Edition, 2001 Prentice Hall, ISBN 0-13-085236-8.

[6] Smith, M., *Application Specific Integrated Circuits*, Addison-Wesley, 1999, ISBN 0-201-50022-1.

[7] Doane, D., and Franzon, P., *Multichip Module Technologies and Alternatives, The Basics*, Van Nostrand Reinhold, 1993, ISBN 0442091236-5.

[8] IPC, http://www.ipc.org

[9] Sears, F., Zemansky, M., and Young, H., *University Physics*, Seventh Edition, Addison-Wesley, 1987, ISBN 0-201-06694-7.

[10] Smithson, G., "Practical RF printed circuit board design," IEE Training Course: "How to Design RF Circuits" (Ref. No. 2000/027), IEE, 5 April 2000, pp. 11/1–11/6.

[11] Sharawi, M. S., "Practical issues in high speed PCB design," *IEEE Potentials*, Vol. 23, Issue 2, April–May 2004, pp. 24–27.

[12] Verma, A., "Optimizing test strategies during PCB design for boards with limited ICT access," 27th International IEEE/SEMI Annual Electronics Manufacturing Technology Symposium (IEMT 2002), 17–18 July 2002, pp. 364–371.

[13] Reeser, S., "Design for in-circuit testability," 11th International IEEE/CHMT Electronics Manufacturing Technology Symposium, 16–18 September 1991, pp. 325–328.

[14] Ghose, A. K., Mandal, S. K., and Deb, G. K., "PCB Design with Low EMI," *Proceedings of the International Conference on Electromagnetic Interference and Compatibility*, 6–8 December 1995, pp. 69–76.

[15] John, W., "Remarks to the solution of EMC-problems on printed-circuit-boards," *Proceedings of the 7th International Conference on Electromagnetic Compatibility*, 28–31 August 1990, pp. 68–72.

[16] XiaoKun Zhu, Bo Qi, Xin Qu, JiaJi Wang, Taekoo Lee, and Hui Wang, "Mechanical test and analysis on reliability of lead-free BGA assembly," *Proceedings of the 6th International Conference on Electronic Packaging Technology*, 20 Aug.–2 Sept. 2005, pp. 498–502.

[17] European Union, Directive 2002/96/EC on Waste Electrical and Electronic Equipment (WEEE).

[18] European Union, Directive 2002/95/EC on the Restriction of Use of Certain Hazardous Substances.

[19] Deubzer, O., Hamano, H., Suga, T., and Griese, H., "Lead-free soldering-toxicity, energy and resource consumption," *Proceedings of the 2001 IEEE International Symposium on Electronics and the Environment*, 7–9 May 2001, pp. 290–295.

Student Exercises

The exercises for this chapter are based on the PCB case study designs. The aim will be to design, manufacture, and test these PCBs, both separately and as a complete system. To achieve this goal, it will be necessary to act as a team. The structure of the team can be decided upon as considered most appropriate, but the following roles should be adopted:

- **Project management**: Coordinating the team to develop and administer the processes to obtain a working PCB design.
- **Schematic entry**: Using the circuit designs provided in Appendix F (see the last paragraph of the Preface for instructions regarding how to access this online content). Develop the circuit schematics in the PCB design tool of choice.
- **Layout**: Develop the PCB layout from the circuit schematic. This step also includes the manufacture of the PCB.
- **Simulation and test**: Developing a suitable test procedure (using simulation if possible), and using suitable test equipment on the manufactured PCB.
- **System interfacing**: Developing a test procedure to test the PCB when integrated into the overall electronic circuit. (This role is to be taken into Question 3 of the exercise.)

1. Identify the possible routes to manufacturing the required PCBs for both *one-off* prototypes. For the chosen manufacturing process, identify the materials used and the required layout design rules.
2. For the circuit designs identified in the case studies, obtain the following information:

   - User guides and relevant information on the PCB design tools to be used
   - Component datasheets
   - Relevant information on the PCB manufacturing facility to be used.

   Using this information, create a suitable information resource center based on HTML pages to operate on a local intranet site. The site is to be readily accessible by those requiring the information.
3. *For each PCB*: Working in teams, design, fabricate, and test each PCB in turn to create the required overall system. Team members should change roles for each PCB so that each member can practice each step.
4. *For the overall system*: When integrating all PCBs into the overall system, assign one person to develop and run board integration tests. Do not assign

a project manager; instead, so each team member is to take on one or more specific roles of a project manager, listed below:

- CPLD development board integration test
- LCD display and keypad board integration test
- PC interface board integration test
- digital I/O board integration test
- analogue I/O board integration test
- overall system test

5. Develop a new PCB design that replaces the digital I/O board with a board that uses suitably placed LEDs (yellow, red, green, blue) to create the lights on a Christmas tree. The CPLD is to be used to switch the LEDs ON and OFF, and to enable a user to set different lighting arrangements from a PC. An example board arrangement is shown in the following figure.

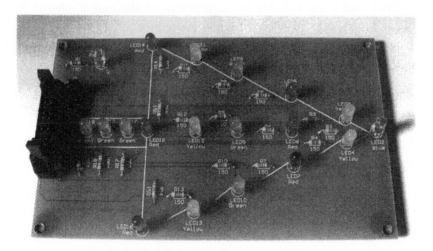

Figure: Example Christmas tree lights board

CHAPTER 4

Design Languages

4.1 Introduction

Design languages provide the means by which to describe the operation of both software programs and hardware. These descriptions, usually text based, are developed and stored as ASCII text files within the computer on which the descriptions are being developed. Over the years, a large number of languages have been developed. Some are still in use today, while others have become obsolete.

Design languages are of two types, software programming languages (SPL) and hardware description languages (HDL). At one time, designers were either software or hardware designers, and design teams were clearly distinguished by these separate roles. Today, however, designers are involved in both software and hardware design and need skills in both areas, although they may be specialized.

Attempting to identify and introduce all the design languages developed and in use would be a book in its own right. This chapter will identify and introduce a number of key languages used in both hardware and software design. Figure 4.1 identifies the languages to be identified and discussed.

4.2 Software Programming Languages

4.2.1 Introduction

Software programming languages (SPLs) allow a software designer to create executable software applications that will operate on a suitable processor. The target processor will be one of three types: microprocessor (μP), microcontroller (μC), or digital signal processor (DSP).

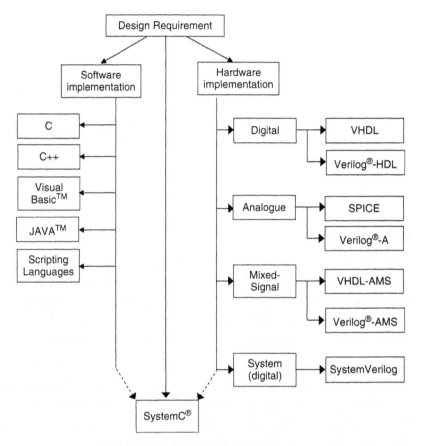

Figure 4.1: Design languages

The *microprocessor* is a software-programmable, integrated circuit built around a central processing unit (CPU) and based on an instruction set that the software program uses to perform a set of required tasks. The instruction set is one of two types: CISC (complex instruction set computer) or a RISC (reduced instruction set computer). As a general-purpose processor that can be designed to undertake a wide range of tasks, the microprocessor architecture is not necessarily optimized for specific tasks.

The *microcontroller* is a type of microprocessor that contains additional circuitry such as memory and communications ports (such as a UART, universal asynchronous receiver transmitter, for RS-232 communications) along with the CPU, and is used in embedded system applications. It does not have the generality of the general-purpose microprocessor, but rather is a self-contained, low-cost "computer on a chip."

The *digital signal processor* is a specialized form of microprocessor used in real-time digital signal processing operations. Although such operations can be performed on a microprocessor, DSP architecture is optimized for the fast computations typically undertaken.

Essentially, software is developed for one of two target areas:

- as a software application to run on a workstation or PC, executing on a suitable operating system (UNIX™, Windows®, Mac OS® or Linux®)

- as embedded software to run on a processor within an embedded system. Examples of embedded systems include control, automotive, and aerospace applications. The processor runs the embedded software program as a stand-alone entity rather than through one of the above software operating systems.

For software applications to run on workstations or PCs, there are a number of software programming languages and supporting development environments to aid the designer. Some supporting development environments are a collection of separate software tools that are executed by the designer separately, others are a collection of software tools contained within a single integrated design environment (IDE).

For embedded software to run on a processor within an embedded system, the choice of programming language and supporting development environments reduces. The C language is most commonly used for embedded software development.

4.2.2   C

The C programming language evolved from two previous programming languages, BCPL and B [1]. BCPL was developed by Martin Richards in 1967. B was then developed by Ken Thompson using many of the features found in BCPL. C evolved from B, and was developed by Dennis Ritchie at Bell Laboratories (USA) and originally implemented in 1972. Initially designed for the UNIX™ operating system, C can now be compiled on almost any computer (UNIX™, Windows®, and Linux® operating systems) and is one of the most commonly used programming languages. Most operating systems, including Microsoft® Windows®, are written in C and/or its extension C++. C is also used to develop the software code run on the majority of processors for use in embedded system applications. The standard for C is the ANSI/ISO Standard C [2]. The standard was first introduced in 1989 and updated in 1999. C is hardware independent, and with careful code design, the same source code can be portable to most computers.

```
1    /*****************************************************/
2    /* This program simply outputs a line of arbitrary text */
3    /*****************************************************/
4
5    #include <stdio.h>
6
7    void main (void) {
8          printf("Hello World\n");
9    }
10
11   /*****************************************************/
12   /* End of File                                    */
13   /*****************************************************/
```

Figure 4.2: "Hello World" program in C

Consider an example of the "Hello World" program written in C. Figure 4.2 shows the program source code and the corresponding line numbers are added for information purposes only.

This program introduces a number of features of C. The first three lines are comments. A comment is a piece of code that is ignored by the C compiler. Comments are used to add useful descriptions of the functionality of the source code, and enable easier reading of the source code by the author as well as by readers. Careful and comprehensive commenting of the program source code is essential to good programming practice.

The fourth, sixth, and tenth lines are left blank for readability purposes.

The fifth line is a directive to the C preprocessor. Lines that begin with the number sign, # (also called a hash character), are processed by the C preprocessor before the program is compiled.

The seventh line is the beginning of the program and is known as the main function. A C program is essentially a number of functions that interact with each other in a predefined manner. At the end of this line is an opening curly bracket, {, and on the last line is a closing curly bracket, }. Curly brackets are used to group a number of statements together. In this case, they are used to mark the beginning and the end of the program, but they can also be used to group statements that are part of other statements such as an if statement or a while statement.

The eighth line is the statement that outputs information using the printf statement. Any text that appears between the quotation marks, '' '', will be printed to the standard output (i.e., the computer display screen). The last two characters of the `printf` statement are \n. This indicates a new line.

The last three lines are comments.

C program development requires a program development environment, the language, and a C standard library. The program development environment provides the software toolset to allow the designer to enter the design software source code, to undertake the phases necessary for the source code to the executed, to accommodate project management, and to enable suitable software code debugging tools. C programs are executed in six phases:

1. *Source code editing*, in which the designer creates and edits the C source code file using a suitable stand-alone text editor or an editor built into an IDE, such as Microsoft® Visual C++ [3].

2. *Preprocessing* is undertaken prior to program compilation and uses specific preprocessor directives that identify needed actions. Such actions include the replacement of specific text characters within the source code and the inclusion of other files include the source code file.

3. *Compilation* uses a compiler program to translate the C source code into machine language code (also called object code) for the particular processor used in the computer system on which the program will run.

4. *Linking*: C programs usually include references to functions defined elsewhere within libraries of functions created elsewhere. The object code created by the C compiler, then, contain gaps for the referenced functions. A linker links the object code with the code for the referenced functions to create an executable image that can then be run.

5. *Loading* places the executable image in memory for execution.

6. *Execution* runs (executes) the executable image on the processor used in the computer system on which the program will run.

A C source code file carries the file extension ".c."

4.2.3 C++

C++ is an extension to the C language that allows an object-oriented programming approach to application development [1]. C++ uses the concepts *classes* and *objects*. Unlike software programming languages such as ANSI standard C, which are procedural in nature, object-oriented programming (OOP) languages such as C++ (and JAVA™) are based on objects. OOP is a design philosophy that identifies and uses the relationship between data (variables, constants, and types) and processes (procedures and functions). Object-oriented design identifies objects, data, and processes that relate to these objects.

Objects can be seen in everyday life and surround us. For example, a motor car that someone owns is an object that is used for a particular purpose. It has particular attributes that are specific to the car (such as the color and mileage), but it also has attributes that are common to all other cars of the make and model (such as manufacturer, engine, fuel requirements).

A vast number of attributes can be identified for any given motor car. In fact, any object is made up of smaller objects that combine to enable its functionality.

For the purpose of this explanation (only), people can also be considered as objects in the world, in that we all have individual attributes (height, age, hair color), and common attributes (one head, two arms, opposable thumbs). In general, then:

- An object can undertake a number of operations, referred to as methods.

- An object has an internal state that might or might not be available to the user of the object, either directly or through the use of the methods.

- An object is to be considered as a black box, which means its internal details are hidden from the user. The user will use the object by applying an input and then receiving an output. How the input is manipulated to form the output is hidden from the user.

- An object is created from a class. The class defines the actions that the object can undertake and the states it can maintain. A class is a template used to create an object.

- An object will have a set of attributes that are particular to the object.

Because C++ is a superset of C, C programs are compiled with a C++ compiler. There are two points to note about C++:

1. Some language additions allow programs to be written in the same manner as a standard C program (i.e., procedural) but they must be kept in mind. The key points are:

   - single-line comments

   - I/O streams

   - declarations in C++

- creating new data types in C++

- function prototypes and type checking

- inline functions

- function overloading

2. Some language additions allow an OOP approach using classes and objects.

The most noticeable aspect of C++ programming are the comments. Both single- and multiline comments commence with a /* and end with a */. In C++, single-line comments can also commence with a //. The second most noticeable aspect of C++ programming is the manner in which input to the file and output from the program is dealt with in the code. In C, input and output is provided with scanf and printf. In C++, streams are used to handle the input (cin) and output (cout). These aspects are shown in the sample C++ source code shown in Figure 4.3. This program prompts the user to enter two integer numbers and calculates the sum and difference. A C++ source code carries the a file extension ".cc."

4.2.4 JAVA™

Java™ was developed in the mid-1990s by developers at Sun Microsystems and was first released in 1995 [1, 4]. The software development kit (the *Java*™*Development Kit*, JDK) is freely available for download via the Internet from the Sun Microsystems [5]. The current release is Java 2 version 5.0, but both Java 2 and Java 1.1 remain in common use.

The development of this object-oriented language was undertaken to overcome the limitations posed by the C++ object-oriented language. It is now widely used in a range of computing applications, in particular for Internet-based software systems. It is an object-oriented and platform neutral language in that:

- *Object-oriented* allows a programmer to follow an object-oriented programming (OOP) approach to software development in which objects are used and work together in the overall system and are created from templates referred to as classes.

- *Platform neutral* allows a program written on one operating system to be run on any other operating system without modification. The source code is compiled into a format referred to as bytecode, and this bytecode is then run using a Java interpreter.

```
//----------------------------------------------------------------
// C++ source code to prompt a user to enter two integer numbers, calculate
// the sum and difference and to display the results to the standard output.
//----------------------------------------------------------------
#include <iostream.h>

void message1();
void message2(void);

int sumFunction(int x, int y);
int diffFunction(int x, int y);

int  a = 0;
int  b = 0;
int  sum = 0;

main() {

   cout << "\n-------------------------------------------\n\n";
   message1();
   cin >> a;
   message2();
   cin >> b;
   cout << "\n-------------------------------------------\n\n";
   sum = sumFunction(a, b);
   cout << "The sum of " << a << " and " << b << " is\t\t\t" << sum << "\n";
   cout << "The difference between " << a << " and " << b << " is\t" << diffFunction(a, b) << "\n";
   cout << "\n-------------------------------------------\n";
   return 0;
}
//----------------------------------------------------------------
// Function prototype to prompt the user to input 'a'
void message1() {
                     cout << "Enter an integer number for a ...";
}
//----------------------------------------------------------------
// Function prototype to prompt the user to input 'b'
void message2(void) {
                     cout << "Enter an integer number for b ...";
}
//----------------------------------------------------------------
// Function prototype to calculate the sum of two numbers
int sumFunction(int x, int y) {
                     return(a + b);
}
//----------------------------------------------------------------
// Function prototype to calculate the difference betweem two numbers
int differenceFunction(int x, int y) {
                     return(a - b);
}
//----------------------------------------------------------------
// End of File
//----------------------------------------------------------------
```

Figure 4.3: C++ source code for input to and output from a program

The Java™ language can be used to develop programs for two types of use:

- when used as an *application* running on an operating system (Microsoft® Windows®, Mac OS®, UNIX™, and LINUX®)

- when used within an Internet browser (such as Microsoft® Internet Explorer), referred to as applets and called from within HTML code viewed in the browser

```
1    import java.lang.*;
2
3    public class HelloWorld {
4
5            public static void main(String[] arguments) {
6
7                    System.out.println("Hello World");
8
9            }
10   }
```

Figure 4.4: "Hello World" program in Java™

Consider an example of the "Hello World" program (application) written in Java™. Figure 4.4 shows the program source code (in the right column of the table), with the corresponding line numbers (in the left column of the table) added for informational purposes only.

This contains a single class called HelloWorld, and within this class is a single method called main. Before the class is written, a package called java.lang is imported. This package contains all of the classes for creating user interfaces and for painting graphics and images. *It is not strictly required for the above source code to work*, but is included to identify the use of predeveloped classes. The program could be written without this line. In this program source code, then:

- The first line

  ```
  import java.lang.*;
  ```

  imports a package called java.lang. The .* means to import all classes within the package. This is used to import packages (groups of classes) as well as individual classes. *It is not strictly required for the above source code to work* as this package is automatically available, but is included to identify the use of predeveloped classes. The program could be written without this line.

- The second, fourth, sixth, and eighth lines are blank lines used to aid readability.

- The third line

  ```
  public class HelloWorld {
  ```

  is the start of the class declaration. The class name HelloWorld matches the file name (without the extension). It is a public class. There may be only one `public` class, which may contain a number of inner `private` classes. A class is a collection of methods and properties.

- The fifth line

```
public static void main(String[] arguments) {
```

is the first line of a method called `main`. A class must contain a `main` method in a Java application for the Java interpreter to run. A `static main` method is first called when an object is created.

- The seventh line is a statement to output a string of text to the system display (the monitor).

- The ninth line is a closing bracket around the `main` method.

- The tenth line is a closing bracket around the `HelloWorld` class.

The basic procedure for creating and running a Java™ application includes three steps:

1. Create the Java™ source code using a suitable text editor. The source code file should have the file extension `.java`.

2. Compile the Java™ source code into Java™ bytecode using javac. This bytecode is machine independent and may be run on Windows®, Linux, Mac OS®, or UNIX™ operating systems. Theoretically, then, all the features used on one operating system should work on the other operating systems, although in practice one must identify any differences between the operating systems. The bytecode carries the file extension `.class` signifying the Java™ class file that contains it.

3. Run the Java™ bytecode. This action runs the class file (note that the .class extension is not included). The command java <class file> runs the Java™ interpreter on the identified file.

4.2.5 Visual Basic™

Visual Basic™ (VB) is a programming language developed by Microsoft® for Windows®-based software applications [6]. Visual Basic™ is also the name of the programming environment. This modern version of the BASIC (Beginner's All-purpose Symbolic Instruction Code) programming language allows developers to use the Visual Basic™ programming environment to create applications with a graphical user interface (GUI) and to take advantage of the language's OOP features. Visual Basic™ is part of the Microsoft® Visual Studio suite of development tools; in Visual Studio 6.0, the tools available are Visual Basic™ 6.0, Visual C++ 6.0, Visual FoxPro 6.0, and Visual InterDev 6.0.

The widely used version of VB is version 6.0, although VB.net has been developed to replace version 6.0. Additionally, VBA (Visual Basic™ for Applications) is a modified version of VB designed as a macro language for the development of macros in software applications such as Microsoft® Word and Excel.

Visual Basic™ applications are designed in three stages:

1. Identify the appearance of the user interface by choosing the required items from a collection of components such as menus, buttons, text boxes, etc.

2. Write the associated scripts with each of the items in the user interface that defines the behavior of the application.

3. Execute the program. This is undertaken from within the programming environment during design development and debugging, and then by creating a stand-alone executable (*.exe*) file.

Consider an example of an application that prints the "Hello World" message written in Visual Basic™. Figure 4.5 shows the program development within the programming environment The center windows show the user interface as it appears in the programming environment (top) and the associated scripts (bottom).

Figure 4.6 shows the window that appears when the program is executed. This application has one label (center of window) to display messages and a single item in a pull-down menu (top of window). The script for this example is held in a single form (*.frm*) file.

The script code created by the designer is shown in Figure 4.7. This consists of two private subroutines:

1. `Private Sub Form_Load()`
2. `Private Sub FileClose_Click()`

The form name is `Form1`. The first subroutine (`Form_Load()`) identifies the actions to be undertaken when the program initially starts. In this application, two actions are taken. The first centers the window on the computer display screen, and the

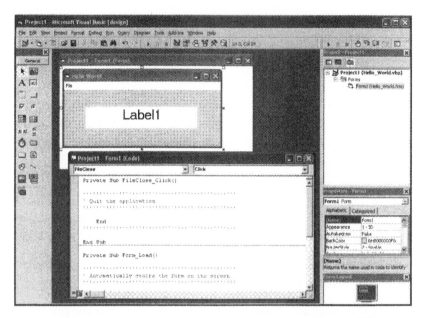

Figure 4.5: "Hello World" program in Visual Basic™

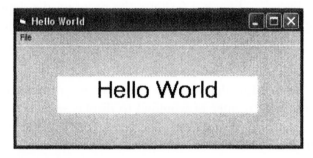

Figure 4.6: "Hello World" program as it appears to the user

second updates the label (Form1.label1) with the required "Hello World" message. The second subroutine identifies the actions when the menu (Figure 4.6, top of window) action (FileClose_Click()) is performed. This ends the application and closes the window.

```
Private Sub FileClose_Click()

'''''''''''''''''''''''''''''''''''''''''''''''
' Quit the application
'''''''''''''''''''''''''''''''''''''''''''''''

    End
'''''''''''''''''''''''''''''''''''''''''''''''

End Sub

Private Sub Form_Load()

'''''''''''''''''''''''''''''''''''''''''''''''
' Automatically centre the form on the screen
'''''''''''''''''''''''''''''''''''''''''''''''

    Left = (Screen.Width - Width) \ 2
    Top = (Screen.Height - Height) \ 2

'''''''''''''''''''''''''''''''''''''''''''''''
' Update label1 with the required text
'''''''''''''''''''''''''''''''''''''''''''''''

    Form1.Label1 = "Hello World"

'''''''''''''''''''''''''''''''''''''''''''''''

End Sub
```

Figure 4.7: "Hello World" program script

4.2.6 Scripting Languages

Scripting languages provide a high-level application programming interface that enables applications to be created and tested quickly [7]. Unlike languages such as C and C++ that are compiled before an executable image of the program is run, programs written in a scripting language are interpreted as they run, thereby removing the step of having to compile a program whenever a change is made. Programs using scripting languages can be found in many workstation or PC applications, as well as Internet-based applications. For example, scripting language applications are commonly used to *glue together* other applications to form a single user interface for a range of existing applications.

The main scripting languages in use today are described below:

- *Javascript* was created by Brendan Eich in 1995 as a Web scripting language for creating interactive web pages on via a suitable Internet browser tool. It was originally called LiveScript and was incorporated into the Netscape Internet browser. Javascript runs on the user's machine (client-side scripting) and allows more operations than possible with HTML alone; the Javascript is placed within the HTML document [7].

- *PERL* (Practical Extraction and Report Language) was developed in the late 1980s by Larry Wall as a simple text processing language. It is also used in a wide range of applications such as file manipulation and processing, interacting with the operating system, and establishing network connections. It originated on the UNIX^{TM} operating systems but is now available on all major operating systems and includes OOP coding capabilities [7].

- *Python* is a high-level object-oriented scripting language. Python was invented in 1990 by Guido van Rossum and first appeared in 1991. It has a wide range of applications from system utility actions through Internet scripting and database access [8].

- *Tcl/Tk* is the tool command language (Tcl) commonly used with an associated toolkit called Tk. This high-level scripting language was created by Professor John Ousterhout and runs on the Windows®, Linux®, Mac OS®, and UNIX^{TM} operating systems. It is machine independent in that the same code (which is stored in an ASCII text file) can be transported across platforms [9].

- *PHP* is a recursive acronym that stands for hypertext preprocessor. It is a server-side scripting language used for creating dynamic web pages. Server-side scripting means that the execution of all PHP source code takes place on the server on which the PHP file is hosted. The output after the PHP source code has been executed is HTML on the user's web browser [10].

- *VBA* (Visual Basic^{TM} for Applications) is a modified version of Visual Basic^{TM} designed as a macro language for the development of macros in software applications such as Microsoft® Word and Excel.

- *VBScript* was developed by Microsoft® as an alternative to Javascript. It runs on the client-side computer and only with Microsoft® Internet Explorer.

4.2.7 PHP

A PHP file can be created using any text editor and saving the file with the extension .php. A PHP file can contain text as well as HTML tags and scripts. When a PHP file is parsed, the PHP parser looks for opening and closing tags indicating that everything between them is to be executed as PHP code. Everything that appears outside of these tags is ignored. The most common, and recommended, syntax for the opening tag is <?php. The syntax for the closing tag in PHP is ?>. Every command in PHP must end with a semicolon, ;. The most basic command in PHP is the echo command, used in the following manner: echo ''some text''. The echo command simply outputs whatever text is placed within the quotation marks (single or double).

Comments can be made in PHP using either of two methods.

1. A single line comment is made by using two forward slashes, //. Everything after this and until the end of the line will be a comment.

2. A multiline comment is made by beginning with /* and ending with */. Everything that appears between these delimiters is a comment.

Consider an example of the "Hello World" application written in PHP. Figure 4.8 shows the program source code (in the right column of the table) and the corresponding line numbers (in the left column of the table) added for informational purposes only. The PHP code here is embedded within an HTML document for browsing on a suitable Internet browser tool such as Micosoft® Internet Explorer.

In Figure 4.8, normal HTML tags are used. As these are not within the opening and closing PHP tags, PHP simply ignores them. Then the opening tag is used on

```
1    <html>
2    <head>
3    <title>PHP Example 1</title>
4    </head>
5
6    <body>
7
8    <?php
9         echo 'Hello World';
10   ?>
11
12   </body>
13   </html>
```

Figure 4.8: "Hello World" using a PHP script

line 8, and the echo statement is used to output the text Hello World on line 9. The closing tag on line 10 concludes the PHP section of the code. Figure 4.9 shows the file viewed in Micosoft® Internet Explorer.

Another way to create a PHP file is to output every element of the HTML file using echo statements. This can be useful if the need arose to create totally different pages depending on a certain event. Figure 4.10 shows a PHP file written in this way; this file outputs the same page as the file in Figure 4.8.

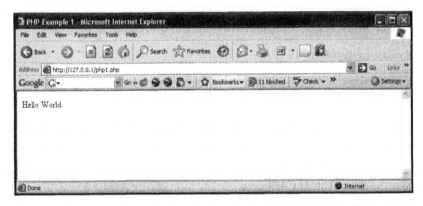

Figure 4.9: "Hello World" using a PHP script viewed in Micosoft® Internet Explorer

```
1    <?php
2
3    /* This is a multi-line comment.
4       It only ends when the closing delimiter
5       is used.
6    */
7
8        echo '<html>';
9
10       echo '<head><title> PHP Example 2</title></head>';
11       echo '<body>';
12
13   //  This is a single line comment
14       echo 'Hello World';
15
16       echo '</html>';
17   ?>
```

Figure 4.10: Alternative "Hello World" using a PHP script

4.3   Hardware Description Languages

4.3.1   *Introduction*

Hardware description language (HDL) design is based on the creation and use of textural based descriptions of a digital logic circuit or system. By using a particular HDL (the two IEEE standards in common use in industry and academia are Verilog®-HDL [11] and VHDL [12]), the description of the circuit can be created at different levels of abstraction from the basic logic gate description according to the language syntax (the grammatical arrangement of the words and symbols used in the language) and semantics (the meaning of the words and symbols used in the language).

Hardware circuit or system designs created using HDL is generated at different levels of abstraction. Starting at the highest level (i.e., furthest from the circuit detail), the system idea or concept is the initial high-level description of the design that provides the design specification. The algorithm level describes the behavioral of the design in mathematical terms. Neither the system idea nor the algorithm describes how the behavior of the design is to be implemented. The algorithm structure in hardware is described by the architecture, which identifies the high-level functional blocks to use and how the functions are connected. The algorithm and architecture levels describe the behavior of the design to be verified in simulation.

The next level down from the architecture is the register transfer level (RTL), which describes the storage (in registers) and flow of data around a design, along with logical operations performed on the data. This level is usually used by synthesis tools that describe the design structure (the netlist of the design in terms of logic gates and interconnect wiring between the logic gates). The logic gates are themselves implemented using transistors. The HDL may also support switch level descriptions that model the transistor operation as a switch (ON/OFF).

When designing with HDLs, the designer chooses what language to use and at what level of design abstraction to work. When choosing language, the following aspects must be considered:

- the availability of suitable electronic design automation (EDA) tools to support the use of the language (including design management capabilities and availability of tool use within a project)

- previous knowledge

- personal preferences

- availability of simulation models

- synthesis capabilities

- commercial issues

- design re-use

- requirements to learn a new language and the capabilities of the language

- supported design flows within an organization

- existence of standards for the language

- access to the standards for the language

- readability of the resulting HDL code

- ability to create the levels of design abstraction required and language or EDA tool support for these abstraction levels

- access to design support tools for the language, such as the existence of automatic code checking tools and documentation generation tools

4.3.2 VHDL

Very high-speed integrated circuit hardware description language—VHSIC HDL or VHDL—began life in 1980 under a United States Department of Defense (DoD) requirement for the design of digital circuits following a common design methodology, providing the ability for self-documentation and re-use with new technologies. VHDL development commenced in 1983, and the language became an IEEE standard in 1987 (IEEE Std 1076-1987). The language was revised in 1993, 2000, and 2002, the latest release being 1076-2002. VHDL also has a number of associated standards relating to modeling and synthesis.

The HDL code is contained in an ASCII text file and therefore is transportable between EDA tools on the same computing system, between computers, between different versions of the EDA tools and between the different engineers within the particular design team.

The HDL code is written to conform to one of three styles:

1. A *structural* description describes the circuit structure in terms of the logic gates used and the interconnect wiring between the logic gates to form a circuit netlist.

2. A *dataflow* description describes the transfer of data from input to output and between signals.

3. A *behavioral* description describes the behavior of the design in terms of the circuit and system behavior using algorithms. This high-level description uses language constructs that resemble a high-level software programming language.

Both the dataflow description and behavioral description use similar language constructs, but in VHDL they differ: a behavioral description uses the language *process* statements, whereas a dataflow description does not.

In VHDL, a design is created initially as an entity declaration and an architecture body. The entity declaration describes the design I/O and includes parameters that customize the entity. The entity can be thought of as a black box with visible I/O connections. The architecture body describes the internal working of the entity and contains any combination of structural, dataflow, or behavioral descriptions used to describe the internal working of the entity.

For example, consider a dataflow level description of a two-input AND gate. This is shown in the right column of Figure 4.11, and the corresponding line numbers are in

```
1   ----------------------------------------------------------------
2   -- And_Gate: Implements a 2-input AND gate.
3   ----------------------------------------------------------------
4
5   LIBRARY IEEE;
6   USE IEEE.STD_LOGIC_1164.ALL;
7
8   ENTITY And_Gate IS
9       Port ( A : IN    STD_LOGIC;
10             B : IN    STD_LOGIC;
11             Z : OUT   STD_LOGIC);
12  END ENTITY And_Gate;
13
14  ARCHITECTURE Dataflow OF And_Gate IS
15
16  BEGIN
17
18        Z <= (A AND B);
19
20  END ARCHITECTURE Dataflow;
21
22  ----------------------------------------------------------------
23  -- End of File
24  ----------------------------------------------------------------
```

Figure 4.11: Two-input AND gate description in VHDL

the left column for informational purposes only. The design has two inputs (*A* and *B*) and one output (*Z*). The code has three main parts:

1. *Top part* identifies the reference libraries to use within the design

2. *Middle part* identifies the design entity

3. *Bottom part* identifies the design architecture.

Comments in VHDL start with --. Lines 1 to 3 and 22 to 24 are comments at the beginning and ending of the file (with a .vhd extension) containing the VHDL code.

- Line 5 identifies the reference library to use in this design (IEEE), and line 6 identifies and the part(s) of the library to use.

- Lines 8 to 12 declare the entity (with a name And_Gate) and the I/O connections (ports).

- Lines 14 to 20 identify the architecture body, using the built-in logical AND operator to define the operation of the design within the architecture.

- Lines 4, 7, 13, 15, 17, 19, and 21 are blank lines to aid code readability.

An example test bench used to simulate the design is shown in Figure 4.12. The structure of the test bench is the same as for a circuit design, except that there are no inputs to or outputs from the test bench. The stimulus to apply to the circuit is defined within the test bench, and an instance of the circuit is placed within the test bench.

4.3.3   Verilog®-HDL

Verilog®-HDL was released in 1983 by Gateway Design System Corporation, together with a Verilog®-HDL simulator. In 1985, the language and simulator were enhanced with the introduction of the Verilog-XL® simulator. In 1989, Cadence Design Systems, Inc. bought the Gateway Design System Corporation, and in early 1990, Verilog®-HDL and Verilog-XL® were separated into two products. Verilog®-HDL, until then a proprietary language, was released into the public domain to facilitate the dissemination of knowledge relating to Verilog®-HDL and to allow Verilog®-HDL to compete with VHDL, which already existed as a nonproprietary language. In 1990, Open Verilog International (OVI) was formed as an industry consortium consisting of computer-aided engineering (CAE) vendors and Verilog®-HDL users to control the

```
-----------------------------------------------------------------
-- Test bench for And_Gate: Implements a 2-input AND gate.
-----------------------------------------------------------------


-----------------------------------------------------------------
-- Libraries and packages to use
-----------------------------------------------------------------

LIBRARY ieee;
USE ieee.std_logic_1164.all;
USE ieee.std_logic_arith.all;
USE ieee.std_logic_unsigned_all;

-----------------------------------------------------------------
-- Test bench Entity
-----------------------------------------------------------------

ENTITY Test_And_Gate_vhd IS
END Test_And_Gate_vhd;

-----------------------------------------------------------------
-- Test bench Architecture
-----------------------------------------------------------------

ARCHITECTURE Behavioural OF Test_And_Gate_vhd IS

        COMPONENT And_Gate
        PORT(
              A : IN std_logic;
              B : IN std_logic;
              Z : OUT std_logic
              );
        END COMPONENT;

        SIGNAL A :  std_logic := '0';
        SIGNAL B :  std_logic := '0';
        SIGNAL Z :  std_logic;

BEGIN

uut: And_Gate PORT MAP(
              A => A,
              B => B,
              Z => Z
);

Testbench_Process : PROCESS
        BEGIN

        Wait for 0 ns;   A <= '0';   B <= '0';
        Wait for 10 ns;  A <= '1';   B <= '0';
        Wait for 10 ns;  A <= '0';   B <= '1';
        Wait for 10 ns;  A <= '1';   B <= '1';
        Wait;

        END PROCESS;
END ARCHITECTURE Behavioural;


-----------------------------------------------------------------
-- End of File
-----------------------------------------------------------------
```

Figure 4.12: VHDL test bench for a two-input AND gate description

```
1     /////////////////////////////////////////////////
2     // Module definition for full-adder.
3     // Design modelled using logic gates
4     /////////////////////////////////////////////////
5
6     module fulladder (A, B, Cin, Sum, Cout);
7
8     input  A, B, Cin;
9     output Sum, Cout;
10
11    xor    g1      (X1, B, Cin);
12    xor    g2      (Sum, X1, A);
13
14    and    g3      (X2, A, B);
15    and    g4      (X3, B, Cin);
16    and    g5      (X4, A, Cin);
17    or     g6      (Cout, X2, X3, X4);
18
19    endmodule
20
21    /////////////////////////////////////////////////
22    // End of File
23    /////////////////////////////////////////////////
```

Figure 4.13: Full-adder description in Verilog®-HDL

language specification. In 1995, Verilog®-HDL was reviewed and adopted as IEEE Standard 1364 (becoming IEEE Std 1364-1995). In 2001, the standard was reviewed, the latest version of the standard now being IEEE Std 1364-2001.

As an example, consider a structural level description of a full-adder design. This is shown in the right column of Figure 4.13 m with the corresponding line numbers in the left column for informational purposes only.

The design is created within a design module, which contains the design defined in terms of logic gate primitives (AND, OR, XOR) and interconnections between the logic gates. These logic gate primitives are defined within the language. The design has three inputs (*A*, *B*, and *Cin*), and 2 outputs (*Sum* and *Cout*).

- Comments in the code start with a // on lines 1 to 4 and 21 to 23.

- The module starts on line 6 and finishes on line 19.

- Line 8 defines the module inputs, and line 9, the module output.

- Lines 11, 12, and 14 to 17 define the circuit in terms of logic gate primitives and the interconnections between the logic gates.

- Lines 5, 7, 10, 13, 18, and 20 are left blank for readability purposes.

An example test fixture for simulating the operation of the full-adder design is shown in Figure 4.14.

4.3.4 Verilog®-A

Verilog®-HDL (sometimes referred to as Verilog®-D for digital) was originally developed to model digital circuits and systems. The need to model analogue circuit behavior led to the development of Verilog®-A, an analogue-only specification providing a unique set of features over the digital modeling language [13]. Features of the language include:

- analogue behavioral descriptions contained in separate analogue blocks

- circuit parameters that can be specified with valid range limits

- control of the simulation time step for accurate simulation

- a full set of mathematical functions and operators describe analogue circuit behavior

- time derivative and integral operators

- circuit noise modeling

- the description of sampled data systems with Z-domain filters and linear continuous time filters with Laplace transforms.

As an example of a Verilog®-A description for an analogue circuit, consider an analogue voltage amplifier with a gain of +2.0. The amplifier is modeled within a module as an ideal amplifier (i.e., infinite input impedance and zero output impedance, along with frequency independence). Figure 4.15 shows a graphical view of the amplifier with input and output voltages. This is a single-ended input, single-ended output voltage amplifier. (No circuit implementation details are included.)

The Verilog®-A description is shown in Figure 4.16. The functionality is the line: V(sigout) <+ 2 * V(sigin);¶, which states that the output voltage is twice (×2) the input voltage. This operation is verified through time domain simulation.

```
/////////////////////////////////////////////////////////////////////
// Module definition for full-adder test fixture
/////////////////////////////////////////////////////////////////////

module test;

reg         A, B, Cin;
wire        Sum, Cout;

fulladder I1 (A, B, Cin, Sum, Cout);

initial

begin

$display("\n--Start of simulation\n");

Cin     = 1'b0; B     = 1'b0; A     = 1'b0;
#5 $display(A, "   ", B, "   ", Cin, "   ", Sum, "   ", Cout);

#5 Cin      = 1'b1; B     = 1'b0; A     = 1'b0;
#5 $display(A, "   ", B, "   ", Cin, "   ", Sum, "   ", Cout);

#5 Cin      = 1'b0; B     = 1'b1; A     = 1'b0;
#5 $display(A, "   ", B, "   ", Cin, "   ", Sum, "   ", Cout);

#5 Cin      = 1'b1; B     = 1'b1; A     = 1'b0;
#5 $display(A, "   ", B, "   ", Cin, "   ", Sum, "   ", Cout);

#5 Cin = 1'b0; B     = 1'b0; A     = 1'b1;
#5 $display(A, "   ", B, "   ", Cin, "   ", Sum, "   ", Cout);

#5 Cin      = 1'b1; B     = 1'b0; A     = 1'b1;
#5 $display(A, "   ", B, "   ", Cin, "   ", Sum, "   ", Cout);

#5 Cin      = 1'b0; B     = 1'b1; A     = 1'b1;
#5 $display(A, "   ", B, "   ", Cin, "   ", Sum, "   ", Cout);

#5 Cin      = 1'b1; B     = 1'b1; A     = 1'b1;
#5 $display(A, "   ", B, "   ", Cin, "   ", Sum, "   ", Cout);

$display("\n--End of simulation\n");

#10 $finish;

end

endmodule

/////////////////////////////////////////////////////////////////////
// End of File
/////////////////////////////////////////////////////////////////////
```

Figure 4.14: Verilog®-HDL test fixture for a full-adder description

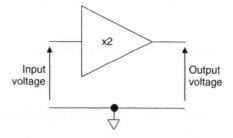

Figure 4.15: Analogue voltage amplifier design with a voltage gain of +2.0

```
//******************************************************
// Verilog-A module for x2 voltage amplifier design
//******************************************************

`include "constants.h"
`include "discipline.h"

module plant_ahdl(sigin, sigout);

input sigin;
output sigout;

electrical sigin, sigout;

analog begin

     V(sigout) <+ 2 * V(sigin);

end

endmodule

//******************************************************
// End of File
//******************************************************
```

Figure 4.16: Verilog®-A amplifier description

Such a description is used for simulation purposes rather than attempting to synthesize the design into analogue circuitry. To simulate the Verilog®-A description, the Spectre® simulator is used [14]. The amplifier design module is instantiated within the Spectre® netlist. In this design, a sine wave input voltage is applied to the amplifier.

```
//++++++++++++++++++++++++++++++++++++++++++++++++++++++++++
// Example Verilog-A design of an analogue amplifier.
// This is simulated using the Spectre simulator.
//++++++++++++++++++++++++++++++++++++++++++++++++++++++++++

global 0
simulator lang=spectre


//++++++++++++++++++++++++++++++++++++++++++++++++++++++++++

adhl_include "/Models/ahdl/veriloga.va"

//++++++++++++++++++++++++++++++++++++++++++++++++++++++++++
// Plant Model

I0   (a b)       plant_ahdl

//++++++++++++++++++++++++++++++++++++++++++++++++++++++++++
// Sine wave input

Vin  (a 0)   vsource dc=0 type=sine fundname="input1" delay=10m \
             sinedc=2.5 ampl=0.5 freq=2 sinephase=0 mag=1 phase=0

//++++++++++++++++++++++++++++++++++++++++++++++++++++++++++
tran tran stop=1  maxstep = 1m
//++++++++++++++++++++++++++++++++++++++++++++++++++++++++++
```

Figure 4.17: Spectre® simulation file for a Verilog®-A amplifier description

Figure 4.17 shows an example Spectre® netlist for simulating the design. Spectre is an analogue and mixed-signal modeling language that provides constructs for DC, AC, transient, and noise analysis in analogue circuits and has a number of features superior to SPICE-based simulation. It provides the features found in SPICE, and the Spectre simulator can simulate designs developed in the native language, along with designs written using SPICE syntax. (SPICE is discussed in detail later in this chapter.)

4.3.5   VHDL-AMS

Two modeling languages are emerging for mixed-signal (analogue and digital) electronic and mixed-technology system modeling, these being Verilog®-AMS [15] and VHDL-AMS [16]. These are extensions to the digital HDL Verilog®-HDL and VHDL, which are widely used as means to model and allow for simulation, documentation, and synthesis of digital circuits and systems from simple Boolean Logic to complex signal processing functions. These extensions from the digital domain are generally referred to as analogue and mixed-signal (AMS) languages for electronic circuits, but the manner in which nondigital electronics are modeled leads to the modeling of nonelectrical and electronic parts using the same model constructs. This provides a common means by which to model mixed-nature, mechatronic systems [17].

VHDL-AMS is the AMS extension to VHDL. This was adopted as a standard in 1999 as IEEE Standard 1076.1-1999. This superset of VHDL supports the description and simulation of continuous and mixed-continuous or discrete time systems. With the ability to model digital, analogue, and mixed-signal electrical and electronic circuits, along with nonelectrical parts, it allows the modeling of mixed-technology, mechatronic systems. Continuous time parts of the system are modeled using ordinary differential and algebraic equations (DAE), in which both conservative and nonconservative systems may be modeled:

- *Conservative systems* use conservation semantics, such as electrical systems obeying Kirchoff's Laws.

- *Nonconservative systems* do not use conservation semantics.

As with VHDL, designs are modeled using entities and architectures. Considering the analogue connections and signals, analogue ports are declared with a simple nature (e.g., electrical) and with any associated quantities (e.g., voltage across the port to a reference point and currents through the port).

Consider a simple electrical resistor-capacitor (RC) network driven by a step voltage source as shown in Figure 4.18. The voltage source (Vsrc) is connected between two nodes in the circuit (the node x1 and the common node). The resistor (R1) is connected between nodes x1 and x2. The capacitor (C1) is connected between nodes x2 and the common node.

The voltage source, resistor, and capacitor used in the design are defined in Figure 4.19. The voltage source produces a step change voltage input that changes at 50 ms and 100 ms.

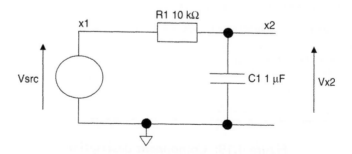

Figure 4.18: RC network

```
-----------------------------------------------------------------
-- Step voltage source
-----------------------------------------------------------------
LIBRARY DISCIPLINES;
USE DISCIPLINES.ELECTROMAGNETIC_SYSTEM.ALL;

ENTITY Source IS
    PORT(TERMINAL pos,neg: ELECTRICAL);
END Source;

ARCHITECTURE behav OF Source IS
    QUANTITY Vsource ACROSS Isource THROUGH pos TO neg;
BEGIN
        IF now < 50 ms or now > 100 ms   USE
          Vsource==0.0;
    ELSE
          Vsource==1.0;
    END USE;
END ARCHITECTURE behav;


-----------------------------------------------------------------
-- 10 kohm resistor
-----------------------------------------------------------------
LIBRARY DISCIPLINES;
USE DISCIPLINES.ELECTROMAGNETIC_SYSTEM.ALL;

ENTITY Resistor IS
    PORT    (TERMINAL pos,neg  : ELECTRICAL);
END Resistor;

ARCHITECTURE behav OF Resistor IS
    QUANTITY Vr ACROSS Ir THROUGH pos TO neg;
BEGIN    Ir == Vr/10000.0;
END behav;


-----------------------------------------------------------------
-- 1 uF capacitor
-----------------------------------------------------------------
LIBRARY DISCIPLINES;
USE DISCIPLINES.ELECTROMAGNETIC_SYSTEM.ALL;

ENTITY Capacitor IS
    PORT    (TERMINAL pos,neg    : ELECTRICAL);
END Capacitor;

ARCHITECTURE behav OF Capacitor IS

    QUANTITY Vc ACROSS Ic THROUGH pos TO neg;

BEGIN
    Ic==1.0e-6 * Vc'dot;
END behav;

-----------------------------------------------------------------
--
-----------------------------------------------------------------
```

Figure 4.19: Component descriptions

```
-------------------------------------------------------------
-- VHDL-AMS test bench for the RC network
-------------------------------------------------------------

LIBRARY DISCIPLINES;
USE DISCIPLINES.ELECTROMAGNETIC_SYSTEM.ALL;

ENTITY TestBench IS
END;

ARCHITECTURE Structure OF TestBench IS
    TERMINAL x1,x2: ELECTRICAL;
BEGIN
    Vsrc: ENTITY Source     (behav) PORT MAP (x1, electrical_ground);
    R1:   ENTITY Resistor   (behav) PORT MAP (x1,x2);
    C1:   ENTITY Capacitor  (behav) PORT MAP (x2, electrical_ground);
END Structure;

-------------------------------------------------------------
-- End of VHDL-AMS test bench
-------------------------------------------------------------
```

Figure 4.20: VHDL-AMS test bench for the RC circuit

These three components are placed within a test bench for simulation purposes. The test bench code is shown in Figure 4.20.

4.3.6 Verilog®-AMS

Verilog®-AMS is the AMS extension to Verilog®-HDL [18]. It provides the extensions to Verilog®-HDL to model mixed-signal (mixed analogue and digital) electronics and mixed-technology (electrical/electronic and nonelectrical/electronic) systems. It encompasses the features of Verilog®-D and Verilog®-A.

4.4 SPICE

Simulation techniques are an essential part of electrical and electronic circuit design, providing an insight into the operation of a designed circuit prior to its being built. This allows circuit design changes and device optimization, along with "what if" scenarios that would be difficult or impossible to undertake on a real circuit. One example is investigating the effects on an analogue amplifier design if transistor parameters were to change because of processing variations.

Electronic circuits and systems can be implemented as:

- printed circuit board (PCB)
- integrated circuit (IC)
- multichip module (MCM)

On a PCB design, simulation is an invaluable input to design verification and can highlight problems that result from component and interconnect placement (e.g., ensuring that signal integrity is maintained). On IC and MCM designs, with complex circuits and systems implemented on (typically) silicon dies and housed within a suitable package, simulation is essential due to the nature of the circuits and the limited ability to access specific parts of the design, with access only via package pins and with potentially hundreds of thousands or millions of transistors within the IC or MCM.

For analogue circuit simulation, SPICE (Simulation Program with Integrated Circuit Emphasis) is the main form of analogue circuit simulation adopted [19, 20]. A range of circuit simulators based on SPICE are available for use (e.g., PSpice and HSPICE). SPICE allows a range of circuit elements to be modeled, connected, and analyzed. The basic analysis methods are:

- DC by DC operating point analysis
- transient by time domain simulation
- AC by frequency domain analysis
- noise by circuit noise analysis over a frequency range (used in conjunction with AC analysis)

The basic (primitive) passive and active circuit elements include:

- resistor
- capacitor
- inductor
- magnetic elements

- bipolar junction transistor (BJT)

- metal oxide semiconductor field effect transistor (MOSFET)

- junction field effect transistor (JFET)

In addition, signal source (voltage and current) and behavioral models (analogue and to a certain extent, digital) circuit elements are utilized.

Consider a resistor. This is defined in SPICE as: `Rname +node -node [model name] value [TC=TC1, [, TC2]]`

which defines the resistor device (`R`) with a unique identifier (name) and with two nodes (+node, −node), an optional model to use (`[model name]`) to modify the resistance calculation value, a resistance value in ohms (`value`) and optional temperature coefficients (`[TC=TC1, [, TC2]]`).

Note also that SPICE syntax is not case sensitive. A simple 10 kΩ resistor (with a name input) connected between two nodes (A and B) is defined as: `Rinput A B 10k`.

A SPICE netlist is created to define the circuit and control the simulation. As an example, consider a simple electrical RC network driven by a step voltage source as shown in Figure 4.21. The voltage source (Vsrc) is connected between two nodes in the circuit (the node x1 and the common node). The resistor (R1) is connected between nodes x1 and x2. The capacitor (C1) is connected between nodes x2 and the common node.

The voltage source produces a step change voltage input that changes at 50 ms and 100 ms. The SPICE netlist for simulation purposes is shown in Figure 4.22.

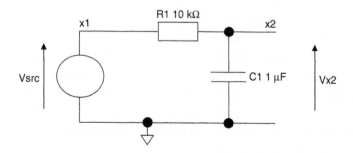

Figure 4.21: RC network

```
* SPICE netlist for RC network

*******************************************
* Set the circuit temperature
*******************************************
.temp 25
*******************************************
* Circuit components
*******************************************
Vin   x1   0    PWL(0,0 50m,0 51m,5 100m,5, 101m,0)
R1    x1   x2   10k
C1    x2   0    1uF
*******************************************
* Run transient analysis
*******************************************
.tran 1ms 200ms
*******************************************
* End simulation
*******************************************
.end
*******************************************
* End of File
*******************************************
```

Figure 4.22: SPICE netlist for RC network

4.5 SystemC®

SystemC® [27] is an ANSI standard C++ class library for supporting the development of electronic systems that are a hybrid of hardware and software. As such, it is used by the developers of complex electronic systems. SystemC® is closely related to the C++ programming language and adheres to terminology used in the ISO/IEC 14882:2003 international standard for the C++ programming language [28]. It is a single unified design and verification language using open-source C++ classes to describe system architectural and other attributes. SystemC® [29] is used for both simulation of hardware providing for simulation performance benefits over RTL level Verilog®-HDL or VHDL design descriptions and for functional verification. In functional verification, then the same platform is used for verification of the software and the entire system.

Early work on SystemC® was undertaken by a number of companies and organisations and is now covered by the IEEE standard 1666™-2005 [30]. This standard provides the definition of the SystemC® class library so that a SystemC® implementation could be developed with reference solely to the standard.

4.6   SystemVerilog

SystemVerilog [27] is a unified hardware description language for design, specification and verification that is used for complex digital ICs that form IC based systems (i.e. system on a chip (SoC) designs) that:

- Have a large number of logic gates (a large *gate count*),

- Are IP based (use IP blocks from one or more sources that are connected to form the overall system),

- Require the use of internal bus arrangements for extensive signal movement around the IC.

With such designs, the verification [28] of such designs starts to dominate the overall system development process.

SystemVerilog [29] was originally developed by Accellera [30] and is now covered by the IEEE standard 1800™-2005 [31]. The motivation for the development of SystemVerilog came from the need to improve the productivity in the design of complex digital ICs that form IC based systems with the above three characteristics. This unified hardware description language forms a major extension to the Verilog® -HDL standard (IEE Std 1364 ™-2005) [32]. It is primarily aimed at IC level design implementation and verification, but includes links into system-level design flow. Through a direct programming interface (DPI), there is a two-way interface between SystemVerilog and C/C++/SystemC functions. Therefore, SystemVerilog designs can be co-simulated with SystemC blocks. This enables a link between system level design and IC implementation/ verification. It also provides features that support the development of hardware models and test fixtures (test benches) using object oriented programming techniques.

4.7   Mathematical Modeling Tools

Mathematical modeling and simulation tools are increasingly used in designing hardware circuits and systems because they allow fast development and interpretation of the algorithms that the hardware is to implement. A number of mathematical tools exist:

- MATLAB® [21, 22]

- Mathematica [23]

- Modelica [24]

- Maple [25]

- Scilab [26]

As an example of such a tool, consider MATLAB® from The Mathworks Inc. It integrates mathematical computing and data visualization tasks that are underpinned with the tool using its own modeling language. MATLAB® is accompanied with a range of toolboxes, blocksets, and other tools that allow a range of engineering and scientific applications. In such an approach, various ideas can be investigated as part of an overall design process to arrive at a final and optimal solution. The toolboxes and blocksets are utilized for:

- data acquisition

- data analysis and exploration

- visualization and image processing

- algorithm prototyping and development

- modeling and simulation

- programming and application development

Examples of the currently available toolboxes and blocksets are shown in Table 4.1.

Simulink® is commonly used by control system designers and increasingly by electronic circuit designers to model the operation of the required circuit or system in a block diagram format. As an example of this, consider a SISO (single input, single

Table 4.1: Example toolboxes within MATLAB®

Communications Blockset	A blockset that builds on the Simulink® system level design environment for modeling the physical layer of a communication system.
Communications Toolbox	A library of MATLAB® functions that supports the design of communication system algorithms and components. It builds on the powerful capabilities of MATLAB® and the Signal Processing Toolbox by providing functions to model the physical layer of a communication system.
Control System Blockset	A collection of algorithms that implement common control system design, analysis, and modeling techniques.
Filter Design HDL Coder	Filter Design HDL Coder allows for the generation of synthesisable and portable HDL code for fixed-point filters that have been designed using the Filter Design toolbox. Both Verilog® -HDL and VHDL code can be generated. It also automatically creates VHDL and Verilog® -HDL test fixtures/test benches for simulating, testing, and verifying the generated HDL code.
Filter Design Toolbox	The Filter Design Toolbox extends the Signal Processing Toolbox. It is a collection of tools that provide techniques for designing, simulating and analysing digital filters with filter architectures and design methods for complex real-time DSP applications
Fuzzy Logic Toolbox	Provides a graphical user interface to support the steps involved in fuzzy logic design.
Signal Processing Toolbox	A collection of MATLAB® functions that provides a customizable framework for analogue and digital signal processing.
Simulink®	An interactive tool for modeling, simulation, and analysis of dynamic, multidomain systems using a graphical, block diagram approach.
Simulink® Fixed Point	Simulink® Fixed Point allows for the design of control and signal processing systems using fixed-point arithmetic.
Simulink® HDL Coder	Simulink®HDL Coder allows for the generation of synthesisable and portable HDL code from Simulink® models, Stateflow® charts and Embedded MATLAB® code. Both Verilog®-HDL and VHDL code can be generated.
Stateflow®	Stateflow®extends Simulink® for developing state machines and flow charts through a design environment. It provides language elements required to describe complex logic in a natural, readable, and understandable form.

output) closed-loop DC motor control system. Here, speed control is required with no steady-state error. The motor is modelled as a first-order system with a Laplace transform and is controlled by a PI (proportional plus integral) controller. Figure 4.23 shows the motor control system block diagram with a PI controller.

An example Simulink® model for this system is shown in Figure 4.24.

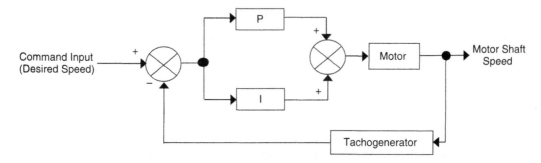

Figure 4.23: Motor control system example with PI control

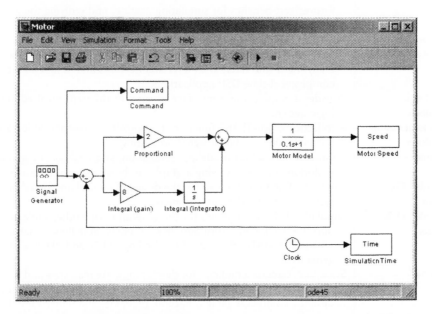

Figure 4.24: Simulink® model for the motor control system example

Therefore, in this model:

- The motor is modeled as a Laplace transform with the transfer function `1/(1 + 0.1s)`.

- The proportional gain is `2`, and the integral gain is `8` (not optimized).

- This is a high-level behavioral model and does not take into account aspects such as value limits, slew rate, and dead-zones.

- The motor model contains the tachogenerator.

- The command input (required speed) and actual motor speed outputs here are considered to be voltages, and the motor shaft speed uses suitable units (e.g., rads/sec).

- The model uses the built-in Simulink® library blocks, and no design hierarchy has been developed.

The motor model is a simple first-order Laplace transform that models the motor and tachogenerator as a single unit. It was created by monitoring the tachogenerator output voltage to a step change in motor speed command input voltage. This is reasonably representative of the motor reaction to larger step changes in command input, but does not model nonideal characteristics such as a motor dead-zone around a null (zero) command input and the need for a minimum command input voltage required for the motor to react to a command input change.

References

[1] Deitel, H. M., and Deitel, P. J., *C, How to Program*, Fourth Edition, Prentice Hall, 2004, ISBN 0-13-122543-X.

[2] American National Standards Institute, INCITS/ISO/IEC 9899-1999 (R2005), Programming languages – C (formerly ANSI/ISO/IEC 9899-1999), http://www.ansi.org

[3] Microsoft® Corporation, Microsoft® Visual C++®, http://www.microsoft.com

[4] Sun Microsystems, *Java Platform, Standard Edition* (J2SE) http://java.sun.com/j2se/

[5] Cadenhead, R., and Lemay, L., *SAMS Teach Yourself Java™ 2 in 21 days*, SAMS, 2004, ISBN 0-672-32628-0.

[6] Microsoft® Corporation, Microsoft® Visual Basic™, http://www.microsoft.com

[7] Barron, D., *The World of Scripting Languages*, Wiley, 2000, ISBN 0-471-99886-9.

[8] Lutz, M., *Programming Python*, Second Edition, O'Reilly, ISBN 0-596-00085-5.

[9] Sastry, V., and Sastry, L., *SAMS Teach Yourself Tcl/Tk in 24 Hours*, SAMS, 2000, ISBN 0-672-31749-4.

[10] Meloni. J. C., *SAMS Teach Yourself PHP, MySQL™ and Apache in 24 Hours, 2003*, ISBN 0-672-32489-X.

[11] The Institute of Electrical and Electronics Engineers, *IEEE Std 1364-2001, IEEE Standard VHDL Language Reference Manual*, http://www.ieee.org

[12] The Institute of Electrical and Electronics Engineers, *IEEE Std 1076-2002, IEEE Standard Verilog Hardware Description Language*, http://www.ieee.org

[13] Open Verilog International, *Verilog-A Language Reference Manual Analog Extensions to Verilog HDL*, Version 1.0, August 1996, http://www.verilog.org/

[14] Cadence Design Systems Inc., USA, http:://www.cadence.com

[15] Accelera Verilog Analog Mixed-Signal Group, http://www.verilog.org/verilog-ams/

[16] The Institute of Electrical and Electronics Engineers, *IEEE Std 1076.1-1999. IEEE Standard VHDL Analog and Mixed-Signal Extensions*http://www.ieee.org

[17] Bradley, D., Seward, D., Dawson, D., and Burge, S., *Mechatronics and the Design of Intelligent Machines and Systems*. Stanley Thornes, 2000, ISBN 0-7487-5443-1.

[18] Pecheux, F., Lallement, C., and Vachoux, A., "VHDL-AMS and Verilog-AMS as Alternative Hardware Description Languages for Efficient Modeling of Multidiscipline Systems," *IEEE Transactions on Computer-Aided Design of Integrated Circuits and Systems*, Vol. 24, No. 2, February 2005.

[19]   *SPICE*: Simulation Program with Integrated Circuit Emphasis, Version 3f5, University of California, Berkeley, USA.

[20]   Tuinenga, P., *SPICE, A Guide to Circuit Simulation and Analysis Using Pspice*. Third Edition, Prentice Hall, 1995, ISBN 0-13-158775-7.

[21]   The Mathworks Inc., http://www.themathworks.com

[22]   Hanselman, D., and Littlefield, B., *Mastering Matlab 6—A Comprehensive Tutorial and Reference*, Prentice Hall, USA, 2001, ISBN 0-13-019468-9.

[23]   Wolfram Research, http://www.wolfram.com/

[24]   Modelica Association, http://www.modelica.org/

[25]   Maplesoft, http://www.maplesoft.com

[26]   Scilab, http://www.scilab.org

[27]   SystemC, http://www.systemc.org

[28]   American National Standards Institute, INCITS/ISO/IEC 14882-2003, Programming languages - C + +, http://www.ansi.org

[29]   Grotker T. et al., "System Design with SystemC", Kluwer Academic Publishers, 2004, ISBN 1-4020-7072-1

[30]   IEEE Std 1666[TM]-2005, IEEE Standard SystemC® Language Reference Manual, IEEE, http://www.ieee.org

[31]   SystemVerilog, http://www.systemverilog.org

[32]   Mintz M. and Ekendahl R., "Hardware Verification with System Verilog: An Object-Oriented Framework", Springer-Verlag New York, May 2007, ISBN 9780387717388

[33]   Sutherland S., Davidmann S. and Flake P., "SystemVerilog for Design: A Guide to Using SystemVerilog for Hardware Design and Modeling", Second Edition, Springer, 2006, ISBN 0-387-3399-1

[34]   Accellera, http://www.accellera.org

[35]   IEEE Std 1800[TM]-2005, IEEE Standard for SystemVerilog - Unified Hardware Design, Specification, and Verification Language, IEEE, http://www.ieee.org

[36]   IEEE Std 1364[TM]-2005, IEEE Standard for Verilog® Hardware Description Language, IEEE, http://www.ieee.org

Student Exercises

4.1   Consider the VHDL design description shown in Figure 4.11. This description is to be placed within a text file on the particular PC or workstation used. Using the C programming language, write a program that will read in this design description and identify the number of inputs and outputs that the design has. This information is to be presented to the user of the program using both the computer visual display unit and as an output text file. Suitably format the information to aid the user of the program.

4.2   Repeat question 4.1, but now use JAVA^{TM}.

4.3   Repeat question 4.1, but now use Visual Basic^{TM}.

4.4   Consider the Verilog$^{®}$-HDL design description shown in Figure 4.13. This description is to be placed within a text file on the particular PC or workstation used. Using the C programming language, write a program that will read in this design description and identify the number of inputs and outputs that the design has, along with the number of each type of logic gate used. This information is to be presented to the user of the program using both the computer visual display unit and as an output text file. Suitably format the information to aid the user of the program.

4.5   Repeat question 4.4, but now use JAVA^{TM}.

4.6   Repeat question 4.4, but now use Visual Basic^{TM}.

4.7   Consider the SPICE design description shown in Figure 4.22. This description is to be placed within a text file on the particular PC or workstation used. Using the C programming language, write a program that will read in this design description and identify the number of circuit nodes that the design has, along with the number of each type of electrical component (resistor, capacitor, etc.) used. This information is to be presented to the user of the program using both the computer visual display unit and as an output text file. Suitably format the information to aid the user of the program.

4.8   Repeat question 4.7, but now use JAVA^{TM}.

4.9   Repeat question 4.7, but now use Visual Basic^{TM}.

4.10   Consider the Simulink$^{®}$ model of the closed-loop control system identified in Figure 4.24. Using this model as a starting point, identify the overall system transfer function (i.e., output / input) using Laplace transforms.

4.11   Using the system transfer function derived in question 4.10, identify the poles and zeros of the system. Comment on the stability of the overall system with the values used for the proportional and integral gains.

CHAPTER 5

Introduction to Digital Logic Design

5.1 Introduction

Although the world that we live in is analogue in nature, in electronic circuits and systems digital circuits are widely used and can be designed to perform many actions that were originally undertaken in analogue circuitry, as well as providing potential benefits over analogue circuit operation. The electronic system shown in Figure 5.1 will perform its operations on signals that are either analogue or digital in nature, using either analogue or digital electronic circuits [1]. Hence, a signal may be one of two types: analogue or digital.

An *analogue signal* is a continuous or discrete-time signal with an amplitude that is continuous in value between a lower and upper limit, but may be either a continuous-time or discrete-time signal.

A *digital signal* is a continuous or discrete-time signal with discrete values between a lower and upper limit. These discrete values are represented by numerical values and are suitable for digital signal processing. If the discrete-time signal has been derived from a continuous-time signal by sampling, then the sampled signal is converted into a digital signal by quantization: quantization produces a finite number of values from a continuous amplitude signal. It is common to use the binary number (i.e., two values, 0 or 1) system to represent a number digitally.

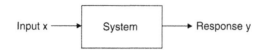

Figure 5.1: Electronic system block diagram

In the digital domain, the choice of implementation technology is essentially whether to use dedicated- (and fixed-) functionality digital logic, to use a software-programmed processor based system (microprocessor, µP; microcontroller, µC; or digital signal processor, DSP), or to use a hardware-configured programmable logic device (PLD) such as the simple programmable logic device (SPLD), complex programmable logic device (CPLD), or the field programmable gate array (FPGA). Memory—random access memory (RAM) or read-only memory (ROM)—is also widely used in many digital electronic circuits and systems. The choices are shown in Figure 5.2.

The initial choice for implementing the digital circuit is between a standard product IC (integrated circuit) and an ASIC (application-specific integrated circuit) [2]:

- **Standard product IC**: an off-the-shelf electronic component that has been designed and manufactured for a given purpose, or range of use, and that is commercially available. It is purchased either from a component supplier or directly from the designer or manufacturer.

- **ASIC**: an integrated circuit that has been specifically designed and manufactured for a particular application.

For many applications, developing an electronic system based on standard product ICs is more cost effective than ASIC design. Undertaking an ASIC design project also requires access to IC design experience, IC computer-aided design (CAD) tools, and a suitable manufacturing and test capability. Whether a standard product IC or ASIC design approach is taken, the type of IC used or developed will be one of four types:

1. **Fixed-functionality**: These ICs have been designed to implement a specific functionality and cannot be changed. The designer uses a set of these ICs to implement a given overall circuit functionality. Changes to the circuit requires a complete redesign of the circuit and the use of different fixed-functionality ICs.

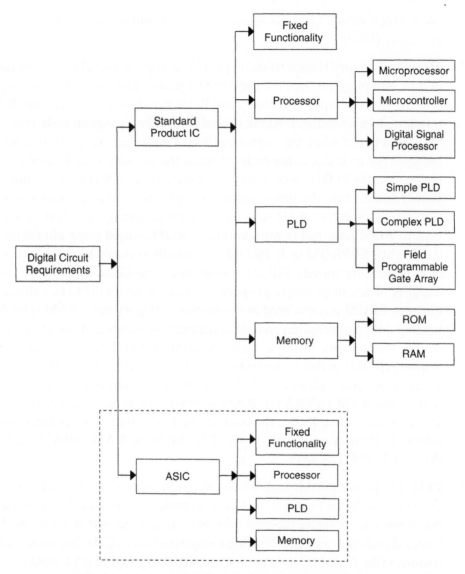

Figure 5.2: Technology choices for digital circuit design

2. **Processor**: Most people are familiar with processors in everyday use; the heart of the PC is a microprocessor. This component runs a software program to implement a required functionality. By changing the software program, the processor will operate a different function. The three types of processor

are microprocessor (μP), microcontroller (μC), and digital signal processor (DSP).

3. **Memory**: Memory is used to store, provide access to, and allow modification of data and program code for use within a processor-based electronic circuit/ system. The two types of memory are ROM (read-only memory) and RAM (random access memory). ROM is used for holding program code that must be retained when the memory power is removed; this is *nonvolatile storage*. The code can either be fixed when the memory is fabricated (mask programmable ROM), electrically programmed once (PROM, programmable ROM) or electronically programmed multiple times. Multiple programming capacity requires the ability to erase prior programming, which is available with EPROM (electrically programmable ROM, erased using ultraviolet [UV] light), EEPROM or E^2PROM (electrically erasable PROM), or flash (also electrically erased). PROM is sometimes considered to be in the same category of circuit as simple programmable logic device (SPLD), although in this text, PROM is considered in the memory category only. RAM is used for holding data and program code that require fast access and the ability to modify the contents during normal operation. RAM differs from read-only memory (ROM) in that it can be both read from and written to in the normal circuit application. However, *flash* memory can also be referred to as nonvolatile RAM (NVRAM). RAM is considered to provide a *volatile storage* since, unlike ROM, the contents of RAM are lost when the power is removed. There are two main types of RAM: static RAM (SRAM) and dynamic RAM (DRAM).

4. **PLD**: The programmable logic device, the main focus of this book, is an IC that contains digital logic cells and programmable interconnect [3, 4] to enable the designer to configure the logic cells and interconnect within the IC itself to form a digital electronic circuit within a single packaged IC. In this, the hardware resources (the available hardware for use) are configured to implement the required functionality. By changing the hardware configuration, the PLD performs a different function. Three types of PLD are available: the simple programmable logic device (SPLD), the complex programmable logic device (CPLD), and the field programmable gate array (FPGA).

Both the processor and PLD enable the designer to implement and change the functionality of the IC by either changing the software program or the hardware configuration. To avoid potential confusion, in this book the following terms will be used to differentiate the PLD from the processor:

- The PLD will be *configured* using a hardware configuration.

- The processor will be *programmed* using a software program.

An ASIC can be designed to create any one of the four standard product IC forms (fixed-functionality, processor, memory, or PLD). A standard product IC is designed in the same manner as an ASIC, so anyone who has access to an ASIC design, fabrication, and test facility can create an equivalent to a standard product IC (given that patent and legal issues of IP [intellectual property] for existing designs and devices are taken into account).

No matter how complex the digital circuit design, and the types of operations it is required to undertake, the operation is based on a small number of basic combinational and sequential logic circuit elements that are connected to form the required circuit operation:

- **Combinational logic**: A combinational logic circuit is defined by a Boolean expression, and the output from the circuit (in logic terms) is a function of the logic input values, the logic gates used (AND, OR, etc.), and the way in which the logic gates are connected [5, 6]. The output becomes a final value when the inputs change after a finite time, which is the time required for the logic values to propagate through the circuit given signal propagation delays in each of the logic gates and any delays in the interconnections between the logic gates. The basic combinational logic circuit elements (gates) are:

  ○ AND gate

  ○ NAND gate

  ○ OR gate

  ○ NOR gate

  ○ exclusive-OR (EX-OR) gate

  ○ exclusive-NOR (EX-NOR) gate

  ○ inverter

  ○ buffer

- **Sequential logic**: In a sequential logic circuit, the output from the circuit becomes a value based on the logic input values, the logic gates used, the way in which the logic gates are connected, and on the current state of the circuit [5, 6]. In a *synchronous sequential logic* circuit, the output change occurs either on the edge of a clock signal change (from 0 to 1 or from 1 to 0) or on a clock signal level (logic 0 or 1). However, an *asynchronous sequential logic* circuit does not use a clock input. In the sequential logic circuit, the circuit will hold or remember its current value (state) and will change state only on clock or data changes. A sequential logic circuit might also contain additional control signals to reset or set the circuit into a known state either when the circuit is initially turned on or during normal circuit operation. The basic sequential logic circuit elements (gates) are:

  - S-R flip-flop
  - J-K flip-flop
  - toggle flip-flop
  - D-latch
  - D-type flip-flop

5.2 Number Systems

5.2.1 Introduction

In everyday life, we use the decimal number system (base, or radix, 10), which allows the creation of numbers with digits in the set: 0, 1, 2, 3, 4, 5, 6, 7, 8, 9. The ten possible digits are combined to create integer and real numbers. However, base 10 is not the only number system. Digital circuits and systems use the binary (base, or radix, 2) number system, which allows for the creation of numbers with digits in the set: 0, 1.

The 0 and 1 numbers are logic levels (0 = logic 0, 1 = logic 1), which are created by voltages in a circuit:

- In **positive logic**, 0 is formed by a low voltage level, and 1 is formed by a high voltage level.

- In **negative logic**, 0 is formed by a high voltage level, and 1 is formed by a low voltage level.

In this text, only positive logic will be used and will use the voltage levels shown in Table 5.1.

**Table 5.1: Typical voltage levels
representing positive logic**

Logic level	+5 V logic	+3.3 V logic
0	+5.0 V	+3.3 V
1	0 V	0 V

Decimal and binary number systems are only two of four number systems used in digital circuits and systems:

1. decimal (base 10)

2. binary (base 2)

3. octal (base 8)

4. hexadecimal (base 16)

As some point in the design and analysis of a digital circuit, it will be necessary to convert between the different number systems to view and manipulate values propagating through the design. Such conversion is typically undertaken to aid the interpretation and understanding of the design operation.

In addition, a binary number can have different meanings as different binary coding can be chosen for different design requirement. The main binary coding schemes used are:

1. unsigned (or straight) binary

2. signed binary (1s complement or 2s complement)

3. Gray code

4. binary coded decimal (BCD)

Unsigned binary numbers are used to represent positive numbers only. Signed binary numbers are used to represent positive and negative numbers that are coded to allow arithmetic using either 1s complement or 2s complement notation. Twos complement notation is more commonly used and will be considered in this text. Gray code allows for a one-bit change when moving from one value to the next (or previous) value. BCD provides a simple conversion between binary and decimal numbers.

All four binary coding schemes are fully discussed in the following sections.

5.2.2 Decimal–Unsigned Binary Conversion

The conversion between decimal and binary involves identifying the particular decimal value for the particular binary code (or vice versa). Both decimal-to-binary and binary-to-decimal conversion is common and a binary number will be needed to represent each decimal number. If both the decimal and binary numbers are unrestricted in size, then an exact conversion is possible.

In unsigned (or straight) binary, the numbers represented by the binary code will be positive numbers only. Each digit in the binary number will contribute to the magnitude of the value. For example, consider the decimal value 8_{10}. In unsigned binary, this is represented by 1000_2. Each digit in the decimal number has a value in the set of (0, 1, 2, 3, 4, 5, 6, 7, 8, 9). Each digit in the binary number is in the set of (0, 1). A binary digit is referred to as a bit (*bi*nary dig*it*).

The magnitude of the decimal number is the sum of the product of the value of each digit in the number (d) and its position (n). The position immediately to the left of the decimal point is position zero (0). The value of the digit has a weight of 2^n where n is the position number. Moving left from position 0 (in the integer part of the number), the position increments by 1. Moving right from position zero (into the fractional part of the number), the position decrements by 1. Therefore, the magnitude of the number is given by:

Magnitude = $(d_n.10^n) + (d_{n-1}.10^{n-1}) + (d_{n-2}.10^{n-2}) + \cdots + (d_0.10^0) + (d_{-1}.10^{-1}) + \cdots + (d_{-n}.10^{-n})$

Here, the decimal number is written as:

$d_n d_{n-1} d_{n-2} \ldots d_0.d_{-1} \ldots . d_{-n}$

Some example decimal numbers are:

$\mathbf{8_{10}}$ is $[(8 \times 10^0)]_{10}$

$\mathbf{18_{10}}$ is $[(1 \times 10^1) + (8 \times 10^0)]_{10}$

$\mathbf{218_{10}}$ is $[(2 \times 10^2) + (1 \times 10^1) + (8 \times 10^0)]_{10}$

$\mathbf{218.3_{10}}$ is $[(2 \times 10^2) + (1 \times 10^1) + (8 \times 10^0) + (3 \times 10^{-1})]_{10}$

$\mathbf{218.37_{10}}$ is $[(2 \times 10^2) + (1 \times 10^1) + (8 \times 10^0) + (3 \times 10^{-1}) + (7 \times 10^{-2})]_{10}$

The binary number is a base 2 number whose magnitude is the sum of the product of the value of each digit in the number (b) and its position (n). Moving left from position 0 (in the integer part of the number), the position increments by 1. The value of the digit has a weight of 2^n where n is the position number. Moving right from position zero (into the fractional part of the number), the position decrements by 1. This allows the creation of numbers with digits in the set: 0, 1. Therefore, in general the magnitude of the number (as a decimal number) is given by:

Magnitude $= (b_n.2^n) + (b_{n-1}.2^{n-1}) + (b_{n-2}.2^{n-2}) + \cdots + (b_0.2^0) + (b_{-1}.2^{-1}) + \cdots + (b_{-n}.2^{-n})$

Here, the binary number is written as $b_n b_{n-1} b_{n-2} \ldots b_0.b_{-1} \ldots b_{-n}$. Some example binary numbers are:

1_2

10_2

101_2

101.1_2

101.01_2

The decimal number equivalent for a binary number can be created by taking the binary number and calculating its magnitude (as a decimal number):

Magnitude $= (b_n.2^n) + (b_{n-1}.2^{n-1}) + (b_{n-2}.2^{n-2}) + \cdots + (b_0.2^0) + (b_{-1}.2^{-1}) + \cdots + (b_{-n}.2^{-n})$

Some example binary numbers are:

$\mathbf{1_{10}}$ is $[(1 \times 2^0)]_{10} = 1_{10}$

$\mathbf{10_{10}}$ is $[(1 \times 2^1) + (0 \times 2^0)]_{10} = 2_{10}$

$\mathbf{101_{10}}$ is $[(1 \times 2^2) + (0 \times 2^1) + (1 \times 2^0)]_{10} = 5_{10}$

$\mathbf{101.1_{10}}$ is $[(1 \times 2^2) + (0 \times 2^1) + (1 \times 2^0) + (1 \times 2^{-1})]_{10} = 5.5_{10}$

$\mathbf{101.01_{10}}$ is $[(1 \times 2^2) + (0 \times 2^1) + (1 \times 2^0) + (0 \times 2^{-1}) + (1 \times 2^{-2})]_{10} = 5.25_{10}.$

The binary number equivalent of a decimal number is created by dividing the decimal number by 2 until the result of the division is 0. The remainder of the total division forms the binary number digits, the remainder from the first division forms the least significant bit (LSB) of the binary number, and the remainder from the last division forms the most significant bit (MSB) of the binary number.

Consider the number 8_{10}. The conversion procedure is:

Action	Division	Remainder	Binary number digit
Start with the decimal number (d = 8)			
Divide by 2	d/2 = 8/2 = 4	0	$b_0 = 0$
Divide by 2	d/2 = 4/2 = 2	0	$b_1 = 0$
Divide by 2	d/2 = 2/2 = 1	0	$b_2 = 0$
Divide by 2	d/2 = 1/2 = 0	1	$b_3 = 1$

The binary number can be read as: $8_{10} = (b_3b_2b_1b_0)_2 = 1000_2$.

Consider now the number 218_{10}. The conversion procedure is:

Action	Division	Remainder	Binary number digit
Start with the decimal number (d = 218)			
Divide by 2	d/2 = 218/2 = 109	0	$b_0 = 0$
Divide by 2	d/2 = 109/2 = 54	1	$b_1 = 1$
Divide by 2	d/2 = 54/2 = 27	0	$b_2 = 0$
Divide by 2	d/2 = 27/2 = 13	1	$b_3 = 1$
Divide by 2	d/2 = 13/2 = 6	1	$b_4 = 1$
Divide by 2	d/2 = 6/2 = 3	0	$b_5 = 0$
Divide by 2	d/2 = 3/2 = 1	1	$b_6 = 1$
Divide by 2	d/2 = 1/2 = 0	1	$b_7 = 1$

The binary number can be read as: $218_{10} = (b_7b_6b_5b_4b_3b_2b_1b_0)_2 = 11011010_2$.

5.2.3 Signed Binary Numbers

Unsigned (or straight) binary numbers are used when the operations use only positive numbers and the result of any operations is a positive number. However, in most DSP tasks, both the number and the result can be either positive or negative, and the unsigned binary number system cannot be used. The two coding schemes used to achieve this are the 1s complement and 2s complement.

The 1s complement of a number is obtained by changing (or inverting) each of the bits in the binary number (0 becomes a 1 and a 1 becomes a 0):

Original binary number: 10001100

1s complement: 01110011

The 2s complement is formed by adding 1 to the 1s complement:

Original binary number: 10001100

1s complement: 01110011

2s complement: 01110100

The MSB of the binary number is used to represent the sign (0 = positive, 1 = negative) of the number, and the remainder of the number represents the magnitude. It is therefore essential that the number of bits used is sufficient to represent the required range, as shown in Table 5.2. Here, only integer numbers are considered.

Twos complement number manipulation is as follows:

- To create a positive binary number from a positive decimal number, create the positive binary number for the magnitude of the decimal number where the MSB is set to 0 (indicating a positive number).

- To create a negative binary number from a negative decimal number, create the positive binary number for the magnitude of the decimal number where the MSB is set to 0 (indicating a positive number), then invert all bits and add 1 to the LSB. Ignore any overflow bit from the binary addition.

- To create a negative binary number from a positive binary number, where the MSB is set to 0 (indicating a positive number), invert all bits and add 1 to the LSB. Ignore any overflow bit from the binary addition.

- To create a positive binary number from a negative binary number, where the MSB is set to 1 (indicating a negative number), invert all bits and add 1 to the LSB. Ignore any overflow bit from the binary addition.

The 2s complement number coding scheme is widely used in digital circuits and system design and so will be explained further. Table 5.3 shows the binary representations of decimal numbers for a four-bit binary number. In the unsigned binary number coding scheme, the binary number represents a positive decimal

Table 5.2: Number range

Number of bits	Unsigned binary range	2s complement number range
4	0_{10} to $+15_{10}$	-8_{10} to $+7_{10}$
8	0_{10} to $+255_{10}$	-128_{10} to $+127_{10}$
16	0_{10} to $+65,535_{10}$	$-32,768_{10}$ to $+32,767_{10}$

Table 5.3: Decimal to binary conversion

Decimal number	4-bit unsigned binary number	4-bit 2s complement signed binary number
+15	1111	—
+14	1110	—
+13	1101	—
+12	1100	—
+11	1011	—
+10	1010	—
+9	1001	—
+8	1000	—
+7	0111	0111
+6	0110	0110
+5	0101	0101
+4	0100	0100
+3	0011	0011
+2	0010	0010
+1	0001	0001
0	0000	0000
−1	—	1111
−2	—	1110
−3	—	1101
−4	—	1100
−5	—	1011
−6	—	1010
−7	—	1001
−8	—	1000

number from 0_{10} to $+15_{10}$. In the 2s complement number coding scheme, the decimal number range is -8_{10} to $+7_{10}$.

In this, the most negative 2s complement number is 1_{10} greater in magnitude than the most positive 2s complement number. The number range for an n-bit number is: -2^N to $+(2^N - 1)$.

Table 5.4: 2s complement addition and subtraction

Arithmetic operation	Polarity of input A	Polarity of input B	Action
Addition(A + B)	**Augend**	**Addend**	Add the augend to the addend and disregard any overflow.
	Positive	Positive	
	Positive	Negative	
	Negative	Positive	
	Negative	Negative	
Subtraction(A − B)	**Minuend**	**Subtrahend**	Negate (invert) the subtrahend, add this to the minuend, and disregard any overflow.
	Positive	Positive	
	Positive	Negative	
	Negative	Positive	
	Negative	Negative	

Addition and subtraction are undertaken by addition and if necessary inversion (creating a negative number from a positive number and vice versa). Table 5.4 shows the cases for addition and subtraction of two numbers (A and B). It is essential to ensure that the two numbers have the same number of bits, the MSB represents the sign of the binary number, and the number of bits used is sufficient to represent the range of possible inputs and the range of possible outputs.

Figure 5.3 shows an arrangement where two inputs are either added or subtracted, depending on the logic level of a control input. This arrangement requires an adder, a

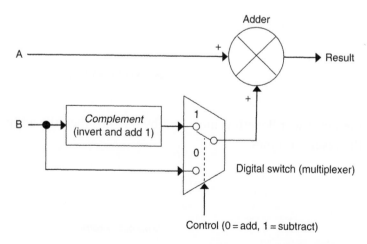

Figure 5.3: Addition and subtraction (2's complement arithmetic)

complement (a logical inversion of the inputs bits and add 1, disregarding any overflow), and a digital switch (multiplexer).

Input numbers in the range -8_{10} to $+7_{10}$ are represented by four bits in binary. However, the range for the result of an addition is -16_{10} to $+14_{10}$, and the range for the result of a subtraction is -15_{10} to $+15_{10}$. The result requires five bits in binary to represent the number range (one bit more than the number of bits required to represent the inputs), so the number of bits to represent the inputs will be increased by one bit before the addition or subtraction:

- In an **unsigned binary number**, to increase the wordlength (number of bits) by one bit, append a 0 to the number as the new MSB:

$0010_2 = 00010_2$

$1010_2 = 01010_2$

- In a **2s complement number**, to increase the wordlength by one bit, then append a bit with the same value as the original MSB to the number as the new MSB:

$0010_2 = 00010_2$

$1010_2 = 11010_2$

Consider the addition of $+2_{10}$ to $+3_{10}$ using 2s complement numbers. The result should be $+5_{10}$. The two input numbers can be represented by three bits, but if 3-bit addition is undertaken, the result will be in error:

```
0  1  0           +2₁₀
0  1  1  +        +3₁₀
_____           _____
1  0  1           -3₁₀   INCORRECT RESULT
```

If, however, the input wordlength is increased by one bit, then the addition is undertaken, the result becomes:

```
0  0  1  0           +2₁₀
0  0  1  1  +        +3₁₀
_____           _____
0  1  0  1           +5₁₀   CORRECT RESULT
```

Consider the subtraction of $+3_{10}$ from -2_{10}. The result should be -5_{10}. The two input numbers can be represented by three bits, but if 3-bit addition is undertaken, then the result will be in error:

```
    1 1 0        -2₁₀
    1 0 1 +      -3₁₀      (Subtrahend is complemented)
  ─────────
  1 0 1 1        +3₁₀   INCORRECT RESULT
  ↑
Overflow is ignored
```

If, however, the input wordlength is increased by one bit, then the addition is undertaken, the result becomes:

```
    1 1 1 0      -2₁₀
    1 1 0 1 +    -3₁₀      (Subtrahend is complemented)
  ───────────
  1 1 0 1 1      -5₁₀   CORRECT RESULT
  ↑
Overflow is ignored
```

5.2.4   Gray Code

The Gray code provides a binary code that changes by one bit only when it changes from one value to the next. The Gray code and the decimal number equivalent of the binary number (in unsigned binary) are shown in Table 5.5. This is no longer a straight binary count sequence.

The Gray code is often used in position control systems which represent either a rotary position as in the output shaft of an electric motor or a linear position as in the position of a conveyor belt. Figure 5.4 shows the Gray code used on a sensor to identify the position of an object that can move left and right. Each code represents a point of position or span of distance in length. The Gray code removes the potential for errors when changing from sensing one position to the next position that could occur in a binary code when more than one bit could change. If there is a time delay in the circuitry that senses the individual bits, and the delay for sensing each bit is different, the result will be a short but finite time during which the position code would be wrong. If the circuitry that uses this position signal detects this wrong position code, it will react to a wrong position, and the result would be an erroneous operation of the circuit.

Table 5.5: Gray code

Decimal number	4-bit Gray code ($d_3d_2d_1d_0$)
0	0000
1	0001
3	0011
2	0010
6	0110
7	0111
5	0101
4	0100
12	1100
13	1101
15	1111
14	1110
10	1010
11	1011
9	1001
8	1000

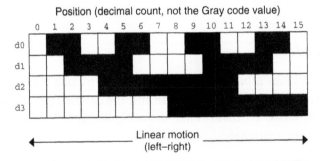

Figure 5.4: Gray code position sensing example

5.2.5 Binary Coded Decimal

Binary coded decimal (BCD) provides a simple conversion between a binary number and the decimal number. For a decimal number, each digit is represented by four bits. For example, the number 12_{10} is represented by 00010010_2.

$$00010010_2 = 0001_2/0010_2$$

$$= (12)_{10}$$

If the MSBs are 0, they might also be left out, so the BCD number could also be represented as 10010_2. This particular BCD code is referred to as 8421 BCD (or straight binary coding) because the binary number is a direct representation of the decimal value for decimal values 0_{10} to 9_{10}. Decimal values 10_{10} to 15_{10} are not represented in the four bits. Other BCD codes can also be implemented.

It is important to understand that a BCD is not the same as a straight binary (unsigned binary) count. For example, consider the number 12_{10}:

$12_{10} = 10010_2$, BCD

$12_{10} = 1100_2$, straight binary

5.2.6 Octal-Binary Conversion

The octal number is a number to the base (or radix) 8, and the magnitude of the number is the sum of the product of the value of each digit in the number (o) and its position (n). This allows the creation of numbers with digits in the set: 0, 1, 2, 3, 4, 5, 6, 7.

The position immediately to the left of the decimal point is zero (0). Moving left from position 0 (in the integer part of the number), the position increments by 1. The value of the digit has a weight of 8^n where n is the position number. Moving right from position 0 (into the fractional part of the number), the position decrements by 1. The eight possible digits are combined to create integers and real numbers. Table 5.6 shows the conversion table.

The magnitude of the number (as a decimal number) is given by:

$$\text{Magnitude} = (o_n.8^n) + (o_{n-1}.8^{n-1}) + (o_{n-2}.8^{n-2}) + \cdots + (o_0.8^0) + (o_{-1}.8^{-1}) + \cdots + (o_{-n}.8^{-n})$$

Here, the octal number is written as $o_n o_{n-1} o_{n-2} \ldots o_0.o_{-1} \ldots . o_{-n}$ (using the decimal equivalent of the octal number).

Some example octal numbers are:

$\mathbf{7_8}$ is $[(7 \times 8^0)]_{10}$

$\mathbf{17_8}$ is $[(1 \times 8^1) + (7 \times 8^0)]_{10}$

$\mathbf{267_8}$ is $[(2 \times 8^2) + (6 \times 8^1) + (7 \times 16^0)]_{10}$

Table 5.6: Octal–decimal–unsigned 4-bit binary conversion

Octal number	Decimal number	4-bit unsigned binary number
0	0	0000
1	1	0001
2	2	0010
3	3	0011
4	4	0100
5	5	0101
6	6	0110
7	7	0111
10	8	1000
11	9	1001
12	10	1010
13	11	1011
14	12	1100
15	13	1101
16	14	1110
17	15	1111

217.5$_8$ is $[(2 \times 8^2) + (1 \times 8^1) + (7 \times 8^0) + (5 \times 8^{-1})]_{10}$

217.57$_8$ is $[(2 \times 8^2) + (1 \times 8^1) + (7 \times 8^0) + (5 \times 8^{-1}) + (7 \times 8^{-2})]_{10}$.

For binary numbers, each octal number represents three bits. Therefore a 6-bit binary number is represented by two octal numbers, an 8-bit binary number is represented by three octal numbers, a 9-bit binary number is also represented by three octal numbers, a 16-bit binary is represented by six octal numbers, and so on. For example, 7_8 is 111_2 and 17_8 is 001111_2:

$$0 \quad 0 \quad 1 \quad 1 \quad 1 \quad 1_2$$
$$1 \qquad\qquad 7_8$$

Some example octal numbers are:

7$_8$ is 111_2

17$_8$ is 001111_2

267$_8$ is 010110111_2

217.5$_8$ is 010001111.101$_2$

217.57$_8$ is 010001111.101111$_2$.

The decimal number equivalent for an octal number is created by calculating the magnitude of the octal number as a decimal number:

$$\text{Magnitude} = (o_n.8^n) + (o_{n-1}.8^{n-1}) + (o_{n-2}.8^{n-2}) + \cdots + (o_0.8^0) + (o_{-1}.8^{-1}) + \cdots + (o_{-n}.8^{-n})$$

Converting from decimal to octal is accomplished in a similar manner as converting from decimal to binary, except now dividing by 8 rather than 2. Consider the number 7_{10}. The conversion procedure is:

Action	Division	Remainder	Octal number digit
Start with the decimal number (d = 7) Divide by 8	d/2 = 7/8 = 0	7	o$_0$ = 7

The octal number can be read as: $7_{10} = (o_0)_8 = 7_8$.

Consider the number 100_{10}. The conversion procedure is:

Action	Division	Remainder	Octal number digit
Start with the decimal number (d = 100) Divide by 8	d/2 = 100/8 = 12	4	o$_0$ = 4
Divide by 8	d/2 = 12/8 = 1	4	o$_1$ = 4
Divide by 8	d/2 = 1/8 = 0	1	o$_2$ = 1

The octal number can be read as: $100_{10} = (o_2 o_1 o_0)_8 = 144_8$.

5.2.7. Hexadecimal-Binary Conversion

The hexadecimal number is a number to the base (or radix) 16, and its magnitude is the sum of the product of the value of each digit in the number (h) and its position (n). This allows the creation of numbers with digits in the set: 0, 1, 2, 3, 4, 5, 6, 7, 8, 9, A, B, C, D, E, F.

The position immediately to the left of the decimal point is zero (0). Moving left from position 0 (in the integer part of the number), the position increments by 1. The value of the digit has a weight of 16^n where n is the position number. Moving right from position zero (into the fractional part of the number), the

Table 5.7: Hexadecimal–decimal–unsigned four-bit binary conversion

Hexadecimal number	Decimal number	4-bit unsigned binary number
0	0	0000
1	1	0001
2	2	0010
3	3	0011
4	4	0100
5	5	0101
6	6	0110
7	7	0111
8	8	1000
9	9	1001
A	10	1010
B	11	1011
C	12	1100
D	13	1101
E	14	1110
F	15	1111

position decrements by 1. The sixteen possible digits are combined to create integers and real numbers. In a decimal equivalent number, the hexadecimal digits A_{16} to F_{16} are the numbers 10_{10} to 15_{10}. Table 5.7 shows the conversion table.

The magnitude of the number (as a decimal number) is given by:

$$\text{Magnitude} = (h_n.16^n) + (h_{n-1}.16^{n-1}) + (h_{n-2}.16^{n-2}) + \cdots + (h_0.16^0) + (h_{-1}.16^{-1}) + \cdots + (h_{-n}.16^{-n})$$

Here, the hexadecimal number is written as $h_n h_{n-1} h_{n-2} \ldots h_0.h_{-1} \ldots . h_{-n}$ (using the decimal equivalent of the hexadecimal number).

Some example hexadecimal numbers are:

$\mathbf{8_{16}}$ is $[(8 \times 16^0)]_{10}$

$\mathbf{A8_{16}}$ is $[(10 \times 16^1) + (8 \times 16^0)]_{10}$

$\mathbf{2A8_{16}}$ is $[(2 \times 16^2) + (10 \times 16^1) + (8 \times 16^0)]_{10}$

$\mathbf{218.F_{16}}$ is $[(2 \times 16^2) + (1 \times 16^1) + (8 \times 16^0) + (15 \times 16^{-1})]_{10}$

$\mathbf{218.F7_{16}}$ is $[(2 \times 16^2) + (1 \times 16^1) + (8 \times 16^0) + (15 \times 16^{-1}) + (7 \times 16^{-2})]_{10}$.

For binary numbers, each hexadecimal number represents four bits. Therefore, an 8-bit binary number is represented by two hexadecimal numbers, a 16-bit binary is represented by four hexadecimal numbers, and so on. For example, 8_{16} is 1000_2 and $A8_{16}$ is 10101000_2.

$$\underbrace{1\ 0\ 1\ 0}_{A}\ \underbrace{1\ 0\ 0\ 0_2}_{8_{16}}$$

Some example hexadecimal numbers are:

8_{16} is 1000_2

$A8_{16}$ is 10101000_2

$2A8_{16}$ is 001010101000_2

$218.F_{16}$ is 001000011000.1111_2

$218.F7_{16}$ is 001000011000.11110111_2.

The decimal number equivalent for a hexadecimal number is created by calculating the magnitude of the hexadecimal number, using the decimal equivalent for hexadecimal numbers A to F, as a decimal number:

$$\text{Magnitude} = (h_n.16^n) + (h_{n-1}.16^{n-1}) + (h_{n-2}.16^{n-2}) + \cdots + (h_0.16^0) + (h_{-1}.16^{-1}) + \cdots + (h_{-n}.16^{-n})$$

Converting from decimal to hexadecimal is accomplished in a similar manner to converting from decimal to binary, except now dividing by 16 rather than 2, and using the letters A to F for decimal remainder values of 10 to 15. Consider the number 7_{10}. The conversion procedure is:

Start with the number (d)	Division	Remainder	Hexadecimal number digit
Start with the decimal number (d = 7) Divide by 16	d/16 = 7/16 = 0	7	h_0 = 7

The hexadecimal number can be read as: $7_{10} = (h_0)_{16} = 7_{16}$.

Consider the number 100_{10}. The conversion procedure is:

Action	Division	Remainder	Hexadecimal number digit
Start with the decimal number (d = 100)			
Divide by 16	d/16 = 100/16 = 6	4	h_0 = 4
Divide by 16	d/16 = 6/16 = 0	6	h_1 = 6

The hexadecimal number can be read as: $100_{10} = (h_1 h_0)_{16} = 64_{16}$.

Consider the number 255_{10}. The conversion procedure is:

Start with the number (d)	Division	Remainder	Hexadecimal number digit
Start with the decimal number (d = 255)			
Divide by 16	d/16 = 255/16 = 15	15	h_0 = F
Divide by 16	d/16 = 15/16 = 0	15	h_1 = F

The hexadecimal number can be read as: $255_{10} = (h_1 h_0)_{16} = FF_{16}$.

Converting from hexadecimal to octal, or vice-versa, is accomplished by converting the number to either a binary or decimal equivalent and from that to the octal to hexadecimal number.

A summary table for the number systems is shown in Table 5.8. Here, unsigned decimal numbers from 0_{10} to 15_{10} are considered.

Both binary and decimal numbers can only be integers or real numbers. Table 5.9 shows the binary and decimal numbers for a real number represented by 40 bits in binary, with 24 bits representing the integer part of the number and 16 bits representing the fractional part of the number.

Table 5.8: Number systems summary

Decimal	Unsigned binary	Octal	Hexadecimal	BCD
0	0000	0	0	0000
1	0001	1	1	0001
2	0010	2	2	0010
3	0011	3	3	0011
4	0100	4	4	0100
5	0101	5	5	0101
6	0110	6	6	0110
7	0111	7	7	0111
8	1000	10	8	1000
9	1001	11	9	1001
10	1010	12	A	00010000
11	1011	13	B	00010001
12	1100	14	C	00010010
13	1101	15	D	00010011
14	1110	16	E	00010100
15	1111	17	F	00010101

Table 5.9: Decimal-binary conversion table, with the positive position to the left of the decimal point and the negative position to the right of the decimal point

Binary location	Unsigned binary number	Binary weighting	Decimal value
23	100000000000000000000000.0000000000000000	2^{23}	8,388,608
22	010000000000000000000000.0000000000000000	2^{22}	4,194,304
21	001000000000000000000000.0000000000000000	2^{21}	2,097,152
20	000100000000000000000000.0000000000000000	2^{20}	1,048,576
19	000010000000000000000000.0000000000000000	2^{19}	524,288
18	000001000000000000000000.0000000000000000	2^{18}	262,144
17	000000100000000000000000.0000000000000000	2^{17}	131,072
16	000000010000000000000000.0000000000000000	2^{16}	65,536
15	000000001000000000000000.0000000000000000	2^{15}	32,768
14	000000000100000000000000.0000000000000000	2^{14}	16,384
13	000000000010000000000000.0000000000000000	2^{13}	8,192
12	000000000001000000000000.0000000000000000	2^{12}	4,096
11	000000000000100000000000.0000000000000000	2^{11}	2,048
10	000000000000010000000000.0000000000000000	2^{10}	1,024
9	000000000000001000000000.0000000000000000	2^9	512
8	000000000000000100000000.0000000000000000	2^8	256
7	000000000000000010000000.0000000000000000	2^7	128

(continued)

Table 5.9 (*Continued*)

Binary location	Unsigned binary number	Binary weighting	Decimal value
6	0000000000000001000000.0000000000000000	2^6	64
5	0000000000000000100000.0000000000000000	2^5	32
4	0000000000000000010000.0000000000000000	2^4	16
3	0000000000000000001000.0000000000000000	2^3	8
2	0000000000000000000100.0000000000000000	2^2	4
1	0000000000000000000010.0000000000000000	2^1	2
0	0000000000000000000001.0000000000000000	2^0	1

Decimal point (.)

−1	0000000000000000000000.1000000000000000	2^{-1}	0.5
−2	0000000000000000000000.0100000000000000	2^{-2}	0.25
−3	0000000000000000000000.0010000000000000	2^{-3}	0.125
−4	0000000000000000000000.0001000000000000	2^{-4}	0.0625
−5	0000000000000000000000.0000100000000000	2^{-5}	0.03125
−6	0000000000000000000000.0000010000000000	2^{-6}	0.015625
−7	0000000000000000000000.0000001000000000	2^{-7}	0.0078125
−8	0000000000000000000000.0000000100000000	2^{-8}	0.00390625
−9	0000000000000000000000.0000000010000000	2^{-9}	0.001953125
−10	0000000000000000000000.0000000001000000	2^{-10}	0.0009765625
−11	0000000000000000000000.0000000000100000	2^{-11}	0.00048828125
−12	0000000000000000000000.0000000000010000	2^{-12}	0.00024414063
−13	0000000000000000000000.0000000000001000	2^{-13}	0.00012207031
−14	0000000000000000000000.0000000000000100	2^{-14}	0.000061035156
−15	0000000000000000000000.0000000000000010	2^{-15}	0.000030517578
−16	0000000000000000000000.0000000000000001	2^{-16}	0.000015258789

5.3   Binary Data Manipulation

5.3.1   Introduction

A digital circuit or system utilizes and manipulates binary data to perform a required operation. Essentially, groups of bits of data are converted from one value to another at a particular point in time. Software-programmed processors typically manipulate groups of 8, 16, 32, 64, or 128 bits of data, although a custom design could manipulate as many bits as required.

Binary data is manipulated using the following:

- *Boolean logic* provides a means to display the operations on input signals and produce a result in mathematical terms using AND, NAND, OR, NOR, EX-OR, EX-NOR, and NOT logical operations.

- *Truth tables* provide a means to display the operations on input signals and produce a result in table format.

- *Karnaugh maps* provide a means to display the operations on input signals and produce a result on a K-map, which allows logic values to be grouped together with loops.

- *Circuit schematics* provide a graphical representation of the Boolean logic expression using logic gate symbols for the logical operations and the connections between the terminals.

Boolean logic, truth tables, Karnaugh maps, and circuit schematics are used in the design and analysis of digital circuits and systems, and the designer must move between these different representations of circuit and system operation many times during the design process. However, these tools are really only suited for design by hand (as it were) for small circuits; for more complex circuits and systems, hardware description languages (HDL) are more commonly used. Understanding Boolean logic, truth tables, and Karnaugh maps, however, will provide the designer with the necessary skills to design, develop, and debug circuit and system designs of any size and complexity.

5.3.2   Logical Operations

A digital circuit or system will consist of a number of operations on logic values. The basic logical operators are the:

- AND
- NAND
- OR
- NOR

- exclusive-OR (EX-OR)

- exclusive-NOR (EX-NOR)

- NOT

Considering two inputs (here called A and B) to a logical operator, the AND, OR, and EX-OR operators provide different results:

- The AND operator provides an output when *both* A and B are at the required values.

- The NAND operator provides an output that is the inverse of the AND operator.

- The OR operator provides an output when *either or both* A and B are at the required values.

- The NOR operator provides an output that is the inverse of the OR operator

- The EX-OR operator provides an output when *either but not both* A and B are at the required values.

- The EX-NOR (or equivalence) operator provides an output that is the inverse of the EX-OR operator.

The NOT operator provides an output that is the logical inverse of the input.

In addition, the BUFFER will provide an output that is the same logic level value as the input. The BUFFER is essentially two NOT gates in series.

These logical operators function in electronic hardware as logic gates. Two inputs (A and B) to the logic gate were considered above, but more inputs are possible to certain logic gates.

5.3.3   Boolean Algebra

Boolean algebra (developed by George Boole and Augustus De Morgan) forms the basic set of rules that regulate the relationship between true-false statements in logic. Applied to digital logic circuits and systems, the true-false statements regulate the relationship between the logic levels (logic 0 and 1) in digital logic circuits and systems. The relationships are based on variables and constants:

- The identifier for the AND logical operator is . (the dot).

- The identifier for the OR logical operator is + (the mathematical addition symbol).

- The identifier for the NOT logical operator is ‾ (a bar across the expression).

- The identifier for the EX-OR logical operator is ⊕ (an encircled addition symbol).

Figure 5.5 shows the Boolean logic expression for each of these operators.

Each of the operators can be combined to create more complex Boolean logic expressions. For example, if a circuit has four inputs (A, B, C, and D) and one output (Z), then if Z is a logic 1 when (A *and* B) is a logic 1 *or* when (C *and* D) is a logic 1, the Boolean expression is:

$$Z = (A.B) + (C.D)$$

Here, parentheses are used to group the ANDed variables and to indicate precedence among various operations—similar to their use in other mathematical expressions. The AND logical operator has a higher precedence than the OR logical operator and so would be naturally grouped together in this way.

Boolean expression	Meaning
Z = A . B	Z is A AND B
Z = $\overline{A . B}$	Z is A NAND B
Z = A + B	Z is A OR B
Z = $\overline{A + B}$	Z is A NOR B
Z = A ⊕ B	Z is A XOR B
Z = $\overline{A ⊕ B}$	Z is A XNOR B
Z = $\overline{A}$	Z is NOT A

Figure 5.5: Boolean expressions for the basic logic operators

A Boolean expression written using Boolean algebra can be manipulated according to a number of theorems to modify it into a form that uses the right logic operators (and therefore the right type of logic gate) and to minimize the number of logic gates. The theorems of Boolean algebra fall into three main categories:

1. Logical operations on constants

2. Logic operations on one variable

3. Logic operations on two or more variables.

Table 5.10 summarizes the logical operations on constants. Each constant value can be either a logic 0 or 1. The result is either a logic 0 or 1 according to the logic operator. A bar above the constant indicates a logical *inversion* of the constant.

**Table 5.10: Logical operations
on constants**

NOT	AND	OR
$\overline{0} = 1$	$0.0 = 0$	$0+0 = 0$
$\overline{1} = 0$	$0.1 = 0$	$0+1 = 1$
	$1.0 = 0$	$1+0 = 1$
	$1.1 = 1$	$1+1 = 1$

Table 5.11 summarizes the logical operations on one variable (A). The operation is performed on the variable alone or on a variable and a constant value. Each variable and constant value can be either a logic 0 or 1. The result is either a logic 0 or 1 according to the logic operator.

**Table 5.11: Logical operations
on one variable**

NOT	AND	OR
$\overline{\overline{A}} = A$	$A.0 = 0$	$A+0 = A$
	$A.1 = A$	$A+1 = 1$
	$A.A = A$	$A+A = A$
	$A.\overline{A} = 0$	$A+\overline{A} = 1$

A bar above the variable indicates a logical inversion of the variable. A double bar indicates a logical inversion followed by another logical inversion. Using the circuit symbol for the NOT gate (the symbol is a triangle with a circle at the end—see Figure 5.8), this effect is shown in Figure 5.6. Logically, a double inversion of a signal has no logical effect.

In practice, the logic gates used to create each of the inversions would create a propagation delay of the value of the variable as it passes through each logic gate. However, a double inversion produces a logic buffer, as shown in Figure 5.7.

The buffer can be used to allow for a signal to drive a large electrical load.

Table 5.12 summarizes the logical operations on two or more variables. Here, two (*A* and *B*) or three variables (*A*, *B*, and *C*) are considered. Each variable value can be either a logic 0 or 1. The result is either a logic 0 or 1 according to the logic operator.

Table 5.12: Logical operations on two or three variables

Commutation Rule	$A+B = B+A$ $A.B = B.A$
Absorption Rule	$A + A.B = A$ $A.(A+B) = A$
Association Rule	$A+(B+C) = (A+B)+C = (A+C)+B = A+B+C$ $A.(B.C) = (A.B).C = (A.C).B = A.B.C$
De Morgan's Theorems	$\overline{A+B} = \overline{A}.\overline{B}$ $\overline{A.B} = \overline{A}+\overline{B}$
Distributive Laws	$A.(B+C) = A.B + A.C$ $A+(B.C) = (A+B).(A+C)$
Minimization Theorems	$A.B + A.\overline{B} = A$ $(A+B).(A+\overline{B}) = A$ $A + \overline{A}.B = A + B$ $A.(\overline{A}+B) = A.B$

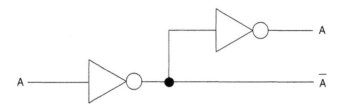

Figure 5.6: Inverting a variable

Figure 5.7: Logic buffer schematic symbol

The commutation rule states that there is no significance in the order of placement of the variables in the expression. The absorption rule is useful for simplifying Boolean expressions, and the association rule allows variables to be grouped together in any order. De Morgan's theorems are widely used in digital logic design as they allow for AND logical operators to be related to NOR logical operators and OR logical operators to be related to NAND logical operators, which allows Boolean expressions to take different forms and thereby be implemented using different logic gates. The distributive laws allow a process similar to factorization in arithmetic, and the minimization theorems allow Boolean expressions to be reduced to a simpler form.

5.3.4 Combinational Logic Gates

Each logic gate that implements the logical operators is represented by a circuit symbol. The commonly used symbols are shown in Figure 5.8. Here, for each logic gate, the inputs are A or A and B, and the output is Z.

An alternative set of logic symbols, IEEE/ANSI standard 91-1984 (*Graphics Symbols for Logic Functions* [7, 8]), is shown in Figure 5.9.

Figures 5.8 and 5.9 use only two-input logic gates for the AND, NAND, OR, and NOR gates, but it is common to use these logic gates with more than two inputs. For example, up to six inputs are available for use in many PLD and ASIC design libraries.

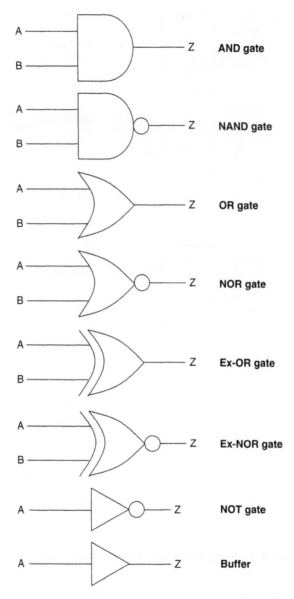

Figure 5.8: Logical operator circuit symbols

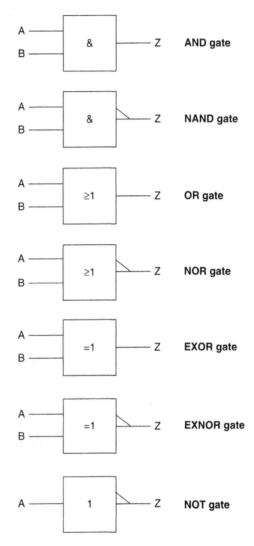

Figure 5.9: Sample IEEE/ANSI standard logic symbols

5.3.5 Truth Tables

The truth table displays the logical operations on input signals in a table format. Every Boolean expression can be viewed as a truth table. The truth table identifies all possible input combinations and the output for each. It is common to create the table so that the input combinations produce an unsigned binary up-count.

The truth table for the AND gate is shown in Table 5.13. Here, the output Z is a logic 1 only when *both* inputs A and B are logic 1.

Table 5.13: AND gate truth table

A	B	Z
0	0	0
0	1	0
1	0	0
1	1	1

The truth table for the NAND gate is shown in Table 5.14. Here, the output Z is a logic 0 only when *both* inputs A and B are logic 1. This is the logical inverse of the AND gate.

Table 5.14: NAND gate truth table

A	B	Z
0	0	1
0	1	1
1	0	1
1	1	0

The truth table for the OR gate is shown in Table 5.15. Here, the output Z is a logic 1 when *either or both* inputs A and B are logic 1.

Table 5.15: OR gate truth table

A	B	Z
0	0	0
0	1	1
1	0	1
1	1	1

The truth table for the NOR gate is shown in Table 5.16. Here, the output Z is a logic 0 when *either or both* inputs A and B are logic 1. This is the logical inverse of the OR gate.

The truth table for the EX-OR gate is shown in Table 5.17. Here, the output Z is a logic 1 when *either but not both* inputs A and B are logic 1.

Table 5.16: NOR gate truth table

A	B	Z
0	0	1
0	1	0
1	0	0
1	1	0

Table 5.17: EX-OR gate truth table

A	B	Z
0	0	0
0	1	1
1	0	1
1	1	0

The truth table for the EX-NOR gate is shown in Table 5.18. Here, the output Z is a logic 0 when *either but not both* inputs A and B are logic 1. This is the logical inverse of the EX-OR gate.

Table 5.18: EX-NOR gate truth table

A	B	Z
0	0	1
0	1	0
1	0	0
1	1	1

The truth table for the NOT gate (inverter) is shown in Table 5.19. This gate has one input only. The output Z is the logical inverse of the input A.

Table 5.19: NOT gate truth table

A	Z
0	1
1	0

The truth table for the BUFFER is shown in Table 5.20. This gate has one input only. The output Z is the same logical value as that of the input A.

Another way to describe a digital circuit or system is by using a suitable HDL such as VHDL [9, 10]. This describes the operation of the circuit or system at different levels

Table 5.20: BUFFER truth table

A	Z
0	0
1	1

of design abstraction. An example VHDL description for each of the basic logic gates using the built-in logical operators in VHDL is shown in Figure 5.10. The syntax and semantics of the language will be provided in Chapter 6. It is sufficient at this point to

```
LIBRARY IEEE;
USE IEEE.STD_LOGIC_1164.ALL;
ENTITY And_Gate IS
     PORT(  A : IN    STD_LOGIC;
            B : IN    STD_LOGIC;
            Z : OUT STD_LOGIC);
END ENTITY And_Gate;
ARCHITECTURE Dataflow OF And_Gate IS
BEGIN
        Z <= A AND B;
END ARCHITECTURE Dataflow;
```
AND gate

Z = (A.B)

```
LIBRARY IEEE;
USE IEEE.STD_LOGIC_1164.ALL;
ENTITY Nand_Gate IS
     PORT(  A : IN    STD_LOGIC;
            B : IN    STD_LOGIC;
            Z : OUT STD_LOGIC);
END ENTITY Nand_Gate;
ARCHITECTURE Dataflow OF Nand_Gate IS
BEGIN
        Z <= A NAND B;
END ARCHITECTURE Dataflow;
```
NAND gate

Z = /(A.B)

```
LIBRARY IEEE;
USE IEEE.STD_LOGIC_1164.ALL;
ENTITY Or_Gate IS
     PORT(  A : IN    STD_LOGIC;
            B : IN    STD_LOGIC;
            Z : OUT  STD_LOGIC);
END ENTITY Or_Gate;
ARCHITECTURE Dataflow OF Or_Gate IS
BEGIN
        Z <= A OR B;
END ARCHITECTURE Dataflow;
```
OR gate

Z = (A+B)

```
LIBRARY IEEE;
USE IEEE.STD_LOGIC_1164.ALL;
ENTITY Nor_Gate IS
     PORT(  A : IN    STD_LOGIC;
            B : IN    STD_LOGIC;
            Z : OUT  STD_LOGIC);
END ENTITY Nor_Gate;
ARCHITECTURE Dataflow OF Nor_Gate IS
BEGIN
        Z <= A NOR B;
END ARCHITECTURE Dataflow;
```
NOR gate

Z = /(A+B)

Figure 5.10: VHDL code examples for the logic gates in Figure 5.8

| ```
LIBRARY IEEE;
USE IEEE.STD_LOGIC_1164.ALL;

ENTITY Xor_Gate IS
    PORT(  A : IN    STD_LOGIC;
           B : IN    STD_LOGIC;
           Z : OUT   STD_LOGIC);
END ENTITY Xor_Gate;

ARCHITECTURE Dataflow OF Xor_Gate IS

BEGIN


    Z <= A XOR B;


END ARCHITECTURE Dataflow;
``` | **EX-OR gate**<br><br>$Z = (A \oplus B)$ |
| ```
LIBRARY IEEE;
USE IEEE.STD_LOGIC_1164.ALL;

ENTITY Xnor_Gate IS
    PORT(  A : IN    STD_LOGIC;
           B : IN    STD_LOGIC;
           Z : OUT   STD_LOGIC);
END ENTITY Xnor_Gate;

ARCHITECTURE Dataflow OF Xnor_Gate IS

BEGIN

    Z <= A XNOR B;
END ARCHITECTURE Dataflow;
``` | **EX-NOR gate**<br><br>$Z = /(A \oplus B)$ |
| ```
LIBRARY IEEE;
USE IEEE.STD_LOGIC_1164.ALL;

ENTITY Not_Gate IS
    PORT(  A : IN    STD_LOGIC;
           Z : OUT   STD_LOGIC);
END ENTITY Not_Gate;

ARCHITECTURE Dataflow OF Not_Gate IS

BEGIN
    Z <= NOT A;
END ARCHITECTURE Dataflow;
``` | **NOT gate**<br><br>$Z = /A$ |
| ```
LIBRARY IEEE;
USE IEEE.STD_LOGIC_1164.ALL;

ENTITY Buffer_Gate IS
    PORT(  A : IN    STD_LOGIC;
           Z : OUT   STD_LOGIC);
END ENTITY Buffer_Gate;

ARCHITECTURE Dataflow OF Buffer_Gate IS

BEGIN
    Z <= A;
END ARCHITECTURE Dataflow;
``` | **Buffer**<br><br>$Z = A$ |

Figure 5.10: (Continued)

note that HDLs exist and for VHDL the basic structure of a VHDL text based description is of the form shown in Figure 5.10.

The EX-OR gate has the Boolean expression:

$$Z = A \oplus B$$

From the truth table for the EX-OR gate, then, a Boolean expression in the *first canonical form* (the *first canonical from* is a set of minterms that are AND logical operators on the variables within the expression with the output of the AND logical operators being logically ORed together) can be written as:

$$Z = (\overline{A}.B) + (A.\overline{B})$$

Therefore, the EX-OR gate can be made from AND, OR, and NOT gates as shown in Figure 5.11.

The truth table can be created to identify the input-output relationship for any logic circuit that consists of combinational logic gates and that can be expressed by Boolean logic. It is therefore possible to move between Boolean logic expressions and truth tables. Consider a three-input logic circuit (A, B, and C) with one output (Z), as shown in the truth table in Table 5.21. The inputs are written as a binary count starting at 0_{10} and incrementing to 7_{10}. The output Z is only a logic 1 when inputs A, B, and C are logic 1. This can be written as a Boolean expression

$$Z = A.B.C$$

Here, where the output Z is a logic 1, the values of inputs A, B, and C are ANDed together. Where a variable is a logic 1, then the variable is used. When the variable is a logic 0, then the inverse (NOT) of the variable is used.

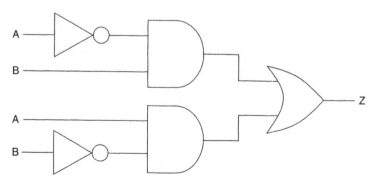

Figure 5.11: EX-OR gate using discrete logic gates

Table 5.21: Three-input logic circuit truth table: Z = A.B.C

A	B	C	Z
0	0	0	0
0	0	1	0
0	1	0	0
0	1	1	0
1	0	0	0
1	0	1	0
1	1	0	0
1	1	1	1

Table 5.22: Three-input logic circuit truth table: Z = NOT (A+B+C)

A	B	C	Z
0	0	0	1
0	0	1	0
0	1	0	0
0	1	1	0
1	0	0	0
1	0	1	0
1	1	0	0
1	1	1	0

Consider now another three-input logic circuit (inputs A, B, and C) with one output (Z), shown in Table 5.22. The inputs are written as a binary count starting at 0_{10} and incrementing up to 7_{10}. The output Z is only a logic 1 when inputs A, B, and C are logic 0. This can be written as a Boolean expression

$$Z = \overline{A}.\overline{B}.\overline{C}$$

Here, where the output Z is a logic 1, the values of inputs A, B, and C are ANDed together. Where a variable is a logic 1, then the variable is used. When the variable is a logic 0, then the inverse (NOT) of the variable is used. The expression identified for the truth table in Table 5.22 can be modified using rules and laws identified in Table 5.12:

$$Z = \overline{A}.\overline{B}.\overline{C}$$

$$Z = \overline{\overline{\overline{A}.\overline{B}.\overline{C}}}$$

$$Z = \overline{\overline{\overline{A}}+\overline{\overline{B}}+\overline{\overline{C}}}$$

$$Z = \overline{A+B+C}$$

The original expression was manipulated by first double-inverting the expression (which logically makes no change), then breaking one of the inversions (the inversion closest in space to the variables) and changing the AND operator to an OR operator (the second De Morgan theorem). This leaves a NOR expression with double-inverted variables. The double-inversion on each input is then dropped.

Now, combining the operations in Table 5.21 and Table 5.22 produces a more complex operation as shown in Table 5.23.

The Boolean expression for this is:

$$Z = (A.B.C) + (\overline{A} + \overline{B} + \overline{C})$$

Each of the ANDed expressions is ORed together. Parentheses group each expression to aid readability of the expression. In this form of expression, the *first canonical form*, a set of minterms (minimum terms) that are AND logical operators are created (one for each line of the truth table where the output is a logic 1). The outputs for each of the AND logical operators are ORed together. This is also referred to as a *sum of products*. A circuit schematic for this circuit is shown in Figure 5.12.

The *second canonical form* is an alternative to the first canonical form. In the second, a set of maxterms that are OR logical operators on the variables within the expression are created (one for each line of the truth table where the output is a logic 0). The outputs for each of the OR logical operators are ANDed together. This is also referred to as a *product of sums*.

Using these approaches, any Boolean logic expression can be described, analyzed, and possibly minimized.

Table 5.23: Three-input logic circuit truth table: complex logic gate

A	B	C	Z
0	0	0	1
0	0	1	0
0	1	0	0
0	1	1	0
1	0	0	0
1	0	1	0
1	1	0	0
1	1	1	1

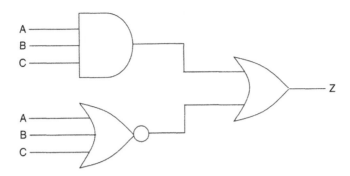

Figure 5.12: Circuit schematic for Boolean expression in Table 5.23

5.4   Combinational Logic Design

5.4.1   Introduction

Using the previous ideas, combinational logic circuits can be combined using either the first canonical form (sum of products) or the second canonical form (product of sums). However, in this text only the first canonical form only will be considered, taking into account logic level 0 or 1 and propagation (time) delays in the cells.

Within a logic gate is an analogue circuit consisting of transistors—either bipolar, using NPN and PNP bipolar junction transistors, or CMOS (complementary metal oxide semiconductor), using n-channel MOS and p-channel MOS transistors. Logic gates in CMOS are of three different circuit architectures at the transistor level [11]: static CMOS, dynamic CMOS, and pass transistor logic CMOS. Today, static CMOS logic is by far the dominant type used. It is built on a network of pMOS and nMOS transistors connected between the power supplies, as shown in Figure 5.13.

The input signals are connected to the gates of the transistors, and the output is taken from the common connection between the transistor networks. The transistors will act as switches, with the switch connections between the drain and source of the transistor. Switch control is via a gate voltage:

- An *nMOS* transistor will be switched ON when high voltage (logic 1) is applied to the transistor gate. Low voltage (logic 0) will turn the switch OFF.

- A *pMOS* transistor will be switched ON when low voltage (logic 0) is applied to the transistor gate. High voltage (logic 1) will turn the switch OFF.

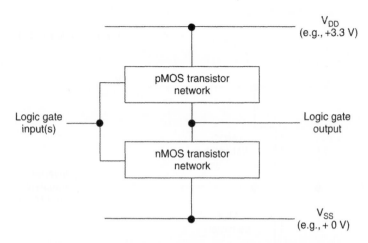

Figure 5.13: Static CMOS logic gate architecture

In the transistor network, a series connection of nMOS transistors will produce an AND effect (i.e., both transistors must be switched ON for the combined effect to be ON). A parallel connection of nMOS transistors will produce an OR effect (i.e., any single transistor must be switched ON for the combined effect to be ON). For the pMOS transistor network, a series connection of nMOS transistors requires a parallel connection of pMOS transistors. A parallel connection of nMOS transistors requires a series connection of pMOS transistors.

The inverter is the most basic logic gate and, in static CMOS, consists of one nMOS and one pMOS transistor. The basic arrangement is shown in Figure 5.14.

The logic gate has both static (DC) and dynamic (time-related) characteristics. Both the voltage (at the different points in the circuit with reference to the common, 0 V, node) and the currents (in particular the power supply current) must be considered.

The static characteristics of the inverter are shown in Figure 5.15; in this case, the static (DC) voltages are not time related. Two graphs are shown. The top graph plots the input voltage (V_{IN}) against the output voltage (V_{OUT}). This shows the operating regions (off, saturation, linear) that each transistor will go through during the input and output voltage changes. A logic 0 is a voltage level of V_{SS}, and a logic 1 is a voltage level of V_{DD}. The bottom graph plots the input voltage (V_{IN}) against the current drawn from the power supply (I_{DD}), showing that the current drawn from the power supply peaks during changes in the input and output voltages.

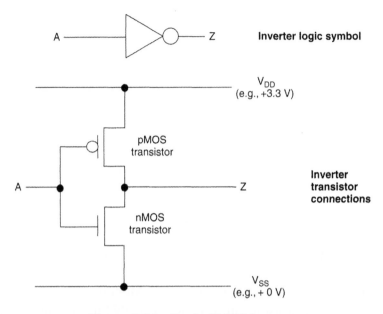

Figure 5.14: Static CMOS inverter

The dynamic characteristics of the inverter are shown in Figure 5.16. These show the operation of the inverter to changes of the inputs and outputs in time. The top graph shows the input test signal, which in this case is a step change for a 0-1-0 logic level change with an instantaneous change in logic value in time. The two voltage levels are V_{OL} and V_{OH}:

- V_{OL} defines the maximum output voltage from the logic gate that would produce a logic 0 output.

- V_{OH} defines the minimum output voltage from the logic gate that would produce a logic 1 output.

- The middle graph shows the output, which changes from a 1 to a 0 and a 0 to a 1 in a finite time. Two values for the propagation time delay are defined, t_{PHL} and t_{PLH}:

  - t_{PHL} defines a propagation time delay from a high level (1) to a low level (0) between the start of the input signal change and the 50 percent change in output.

  - t_{PLH} defines a propagation time delay from a low level (0) to a high level (1) between the start of the input signal change and the 50 percent change in output.

- The bottom graph shows the output, which changes from a 1 to a 0 and a 0 to a 1 in a finite time. Two values for the rise and fall times are defined, t_{FALL} and t_{RISE}:

  o t_{FALL} defines a fall time from a high level (1) to a low level (0) between the 90 percent and 10 percent levels between the high and low levels.

  o t_{RISE} defines a rise time from a low level (0) to a high level (1) between the 10 percent and 90 percent levels between the low and high levels.

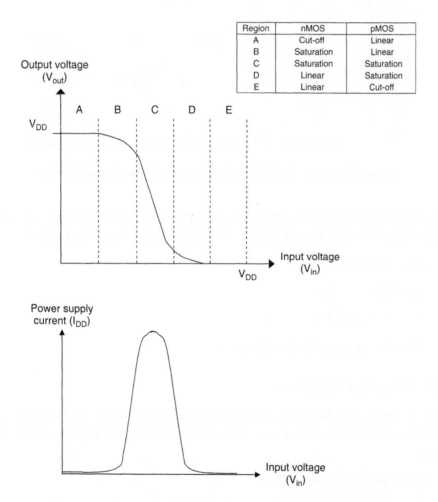

Region	nMOS	pMOS
A	Cut-off	Linear
B	Saturation	Linear
C	Saturation	Saturation
D	Linear	Saturation
E	Linear	Cut-off

Figure 5.15: Static CMOS inverter—static characteristics

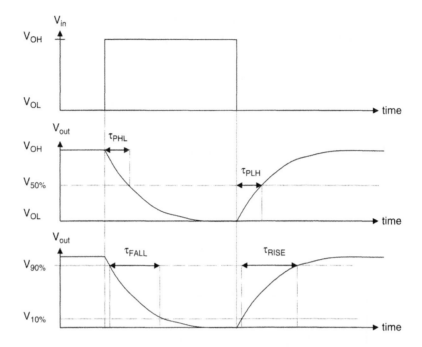

Figure 5.16: Static CMOS inverter—dynamic characteristics

Having considered the static CMOS inverter operation, the logical operation of more complex logic gates will be considered through the following four examples:

1. Two-input multiplexer

2. One-bit half-adder

3. One-bit full-adder

4. Partial odd/even number detector

Example 1: Two-Input Multiplexer

Consider a circuit that has two data inputs (A and B) and one data output (Z). An additional control input, Select, is used to select which input appears at the output, such that:

*   when Select = 0, A → Z

*   when Select = 1, B → Z

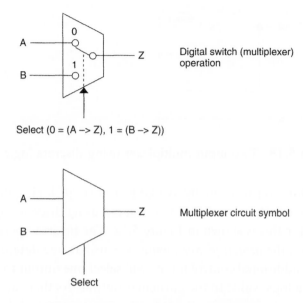

Digital switch (multiplexer)
operation

Select (0 = (A –> Z), 1 = (B –> Z))

Multiplexer circuit symbol

Figure 5.17: Two-input multiplexer

This circuit is the multiplexer, and the circuit symbol is shown in Figure 5.17.

In general, the multiplexer can have as many data inputs as required, and the number of control signals required will reflect the number of data inputs. For the two-input multiplexer, the truth table has three inputs (for eight possible combinations) and the output as shown in Table 5.24. The Boolean expression can be created for this and a reduced form would be:

$$Z = (\overline{Select}.A) + (Select.B)$$

Table 5.24: Two-input multiplexer truth table

Select	A	B	Z
0	0	0	0
0	0	1	0
0	1	0	1
0	1	1	1
1	0	0	0
1	0	1	1
1	1	0	0
1	1	1	1

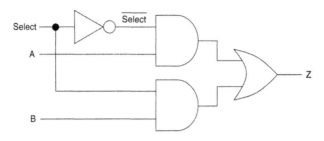

Figure 5.18: Two-input multiplexer using discrete logic gates

Although the multiplexer is normally available as a single logic gate in an ASIC or PLD reference library, the circuit could be created using discrete logic gates. The circuit schematic for this is shown in Figure 5.18. The inverse operation of the *multiplexer* (mux) is the *demultiplexer* (demux). This has one data input and multiple data outputs. The additional control inputs will select one output to be active and will pass the input data logic value to the particular output. As there are multiple outputs in the *demultiplexer*, the remaining outputs (those outputs which have not been selected) will output a logic '0' value.

Example 2: One-Bit Half-Adder

The half-adder is an important logic design created from basic logic gates, as shown in Figure 5.19. This is a design with two inputs (A and B) and two outputs (Sum and Carry-out, Cout). This cell adds the two binary input numbers to produce sum and carry-out terms.

The truth table for this design is shown in Table 5.25.

From viewing the truth table, the Sum output is only a logic 1 when *either but not both* inputs are logic 1:

$$\text{Sum} = (\overline{A}.B) + (A.\overline{B})$$

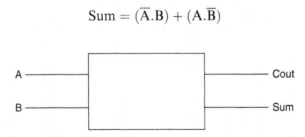

Figure 5.19: One-bit half-adder cell

Table 5.25: One-bit half-adder cell truth table

A	B	Sum	Cout
0	0	0	0
0	1	1	0
1	0	1	0
1	1	0	1

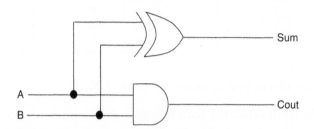

Figure 5.20: One-bit half-adder circuit schematic

This is actually the EX-OR function, so:

$$\text{Sum} = (A \oplus B)$$

From viewing the Cout output in the truth table, the output is logic 1 only when *both* inputs are logic 1 (i.e., A AND B):

$$\text{Cout} = (A.B)$$

This can be drawn as a circuit schematic as shown in Figure 5.20.

Example 3: One-Bit Full-Adder

The full-adder extends the concept of the half-adder by providing an additional carry-in (Cin) input, as shown in Figure 5.21. This is a design with three inputs (A, B, and Cin) and two outputs (Sum and Cout). This cell adds the three binary input numbers to produce sum and carry-out terms.

Figure 5.21: One-bit full-adder cell

Table 5.26: One-bit full-adder cell truth table

A	B	Cin	Sum	Cout
0	0	0	0	0
0	0	1	1	0
0	1	0	1	0
0	1	1	0	1
1	0	0	1	0
1	0	1	0	1
1	1	0	0	1
1	1	1	1	1

The truth table for this design is shown in Table 5.26.

From viewing the truth table, the Sum output is only a logic 1 when *one or three* (but not two) of the inputs is logic 1. The Boolean expression for this is (in reduced form):

$$\text{Sum} = \text{Cin} \oplus (A \oplus B)$$

From viewing the truth table, the Cout output is only a logic 1 when *two or three* of the inputs is logic 1. The Boolean expression for this is (in reduced form):

$$\text{Cout} = (A.B) + (\text{Cin}.(A \oplus B))$$

This can be drawn as a circuit schematic as shown in Figure 5.22.

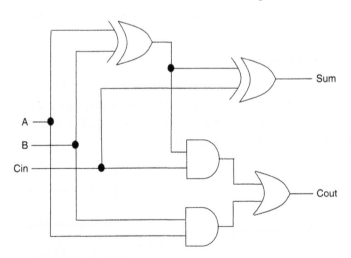

Figure 5.22: One-bit full-adder circuit schematic

Any number of half- and full-adder cells can be connected together to form an *n*-bit addition. Figure 5.23 shows the connections for a four-bit binary adder. In this design, there is no Cin input. Inputs A and B are four bits wide, and bit 0 (A(0) and B(0)) are the LSBs.

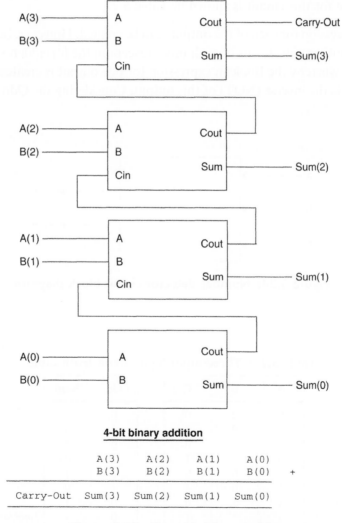

4-bit binary addition

	A(3)	A(2)	A(1)	A(0)	
	B(3)	B(2)	B(1)	B(0)	+
Carry-Out	Sum(3)	Sum(2)	Sum(1)	Sum(0)	

Figure 5.23: Four-bit binary adder

Example 4: Partial Odd/Even Number Detector

Consider a circuit that receives a three-bit unsigned binary number (A, B, and C where A is the MSB and C is the LSB) and is to detect when the number is ODD or EVEN. The circuit will have two outputs (Odd and Even), as shown in Figure 5.24. The Odd output is a logic 1 when the input number (in decimal) is 1, 3, or 5 *but not* 7. The input 7 is to be considered a forbidden input in this circuit. The Even output is a logic 1 when the input number (in decimal) is 0, 2, 4, 6.

The truth table for this circuit is shown in Table 5.27.

A Boolean expression for each of the outputs can be created. However, because the Odd and Even outputs are inversions of each other (except in the forbidden state), a circuit can be created whereby the Boolean expression for one output is created and the second output is the inverse (NOT) of this output. Considering the Odd output (with

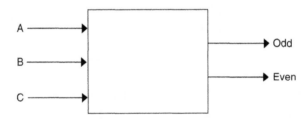

Figure 5.24: Number detector circuit block diagram

Table 5.27: Three-input logic circuit truth table

A	B	C	Odd	Even	
0	0	0	0	1	
0	0	1	1	0	
0	1	0	0	1	
0	1	1	1	0	
1	0	0	0	1	
1	0	1	1	0	
1	1	0	0	1	
1	1	1	0	1	*Forbidden input*

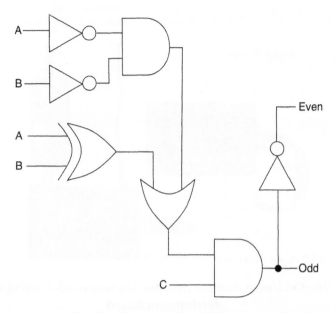

Figure 5.25: Circuit schematic for odd/even number detector

three 1s, compared to four in the Even output, making it a smaller Boolean expression), then the Boolean expression for the Odd and Even outputs would be:

$$Odd = (\overline{A}.\overline{B}.C) + (\overline{A}.B.C) + (A.\overline{B}.C)$$

$$Odd = (\overline{A}.\overline{B}.C) + C.((\overline{A}.B) + (A.\overline{B}))$$

$$Odd = (\overline{A}.\overline{B}.C) + C.(A \oplus B)$$

$$Odd = C.((\overline{A}.\overline{B}) + (A \oplus B))$$

$$Even = \overline{Odd}$$

The circuit schematic for this design is shown in Figure 5.25.

A problem with this circuit is that when the odd number input 7 is applied, the circuit produces a logic 0 on Odd and a logic 1 on Even, which is incorrect. If this circuit is to be used, then the input 7 must be taken into account and the circuit redesigned, or the input 7 must never be applied by design.

If the input 7 is considered in the creation of the Boolean logic expression for the Odd output, then the logic for the Odd output simply becomes the value for

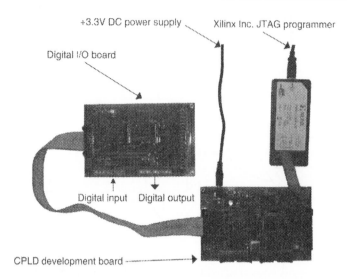

Figure 5.26: Odd/even number detector implementation using the CPLD development board

the C input. This design can be implemented in the CPLD development system (refer to Chapter 3 and Appendix F—See the last paragraph of the Preface for instructions regarding how to access this online content) using a CPLD development board and a digital I/O board.

The basic arrangement is shown in Figure 5.26. Here, the Coolrunner™-II CPLD on the CPLD development board is configured with the digital logic circuit, and the digital I/O board is interfaced to external test and measurement equipment. The CPLD is configured using the pins identified in Table 5.28.

Table 5.28: Odd/even number detector CPLD pin assignment

Signal name	CPLD pin number	Digital I/O board identifier	Comment
A	13	B0 (input bit 0)	CPLD input, design A input
B	14	B1 (input bit 1)	CPLD input, design B input
C	15	B2 (input bit 2)	CPLD input, design C input
Odd	3	A0 (output bit 0)	CPLD output, design Odd output
Even	4	A1 (output bit 1)	CPLD output, design Even output
Input buffer enable	12	OE2 (input enable)	CPLD output, tie to logic 0 in CPLD design
Output buffer enable	2	OE1(output enable)	CPLD output, tie to logic 0 in CPLD design

Here, the five design I/Os are defined and are connected to the relevant CPLD pins to connect to Header A for the digital I/O board. In addition, the CPLD design must also incorporate two additional outputs to enable the tristate buffers used on the digital I/O board. Here, the enable (OE, output enable) pins on the tristate buffers must be tied to logic 0 to enable the buffers.

External circuitry is connected to the digital I/O board to provide the logic levels for inputs A, B, and C and to monitor the outputs Odd and Even where:

- logic 0 = 0 V

- logic 1 = +3.3 V

The CPLD is programmed from using an appropriate JTAG (Joint Test Action Group) programmer.

5.4.2   NAND and NOR logic

Logical operations using AND, OR, and NOT logic gates can also be undertaken using either NAND or NOR logic gates. A Boolean expression using AND, OR, and NOT logic can be manipulated to produce NAND and NOR logic. For example, the Boolean expression:

$$Z = (A.B)$$

can also be expressed as:

$$Z = \overline{(\overline{A} + \overline{B})}$$

Figure 5.27 shows the two logic gate implementations for these Boolean expressions.

Similarly, the Boolean expression:

$$Z = (A + B)$$

can also be expressed as:

$$Z = \overline{(\overline{A} . \overline{B})}$$

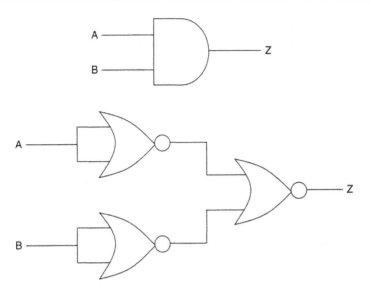

Figure 5.27: NOR implementation for the AND gate

Figure 5.28 shows the two logic gate implementations for these Boolean expressions.

If only NAND and NOR gates are available, any Boolean logic expression can be implemented through such manipulation.

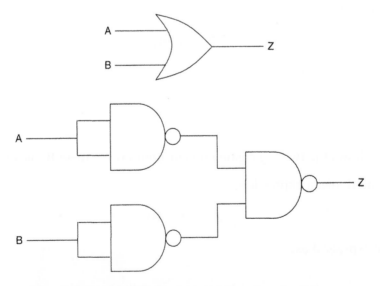

Figure 5.28: NAND implementation for the OR gate

5.4.3 Karnaugh Maps

The Karnaugh map (or K-map) provides a means to display logical operations on input signals as a map showing the output values for each of the input values. This allows groups of logic values to be looped together with suitably sized loops to minimize the resulting Boolean logic expression. The size of the Karnaugh map depends on the number of inputs to the combinational logic circuit. Karnaugh maps for two-, three-, and four-input circuits are shown in Figure 5.29:

- A *two-input Karnaugh map* contains four cells, one cell for each possible input combination (2^n where n is the number of inputs). Here, the inputs are named A and B.

- A *three-input Karnaugh map* contains eight cells, one cell for each possible input combination (2^n where n is the number of inputs). Here, the inputs are named A, B, and C.

- A *four-input Karnaugh map* contains sixteen cells, one cell for each possible input combination (2^n where n is the number of inputs). Here, the inputs are named A, B, C, and D.

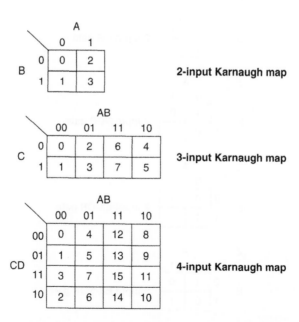

Figure 5.29: Two-, three-, and four-input Karnaugh maps

The Karnaugh map has a direct correspondence with the truth table for a Boolean logic expression. Each K-map cell is filled with the logic value of the output (0 or 1) for the corresponding input combination. In Figure 5.29, the cells are filled with (for reference purposes) the decimal number equivalent for the unsigned binary value of the input combination (A is the MSB of the binary input value). Note the values and locations of the values within the cells.

The Karnaugh maps for the two-input logic gates in Figure 5.8 are shown in Figure 5.30.

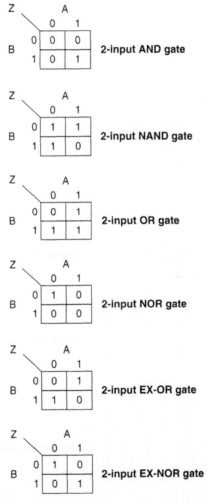

Figure 5.30: Truth table for two-input logic gates

The Karnaugh map can then be analyzed, and loops of output logic levels within the cells can be created. In the first canonical form, logic 1s are grouped together. In the second canonical form, logic 0s are grouped together. In this text, the first canonical form will be considered, so in the Karnaugh map, logic 1s are grouped together.

The larger the loop, the smaller the resulting Boolean logic expression (with fewer variables to be considered). The variables in the loop will be ANDed together, and each group will be ORed together.

For a *two-input* Karnaugh map, then:

- A group of one logic 1 will result in the ANDing of two variables.

- A group of two logic 1s will result in one variable.

- A group of four logic 1s will result in a constant logic 1.

For a *three-input* Karnaugh map, then:

- A group of one logic 1 will result in the ANDing of three variables.

- A group of two logic 1s will result in the ANDing of two variables.

- A group of four logic 1s will result in one variable.

- A group of eight logic 1s will result in a constant logic 1.

For a *four-input* Karnaugh map, then:

- A group of one logic 1 will result in the ANDing of four variables.

- A group of two logic 1s will result in the ANDing of three variables.

- A group of four logic 1s will result in the ANDing of two variables.

- A group of eight logic 1s will result in one variable.

- A group of sixteen logic 1s will result in a constant logic 1.

Consider the two-input AND gate: it has only one logic 1, so only a loop of 1 can be created, as shown in Figure 5.31. Where the input variable is a logic 1, the variable is used. When the input variable is a logic 0, the inverse (NOT) of the variable is used.

Consider now the two-input OR gate: it has three logic 1s, two loops of two can be created, as shown in Figure 5.32. Where the input variable is a logic 1, the variable is used. When the input variable is a logic 0, the inverse (NOT) of the variable is used.

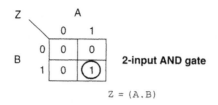

Figure 5.31: Two-input AND gate

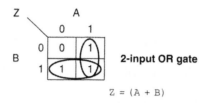

Figure 5.32: Two-input OR gate

When a group of two is created, one of the variables can be a logic 0 or a logic 1 and so can be dropped from the resulting Boolean logic expression. The vertical group of two retains the variable A but drops the variable B. The horizontal group of two retains the variable B but drops the variable A.

The grouping of logic 1s follows the following rules:

1. Loops of 2^n adjacent cells can be made where n is an integer number starting at 0.

2. All cells containing a 1 (first canonical form; or 0 in the second canonical form) must be covered.

3. Loops can overlap provided they contain at least one unlooped cell.

4. Loops must be square or rectangular (diagonal or L-shaped loops are not permitted).

5. Any loop that has all of its cells included in other loops is redundant.

6. The edges of a map are considered to be adjacent—a loop can leave the side of the Karnaugh map and re-enter at the other side, or leave from the top of the Karnaugh map and return at the bottom, as shown in Figure 5.33.

One potential problem with combinational logic arises from hazards. Here, because of the finite time for a signal change to propagate through the combinational logic (due to any logic gate delays and interconnect delays), there is potential for erroneous

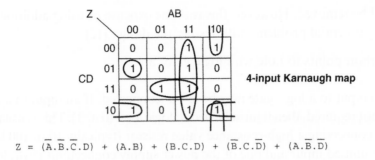

$$Z = (\overline{A}.\overline{B}.\overline{C}.D) + (A.B) + (B.C.D) + (\overline{B}.C.\overline{D}) + (A.\overline{B}.\overline{D})$$

Figure 5.33: Adjacent cells in a Karnaugh map

output during the time that the change occurs. This results from different time delays in different paths within the combinational logic. Although the final output would be correct, an erroneous output (i.e., wrong logic level) can occur during the change, which would cause problems if detected and used.

If the digital circuit or system can be designed so that the output from the combinational logic with a hazard is only used after it is guaranteed correct, then the hazard, although not eliminated, will not cause a problem in the design.

A way to eliminate hazards using the Karnaugh map is to ensure that all loops are joined together. Although this will introduce a redundant term (see Figure 5.34), the

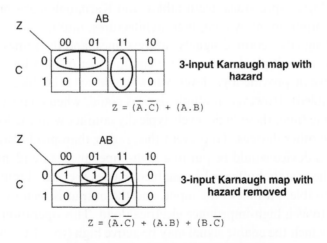

$$Z = (\overline{A}.\overline{C}) + (A.B)$$

$$Z = (\overline{A}.\overline{C}) + (A.B) + (B.\overline{C})$$

Figure 5.34: Eliminating hazards

hazard will be removed. However, this is at the expense of using additional logic and introducing potential problems with testing the design [12].

Two important points to note with logic gates are:

1. No input to a logic gate may be left unconnected. If an input to a logic gate is not required, then it must be tied to logic level (0 or 1). This is usually achieved by connecting a high-resistance value resistor (typically 10 to 100 kΩ) between the unused input and one of the power supply connections (V_{DD} for logic 1, V_{SS} or GND for logic 0). In some ICs, specific inputs might be designed for use only under specific circumstances and with a pull-up (to logic 1) or pull-down (to logic 0) component integrated into the IC circuitry. Such integrated pull-up or pull-down components alleviate the need for the designer to place resistors on the PCB and so reduce the PCB design requirements.

2. Where a logic gate only produces a logic 0 or 1 output, then no two or more logic gate outputs are to be connected unless the implementation technology (circuitry within the logic gate) allows it. Certain logic gate outputs can be put into a high-impedance state, which stops the output from producing a logic output and instead turns the output into a high-impedance electrical load. Circuits with a high-impedance output are used where multiple devices are to be connected to a common set of signals (a bus) such as a microprocessor data bus.

Whereas the previous logic gates considered in the design of digital circuits using Boolean logic expressions, truth tables, and Karnaugh maps provided only a logic 0 or 1 output, in many computer architectures, multiple devices share a common set of signals—control signals, address lines, and data lines. In a computer architecture where multiple devices share a common set of data lines, these devices can either receive or provide logic levels when the device is enabled (and all other devices are disabled). However, multiple devices could, when enabled, provide logic levels at the same time; these logic levels typically conflict with the logic levels provided by the other devices. To prevent this, rather than producing a logic level when disabled, a device would be put in a high-impedance state (denoted by the character z). The tristate buffer, when enabled, passes the logic input level to the output; when disabled, it blocks the input, and the output is seen by the circuit that it is connected to as a high-impedance electrical load. This operation is shown in Figure 5.35, in which the enable signal may be active high (top, 1 to enable the buffer) or active low (bottom, 0 to enable the buffer).

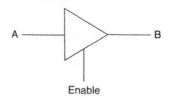

Enable	A	B
0	0	Z
0	1	Z
1	0	0
1	1	1

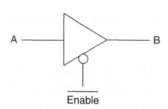

$\overline{\text{Enable}}$	A	B
0	0	0
0	1	1
1	0	Z
1	1	Z

The bar above the *Enable* input name
indicates that the input is active low

Figure 5.35: Tristate buffer symbol

5.4.4 *Don't Care* Conditions

In some situations, certain combinations of input might not occur, so the designer could consider that these conditions are not important. They are referred to as *Don't care* conditions. As such, the output in these conditions could be either a logic 0 or a logic 1, so the designer is free to choose the output value that results in the simpler output logic (i.e., using fewer logic gates).

5.5 Sequential Logic Design

5.5.1 Introduction

Sequential logic circuits are based on combinational logic circuit elements (AND, OR, etc.) working alongside sequential circuit elements (latches and flip-flops). A generic sequential logic circuit is shown in Figure 5.36. Here, the circuit inputs are applied to and the circuits outputs are derived from a combinational logic block. The sequential logic circuit elements store an output from the combinational logic that is fed back to the combinational logic input to constitute the present state of the circuit. The output from the combinational logic that forms the inputs to the sequential logic circuit elements constitutes the next state of the circuit. These sequential logic circuit

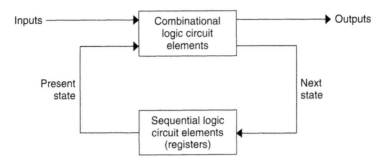

Figure 5.36: Generic sequential logic circuit (counter or state machine)

elements are grouped together to form registers. The circuit changes state from the present state to the next state on a clock control input (as happens in a synchronous sequential logic circuit). Commonly the D-latch and D-type flip-flop are used (rather than other forms of latch and flip-flop such as the S-R, toggle, and J-K flip-flops), and they will be discussed in this text. The output from the circuit is taken from the output of the combinational logic circuit block.

In general, sequential logic circuits may be asynchronous or synchronous:

1. **Asynchronous sequential logic**. This form of sequential logic does not use a clock input signal to control the timing of the circuit. It allows very fast operation of the sequential logic, but its operation is prone to timing problems where unequal delays in the logic gates can cause the circuit to operate incorrectly.

2. **Synchronous sequential logic**. This form of sequential logic uses a clock input signal to control the timing of the circuit. The timing of changes in states in the sequential logic is designed to occur either on the edge of the clock input when flip-flops are used, or at a particular logic level, as when latches are used. State changes that occur on the edge of the clock input, as when flip-flops are used, occur either on a 0 to 1 rise, referred to as positive edge triggered, or on a 1 to 0 fall, referred to as negative edge triggered.

In this text, only synchronous sequential logic will be considered.

An alternative view for the generic sequential logic circuit in Figure 5.36, is shown in Figure 5.37. Here, the combinational logic is separated into input and output logic. Both views are commonly used in the description of sequential logic circuits.

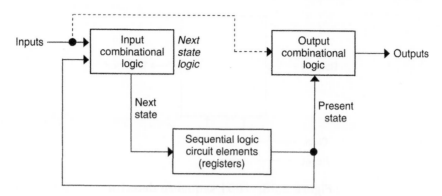

Figure 5.37: Alternative view for the generic sequential logic circuit

In designing the synchronous sequential logic circuit (from now on simply referred to as the sequential logic circuit), the designer must consider both the type of sequential logic circuit elements (latch or flip-flop) and the combinational logic gates. The design uses the techniques previously discussed—Boolean logic expressions, truth tables, schematics, and Karnaugh maps—to determine the required input combinational logic (the next state logic) and determine the required output combinational logic.

The sequential logic circuit will form one of two types of machines:

1. In the **Moore machine**, the outputs are a function only of the present state only.

2. In the **Mealy machine**, the outputs are a function of the present state and the current inputs.

In addition, the sequential logic circuit will be designed either to react to an input or to be autonomous. In an autonomous sequential logic circuit, there are no inputs (apart from the clock and reset/set) to control the operation of the circuit, so the circuit moves through states under the control of only the clock input. An example of an autonomous sequential logic circuit is a straight binary up-counter that moves through a binary count sequence taking the outputs directly from the sequential logic circuit element outputs. A sequential logic circuit can also be designed to react to an input: a sequential logic circuit that reacts to an input is called a state machine in this text.

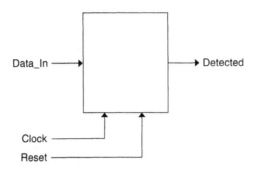

Figure 5.38: 1001 sequence detector

Sequential logic circuit design follows a set design sequence aided by:

- **state transition diagram**, which provides a graphical means to view the states and the transitions between states

- **state transition table**, similar in appearance to a combinational logic truth table, which identifies the current state outputs and the possible next state inputs to the sequential logic circuit elements.

As an example, consider a circuit that is to detect the sequence 1001 on a serial bit-stream data input and produce a logic 1 output when the sequence has been detected, as shown in Figure 5.38. The state machine will have three inputs—one Data_In that is to be monitored for the sequence and two control inputs, Clock and Reset—and one output, Detected. Such a state machine could be used in a digital combinational lock circuit.

An example *state transition diagram* for this design is shown in Figure 5.39. The circuit is to be designed to start in *State 0* and has five possible states. With these five states, if D-type flip-flops are to be used, then there will need to be a need for three flip-flops (producing eight possible states although only five will be used when each state is to be represented by one value of a straight binary count sequence 0, 1, 2, 3, 4, 0, etc.). The arrangement for the *state transition diagram* is:

1. The *circles* identify the states. The name of the state (the *state identifier*) and the outputs for each state are placed within the circle. Each state is referred to as a *node*.

2. The transition between states uses a *line* with the arrow end identifying the direction of movement. Each line starts and ends at a node.

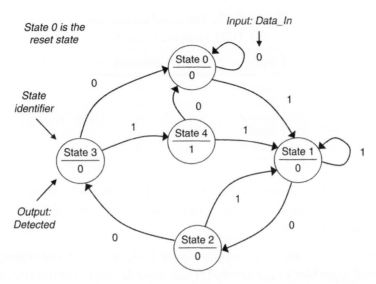

Figure 5.39: "1001" sequence detector state transition diagram (Moore machine)

3. Each line is accompanied by an *identifier* that identifies the logical value of the input (here *Data_In*) that controls the state machine to go to the next particular state.

This form of the state transition diagram is for a *Moore machine* and in this form the outputs for each state are identified within the circles. The alternative to the *Moore machine* is the *Mealy machine*. In the *Mealy machine*, the outputs for a particular state are identified on the lines connecting the states along with the *identifier*.

The state transition table (also referred to as a present state/next state table) for the 1001 sequence detector state diagram is shown in Table 5.29. Each possible input

Table 5.29: State transition table for the 1001 sequence detector

	Data_In = 0	Data_In = 1
Present state	Next state	Next state
State 0	State 0	State 1
State 1	State 1	State 2
State 2	State 3	State 1
State 3	State 0	State 4
State 4	State 0	State 1

Table 5.30: Detected output for the 1001 sequence detector

State	Detected
State 0	0
State 1	0
State 2	0
State 3	0
State 4	1

condition has its own column, and each row contains the present state and the next state for each possible input condition. The Detected output is defined in the truth table shown in Table 5.30.

Using the circuit architecture shown in Figure 5.37, the input and output combinational logic blocks are created. Each state is created using the outputs from the sequential logic circuit element block. Flip-flops form a register whose outputs produce a binary value that defines one of the states. It is common to create the states as a straight binary count. Using n-flip-flops, 2^n states are possible in the register output. However, any count sequence could be used. For example, one-hot encoding uses n-flip-flops to represent n states. In the one-hot encoding scheme, to change from one state to the next, only two flip-flop outputs will change (the first from a 1 to a 0, and the second from a 0 to a 1). The advantage of this scheme is less combinational logic to create the next state values.

5.5.2 Level Sensitive Latches and Edge-Triggered Flip-Flops

The two sequential logic circuit elements are the latch and the flip-flop. These elements store a logic value (0 or 1). The basic latches and flip-flops are:

Latches:

- D-latch

- S-R latch (set-reset latch)

Flip-flops:

- S-R flip-flop (set-reset flip-flop)

- J-K flip-flop

- T-flip-flop (toggle flip-flop)

- D-type flip-flop

Each latch and flip-flop has its own particular characteristics and operation requirements. In this text, only the D-latch and the D-type flip-flop will be considered.

5.5.3   The D Latch and D-Type Flip-Flop

The basic D latch circuit symbol, shown in Figure 5.40, includes two inputs, the data input (D, value to store) and the control input (C). There is one output (Q).

In the D latch, when the C input is at a logic 1, the Q output is assigned the value of the D input. When the C input is a logic 0, the Q output holds its current value even when the D input changes. In addition, many D latches also include a logical inversion of the Q output (the NOT-Q output) as an additional output.

Latches are normally designed as part of normal circuit operation. However, a problem can occur when writing HDL code in that a badly written design will create unintentional latches. When a design description is synthesized, the synthesis tool will *infer* latches. In VHDL, two common coding mistakes that result in inferred latches are:

- an *If* statement without an *Else* clause

- a register description without a clock rising or falling edge construct

An example of an *If* statement without an *Else* clause in VHDL is shown in Figure 5.41. Here, a circuit has two input signals (Data_In and Enable) and one output signal (Data_Out). The output is the logical value of the Data_In input when the Enable input is a logic 1. The operation of the circuit is defined on lines 20 to 30 of the code.

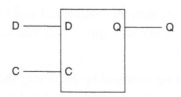

Figure 5.40: D latch

```
1    ------------------------------------------------
2
3    LIBRARY IEEE;
4    USE IEEE.STD_LOGIC_1164.ALL;
5
6    ------------------------------------------------
7
8    ENTITY Inferred_Latch is
9        PORT ( Enable   : IN   STD_LOGIC;
10               Data_In  : IN   STD_LOGIC;
11               Data_Out : OUT  STD_LOGIC);
12   END Inferred_Latch;
13
14   ------------------------------------------------
15
16   ARCHITECTURE Behavioural OF Inferred_Latch IS
17
18   BEGIN
19
20   Enable_Process: PROCESS (Enable, Data_In)
21
22   BEGIN
23
24       IF (Enable = '1') THEN
25
26             Data_Out <= Data_In;
27
28         END IF;
29
30   END PROCESS Enable_Process;
31
32   END Behavioural;
33
34   ------------------------------------------------
```

Figure 5.41: *If* statement without an *Else* clause

The RTL schematic for this design, as synthesized and viewed as a schematic within the Xilinx® ISE™ tools, is shown in Figure 5.42. The latch is the LD symbol in the middle of the schematic view.

This unintentional latch can be removed by including the *Else* clause, as shown in Figure 5.43. Here, when the Enable input is a 0, then the Data_Out output is a logic 0 also. The schematic for this design, as synthesized and viewed as a RTL schematic

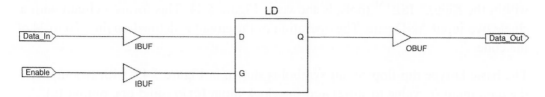

Figure 5.42: Schematic of inferred latch design

```
1       ------------------------------------------------
2
3       LIBRARY IEEE;
4       USE IEEE.STD_LOGIC_1164.ALL;
5
6       ------------------------------------------------
7
8       ENTITY Inferred_Latch is
9           PORT ( Enable   : IN    STD_LOGIC;
10                  Data_In  : IN    STD_LOGIC;
11                  Data_Out : OUT   STD_LOGIC);
12      END Inferred_Latch;
13
14      ------------------------------------------------
15
16      ARCHITECTURE Behavioural OF Inferred_Latch IS
17
18      BEGIN
19
20      Enable_Process: PROCESS(Enable, Data_In)
21
22      BEGIN
23
24          IF (Enable = '1') THEN
25
26                  Data_Out <= Data_In;
27
28          ELSE
29
30                  Data_Out <= '0';
31
32          END IF;
33
34      END PROCESS Enable_Process;
35
36      END Behavioural;
37
38      ------------------------------------------------
```

Figure 5.43: *If* statement with an *Else* clause

within the Xilinx® ISE™ tools, is shown in Figure 5.44. This forms a circuit with a single two-input AND gate. The operation of the circuit is defined on lines 20 to 34 of the code.

The basic D-type flip-flop circuit symbol is shown in Figure 5.45, with two inputs—the data input (D, value to store) and the clock input (CLK)—and one output (Q).

In the D-type flip-flop, when the CLK input changes from a 0 to a 1 (positive edge triggered) or from a 1 to a 0 (negative edge triggered), the Q output is assigned the value of the D input. When the CLK input is steady at a logic 0 or a 1, the Q output holds its current value even when the D input changes.

It is common, however, for the flip-flop to have a reset or set input to initialize the output Q to either logic 0 (reset) or logic 1 (set). This reset/set input can either be asynchronous (independent of the clock input) or synchronous (occurs in a clock edge) and active high (reset/set occurs when the signal is a logic 1) or active low (reset/set occurs when the signal is a logic 0). The circuit symbol for the D-type flip-flop with active low reset is shown in Figure 5.46.

The circle on the reset input indicates an active low reset: if no circle is used, then the flip-flop is active high reset. Many D-type flip-flops also include a logical inversion of the Q output (the NOT-Q output) as an additional output. The circuit symbol for the D-type flip-flop with a NOT-Q output is shown in Figure 5.47.

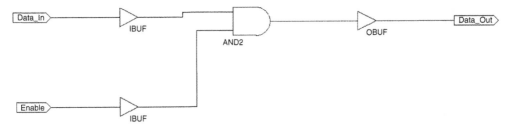

Figure 5.44: Schematic of the design with an *Else* clause

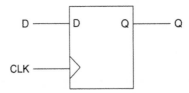

Figure 5.45: D-type flip-flop

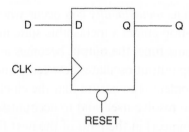

Figure 5.46: D-type flip-flop with active low reset

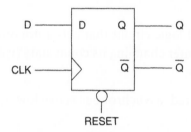

Figure 5.47: D-type flip-flop with active low reset and not-Q output

When a D-input change is to be stored in the flip-flop, specific timing requirements must be considered for the inputs of the flip-flop in both set-up (how long *before* the clock input must the D input be static?) and hold (how long *after* the clock input must the D input be static?). This is shown in Figure 5.48. If these times are violated, then problems with the flip-flop operation will occur.

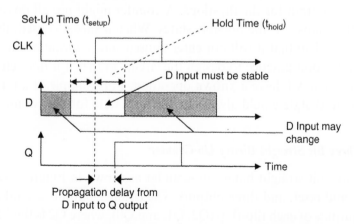

Figure 5.48: D-type flip-flop set-up and hold times

A potential problem known as *metastability* can occur when flip-flop set-up and hold times are violated. The flip-flop enters a metastable state in which the output is unpredictable until, after some time, the output becomes a logic 0 or 1. In the metastable state, the flip-flop output oscillates between 0 and 1. A simple way to design a circuit that avoids this problem is to ensure that the clock period is long enough to allow the metastable state to resolve itself and to account for signal delays (resulting from logic gates and interconnect) in the path of the next flip-flop in the circuit.

5.5.4   Counter Design

The counter is a sequential logic circuit that acts autonomously to perform the functions of a number counter changing its count state (value) on a clock edge. In the following discussions, then:

- Positive edge triggered, asynchronous active low or high D-type flip-flops will be used.

- All flip-flops will have a common reset input.

- All flip-flops will have a common clock input.

In addition, the output from the counter either can be taken directly from the Q outputs of each flip-flop, or can be decoded using output combinational logic to form specific outputs for specific states of the counter.

Because the counter uses flip-flops, for *n*-flip-flops, there will be 2^n possible combinations of output for the flip-flops. A counter might use all possible states or might use only a subset of the possible states. When a subset is used, the counter should be designed so that it will not enter unused states during normal operation. In addition, it is good practice to design the circuit so that if it does enter one of the unused states, it will have a known operation. For example, if an unused state is entered, the next state would always be the reset state for the counter.

Example 1: Three-Bit Straight Binary Up-Counter

Consider a three-bit straight binary up-counter as shown in Figure 5.49, using two inputs, clock and reset, and three outputs. The counter outputs are taken directly from the Q outputs of each flip-flop (Q2, Q1, and Q0), where Q2 is the MSB and Q0 is the LSB.

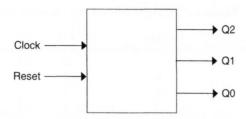

Figure 5.49: Three-bit straight binary up-counter

The design process begins by creating the state transition diagram (Figure 5.50) and the state transition table. The counter is designed to reset (i.e., when reset is a logic 0) to a count of 000_2 (0_{10}), which will be state 0. When the reset is removed (i.e., when reset becomes a logic 1), then the counter will count through the sequence 0, 1, 2, 3, 4, 5, 6, 7, 0, etc. This means that when the counter output reaches 111_2 (state 7), it will automatically wrap around back to 000_2.

Each state in the counter will be encoded by the Q outputs of the D-type flip-flops, as shown in Table 5.31, so that it produces the required straight binary count sequence.

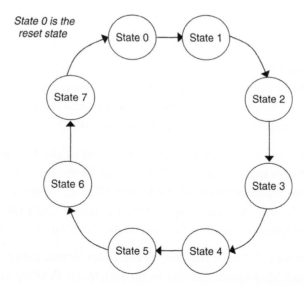

Figure 5.50: Three-bit straight binary up-counter state transition diagram

Table 5.31: Three-bit straight binary up-counter state encoding

State	Q2	Q1	Q0
State 0	0	0	0
State 1	0	0	1
State 2	0	1	0
State 3	0	1	1
State 4	1	0	0
State 5	1	0	1
State 6	1	1	0
State 7	1	1	1

Table 5.32: Three-bit straight binary up-counter state transition table

Present state				Next state		
State name	Current Q outputs			Current D inputs		
	Q2	Q1	Q0	D2	D1	D0
State 0	0	0	0	0	0	1
State 1	0	0	1	0	1	0
State 2	0	1	0	0	1	1
State 3	0	1	1	1	0	0
State 4	1	0	0	1	0	1
State 5	1	0	1	1	1	0
State 6	1	1	0	1	1	1
State 7	1	1	1	0	0	0

The state transition table for the counter can then be created (Table 5.32). For the next state logic, the Q output for each flip-flop in the next state is actually the D input for each flip-flop in the current state. In this view of the state transition table, the current Q outputs and the current D inputs (next state Q outputs) are defined.

The Boolean logic expression can be created for each of the D inputs so that the counter of the form shown in Figure 5.51 is created. Here, the next state logic for each flip-flop (Dff*n*) uses a combination of the Q and NOT-Q outputs from each flip-flop. Manipulation of the Boolean logic expression, the use of truth tables, and Karnaugh maps allow the designer to create a Boolean logic expression of a required form.

An example of the logic for each flip-flop D input developed using the Karnaugh map is shown in Figure 5.52. Figure 5.53 shows the schematic developed for the counter in which each D-type flip-flop only has a Q output, and the NOT-Q output is created

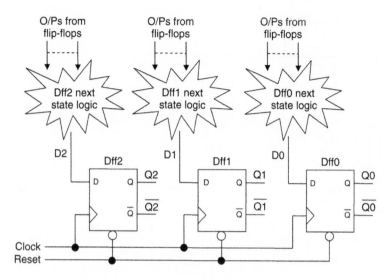

Figure 5.51: Three-bit counter structure

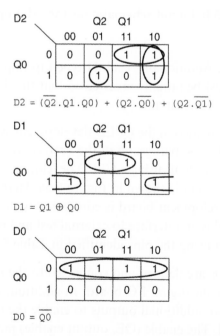

$$D2 = \overline{(Q2.Q1.Q0)} + (Q2.\overline{Q0}) + (Q2.\overline{Q1})$$

$$D1 = Q1 \oplus Q0$$

$$D0 = \overline{Q0}$$

Figure 5.52: Three-bit up-counter D-input Boolean expressions

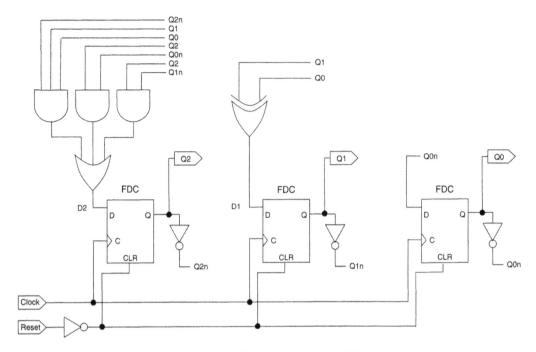

Figure 5.53: Circuit schematic for three-bit up-counter

using a discrete inverter. Additionally, each D-type flip-flop has an asynchronous active high reset that must be initially inverted so that the design reset input sees an asynchronous active low reset circuit.

This design can be implemented in the CPLD development system (refer to Chapter 3 and Appendix F—See the last paragraph of the Preface for instructions regarding how to access this online content) using a CPLD development board and a digital I/O board. The basic arrangement is shown in Figure 5.54. Here, the Coolrunner^{TM}-II CPLD on the CPLD development board is configured with the digital logic circuit, and the digital I/O board is an interface to external test and measurement equipment. The CPLD is configured using the pins identified in Table 5.33.

Here, the five design I/Os are defined and connected to the relevant CPLD pins to connect to Header A for the digital I/O board. In addition, the CPLD design must also incorporate two additional outputs to enable the tristate buffers used on the digital I/O board. The enable (OE, output enable) pins on the tristate buffers must be tied to logic 0 to enable the buffers.

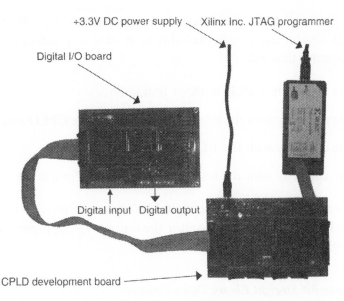

Figure 5.54: Three-bit up-counter implementation using the CPLD development board

Table 5.33: Three-bit up-counter CPLD pin assignment

Signal name	CPLD pin number	Digital I/O board identifier	Comment
Clock	13	B0 (input bit 0)	CPLD input, design Clock input
Reset	14	B1 (input bit 1)	CPLD input, design Reset input
Q0	3	A0 (output bit 0)	CPLD output, design Q0 output
Q1	4	A1 (output bit 1)	CPLD output, design Q1output
Q2	5	A2 (output bit 2)	CPLD output, design Q2 output
Input buffer enable	12	OE2 (input enable)	CPLD output, tie to logic 0 in CPLD design
Output buffer enable	2	OE1 (output enable)	CPLD output, tie to logic 0 in CPLD design

An alternative to using the digital I/O board for the clock and reset inputs is to use the clock and reset inputs available on the CPLD development board:

- **Clock**—a 50 MHz clock connected to CPLD pin 38

- **Reset**—a reset push switch (SW1) that produces a low signal when activated (pressed) via the MAX811-S voltage monitor IC connected to CPLD pin 143.

This removes the need to generate external clock and reset inputs. If the 50 MHz clock is used, a clock divider circuit will probably be needed to produce a lower frequency clock to the counter that will:

- operate within the CPLD without timing errors

- operate correctly with the interconnect between the CPLD and the I/O buffers

- operate correctly with the I/O buffers

- operate correctly with the test and measurement equipment used

External circuitry is connected to the digital I/O board to provide the logic levels for inputs clock and reset, and to monitor the outputs Q2, Q1, and Q0 where logic $0 = 0$ V and logic $1 = +3.3$ V.

Example 2: Three-Bit Straight Binary Down-Counter

Consider a three-bit straight binary down-counter, as shown in Figure 5.55 with two inputs, clock and reset. The counter outputs are taken directly from the Q outputs of each flip-flop (Q2, Q1, and Q0), where Q2 is the MSB and Q0 is the LSB. This is similar to the up-counter, except now the binary count is downward.

The design process begins by creating the state transition diagram (Figure 5.56) and the state transition table. The counter is designed to reset to a count of 000_2 (0_{10}). When the reset is removed (i.e., becomes a logic 1), the counter will count through the sequence 0, 7, 6, 5, 4, 3, 2, 1, 0, etc. This means that when the counter output reaches 000_2, it will automatically wrap around back to 111_2.

Each state in the counter is encoded by the Q outputs of the D-type flip-flops, as shown in Table 5.34.

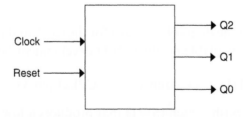

Figure 5.55: Three-bit straight binary down-counter

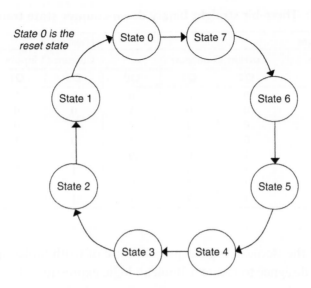

Figure 5.56: Three-bit straight binary down-counter state transition diagram

Table 5.34: Three-bit straight binary down-counter state encoding

State	Q2	Q1	Q0
State 0	0	0	0
State 1	0	0	1
State 2	0	1	0
State 3	0	1	1
State 4	1	0	0
State 5	1	0	1
State 6	1	1	0
State 7	1	1	1

The state transition table for the counter can then be created (Table 5.35). For the next state logic, the Q output for each flip-flop is actually the D input for each flip-flop in the current state. In this view of the state transition table, the current Q outputs and the current D inputs (next state Q outputs) are defined.

The Boolean logic expression can be created for each of the D inputs so that the counter of the form shown in Figure 5.51 is created. The next state logic for each flip-flop uses a combination of the Q and NOT-Q outputs from each flip-flop.

Table 5.35: Three-bit straight binary down-counter state transition table

Present state				Next state		
State name	Current Q outputs			Current D inputs		
	Q2	Q1	Q0	D2	D1	D0
State 0	0	0	0	1	1	1
State 1	0	0	1	0	0	0
State 2	0	1	0	0	0	1
State 3	0	1	1	0	1	0
State 4	1	0	0	0	1	1
State 5	1	0	1	1	0	0
State 6	1	1	0	1	0	1
State 7	1	1	1	1	1	0

Manipulation of the Boolean logic expression, use of truth tables, and Karnaugh maps allows the designer to create a Boolean logic expression of a required form.

An example of the logic for each flip-flop D input developed using the Karnaugh map is shown in Figure 5.57; Figure 5.58 shows a schematic for the counter. Each

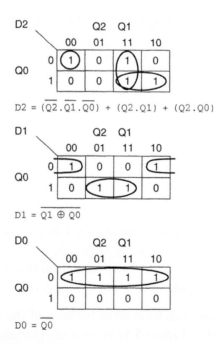

$$D2 = (\overline{Q2.Q1.Q0}) + (Q2.Q1) + (Q2.Q0)$$

$$D1 = \overline{Q1} \oplus Q0$$

$$D0 = \overline{Q0}$$

Figure 5.57: Three-bit down-counter D input Boolean expressions

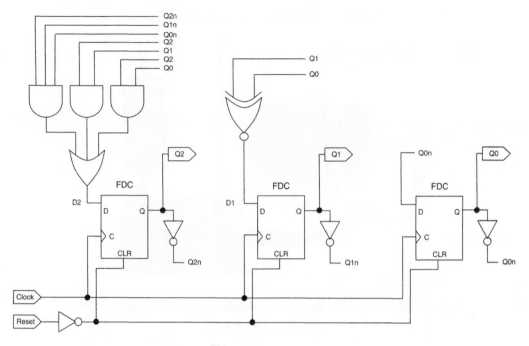

Figure 5.58: Xilinx ISE™ schematic for three-bit down-counter

D-type flip-flop has only a Q output, and the NOT-Q output is created using a discrete inverter. Additionally, each D-type flip-flop has an asynchronous active high reset that must be initially inverted so that the design reset input sees an asynchronous active low reset circuit.

This design can be implemented in the CPLD development system (refer to Chapter 3 and Appendix F—See the last paragraph of the Preface for instructions regarding how to access this online content) using a CPLD development board and a digital I/O board. The basic arrangement is shown in Figure 5.59. Here, the Coolrunner™-II CPLD on the CPLD development board is configured with the digital logic circuit, and the digital I/O board interfaces to external test and measurement equipment. The CPLD is configured using the pins identified in Table 5.36.

The five design I/Os are defined and connected to the relevant CPLD pins to connect to Header A for the digital I/O board. In addition, the CPLD design must also incorporate two additional outputs to enable the tristate buffers used on the digital I/O board. The enable (OE, output enable) pins on the tristate buffers must be tied to logic 0 to enable the buffers.

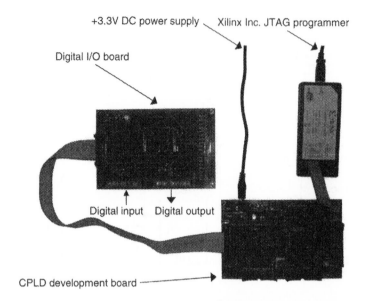

Figure 5.59: Three-bit up-counter implementation using the CPLD development board

Table 5.36: Three-bit up-counter CPLD pin assignment

Signal name	CPLD pin number	Digital I/O board identifier	Comment
Clock	13	B0 (input bit 0)	CPLD input, design Clock input
Reset	14	B1 (input bit 1)	CPLD input, design Reset input
Q0	3	A0 (output bit 0)	CPLD output, design Q0 output
Q1	4	A1 (output bit 1)	CPLD output, design Q1output
Q2	5	A2 (output bit 2)	CPLD output, design Q2 output
Input buffer enable	12	OE2 (input enable)	CPLD output, tie to logic 0 in CPLD design
Output buffer enable	2	OE1 (output enable)	CPLD output, tie to logic 0 in CPLD design

An alternative to using the digital I/O board for the clock and reset inputs is to use the clock and reset inputs available on the CPLD development board:

- **Clock**—a 50 MHz clock connected to CPLD pin 38

- **Reset**—a reset push switch (SW1) that produces a low signal when activated (pressed) via the MAX811-S voltage monitor IC connected to CPLD pin 143.

This removes the need to generate external clock and reset inputs. If the 50 MHz clock is used, then a clock divider circuit is probably needed to produce a lower frequency clock to the counter that will:

- operate within the CPLD without timing errors

- operate correctly with the interconnect between the CPLD and the I/O buffers

- operate correctly with the I/O buffers

- operate correctly with the test and measurement equipment used

External circuitry is connected to the digital I/O board to provide the logic levels for inputs clock and reset, and to monitor the outputs Q2, Q1, and Q0 where logic $0 = 0$ V and logic $1 = +3.3$ V.

Example 3: Divide-by-5 Circuit

Consider a circuit that receives a clock signal and produces a single output pulse on every fifth clock input pulse. This simple divide-by-5 circuit, shown in Figure 5.60, can be used in a clock divider circuit.

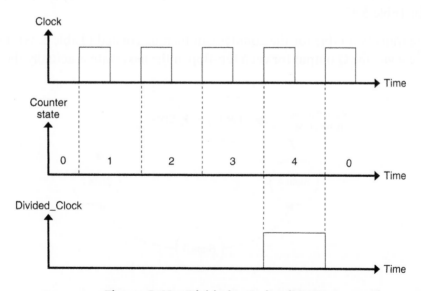

Figure 5.60: Divide-by-5 circuit I/O

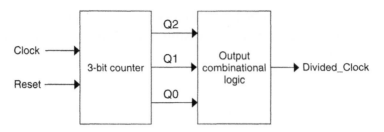

Figure 5.61: Divide-by-5 circuit

To create this output signal (Divided_Clock), a counter with five count states (0, 1, 2, 3, 4) is created and the output decoded using combinational logic so that on state 4 of the count, the output is a logic 1 only. This ensures that when the counter is reset (either at power on or by an external circuit), the output will be a logic 0. This arrangement is shown in Figure 5.61.

The design process begins by creating the state transition diagram (Figure 5.62) and the state transition table. The counter is designed to reset (into state 0) to a count value of 000_2 (0_{10}). When the reset is removed (i.e., becomes a logic 1), then the counter will count through the sequence 0, 1, 2, 3, 4, 0, etc. When the counter output reaches 100_2, it will automatically wrap around back to 000_2.

Each state in the counter is encoded by the Q outputs of the D-type flip-flops, as shown in Table 5.37.

The state transition table for the counter can then be created (Table 5.38). For the next state logic, the Q output for each flip-flop in the next state is actually the D input

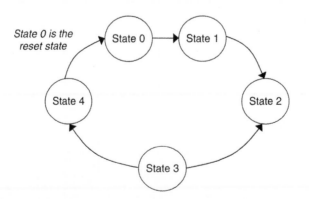

Figure 5.62: Divide-by-5 circuit state transition diagram

Table 5.37: Divide-by-5 circuit state encoding

State	Q2	Q1	Q0
State 0	0	0	0
State 1	0	0	1
State 2	0	1	0
State 3	0	1	1
State 4	1	0	0
Unused states			
State 5	1	0	1
State 6	1	1	0
State 7	1	1	1

Table 5.38: Divide-by-5 circuit state transition table

Present state				Next state		
State name	Current Q outputs			Current D inputs		
	Q2	Q1	Q0	D2	D1	D0
State 0	0	0	0	0	0	1
State 1	0	0	1	0	1	0
State 2	0	1	0	0	1	1
State 3	0	1	1	1	0	0
State 4	1	0	0	0	0	0
Unused states						
State 5	1	0	1	0	0	0
State 6	1	1	0	0	0	0
State 7	1	1	1	0	0	0

for each flip-flop in the current state. In this view of the state transition table, the current Q outputs and the current D inputs (next state Q outputs) are defined. The unused states are also shown and are set so that if they are entered, the next state will be state 0.

A Boolean logic expression can be created for each of the D inputs so that a counter of the form shown in Figure 5.51 is created. Here, the next state logic for each flip-flop uses a combination of the Q and NOT-Q outputs from each flip-flop. Manipulation of the Boolean logic expression, the use of truth tables, and Karnaugh maps allow the designer to create a Boolean logic expression of a required form.

An example of the logic for each flip-flop D input developed using the Karnaugh map is shown in Figure 5.63. Figure 5.64 shows the schematic developed for the counter, in

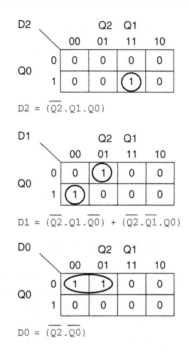

$$D2 = \overline{(Q2.Q1.Q0)}$$

$$D1 = \overline{(Q2.Q1.\overline{Q0})} + \overline{(Q2.\overline{Q1}.Q0)}$$

$$D0 = \overline{(Q2.\overline{Q0})}$$

Figure 5.63: Divide-by-5 D input Boolean expressions

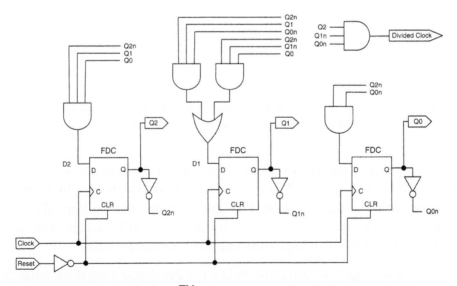

Figure 5.64: Xilinx ISE™ schematic for the divide-by-5 circuit

Table 5.39: Three divide-by-5 circuit output logic decodings

State	Q2	Q1	Q0	Divided_Clock
State 0	0	0	0	0
State 1	0	0	1	0
State 2	0	1	0	0
State 3	0	1	1	0
State 4	1	0	0	1
Unused states				
State 5	1	0	1	0
State 6	1	1	0	0
State 7	1	1	1	0

which each D-type flip-flop only has a Q output, and the NOT-Q output is created using a discrete inverter. Additionally, each D-type flip-flop has an asynchronous active high reset that must be initially inverted so that the design reset input sees an asynchronous active low reset circuit.

The output combinational logic is provided in the truth table shown in Table 5.39. The output is a logic 1 only when the counter is in state 4.

The Boolean logic expression for the Divided_Clock output is given as:

$$\text{Divided\_Clock} = (Q2.\overline{Q1}.\overline{Q0})$$

This design can be implemented in the CPLD development system (refer to Chapter 3 and Appendix F—See the last paragraph of the Preface for instruction regarding how to access this online content) using a CPLD development board and a digital I/O board. The basic arrangement is shown in Figure 5.65. Here, the Coolrunner™-II CPLD on the CPLD development board is configured with the digital logic circuit, and the digital I/O board interfaces to external test and measurement equipment. The CPLD is configured using the pins identified in Table 5.40.

The three design I/Os are defined and are connected to the relevant CPLD pins to connect to Header A for the digital I/O board. The CPLD design must also incorporate two additional outputs to enable the tristate buffers used on the digital I/O board. The enable (OE, output enable) pins on the tristate buffers must be tied to logic 0 to enable the buffers.

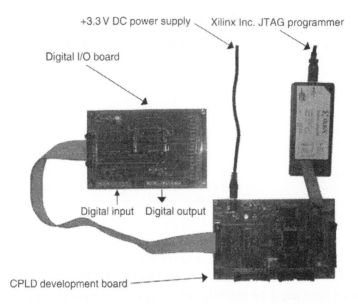

Figure 5.65: Divide-by-5 circuit implementation using the CPLD development board

Table 5.40: Divide-by-5 circuit CPLD pin assignment

Signal name	CPLD pin number	Digital I/O board identifier	Comment
Clock	13	B0 (input bit 0)	CPLD input, design Clock input
Reset	14	B1 (input bit 1)	CPLD input, design Reset input
Divided_Clock	3	A0 (output bit 0)	CPLD output, design Q0 output
Input buffer enable	12	OE2 (input enable)	CPLD output, tie to logic 0 in CPLD design
Output buffer enable	2	OE1 (output enable)	CPLD output, tie to logic 0 in CPLD design

An alternative to using the digital I/O board for the clock and reset inputs is to use the clock and reset inputs available on the CPLD development board:

- **Clock**—a 50 MHz clock connected to CPLD pin 38

- **Reset**—a reset push switch (SW1) that produces a low signal when activated (pressed) via the MAX811-S voltage monitor IC connected to CPLD pin 143.

This removes the need to generate external clock and reset inputs. If the 50 MHz clock is used, a clock divider circuit will probably be necessary to produce a lower frequency clock to the counter that will:

- operate within the CPLD without timing errors
- operate correctly with the interconnect between the CPLD and the I/O buffers
- operate correctly with the I/O buffers
- operate correctly with the test and measurement equipment used

External circuitry is connected to the digital I/O board to provide the logic levels for inputs clock and reset, and to monitor the outputs Q2, Q1, and Q0 where logic $0 = 0$ V and logic $1 = +3.3$ V.

5.5.5 State Machine Design

The sequential logic circuit is designed either to react to an input, called a state machine in this text, or to be autonomous, in which no inputs control circuit operation. Figure 5.66 shows the basic structure of the state machine.

In the following discussions, then:

- Positive edge triggered, asynchronous active low or high D-type flip-flops will be used.
- All flip-flops will have a common reset input.
- All flip-flops will have a common clock input.

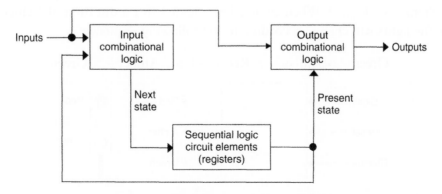

Figure 5.66: State machine structure

In addition, the output from the state machine either can be taken directly from the Q outputs of each flip-flop or can be decoded using output combinational logic to form specific outputs for specific states of the counter.

The state machine uses flip-flops, so for *n*-flip-flops, there are 2^n possible combinations of output. A state machine might use all possible states or might use only a subset of the possible states. When a sub-set is used, the state machine should be designed so that in normal operation, it will not enter the unused states. However, it is good practice to design the circuit so that if it did enter one of the unused states, it will have a known operation. For example, if an unused state is entered, the next state would always be the reset state for the state machine.

The state machine will be based on either a *Moore* machine or *Mealy* machine. In the Moore machine, the outputs will be a function of the present state only. As such, the outputs will be valid whilst the state machine is within this state and will not be valid during state transitions. In the Mealy machine, the output is a function of the present state and current inputs. As such, the output of the Mealy machine will change immediately whenever there is a change on the input whilst the output of the Moore machine would be synchronised to the clock.

Example 1: Traffic Light Sequencer

Consider a state machine design to control a set of traffic lights that moves from green to amber to red and back to green whenever a person pushes a button. This is shown in Figure 5.67. There are three inputs—clock, reset, and change—and three outputs—red, amber, and green.

The light begins on green (when the circuit is reset) and stays in the green state when change is a logic 0. When change becomes a 1 (for a duration of 1 clock cycle), the lights will change according to the following sequence:

Green– > Amber– > Red– > Red_Amber– > Green

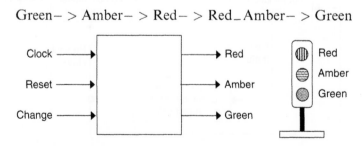

Figure 5.67: Three-bit straight binary up-counter

Table 5.41: Traffic light sequence

State	Green	Amber	Red
Green	**ON**	OFF	OFF
Amber	OFF	**ON**	OFF
Red	OFF	OFF	**ON**
Red_Amber	**ON**	**ON**	OFF

The four states and their corresponding outputs (ON = logic 1, OFF = logic 0) are defined in Table 5.41.

The state machine is designed so that when a change input is detected, the lights will change from green to red and back to green. It will then wait for another change input to be detected. During the light changes, the value of change is considered a *Don't care* condition (i.e., it could be a logic 0 or 1).

The design process begins by creating the state transition diagram (Figure 5.68) and the state transition table. There are four distinct states, so two D-type flip-flops are used (where n = 2, giving 2^n = 4 possible states). The state machine is designed to reset (i.e., when reset is a logic 0) to a count of 00_2 (0_{10}), which is the green state. When the reset is removed (i.e., when reset becomes a logic 1), the

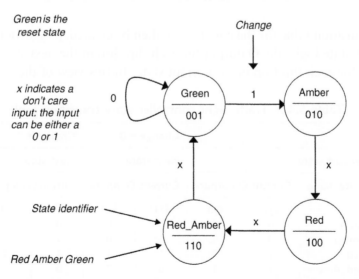

Figure 5.68: Traffic light controller circuit state transition diagram (Moore machine)

state will count through the sequence Green, Amber, Red, Red_Amber, Green, etc. when the *Change* button is pushed, and when the state machanic output reaches 11_2 (state Red_Amber), it will automatically wrap around back to Green. State machine changes are summarized below:

- State is green and input (change) is a logic 0: state remains green.

- State is green and input (change) is a logic 1: state changes to amber.

- State is amber and input (change) is a logic 0 or 1: state changes to red.

- State is red and input (change) is a logic 0 or 1: state changes to red_amber.

- State is red_amber and input (change) is a logic 0 or 1: state changes to green.

Each state in the state machine will be encoded by the Q outputs of the D-type flip-flops, as shown in Table 5.42.

Table 5.42: Divide-by-5 circuit state encoding

State	Q1	Q0
Green	0	0
Amber	0	1
Red	1	0
Red_Amber	1	1

The state transition table for the counter can then be created, as shown in Table 5.43. For the next state logic, the Q output for each flip-flop in the next state is actually the D input for each flip-flop in the current state. In this view of the state transition

Table 5.43: Traffic light controller state transition table

			Change = 0		Change = 1	
Present state			Next state		Next state	
State name	Current Q outputs		Current D inputs		Current D inputs	
	Q1	Q0	D1	D0	D1	D0
Green	0	0	0	0	0	1
Amber	0	1	1	0	1	0
Red	1	0	1	1	1	1
Red_Amber	1	1	0	0	0	0

table, the current Q outputs and the current D inputs (next state Q outputs) are defined. The change input is included in the state transition table, and the state machine moves into one of two possible next states.

The Boolean logic expression can be created for each D input so that a state machine like that shown in Figure 5.66 is created. The next state logic for each flip-flop uses a combination of the Q and NOT-Q outputs from each flip-flop along with the change input. Manipulation of the Boolean logic expression, the use of truth tables, and Karnaugh maps allow the designer to create the required Boolean logic expression.

The logic for each flip-flop D input can be developed using the truth table. As there is only one D input to each flip-flop, but two possible input conditions (depending on the value of change), Boolean logic expressions for each possible input are created and the results ORed together to determine the D input of each flip-flop. This idea is shown in Figure 5.69 The resulting Boolean logic expression should then be minimized.

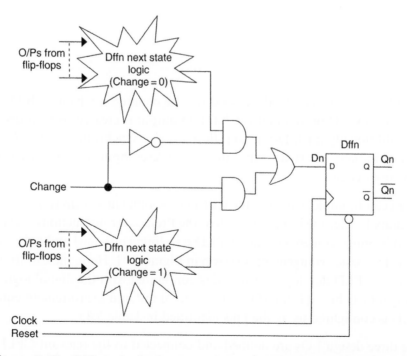

Figure 5.69: ORing the logic expressions to form the flip-flop D input

Table 5.44: Traffic light controller output logic decoding

State	Q1	Q0	Red	Amber	Green
Green	0	0	0	0	1
Amber	0	1	0	1	0
Red	1	0	1	0	0
Red_Amber	1	1	1	1	0

An example for the Boolean logic expressions for each of the flip-flops is as follows:

$$D1 = (Q1 \oplus Q0)$$

$$D0 = \overline{Change}.(Q1.\overline{Q0}) + (Change.\overline{Q0})$$

This shows that the D1 input is actually independent of the change input logic value. The output combinational logic is provided in the truth table shown in Table 5.44.

The Boolean logic expressions for the outputs are given as:

$$Green = (\overline{Q1}.\overline{Q0})$$

$$Amber = (Q0)$$

$$Red = (Q1)$$

Figure 5.70 shows a schematic developed for the counter in which each D-type flip-flop has only a Q output and the NOT-Q output is created using a discrete inverter. Additionally, each D-type flip-flop has an asynchronous active high reset that must be initially inverted so that the design reset input sees an asynchronous active low reset circuit.

This design can be implemented in the CPLD development system (refer to Chapter 3 and Appendix F—See the last paragraph of the Preface for instructions regarding how to access this online content) using a CPLD development board and a digital I/O board. The basic arrangement is shown in Figure 5.71. Here, the Coolrunner™-II CPLD on the CPLD development board is configured with the digital logic circuit, and the digital I/O board is interfaced to external test and measurement equipment. The CPLD is configured using the pins identified in Table 5.45.

Here, the three design I/Os are defined and connected to the relevant CPLD pins to connect to Header A for the digital I/O board. In addition, the CPLD design must

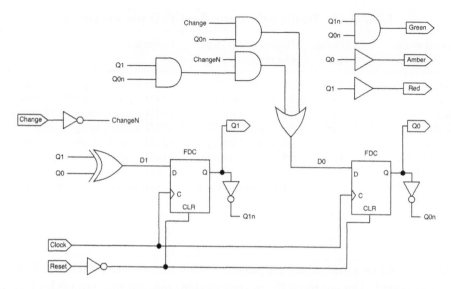

Figure 5.70: Circuit schematic for the traffic light controller

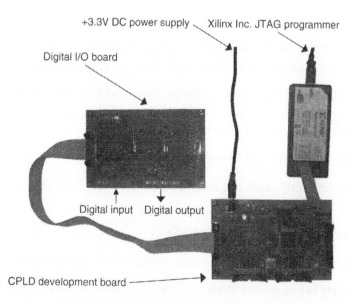

Figure 5.71: Traffic light controller implementation using the CPLD development board

Table 5.45: Traffic light controller CPLD pin assignment

Signal name	CPLD pin number	Digital I/O board identifier	Comment
Clock	13	B0 (input bit 0)	CPLD input, design Clock input
Reset	14	B1 (input bit 1)	CPLD input, design Reset input
Red	3	A0 (output bit 0)	CPLD output, design Red output
Amber	4	A0 (output bit 1)	CPLD output, design Amber output
Green	5	A0 (output bit 2)	CPLD output, design Green output
Input buffer enable	12	OE2 (input enable)	CPLD output, tie to logic 0 in CPLD design
Output buffer enable	2	OE1 (output enable)	CPLD output, tie to logic 0 in CPLD design

also incorporate two additional outputs to enable the tristate buffers used on the digital I/O board. The enable (OE, output enable) pins on the tristate buffers must be tied to logic 0 to enable the buffers.

An alternative to using the digital I/O board for the clock and reset inputs is to use the clock and reset inputs available on the CPLD development board:

- **Clock**—a 50 MHz clock connected to CPLD pin 38

- **Reset**—a reset push switch (SW1) which produces a low signal when activated (pressed) via the MAX811-S voltage monitor IC connected to CPLD pin 143.

This removes the need to generate external clock and reset inputs. If the 50 MHz clock is used, clock divider circuit will likely be needed to produce a lower frequency clock to the counter that will:

- operate within the CPLD without timing errors

- operate correctly with the interconnect between the CPLD and the I/O buffers

- operate correctly with the I/O buffers

- operate correctly with the test and measurement equipment used

External circuitry is connected to the digital I/O board to provide the logic levels for inputs clock, reset, and change, and to monitor the outputs red, amber, and green where logic $0 = 0$ V and logic $1 = +3.3$ V.

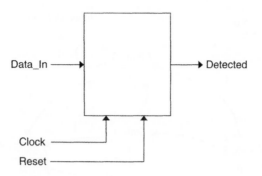

Figure 5.72: 1001 sequence detector

Example 2: 1001 Sequence Detector

Consider the circuit that is to detect the sequence "1001" on a serial bit-stream data input and produce a logic 1 output when the sequence has been detected, as shown in Figure 5.72. The state machine will have three inputs—one Data_In to be monitored for the sequence and two control inputs, Clock and Reset—and one output, Detected. Such a state machine could be used in a digital combinational lock circuit.

The design process begins by creating the state transition diagram (Figure 5.73) and the state transition table. There are five distinct states, so three D-type flip-flops are used (where n = 3, giving $2^n = 8$ possible states, although only five states are used). The state machine is designed to reset (i.e., when Reset is a logic 0) to a count of 000_2 (0_{10}), which will be the state 0 state. When the reset is removed (i.e., when reset becomes a logic 1), then the state machine becomes active. State machine changes are summarized below:

- At state 0 and input (Data_In) is a logic 0: state remains in state 0.

- At state 0 and input (Data_In) is a logic 1: state changes to state 1.

- At state 1 and input (Data_In) is a logic 0: state changes to state 2.

- At state 1 and input (Data_In) is a logic 1: state remains in state 1.

- At state 2 and input (Data_In) is a logic 0: state changes to state 3.

- At state 2 and input (Data_In) is a logic 1: state changes back to state 1.

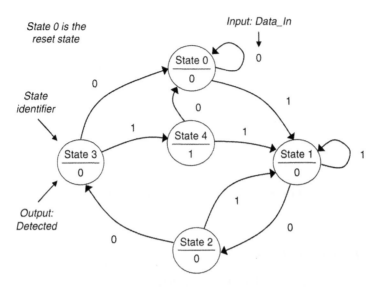

Figure 5.73: 1001 sequence detector state transition diagram (Moore machine)

- At state 3 and input (Data_In) is a logic 0: state changes back to state 0.

- At state 3 and input (Data_In) is a logic 1: state changes to state 4.

- At state 4 and input (Data_In) is a logic 0: state changes back to state 0.

- At state 4 and input (Data_In) is a logic 1: state changes back to state 1.

Whenever an unused state is encountered, the state machine is designed to enter state 0 on the next clock rising edge.

Each state in the counter is encoded by the Q outputs of the D-type flip-flops, as shown in Table 5.46.

The state transition table for the counter can then be created, as shown in Table 5.47. For the next state logic, the Q output for each flip-flop in the next state is actually the D input for each flip-flop in the current state. In this view of the state transition table, the current Q outputs and the current D inputs (next state Q outputs) are defined. The change input is included in the state transition table, and the state machine can move into one of two possible next states.

Table 5.46: 1001 sequence detector state encoding

State	Q2	Q1	Q0
State 0	0	0	0
State 1	0	0	1
State 2	0	1	0
State 3	0	1	1
State 4	1	0	0
Unused states			
State 5	1	0	1
State 6	1	1	0
State 7	1	1	1

Table 5.47: 1001 sequence detector state transition table

Present state				Data_In = 0			Data_In = 1		
				Next state			Next state		
State name	Current Q outputs			Current D inputs			Current D inputs		
	Q2	Q1	Q0	D2	D1	D0	D2	D1	D0
State 0	0	0	0	0	0	0	0	0	1
State 1	0	0	1	0	1	0	0	0	1
State 2	0	1	0	0	1	1	0	0	1
State 3	0	1	1	0	0	0	1	0	0
State 4	1	0	0	0	0	0	0	0	1
Unused states									
State 5	1	0	1	0	0	0	0	0	0
State 6	1	1	0	0	0	0	0	0	0
State 7	1	1	1	0	0	0	0	0	0

An example for the Boolean logic expressions for each of the flip-flops is as follows:

$$D2 = Data\_In.(\overline{Q2}.Q1.Q0)$$

$$D1 = \overline{Data\_In}.\overline{Q2}.(Q1 \oplus Q0)$$

$$D0 = \overline{Data\_In}.(\overline{Q2}.Q1.\overline{Q0}) + Data\_In.((\overline{Q2}.\overline{Q1}) + (Q1.\overline{Q0}))$$

The output combinational logic is provided in the truth table shown in Table 5.48.

Table 5.48: 1001 sequence detector output logic decoding

State	Q2	Q1	Q0	Detected
State 0	0	0	0	0
State 1	0	0	1	0
State 2	0	1	0	0
State 3	0	1	1	0
State 4	1	0	0	1
Unused states				
State 5	1	0	1	0
State 6	1	1	0	0
State 7	1	1	1	0

The Boolean logic expressions for the output is given as:

$$\texttt{Detected} = (\texttt{Q2}.\overline{\texttt{Q1}}.\overline{\texttt{Q0}})$$

Figure 5.74 shows a schematic developed for the counter in which each D-type flip-flop has only a Q output and the NOT-Q output is created using a discrete inverter. Additionally, each D-type flip-flop has an asynchronous active high reset that must be initially inverted so that the design reset input sees an asynchronous active low reset circuit.

5.5.6   *Moore versus Mealy State Machines*

Sequential logic circuit designs create counters and state machines. The state machines are based on either the Moore machine or the Mealy machine, shown in Figure 5.75.

The diagrams shown in Figure 5.75 are a modification of the basic structure identified in Figure 5.36 by separating the combinational logic block into two blocks, one to create the next state logic (inputs to the state register that store the state of the circuit) and the output logic:

- In the **Moore machine**, the outputs are a function only of the present state only.

- In the **Mealy machine**, the outputs are a function of the present state and the current inputs.

The types of circuits considered here will be synchronous circuits in that activity occurs under the control of a clock control input, these are synchronous circuits. A number of possible circuits can be formed to produce the required circuit functionality.

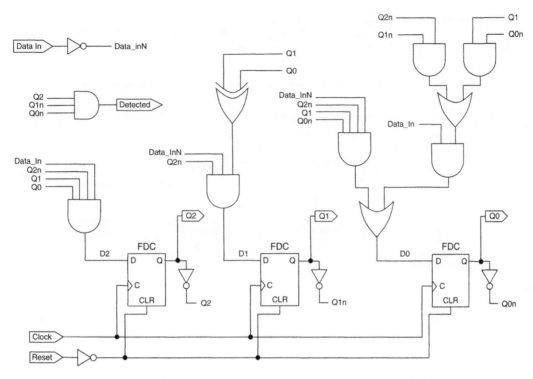

Figure 5.74: Circuit schematic for the 1001 sequence detector

5.5.7 Shift Registers

The D-type flip-flops can be connected so that the Q output of one flip-flop is connected to the D input of the next flip-flop, as shown in Figure 5.76. With a single input (Data_In), a serial bit-stream can be applied to the circuit. Whenever a clock edge occurs, the D input of a flip-flop is stored and presented as the Q output of that flip-flop.

If there are *n*-flip-flops in the circuit, the serial bitstream applied at the input appears at the output (as Data_Out) after *n* clock cycles. The serial bitstream input is available as a serial bitstream output, which is referred to as a serial-in, serial-out shift register because the input is shifted by the clock signal to become the output.

Modifications to this circuit allow parallel input to the shift register (a parallel data load, rather than a serial data load) and parallel output. A shift register that provides for serial input along with serial and parallel output is shown in Figure 5.77.

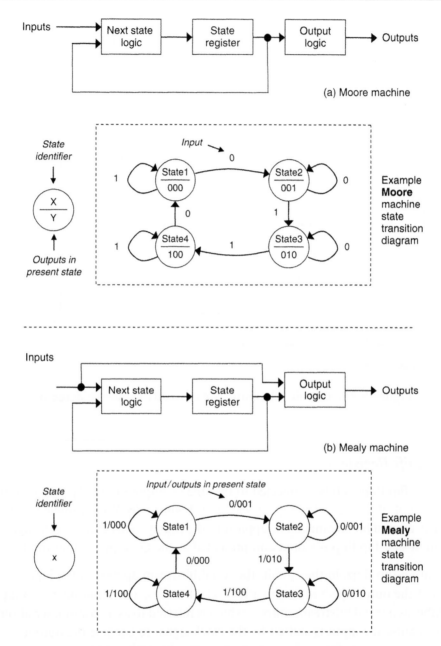

Figure 5.75: Moore and Mealy state machines

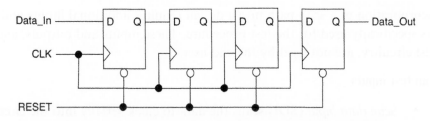

Figure 5.76: Four-bit shift register (serial in, serial out)

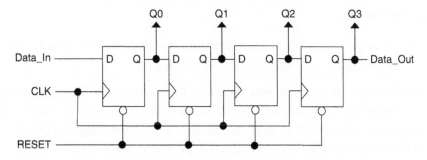

Figure 5.77: Four-bit shift register (serial in, parallel and serial out)

5.5.8   Digital Scan Path

The shift register is used to support circuit and system testing. This arrangement forms a scan path [12]. Scan path testing is the main method to provide access for internal node controllability and observability of digital sequential logic circuits, where:

- *controllability* is the ability to control specific parts of a design to set particular logic values at specific points.

- *observability* is the ability to observe the response of a circuit to a particular circuit stimulus.

In scan path, the circuit is designed for two modes of operation:

- *normal operating mode*, in which the circuit is running according to its required end-user function

- *scan test mode*, in which logic values are serially clocked into circuit flip-flop elements from an external signal source, and the results are serially clocked out for external monitoring.

The incorporation of a scan path into a design requires additional inputs and outputs specifically used for the test procedure. These inputs and outputs, and the scan test circuitry, are not used by the end user.

Scan test inputs:

- *Scan data input* (SDI) scans the data to clock serially into the circuit.

- *Scan enable* (SE) enables the scan path mode.

Scan test output:

- *Scan data out* (SDO) scans the data (results) that are serially clocked out of the scan path for external monitoring.

Using the basic circuit arrangement shown in Figure 5.77, the D-type flip-flops within the sequential logic circuit are put into a serial-in, serial-out shift register as shown in Figure 5.78, showing SDI and SDO. The parallel outputs (Q0, Q1, Q2, and Q3) form inputs to the combinational logic within the design.

A typical scan path test arrangement is shown in Figure 5.79, including the combinational logic block and D-type flip-flops. Each flip-flop has a common clock and reset input. Between the flip-flop D input and the combinational logic (the next state logic), a two-input multiplexer is inserted. The first data input to the multiplexer is the output from the next state logic. The second data input comes from the Q output of a flip-flop. This allows either of these signals to be applied to the D input of the flip-flop using the select input on the multiplexer (connected to SE).

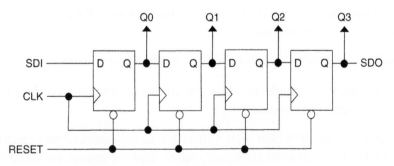

Figure 5.78: Scan test shift register

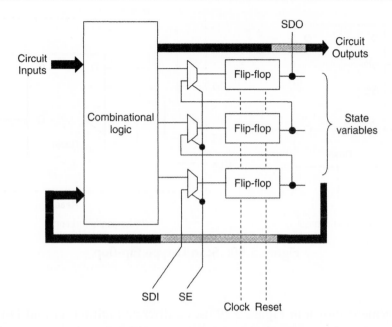

Figure 5.79: Scan path insertion using D-type flip-flops and multiplexers

In normal operating mode, the next state logic is connected to the flip-flop D input. In scan test mode, the Q output from a previous flip-flop is connected to the flip-flop D input. This isolates the flip-flop from the next state logic, and the flip-flops form a shift register of the form shown in Figure 5.78. Test data can therefore be scanned in (using the SDI input), and test results can be scanned out (using the SDO output). An example operation of this scan path follows:

1. The circuit is put into scan test mode (by control of the SE). The test data is serially scanned into the design to set the flip-flop Q outputs to known values (i.e., to put the circuit into a known, initial state) by applying the test data to the SDI pin.

2. The circuit is put back into its normal operating mode and operated for a set number of clock cycles.

3. The circuit is again put into scan test mode. The results of the test are stored on the Q outputs of the flip-flops and serially scanned out and monitored on the SDO pin.

4. The monitored values are compared with the expected values, and the circuit is then checked to see if it has passed (expected values received) or failed (the circuit output is not as expected) the test.

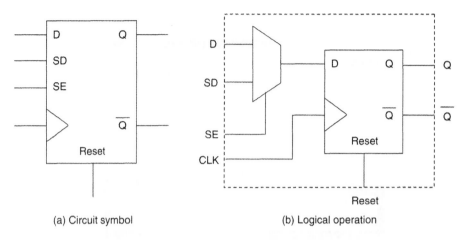

(a) Circuit symbol (b) Logical operation

Figure 5.80: Scan D-type flip-flop

The arrangement shown in Figure 5.79 uses a discrete multiplexer and D-type flip-flop to create the scan path. In many circuits, these functions are combined into a single scan D-type flip-flop circuit element, as shown in Figure 5.80. This has the same logic functionality as a discrete flip-flop and multiplexer arrangement, but is optimized for size and speed of operation. It has two data inputs (D, normal data, and SD, scan data input) and a scan enable (SE) control input to select between normal and scan test modes, in addition to the clock, reset (and/or set) inputs and Q/NOT-Q outputs.

5.6 Memory

5.6.1 Introduction

Memory is used to store, provide access to, and allow modification of data and program code for use within a processor-based electronic circuit or system. The two basic types of memory are ROM (read-only memory, and RAM (random access memory). Memory can be considered for use for one of the following three data or program storage purposes:

1. *Permanent storage* for values that are normally only read within the application and can be changed (if at all) only by removing the memory from the application and reprogramming or replacing it.

2. *Semi-permanent storage* for values that can be read only within the application (as with permanent storage). However, stored values can be modified by reprogramming while the memory remains in the circuit.

3. *Temporary storage* for values needed only for temporary use and requiring fast access or modification (such as data and program code within a computer system that can be removed when no longer needed).

These memories are typically used within a computer architecture of the form shown in Figure 5.81. Here, the ROM and RAM are connected to the other computer functional blocks:

- *ALU*, arithmetic and logic unit

- *I/O*, input/output to external circuitry

- *controller* to provide the necessary timing for the circuitry.

Each of the functional blocks is connected to a common set of data, address, and control lines required to access and manipulate the digital data at specific points in time. Also needed is a power supply for each circuit to implement the functional blocks within the computer.

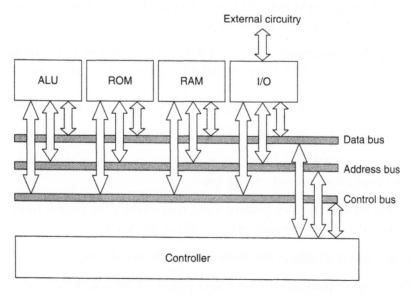

Figure 5.81: Basic computer architecture

The key drivers for memory development are driven by the end-user, who is constantly demanding more functionality at a lower cost. Hence, the key drivers for memory development are:

1. Increased capacity—the amount of data that can be stored within a single memory circuit

2. Increased operating speed—to reduce time to write data to and read data from the memory

3. Lower cost

Memory bandwidth, the amount of information that can be transferred to and from memory per unit time, is an increasingly important aspect to memory design and choice for use. This is driven by the increase in processor performance and demanding applications such as multimedia and communications.

5.6.2   Random Access Memory

RAM (also referred to as read-write memory, RWM) is considered volatile storage because its contents are lost when the power is removed. There are two main types of RAM, static RAM (SRAM) and dynamic RAM (DRAM). In addition, ferromagnetic RAM (FRAM) is also available.

A view of SRAM connections where the SRAM is provided in a dual in-line package (DIP) is shown in Figure 5.82. Here, the SRAM consists of the following connections:

- *Address lines* define the memory location to be selected for reading or writing.

- *Input/output data lines* define the data to write to or read from memory.

- *Write enable* (WE) is a control input that selects between the memory read and write operations (usually active low).

- *Output enable* (OE) is a control input that enables the output buffer for reading data from the memory (usually active low).

- *Chip select* (CS) selects the memory (usually active low).

- *Power supply* provides the necessary power to operate the circuit.

Figure 5.82: SRAM in a DIL package

Where the SRAM is provided as a discrete packaged device, a suitable power supply (V_{DD}/V_{SS}) along with power supply decoupling (capacitors) on the PCB will be needed. Increasingly, SRAMs are provided as macro cells within ICs (such as in the Xilinx® Spartan™-3 FPGA), in which the power supply has already been routed and the SRAM is ready for use.

In some RAM designs, the two *write enable* (WE) and *output enable* (OE) control signals identified above are combined into a single *read/write* (R/W) control signal. This reduces the pin count by one and the logic level of the R/W input will determine if the RAM is written to, or read from.

5.6.3 Read-Only Memory

ROM is used for holding program code that must be retained when the memory power is removed, so it is considered nonvolatile storage. The code can take one of three forms:

1. Fixed when the memory is fabricated—mask-programmable ROM

2. Electrically programmed once—PROM, programmable ROM

3. Electrically programmed multiple times—EPROM (electrically programmable ROM) erased using ultraviolet (UV) light; EEPROM or E^2PROM (electrically erasable PROM); and flash (also electrically erased).

Figure 5.83: Basic ROM in a DIL package

PROM is sometimes considered in the same category of circuit as programmable logic, although in this text, PROM is discussed only in the memory category.

RAM is used for holding data and program code that must be accessed quickly and modified during normal operation. RAM differs from read-only memory (ROM) in that it can be both read from and written to in the normal circuit application. However, flash memory is also referred to as nonvolatile RAM (NVRAM).

A basic ROM design in which ROM is provided in a dual in-line package is shown in Figure 5.83. Here, the ROM consists of the following connections:

- *Address lines* define the memory location to be selected for reading or writing.

- *Output data lines* access the data from memory.

- *Output enable* (OE) is a control input that enables the output buffer for reading data from the memory (usually active low).

- *Chip select* (CS) selects the memory (usually active low).

- *Power supply* provides the necessary power to operate the circuit.

In this view, the data bus is considered to be unidirectional (i.e., output only). Where the ROM may be electrically programmed, then the data and control line arrangement will be more complex.

References

[1] Meade, M., and Dillon, C., *Signals and Systems, Models and Behaviour*, Second Edition, Chapman and Hall, 1991, ISBN 0-412-40110-x.

[2] Smith, M., *Application Specific Integrated Circuits*, Addison-Wesley, 1999, ISBN 0-201-50022-1.

[3] Skahill, K., *VHDL for Programmable Logic*, Addison-Wesley, 1996, ISBN 0-201-89573-0.

[4] Maxfield, C., *The Design Warrior's Guide to FPGAs*, Newnes, 2004, ISBN 0-7506-7604-3.

[5] Stonham, T. J., *Digital Logic Techniques: Principles and practice*, Second Edition, Van Nostrand Reinhold, UK, 1988, ISBN 0-278-00011-8.

[6] Tocci, R. J., Widmer, N. S., and Moss, G. L. K., *Digital Systems*, Ninth Edition, Pearson Education International, USA 2004, ISBN 0-13-121931-6.

[7] The Institute of Electrical and Electronics Engineers, IEEE Standard, 91-1984, *Graphics Symbols for Logic Functions*, IEEE, USA.

[8] Overview of IEEE Standard 91-1984, *Explanation of Logic Symbols*, 1996, Texas Instruments, USA.

[9] The Institute of Electrical and Electronics Engineers, IEEE Standard 1076-2002, *IEEE Standard VHDL Language Reference Manual*, IEEE, USA.

[10] Zwolinski, M., *Digital System Design with VHDL*, Pearson Education Limited, 2000, England, ISBN 0-201-36063-2.

[11] Kang, S., and Leblebici, Y., *CMOS Digital Integrated Circuits Analysis and Design*. McGraw-Hill International Editions, Singapore, 1996, ISBN 0-07-114423-4.

[12] Grout, I. A., *Integrated Circuit Test Engineering Modern Techniques*, Springer, 2006, ISBN 1-84628-023-0.

Student Exercises

5.1   Convert the following decimal numbers to unsigned binary:

- 145_{10}
- 10_{10}
- 21.75_{10}
- $1{,}256.125_{10}$

5.2   Convert the following decimal numbers to 2s complement signed binary:

- 145_{10}
- -10_{10}
- 21.75_{10}
- $-1{,}256.125_{10}$

5.3   Convert the following unsigned binary numbers to decimal:

- 10101010_2
- 01010101_2
- 1100.001101_{10}
- $1111000011110000.00001111_{10}$

5.4   Convert the following unsigned binary numbers to octal:

- 101010010_2
- 101010101_2
- 100.001101_2
- 111100001111.000111_2

5.5   Convert the following unsigned binary numbers to hexadecimal:

- 10101010_2
- 101010101111_2
- 1100.0011_2
- 111100001111.00001111_2

5.6   Consider the Boolean logic expression:

$$Z = \overline{((A + B).C).D}$$

Draw the logic level schematic for this design:

- As it is presented.
- Using two-input NAND logic gates only.
- Using two-input NOR logic gates only.

5.7   Consider the Boolean logic expression:

$$z = A + (B.C)$$

Draw the logic level schematic for this design:

- As it is presented.
- Using two-input NAND logic gates only.
- Using two-input NOR logic gates only.

5.8   Consider the Boolean expression:

$$z = (A.\overline{B}) + B.C$$

Draw the logic level schematic for this design:

- As it is presented.
- Using two-input NAND logic gates only.
- Using two-input NOR logic gates only.

5.9   Design a circuit using combinational logic that will convert a four-bit unsigned binary into Gray code.

5.10   Create the truth table for a five-bit Gray code count.

5.11   Design a circuit using combinational logic that will implement parity checking on a four-bit input. The circuit is to use even parity coding.

5.12   Design a circuit using combinational logic that will implement parity checking on a four-bit input. The circuit is to use odd parity coding.

5.13   Design a synchronous sequential circuit that will produce a four-bit straight binary up-count. The counter is to use positive edge triggered active low asynchronous reset D-type flip-flops.

5.14   Design a synchronous sequential circuit that will produce a four-bit straight binary down-count. The counter is to use positive edge triggered active low asynchronous reset D-type flip-flops.

5.15 Design a synchronous sequential circuit that will produce a four-bit straight binary up/down-count. When a direction input is 0, the counter will count up. When the direction input is 1, the counter will count down. The counter is to use positive edge triggered active low asynchronous reset D-type flip-flops.

5.16 For the range of Xilinx® FPGAs, identify which FPGAs contain the following:

- Hardware multiplier.
- SRAM.

If these resources are available for the designer to use, identify the number and size of each macro within each type of FPGA.

5.17 For the range of Xilinx® CPLDs, identify which CPLDs contain the following:

- Hardware multiplier.
- SRAM.

If these resources are available for the designer to use, identify the number and size of each macro within each type of FPGA.

5.18 For the range of Lattice® Semiconductor FPGAs, identify which FPGAs contain the following:

- Hardware multiplier.
- SRAM.

If these resources are available for the designer to use, identify the number and size of each macro within each type of FPGA.

5.19 For the range of Lattice® Semiconductor CPLDs, identify which CPLDs contain the following:

- Hardware multiplier.
- SRAM.

If these resources are available for the designer to use, identify the number and size of each macro within each type of FPGA.

5.20 Design a synchronous counter that will produce a four-bit Gray code count. The counter is to use active low asynchronous reset D-type flip-flops.

5.21   Design a synchronous counter that will produce a four-bit BCD count. The counter is to use active low asynchronous reset D-type flip-flops.

5.22   Design a combinational logic circuit that will accept an eight-bit BCD value and produce the outputs to display the BCD count on two seven-segment displays (using common cathode displays).

5.23   Design a counter that will control the lights for a traffic light system at the crossroads of two roads. The counter will automatically cycle through each road from red to green for each road in turn.

5.24   Modify the traffic light control system so that now each road will become green if a car is detected on that road. What are the limitations of the implemented system?

CHAPTER 6

Introduction to Digital Logic Design with VHDL

6.1 Introduction

In the past, digital circuits were designed by hand on paper using techniques such as Boolean expressions, circuit schematics, Karnaugh maps, and state transition diagrams. With the increasing use of computer-based design methods and tools, the design process migrated to the computer using electronic design automation (EDA) tools [1]. These are computer-aided design (CAD) tools developed to support the designers of electronic hardware and software systems. Circuit schematic design entry, supported with design simulation tools, became the design entry and validation (through simulation) method available. With the subsequent development and standardization of hardware description languages (HDL), the HDL design entry method using text-based descriptions of circuits is now often the preferred choice of designers [2-4]. HDL design is supported with simulation, as with circuit schematic design entry, and with logic synthesis (normally referred to simply as synthesis), which converts (synthesizes) the HDL design description into a circuit netlist consisting of the required logic gates and interconnection wiring [5, 6]. Many EDA tools also provide a means by which to view the HDL code as a circuit schematic, thereby providing a graphical view of the design hardware. Such graphical views can aid the designer in understanding the circuit operation and for design debugging purposes.

This chapter provides an introduction to design with HDLs, with particular emphasis on the VHDL language. As such, all examples in this chapter and this text book are provided in VHDL.

6.2   Designing with HDLs

Hardware description language (HDL) design entry is based on the creation and use of text-based descriptions of a digital logic circuit or system. Here, using a particular HDL (the two IEEE standards in common use in industry and academia are Verilog®-HDL [7] and VHDL [8]), the description of the circuit can be created at different levels of abstraction from the basic logic gate description according to the language syntax (the grammatical arrangement of the words and symbols used in the language) and semantics (the meaning of the words and symbols used in the language).

Verilog®-HDL and VHDL are both set in the IEEE standards [9]:

- Verilog®-HDL, IEEE Std 1364™-2005

- VHDL, IEEE Std 1076™-2002

Verilog®-HDL was released in 1983 by Gateway Design System Corporation, together with a Verilog®-HDL simulator. In 1985, the language and simulator were enhanced with the introduction of the Verilog-XL® simulator. In 1989, Cadence Design Systems, Inc. brought the Gateway Design System Corporation, and early in 1990, Verilog®-HDL and Verilog-XL® were separated to become two separate products. Verilog®-HDL, until then a proprietary language, was released into the public domain to facilitate the dissemination of knowledge relating to Verilog®-HDL and to allow Verilog®-HDL to compete with VHDL, already existing as a nonproprietary language. Also in 1990, Open Verilog International (OVI) [10] was formed as an industry consortium consisting of computer-aided engineering (CAE) vendors and Verilog®-HDL users to control the language specification. In 1995, Verilog®-HDL was reviewed and adopted as IEEE standard (Std) 1364 (becoming IEEE Std 1364-1995). In 2001 and 2005, the standard was reviewed and the current version is the IEEE Std 1364™-2005.

VHDL (VHSIC HDL, very high-speed integrated circuit hardware description language) began life in 1980 under a United States Department of Defense (DoD) requirement for the design of digital circuits following a common design methodology to provide the ability for self-documentation and reuse with new technologies. VHDL development commenced in 1983, and the language was reviewed in 1987 to become IEEE Std 1076-1987. The language has been revised since in 1993, 2000, and 2002, the latest release being 1076-2002. VHDL also has a number of associated standards relating to modeling and synthesis.

When designing with HDLs, the designer must decide what language to use and at what level of design abstraction to work. When considering the choice of language, a number of factors come into play, including:

- the availability of suitable EDA tools to support the use of the language (including design management capabilities and availability of tool use within a project)

- previous knowledge

- personal preferences

- availability of simulation models

- synthesis capabilities

- commercial issues

- design re-use

- requirements to learn a new language and the capabilities of the language

- supported design flows within an organization

- existence of standards for the language

- access to the standards for the language

- readability of the resulting HDL code

- ability to create the levels of design abstraction required, and language and/or EDA tool support for these abstraction levels

- access to design support tools for the language (such as the existence of automatic code checking tools and documentation generation tools)

Figure 6.1 shows the different levels of design abstraction that are used. One or more levels of abstraction are used in a typical design project.

Starting at the highest level of abstraction (furthest from the circuit detail), the system idea or concept is the initial high-level description of the design that provides the design specification. The algorithm level describes a high-level behavioral description of the design operation at a mathematical description level of behavior. Neither the system idea nor the algorithm describes the way in which the behavior of the design is to be implemented. The algorithm structure in hardware is described

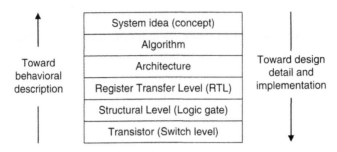

Figure 6.1: Levels of design description abstraction

by the architecture, identifying the high-level functional blocks to use and the way in which the functions are connected together. The algorithm and architecture levels describe the behavior of the design to be verified in simulation.

The next level down from the architecture is the register transfer level (RTL), which describes the storage (in registers) and flow of data around a design, along with logical operations performed on the data. It is this level that is usually used by synthesis tools to convert the design description into a structural level (the netlist of the design in terms of logic gates and interconnect wiring between the logic gates). The logic gates are themselves implemented using transistors. The HDL may also support switch level descriptions that model the transistor operation as a switch (ON/OFF).

A typical design flow for a digital circuit or system using VHDL is shown in Figure 6.2. From the initial design idea, a behavioral description of the design is written in VHDL. This is simulated to verify its operation and determine that the description matches the design idea (with the design idea in the form of a design specification).

Both Verilog®-HDL and VHDL could be used, the choice depending on a number of considerations given for a particular design project. A combination of both languages within a single design project is also common.

In VHDL, the simulation control and test stimulus to apply is created within the test bench. In Verilog®-HDL, the simulation control and test stimulus to apply is created within the test fixture.

When the behavioral level design description has been successfully validated through simulation, the design is translated to RTL code. This might be undertaken manually or automatically, if a suitable behavioral synthesis tool is available. The resulting

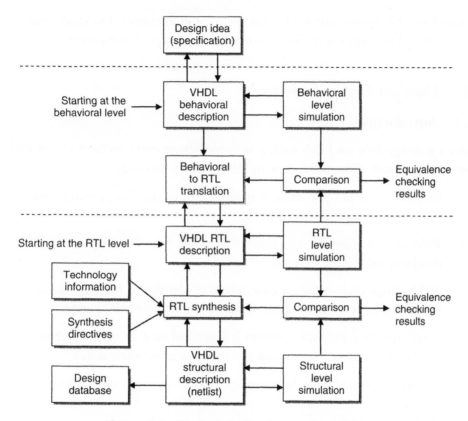

Figure 6.2: Typical design flow using HDLs

RTL VHDL code is then synthesized using the same set of criterion as the behavioral level design, and the results of the simulation are compared for equivalence to prove that the RTL code performs in the same manner as the behavioral level code. If differences are noted, then the RTL code is modified to ensure equivalence.

To summarize: the design entry point is the design idea. In Figure 6.2, the idea is then coded in VHDL as a behavioral description. This is one approach. In many situations, the RTL level code is generated directly from the idea, skipping the behavioral description stage.

The next step is to perform RTL synthesis on the code to produce a design netlist for the target technology (ASIC or PLD). Information about the target technology and user set synthesis directives is required to control the synthesis operation. The pre- and postsynthesis designs are then compared by simulating the postsynthesis netlist and checking this against the RTL level code for equivalence. On successful completion of this step, the netlist is stored in the design database for use.

6.3   Design Entry Methods

6.3.1   Introduction

To enter a design into an EDA tool, a suitable design entry method is required. Typically, tools will allow the following design entry methods:

1. *Circuit schematics* present a graphical view of the design using logic gate symbols and interconnect wiring.

2. *Boolean expressions* can be entered as a text-based description in combinational logic designs.

3. *HDL design entry* allows a description of the digital logic circuit or system operation to be entered in text form using a suitable language.

4. *State transition diagrams* present a graphical view of state machines that identifies the design states and the transitions between states.

The availability of a particular design entry method depends on the EDA tool used.

6.3.2   Schematic Capture

Schematic capture is undertaken by creating a circuit diagram (schematic) showing the logic gate symbols and the interconnections between the symbols. Figure 6.3 shows an example of a circuit schematic.

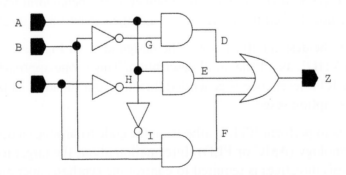

Figure 6.3: Example circuit schematic

This combinational logic circuit includes three primary inputs (A, B, and C), one primary output (Z), and six internal nodes (D, E, F, G, H, and I), as well as seven logic gates (three inverters, two two-input AND gates, one three-input AND gate, and one three-input OR gate). The circuit is represented by the following Boolean expression:

$$Z = (A \cdot \overline{B}) + (A \cdot \overline{C}) + (\overline{A} \cdot B \cdot C)$$

For a small circuit of this form, creating a circuit schematic (initially on paper, then within an EDA tool) is a straightforward task. However, consider a design with hundreds or thousands of logic gates and interconnect wiring: the task of creating, debugging, modifying, and maintaining the schematic becomes immense. Additionally, the schematic is created for a particular implementation technology and is therefore technology dependent. An example of a complex logic schematic is shown in Figure 6.4. In this view, it is not possible to identify particular logic gates, and navigating through the design to identify particular logic gates and interconnect wires would be a time-consuming task. A design description that was initially technology independent would therefore be of greater use as the same description could be used to target a range of implementation technologies. Complex designs also commonly implement a hierarchical design approach in which symbols identifying the input and output connections of blocks of circuitry are connected together in a schematic, and other schematics contain the detailed circuits for the symbols used. In this way, complex designs can be created and validated for use multiple times in a structured manner by multiple designers.

6.3.3 HDL Design Entry

Hardware description language (HDL) design entry is based on the creation and use of text-based descriptions of a digital logic circuit or system using a particular HDL (the two IEEE standards in common use are Verilog®-HDL and VHDL). It is common to adopt a hierarchical design approach to keep a project manageable. The HDL code is written to conform to one of three styles:

1. A *structural* description describes the circuit structure in terms of the logic gates used and the interconnect wiring between the logic gates to form a circuit netlist.

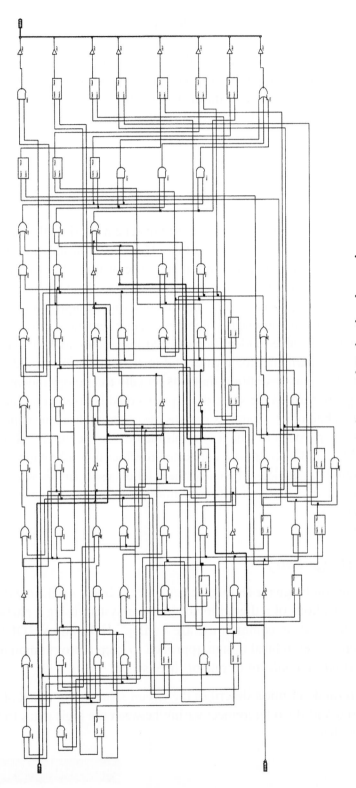

Figure 6.4: Example complex circuit schematic

2. A *dataflow* description describes the transfer of data from input to output and between signals.

3. A *behavioral* description describes the behavior of the design in terms of circuit or system behavior using algorithms. This high-level description uses language constructs that resemble a high-level software programming language.

Both the dataflow and behavioral descriptions can use similar language constructs. They differ in VHDL in that a behavioral description uses language process statements, whereas a dataflow description does not.

6.4   Logic Synthesis

An important feature of design with HDLs is the need for logic synthesis, hereafter called simply synthesis. Synthesis is the means by which an HDL description is converted (translated) to a circuit netlist that identifies the logic gates used and the interconnection wiring. In the process of synthesis, an initial HDL-based design that is technology independent (i.e., does not describe anything relating to the final implementation technology) is converted to a technology dependent netlist. As such, only at the synthesis stage is the design fixed to a particular implementation technology. This is an advantage to the designer as the same initial HDL code can be used to target different technologies, particularly important if a design is to be migrated to a new implementation technology, as often happens in design projects.

The basic synthesis process is shown in Figure 6.5. A synthesis tool requires specific information such as tool set-up routines, technology libraries, and synthesis directives. The tool set-up routines configure the synthesis tool to the particular computing platform on which it is installed. The technology libraries provide specific information relating to the target implementation technology. The synthesis directives are applied by the user to direct the synthesis tool during the design synthesis operation.

Synthesis consists of seven steps, identified in Figure 6.5. The initial HDL description is translated (1) to an RTL level description. This form is optimized (2) and then translated (3) into a logic level description. This is optimized (4) and translated (mapped) to a gate level description (5). This is optimized (6), and the result is translated to the final netlist (7). At each step, the description created is closer to the final netlist. The designer sets up specific synthesis directives to direct the synthesis tool in creating the design netlist. Constraints are typically size, speed, and power consumption. Applying different synthesis constraints typically result in different final netlists.

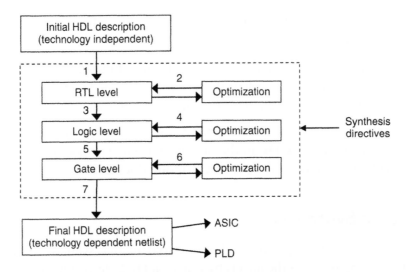

Figure 6.5: Basic synthesis process

These considerations apply to *logic synthesis* (commonly referred to as synthesis). However, there is also *physical synthesis* which relates to the automatic synthesis of design layouts in integrated circuit (IC) design at the silicon level. However, physical synthesis will not be considered further in this text.

When RTL code is synthesized, this is referred to as RTL synthesis. When behavioral level HDL code is synthesized, this is referred to as behavioral synthesis. Additionally, the initial HDL code must be created so that it is synthesizable. An HDL design description can be created as simulated and the correct operation ascertained, but in certain circumstances the HDL code written like this will be unsynthesizable.

As an example of synthesis, Figure 6.6 shows an example VHDL description for a digital circuit. This is written in the dataflow style and adds two numbers (A and B) to form a third number (Z). All numbers are eight bits wide.

The code has three main parts:

1. *Top part* identifies the reference libraries and packages to use within the design.

2. *Middle part* identifies the design entity.

3. *Bottom part* identifies the design architecture.

```
LIBRARY IEEE;
USE IEEE.STD_LOGIC_1164.ALL;
USE IEEE.STD_LOGIC_ARITH.ALL;
USE IEEE.STD_LOGIC_UNSIGNED.ALL;

ENTITY Design1 IS
    PORT ( A : IN    STD_LOGIC_VECTOR (7 downto 0);
           B : IN    STD_LOGIC_VECTOR (7 downto 0);
           Z : OUT   STD_LOGIC_VECTOR (7 downto 0));
END ENTITY design1;

ARCHITECTURE Dataflow OF Design1 IS

BEGIN

            Z (7 downto 0) <= A (7 downto 0) + B (7 downto 0);

END ARCHITECTURE Dataflow;
```

Figure 6.6: Eight-bit adder design in VHDL

The design entity has the general declaration of the form as shown in Figure 6.7.

The items enclosed within the square brackets are optional. The SIGNAL preceding the identifier is often omitted in VHDL design descriptions, although care must be taken when creating VHDL-AMS (for mixed-signal and mixed-technology design) descriptions because SIGNAL identifies digital signals and would be used then.

For VHDL-based designs, note that VHDL is not a case-sensitive language and as such both lower- and uppercase letters can be used. However, it is wise to adopt a consistent style (e.g., that uppercase letters have particular meanings).

```
ENTITY Entity_Name IS
    PORT(
        SIGNAL Signal_Identifier_1   :   Mode   Signal_Type;
        SIGNAL Signal_Identifier_2   :   Mode   Signal_Type;
        SIGNAL Signal_Identifier_3   :   Mode   Signal_Type;
        SIGNAL Signal_Identifier_4   :   Mode   Signal_Type);
END ENTITY Entity_Name;
```

Figure 6.7: General entity declaration used within the VHDL code examples

```
ARCHITECTURE Architecture_Name OF Entity_Name IS

   Signal_Declaration

   Constant_Declaration

   Component_Declaration

BEGIN

   Process_Statement

   Concurrent_Signal_Assignment_Statement

   Component_Instantiation_Statement

END ARCHITECTURE Architecture_Name;
```

Figure 6.8: General architecture declaration used within the VHDL code examples

The architecture body has the general form as shown in Figure 6.8.

This VHDL description does not describe anything relating to the circuit netlist. This description can be synthesized into the netlist (structural level) description and viewed as a schematic. For the above adder design, the resulting technology schematic created using the Xilinx® ISE™ tools is shown in Figure 6.9.

6.5 Entities, Architectures, Packages, and Configurations

6.5.1 Introduction

In VHDL, a design is created initially as an entity declaration and an architecture body. The entity declaration describes the design I/O and includes parameters that customize the design. The entity can be thought of as a black box with the I/O connections visible. The architecture body describes the internal working of the entity and contains any combination of structural, dataflow, or behavioral descriptions needed to describe the internal working of the entity.

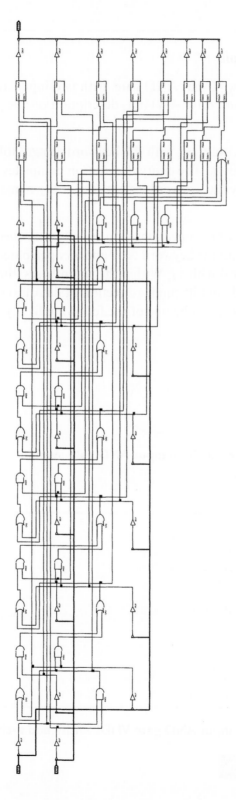

Figure 6.9: Synthesized design schematic for the eight-bit adder design

6.5.2 AND Gate Example

As an example, consider a two-input AND gate, with two inputs (A and B) and one output (Z), as shown in Figure 6.10. A VHDL description of this gate is shown in Figure 6.11.

The entity declaration starts on line 1 with the keyword ENTITY followed by the entity name and the keyword IS. The entity declaration completes on line 5 with the keyword END followed by the keyword ENTITY, the entity name and a semicolon.

Within the entity, the entity I/O connections—ports—are described. The port description starts on line 2 with the keyword PORT and an open parenthesis (. The port description completes on line 4 with) ;. Within the parentheses, the name for each of the I/O connections is stated, and its mode (direction of data transfer) and type of connection identified. The port will be one of the following five types, named for the direction of data transfer:

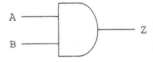

Figure 6.10: Two-input AND gate symbol

```
1    ENTITY And_Gate IS
2        PORT(  A : IN    STD_LOGIC;
3               B : IN    STD_LOGIC;
4               Z : OUT   STD_LOGIC);
5    END ENTITY And_Gate;
6
7    ARCHITECTURE Dataflow OF And_Gate IS
8
9    BEGIN
10
11          Z <= A AND B;
12
13   END ARCHITECTURE Dataflow;
```

Figure 6.11: Two-input AND gate VHDL entity and architecture

1. **IN**. This port can be read but not updated. The driver for the port is external to the entity. IN ports are primarily used for control signals and unidirectional data signals.

2. **OUT**. This port may be updated (assigned a value) but not read. The driver for the port is from within the entity. Because OUT ports cannot be read, they are used as outputs from the entity, but not for internal feedback within the entity.

3. **INOUT**. This bidirectional port may be read (from an external driver) and updated (from an internal driver). It also allows internal feedback of signals within the entity. Specific attributes may also be read.

4. **BUFFER**. This port may be read and updated. BUFFER ports are used when an output from the entity is also used for internal feedback within the entity. They do not allow for bidirectional ports. Attributes may also be read. BUFFER ports will not be considered in this text.

5. **LINKAGE**. This port may be read and updated. LINKAGE ports will not be considered in this text.

Note that blank lines and text indentation is used in the VHDL code to aid readability. The code shown in Figure 6.11 can be abbreviated by removing the optional words in the entity declaration and architecture body. Figure 6.12 shows the same code but with the optional words removed.

A benefit of incorporating the optional words is that the code can be more readable to another person, or even to the designer after some time has elapsed. If the code is

```
1    ENTITY And_Gate IS
2        PORT(  A : IN   STD_LOGIC;
3               B : IN   STD_LOGIC;
4               Z : OUT  STD_LOGIC);
5    END;
6
7    ARCHITECTURE Dataflow OF And_Gate IS
8
9    BEGIN
10
11          Z <= A AND B;
12
13   END;
```

Figure 6.12: Two-input AND gate VHDL entity and architecture

automatically generated, then the addition of these words is automatic and this aspect of code generation is transparent to the user of the code. However, for handwritten code, the addition of the optional words requires a bit more effort to create the code.

The architecture body identifies the functionality internal to the entity and the relationship between the entity ports. VHDL allows multiple architectures to be defined for a single entity, each of which is identified by a unique name. However, only one of the available architectures can be associated with an entity at any one time. In Figure 6.11, the architecture body is defined between lines 7 and 13.

The architecture body starts on line 7 with the keyword ARCHITECTURE, followed by the architecture name, the keyword OF, the entity name (with which the architecture is associated), and the keyword IS. The architecture body is completed on line 13 with the keyword END, followed by the keyword ARCHITECTURE, the architecture name and a semicolon. The details within the architecture are placed after the keyword BEGIN on line 9. The operation of this design appears on one line only, line 11:

```
Z <= A AND B;
```

This line assigns the signal Z with a value (A AND B). AND is the logical AND operator in VHDL. No time delay is associated with this; that is, the logic gate has zero time delay as far as the logic gate is concerned. However, for actual signal assignments in VHDL without the AFTER clause, the right-hand side of the assignment is assigned to the left-hand side after delta time interval. In a physical sense, this is zero time, but for signal scheduling, this delta time is significant. A signal assignment that is scheduled to occur one delta time later will happen before a signal assignment that is scheduled to occur two delta times later. The later signal assignment is not available for use by the earlier signal assignment. However, both signal assignments will occur before the smallest physical time unit used.

For an entity–architecture pair to be used for simulation and synthesis, the reference libraries used within the design must be identified before the entity declaration. In the design shown in Figure 6.11, one reference library is also required:

```
LIBRARY IEEE;
USE IEEE.STD_LOGIC_1164.ALL;
```

The first line of the above code identifies the reference library to use with the keyword LIBRARY followed by the library name IEEE and a semicolon. This clause makes the library accessible. The IEEE library contains IEEE standard design units such as the packages std_logic_1164, numeric_std, and numeric_bit. However, the design units contained within the library are not immediately made accessible. The designer must make the design units (components, declarations, functions, procedures, etc.) visible by using the USE clause. This is done on the second line:

```
USE IEEE.STD_LOGIC_1164.ALL;
```

The line starts with the keyword USE followed by the library IEEE, then package std_logic_1164, all definitions within the package (ALL), and a semicolon. Each part of the statement separated with a full-stop. The package STD_LOGIC_1164 contains a nine value type called STD_LOGIC and STD_LOGIC_VECTOR, along with simple operators such as AND, NOT, etc.

Table 6.1 identifies a number of key libraries and packages required for basic operations. The libraries and packages to be used by an entity will appear immediately before the particular entity declaration. Therefore the libraries and packages to be used by an entity will need to appear before each entity declaration.

Table 6.1: Key libraries and packages

Library	Package	Required for
IEEE	STD_LOGIC_1164	Defines the standard for describing the interconnection data types used in the VHDL language, along with the STD_LOGIC and STD_LOGIC_VECTOR types.
	STD_LOGIC_ARITH	Defines UNSIGNED and SIGNED types, conversion functions, and arithmetic/comparison operations for use with the UNSIGNED and SIGNED types.
	STD_LOGIC_UNSIGNED	Functions to allow the use of STD_LOGIC_VECTOR types as if they were UNSIGNED types.
	STD_LOGIC_SIGNED	Functions to allow the use of STD_LOGIC_VECTOR types as if they were SIGNED types.
	NUMERIC_STD	Arithmetic operations following the IEEE standard. For unsigned and signed arithmetic operations, this is the preferred package in many scenarios to using the STD_LOGIC_ARITH, STD_LOGIC_UNSIGNED and STD_LOGIC_SIGNED packages.
STD	STANDARD	Predefined definitions for the types and functions of the VHDL language
	TEXTIO	File I/O operations
WORK	<SET BY USER >	Current work library

The names given to libraries are logical names rather than physical files and directories on the host computer system. Different EDA tools implement the physical file and directory structures on the host computer system differently.

Every VHDL design unit except for the STANDARD package is assumed automatically to contain the following as part of its context clause:

```
LIBRARY STD, WORK;
USE STD.STANDARD.ALL;
```

Therefore, these need not be explicitly stated by the designer.

The code identified in Figure 6.11 now becomes the code shown in Figure 6.13.

Within VHDL, entity and architecture descriptions (design units) are placed within libraries. These may be either *working* or *resource* libraries. In the VHDL standard, these are both referred to as *design libraries*, where:

- A *working library* contains a particular design that is being created, analyzed or modified by the designer. These are editable (i.e., can be read from and written to). Only one library can be the current working library.

- A *resource library* contains existing designs that can be used. These would normally be accessible for read only by the designer and would not be modified except by the library designer.

```
1    LIBRARY IEEE;
2    USE IEEE.STD_LOGIC_1164.ALL;
3
4    ENTITY And_Gate IS
5        PORT(  A : IN    STD_LOGIC;
6               B : IN    STD_LOGIC;
7               Z : OUT   STD_LOGIC);
8    END ENTITY And_Gate;
9
10   ARCHITECTURE Dataflow OF And_Gate IS
11
12   BEGIN
13
14       Z <= A AND B;
15
16   END ARCHITECTURE Dataflow;
```

Figure 6.13: Two-input AND gate VHDL entity and architecture

This compares with generally used library definitions for EDA/CAD tools that also utilise libraries in the creation and management of designs, where the libraries can be referred to as either a *design library* or *reference library*, where:

- A *design library* would contain a particular design that is being created, analyzed, or modified. These libraries are editable (i.e., they can be read from and written to). A design library used by another designer would be a reference library for the other person.

- A *reference library* would contain existing designs that can be used. These libraries are normally read only access and would not be editable except by the library designer.

Entities and architectures are also referred to as design units. A VHDL design will consist of a number of these design units. When a VHDL design unit has been created as a text file, it must be analyzed (or compiled) before it can be used. When a design unit has been analyzed and found to be error-free, an entry for this design unit is made within a VHDL design library. The *work library* is the name given to the current working library; it is a design library. Previously analyzed design units are placed within reference libraries and may be accessed by the designer.

Packages and configurations are also design units. Packages are VHDL design units that are used to group certain components based on specific design requirements. Configurations are VHDL design units that allow generic components to be configured with specific parameters when the components are used. For example, if a design entity has two or more architectures, a configuration identifies which architecture is to be used.

Aside: In VHDL code examples presented in this text, the IEEE library is used with the following packages made accessible (depending on the particular example):

```
USE IEEE.STD_LOGIC_1164.ALL;
USE IEEE.STD_LOGIC_ARITH.ALL;
USE IEEE.STD_LOGIC_UNSIGNED.ALL;
USE IEEE.NUMERIC_STD.ALL;
```

Most of the examples only require the use of the STD_LOGIC_1164 package, but the other packages are included. The reader is encouraged to identify when particular packages are required in which examples, and to include or remove the packages as and when required. The above packages however are included in the examples for reference. In many instances, then the NUMERIC_STD package is preferred for arithmetic operations.

Table 6.2: Logical operators in VHDL

Logical operation	Operator	Example
AND	AND	Z <= (A AND B);
NAND	NAND	Z <= (A NAND B);
NOR	NOR	Z <= (A NOR B);
NOT	NOT	Z <= NOT (A);
OR	OR	Z <= (A OR B);
XNOR	XNOR	Z <= (A XNOR B);
XOR	XOR	Z <= (A XOR B);

In Figure 6.13, the logical AND operator is used. Within VHDL, no logical operator has precedence over another logical operator and so parentheses are required to group logical operations together within a Boolean expression. Table 6.2 shows the available logical operators available in VHDL, using parentheses to aid readability.

In addition to logical operators, relational operators are also available in VHDL, as shown in Table 6.3.

Table 6.4 shows the arithmetic operators available in VHDL.

Table 6.3: Relational operators in VHDL

Relational operation	Operator	Example
Equal to	=	If (A = B) Then
Not equal to	/=	If (A /= B) Then
Less than	<	If (A < B) Then
Less than or equal to	<=	If (A <= B) Then
Greater than	>	If (A > B) Then
Greater than or equal to	>=	If (A >= B) Then

Table 6.4: Arithmetic operators in VHDL

Arithmetic operation	Operator	Example
Addition	+	Z <= A + B;
Subtraction	−	Z <= A − B;
Multiplication	*	Z <= A * B;
Division	/	Z <= A / B;
Exponentiating	**	Z <= 4 ** 2;
Modulus	MOD	Z <= A MOD B;
Remainder	REM	Z <= A REM B;
Absolute value	ABS	Z <= ABS A;

Table 6.5: Concatenation operator in VHDL

Concatenation operation	Operator	Example
AND	&	Z <= A & B;

Table 6.6: Shift operators in VHDL

Shift operation	Operator	Example
Rotate left logical	rol	Z <= a rol 2;
Rotate right logical	ror	Z <= a ror 2;
Shift left arithmetic	sla	Z <= a sla 2;
Shift left logical	sll	Z <= a sll 2;
Shift right arithmetic	sra	Z <= a sra 2;
Shift right logical	srl	Z <= a srl 2;

A single concatenation operator is also available in VHDL, shown in Table 6.5.

Finally, the shift operators available in VHDL are shown in Table 6.6.

The shift right logical operation shifts an array to the right, drops the rightmost value, and fills the leftmost value with a fill value. The shift left logical operation shifts an array to the left, drops the leftmost value, and fills the rightmost value with a fill value.

The shift right arithmetic operation shifts an array to the right and uses the leftmost element for the left fill. The shift left arithmetic operation shifts an array to the left and uses the rightmost element for the right fill.

Note that in the VHDL reference manual, the **, ABS, and NOT operators are noted as miscellaneous operators, but are placed in the above tables in this text.

6.5.3 Commenting the Code

In the code shown in Figure 6.13, an obvious part missing is the code commenting. Comments in VHDL are included after a double dash (--) either at the beginning of a line or in-line (after) the main code. Note that VHDL is not case sensitive, but using lower- and uppercase characters aids readability.

Figure 6.14 is the code from Figure 6.13, but now with comments added.

```
1    ----------------------------------------------------------------
2    -- 2-input AND gate design
3    -- Dataflow description
4    ----------------------------------------------------------------
5
6    ----------------------------------------------------------------
7    -- Libraries and packages to use
8    ----------------------------------------------------------------
9
10   LIBRARY IEEE;
11   USE IEEE.STD_LOGIC_1164.ALL;
12
13   ----------------------------------------------------------------
14   -- Entity declaration
15   ----------------------------------------------------------------
16
17   ENTITY And_Gate IS
18       PORT(  A : IN   STD_LOGIC;    -- Input A
19              B : IN   STD_LOGIC;    -- Input B
20              Z : OUT  STD_LOGIC);   -- Output Z
21   END ENTITY And_Gate;
22
23   ----------------------------------------------------------------
24   -- Architecture body
25   ----------------------------------------------------------------
26
27   ARCHITECTURE Dataflow OF And_Gate IS
28
29   BEGIN
30
31   ----------------------------------------------
32   -- Z becomes A AND B
33   ----------------------------------------------
34
35         Z <= A AND B;
36
37   END ARCHITECTURE Dataflow;
38
39   ----------------------------------------------------------------
40   -- End of File
41   ----------------------------------------------------------------
```

Figure 6.14: Two-input AND gate VHDL entity and architecture with commenting

6.6   A First Design

6.6.1   Introduction

As a first design exercise, a 2-to-1 multiplexer design will be created. The multiplexer is a multiple-input, single-output circuit whose function is to provide a selection of an input: it is also considered to provide a parallel-to-serial conversion. Each input is selected, and using additional control inputs, the actual signal selected depends on the value of the control signal. Figure 6.15 shows a 2-to-1 multiplexer symbol.

A and B are the logic data inputs, and F is the logic data output. Sel0 is the control logic input. The truth table for this is shown in Figure 6.16.

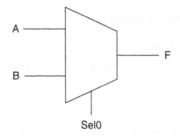

Figure 6.15: Two-to-one multiplexer

Sel0	F
0	F = A
1	F = B

Sel0	A	B	F
0	0	0	0
0	0	1	0
0	1	0	1
0	1	1	1
1	0	0	0
1	0	1	1
1	1	0	0
1	1	1	1

Figure 6.16: Two-to-one multiplexer operation

The Boolean expression for this in terms of basic logic gates is:

$$F = (A \cdot \overline{Sel0}) + (B \cdot Sel0)$$

The multiplexer design can be created in VHDL in a number of ways. For the same combinational logic design, the following coding styles will be considered:

- *Dataflow* describes the transfer of data from input to output and between signals.

- *Behavioral* describes the behavior of the design in terms of circuit or system behavior using algorithms.

- *Structural* describes the circuit structure in terms of the logic gates and interconnect wiring between the logic gates to form a circuit netlist.

6.6.2 Dataflow Description Example

In this example, the built-in logical operators AND, OR, and NOT are used to create the Boolean expression. Figure 6.17 provides an example VHDL code.

```
1    LIBRARY ieee;
2    USE ieee.std_logic_1164.all;
3    USE ieee.std_logic_arith.all;
4    USE ieee.std_logic_unsigned.all;
5
6    ENTITY Two_To_One_Mux_DataFlow is
7        PORT ( A    : IN    STD_LOGIC;
8               B    : IN    STD_LOGIC;
9               Sel0 : IN    STD_LOGIC;
10              F    : OUT   STD_LOGIC);
11   END ENTITY Two_To_One_Mux_DataFlow;
12
13   ARCHITECTURE DataFlow OF Two_To_One_Mux_DataFlow IS
14
15   BEGIN
16
17       F <= ((A AND NOT(Sel0)) OR (B AND Sel0));
18
19   END ARCHITECTURE DataFlow;
```

Figure 6.17: Two-to-one multiplexer dataflow description

The Boolean expression is placed after the BEGIN within the ARCHITECTURE body on line 17:

```
F <= ((A AND NOT(Sel0)) OR (B AND Sel0));
```

Parentheses are used to group the AND operators together and to group the result of the AND operators being ORed together.

In the dataflow description, operations are performed concurrently; that is, all operations are performed at the same time, called concurrent signal assignment. The dataflow description does not use processes (unlike a behavioral description) and their sequential signal assignment statements. Concurrent signal assignment statements are placed outside of process statements.

An example VHDL test bench for this design is shown in Figure 6.18. Here, all possible input combinations are applied changing every 10 ns.

6.6.3   Behavioral Description Example

In this example, the dataflow description is modified to place the Boolean logic expression within a process statement and the two inputs in the sensitivity list. Figure 6.19 provides an example behavioral description for the multiplexer design.

In this example, the difference between the dataflow and behavioral descriptions is that the behavioral description uses the PROCESS statement.

The PROCESS statement starts on line 17 with an optional name for the process (Mux_Process) followed by a colon, then the keyword PROCESS, and in parentheses, the signals that are in the sensitivity list. The process will react to changes in these signals, so it is essential that the sensitivity list contain all signals that will affect the process behavior.

The process finishes on line 23 with the keywords END and PROCESS, followed by a semicolon.

Within the process the behavior of the process is described using sequential statements to be executed in turn. Therefore, the ordering of sequential statements within the process is important because they are executed as they appear.

```
LIBRARY ieee;
USE ieee.std_logic_1164.all;
USE ieee.std_logic_arith.all;
USE ieee.std_logic_unsigned.all;

ENTITY Test_Two_To_One_Mux_DataFlow_vhd IS
END Test_Two_To_One_Mux_DataFlow_vhd;

ARCHITECTURE Behavioural OF Test_Two_To_One_Mux_DataFlow_vhd IS

COMPONENT Two_To_One_Mux_DataFlow
PORT(
        A    : IN   STD_LOGIC;
        B    : IN   STD_LOGIC;
        Sel0 : IN   STD_LOGIC;
        F    : OUT  STD_LOGIC);
END COMPONENT;

SIGNAL A    : STD_LOGIC := '0';
SIGNAL B    : STD_LOGIC := '0';
SIGNAL Sel0 : STD_LOGIC := '0';

SIGNAL F    : STD_LOGIC;

BEGIN

-------------------------------------------------------
-- Instantiate the Unit Under Test (UUT)
-------------------------------------------------------

uut: Two_To_One_Mux_DataFlow PORT MAP(
        A    => A,
        B    => B,
        Sel0 => Sel0,
        F    => F);

Test_Bench_Process : PROCESS

BEGIN

        wait for  0 ns; Sel0 <= '0'; A <= '0'; B <= '0';
        wait for 10 ns; Sel0 <= '0'; A <= '0'; B <= '1';
        wait for 10 ns; Sel0 <= '0'; A <= '1'; B <= '0';
        wait for 10 ns; Sel0 <= '0'; A <= '1'; B <= '1';
        wait for 10 ns; Sel0 <= '1'; A <= '0'; B <= '0';
        wait for 10 ns; Sel0 <= '1'; A <= '0'; B <= '1';
        wait for 10 ns; Sel0 <= '1'; A <= '1'; B <= '0';
        wait for 10 ns; Sel0 <= '1'; A <= '1'; B <= '1';
        wait for 10 ns;

END PROCESS;

END ARCHITECTURE Behavioural;
```

Figure 6.18: Two-to-one multiplexer dataflow description test bench

```
 1    LIBRARY ieee;
 2    USE ieee.std_logic_1164.all;
 3    USE ieee.std_logic_arith.all;
 4    USE ieee.std_logic_unsigned.all;
 5
 6    ENTITY Two_To_One_Mux_Behavioural is
 7        PORT ( A    : IN    STD_LOGIC;
 8               B    : IN    STD_LOGIC;
 9             Sel0 : IN    STD_LOGIC;
10               F    : OUT   STD_LOGIC);
11    END ENTITY Two_To_One_Mux_Behavioural;
12
13    ARCHITECTURE Behavioural OF Two_To_One_Mux_Behavioural IS
14
15    BEGIN
16
17    Mux_Process: PROCESS(A, B, Sel0)
18
19    BEGIN
20
21              F <= ((A AND NOT(Sel0)) OR (B AND Sel0));
22
23    END PROCESS;
24
25    END ARCHITECTURE Behavioural;
```

Figure 6.19: Two-to-one multiplexer behavioral description

Within an architecture are typically two or more processes that operate concurrently and can be thought of a blocks of hardware circuitry running in parallel.

An example VHDL test bench for this design is shown in Figure 6.20. Here, all possible input combinations are applied changing every 10 ns.

6.6.4 *Structural Description Example*

Structural descriptions are based on VHDL netlists and describes the instantiation of components and the interconnections between components. Components range from basic logic gates to complex subsystems of large digital systems designs. Structural designs are therefore hierarchical.

```
LIBRARY ieee;
USE ieee.std_logic_1164.all;
USE ieee.std_logic_arith.all;
USE ieee.std_logic_unsigned.all;

ENTITY Test_Two_To_One_Mux_Behavioural_vhd IS
END Test_Two_To_One_Mux_Behavioural_vhd;

ARCHITECTURE Behavioural OF Test_Two_To_One_Mux_Behavioural_vhd IS

COMPONENT Two_To_One_Mux_Behavioural
PORT(
        A    : IN    STD_LOGIC;
        B    : IN    STD_LOGIC;
        Sel0 : IN    STD_LOGIC;
        F    : OUT   STD_LOGIC);
END COMPONENT;

SIGNAL A    :  STD_LOGIC := '0';
SIGNAL B    :  STD_LOGIC := '0';
SIGNAL Sel0 :  STD_LOGIC := '0';

SIGNAL F    :  STD_LOGIC;

BEGIN

uut: Two_To_One_Mux_Behavioural PORT MAP(
        A      => A,
        B      => B,
        Sel0   => Sel0,
        F      => F);

Test_Bench_Process : PROCESS

BEGIN

        wait for  0 ns;  Sel0 <= '0';  A <= '0';  B <= '0';
        wait for 10 ns;  Sel0 <= '0';  A <= '0';  B <= '1';
        wait for 10 ns;  Sel0 <= '0';  A <= '1';  B <= '0';
        wait for 10 ns;  Sel0 <= '0';  A <= '1';  B <= '1';
        wait for 10 ns;  Sel0 <= '1';  A <= '0';  B <= '0';
        wait for 10 ns;  Sel0 <= '1';  A <= '0';  B <= '1';
        wait for 10 ns;  Sel0 <= '1';  A <= '1';  B <= '0';
        wait for 10 ns;  Sel0 <= '1';  A <= '1';  B <= '1';
        wait for 10 ns;

END PROCESS;

END ARCHITECTURE Behavioural;
```

Figure 6.20: Two-to-one multiplexer behavioral description test bench

The two-to-one multiplexer uses four logic gates in the circuit:

- two-input AND gate (two)

- two-input OR gate

- inverter (NOT gate)

In a structural description, these basic logic gates are first created, then instantiated. Figure 6.21 provides a behavioral description of these logic gates.

These logic gates are used within the multiplexer design by instantiating the gates. Figure 6.22 provides a structural description for the multiplexer design. Within the architecture, and before the BEGIN, the following are declared:

- **Internal signals**, X1, X2, and X3. The type of signal is STD_LOGIC, which represents signals that are internal to the design architecture and are not listed in the entity port declaration. The internals have a type but not a mode.

```
LIBRARY ieee;
USE ieee.std_logic_1164.all;
USE ieee.std_logic_arith.all;
USE ieee.std_logic_unsigned.all;

ENTITY And_Gate is
    PORT ( A     : IN    STD_LOGIC;
           B     : IN    STD_LOGIC;
           Z     : OUT   STD_LOGIC);
END ENTITY And_Gate;

ARCHITECTURE Behavioural OF And_Gate IS

BEGIN

AndGate_Process: PROCESS(A, B)

BEGIN

            Z <= (A AND B);

END PROCESS AndGate_Process;

END ARCHITECTURE Behavioural;
```

2-Input AND gate

Figure 6.21: Basic logic gate entity and architecture

```
LIBRARY ieee;
USE ieee.std_logic_1164.all;
USE ieee.std_logic_arith.all;
USE ieee.std_logic_unsigned.all;

ENTITY Or_Gate is
    PORT ( A    : IN   STD_LOGIC;
           B    : IN   STD_LOGIC;
           Z    : OUT  STD_LOGIC);
END ENTITY Or_Gate;

ARCHITECTURE Behavioural OF Or_Gate IS

BEGIN

OrGate_Process: PROCESS(A, B)

BEGIN
            Z <= (A OR B);

END PROCESS OrGate_Process;

END ARCHITECTURE Behavioural;
```

2-Input OR gate

```
LIBRARY ieee;
USE ieee.std_logic_1164.all;
USE ieee.std_logic_arith.all;
USE ieee.std_logic_unsigned.all;

ENTITY Inverter is
    PORT ( A    : IN   STD_LOGIC;
           Z    : OUT  STD_LOGIC);
END ENTITY Inverter;

ARCHITECTURE Behavioural OF Inverter IS

BEGIN

InverterGate_Process: PROCESS(A)

BEGIN
            Z <= NOT A;

END PROCESS InverterGate_Process;

END ARCHITECTURE Behavioural;
```

Inverter

Figure 6.21: (Continued)

```
LIBRARY ieee;
USE ieee.std_logic_1164.all;
USE ieee.std_logic_arith.all;
USE ieee.std_logic_unsigned.all;

ENTITY Two_To_One_Mux_Structural is
    PORT ( A     : IN    STD_LOGIC;
           B     : IN    STD_LOGIC;
           Sel0  : IN    STD_LOGIC;
           F     : OUT   STD_LOGIC);
END ENTITY Two_To_One_Mux_Structural;

ARCHITECTURE Structural OF Two_To_One_Mux_Structural IS

SIGNAL  X1 : STD_LOGIC;
SIGNAL  X2 : STD_LOGIC;
SIGNAL  X3 : STD_LOGIC;

COMPONENT And_Gate
PORT(
                A : IN   STD_LOGIC;
                B : IN   STD_LOGIC;
                Z : OUT  STD_LOGIC);
END COMPONENT;

COMPONENT Or_Gate
PORT(
                A : IN   STD_LOGIC;
                B : IN   STD_LOGIC;
                Z : OUT  STD_LOGIC);
END COMPONENT;

COMPONENT Inverter
PORT(
                A : IN   STD_LOGIC;
                Z : OUT  STD_LOGIC);
END COMPONENT;

BEGIN

I1: And_Gate
            PORT MAP(A => A, B => X1, Z => X2);

I2: And_Gate
            PORT MAP(A => Sel0, B => B, Z => X3);

I3: Or_Gate
            PORT MAP(A => X2, B => X3, Z => F);

I4: Inverter
            PORT MAP(A => Sel0, Z => X1);

END ARCHITECTURE Structural;
```

Figure 6.22: Two-to-one multiplexer structural description

- **Components**. Each component to be used in the structural description netlist must be declared. This has the same format as the component entity, except now the keyword ENTITY is replaced with the keyword COMPONENT. A component is an entity used within another entity.

After the BEGIN, each of the components placed in the structural design is instantiated. Each instance has a unique name; for example, the first instance is:

```
I1: And_Gate
        PORT MAP(A => A, B => X1, Z => X2);
```

The first line commences with the instance name (I1) followed by a colon and the component name (And_Gate).

The second line, Port Map, identifies how the component is to be connected within the design netlist. This starts with the keywords Port Map, followed by an open parenthesis. The mapping of the component port to the design signals is then declared, and at the end, the line is finished with);. Note that everything is placed, in this example, on one line to save space. A semicolon indicates a new line, so another way of writing this case would be:

```
I1: And_Gate
        PORT MAP( A => A,
                  B => X1,
                  Z => X2
                );
```

Instantiation and port mapping is shown for the AND gate in Figure 6.23.

An example VHDL test bench for this design is shown in Figure 6.24. Here, all possible input combinations are applied changing every 10 ns.

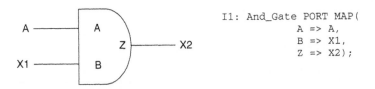

Figure 6.23: Component instantiation and port mapping

```
LIBRARY ieee;
USE ieee.std_logic_1164.all;
USE ieee.std_logic_arith.all;
USE ieee.std_logic_unsigned.all;

ENTITY Test_Two_To_One_Mux_Structural_vhd IS
END Test_Two_To_One_Mux_Structural_vhd;

ARCHITECTURE Behavioural OF Test_Two_To_One_Mux_Structural_vhd IS

COMPONENT Two_To_One_Mux_Structural
PORT(
      A    : IN    STD_LOGIC;
      B    : IN    STD_LOGIC;
      Sel0 : IN    STD_LOGIC;
      F    : OUT   STD_LOGIC);
END COMPONENT;

SIGNAL A    :  STD_LOGIC := '0';
SIGNAL B    :  STD_LOGIC := '0';
SIGNAL Sel0 :  STD_LOGIC := '0';

SIGNAL F :  STD_LOGIC;

BEGIN

uut: Two_To_One_Mux_Structural PORT MAP(
      A    => A,
      B    => B,
      Sel0 => Sel0,
      F    => F);

Test_Bench_Process : PROCESS

BEGIN

      wait for  0 ns;  Sel0 <= '0';  A <= '0';  B <= '0';
      wait for 10 ns;  Sel0 <= '0';  A <= '0';  B <= '1';
      wait for 10 ns;  Sel0 <= '0';  A <= '1';  B <= '0';
      wait for 10 ns;  Sel0 <= '0';  A <= '1';  B <= '1';
      wait for 10 ns;  Sel0 <= '1';  A <= '0';  B <= '0';
      wait for 10 ns;  Sel0 <= '1';  A <= '0';  B <= '1';
      wait for 10 ns;  Sel0 <= '1';  A <= '1';  B <= '0';
      wait for 10 ns;  Sel0 <= '1';  A <= '1';  B <= '1';
      wait for 10 ns;

END PROCESS;

END ARCHITECTURE Behavioural;
```

Figure 6.24: Two-to-one multiplexer structural description test bench

6.7   Signals versus Variables

6.7.1   Introduction

A carrier of values in VHDL can be declared as either a signal or a variable, and this declaration must be carefully made. Signals are declared when a design is intended for simulation and synthesis purposes, whereas variables are mainly used for behavioral modeling and for simulation purposes. The reason for this is that synthesis of variables in not always well defined in synthesis tools, and although possible, attempting to synthesize designs that use variables might result in different results for different synthesis tools.

Signals are used to connect components within a design and to carry information. Signals have hardware significance in that particular timing components are associated with them. The assignment symbol for signals is <=, which has a non-zero time component. Signals can be used in sequential (sequential statements inside processes) and concurrent bodies of VHDL, but can only be declared within concurrent bodies of VHDL.

The <= signal assignment symbol can be scheduled (in time) by the use of the AFTER clause. The delay is either an inertial delay or a transport delay, whose differences are explained by B <= A; and illustrated in Figure 6.25.

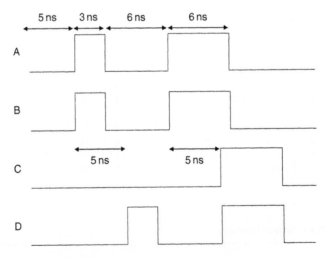

Figure 6.25: Inertial and transport delays

This means B is assigned the value of A after a zero time delay. Note that, for scheduling purposes, this is a delta delay.

```
C <= A AFTER 5 NS;
C <= INERTIAL A AFTER 5 NS;
```

This means C is assigned the value of A after a time delay of 5 ns. This, for example, can be used to model the propagation delay of signals within logic gates or within the interconnect between logic gates. It is useful for simulation purposes. However, a synthesis tool would not be able to understand this time delay (what type of circuit would implement such a delay time?). It is also referred to as an inertial delay. In addition to the 5 ns delay, any input signal pulse of less than 5 ns is suppressed. The reserved INERTIAL is optional and in many cases is omitted by code developers.

```
D <= TRANSPORT A AFTER 5 NS;
```

This means D is assigned the value of A after a time delay of 5 ns. This, can also be used to model the propagation delay of signals within logic gates or within the interconnect between logic gates. It is useful for simulation purposes. However, a synthesis tool would not be able to understand this time delay (what type of circuit would implement such a delay time?). It is referred to as a transport delay. The signal is simply delayed by 5 ns. Pulse widths of less than 5 ns are passed.

The differences between the inertial and transport delays are shown in Figure 6.25. Delays using the AFTER clause are not used where the delay is integral to the operation of the design after design synthesis. In such cases, a design is based on clocks to create designs with a determinable delay or occasionally by creating a string of logic gates where the delays within the logic gates combine to produce a delay. (This is not a well-defined time delay due to manufacturing process variations and because creating designs are difficult to test.)

The operation of the different types of delay can be seen in the VHDL code example shown in Figure 6.26. Here, the design has one input (A) and three outputs (B, C, and D). Each output becomes the value of the input after a time delay. For the inertial delay (output C), an input pulse width of less than 5 ns will be suppressed.

An example VHDL test bench for the design is shown in Figure 6.27.

Variables are used within processes to compute values. Unlike signals, variables do not have an associated time component. They are used to hold immediate values, in

```
LIBRARY ieee;
USE ieee.std_logic_1164.ALL;
-- Example design incorporating "numeric_std" package
USE ieee.numeric_std.ALL;

ENTITY Design1 is
    PORT ( A : IN    STD_LOGIC;
           B : OUT   STD_LOGIC;
           C : OUT   STD_LOGIC;
           D : OUT   STD_LOGIC);
END ENTITY Design1;

ARCHITECTURE  DataFlow OF Design1 IS

BEGIN

    B  <=  A;
    C  <=  A AFTER 5 NS;
    D  <=  TRANSPORT  A AFTER 5 NS;

END ARCHITECTURE DataFlow;
```

Figure 6.26: VHDL dataflow description showing inertial and transport delays

the same sense as software programming languages do. The assignment symbol for variables is :=, which has a zero time component. Variables can only be declared and used in sequential bodies of VHDL (including processes, functions, and procedures) and are local to the body in which they are declared.

6.7.2   Example: Architecture with Internal Signals

Signals internal to the design can be created to allow the output from some parts of the architecture to be used as inputs to other parts of the architecture, and to allow signals that drive ports of mode direction OUT to be used within the architecture itself. Consider the combinational logic circuit shown in Figure 6.28.

The output from the two-input OR gate is used for two purposes:

1. To drive the entity port (of mode OUT)

2. To drive the input of the inverter gate.

```
LIBRARY ieee;
USE ieee.std_logic_1164.ALL;
-- Example incorporating "numeric_std" package
USE ieee.numeric_std.ALL;

ENTITY Test_Design1_vhd IS
END ENTITY Test_Design1_vhd;

ARCHITECTURE Behavioural OF Test_Design1_vhd IS

COMPONENT Design1
PORT(
        A :  IN    STD_LOGIC;
        B :  OUT   STD_LOGIC;
        C :  OUT   STD_LOGIC;
        D :  OUT   STD_LOGIC);
END COMPONENT;

SIGNAL A :  STD_LOGIC := '0';

SIGNAL B :  STD_LOGIC;
SIGNAL C :  STD_LOGIC;
SIGNAL D :  STD_LOGIC;

BEGIN

uut: Design1 PORT MAP(
            A => A,
            B => B,
            C => C,
            D => D);

Test_Bench_Process : PROCESS

BEGIN

        wait for  0 ns;    A <= '0';
        wait for  5 ns;    A <= '1';
        wait for  3 ns;    A <= '0';
        wait for  6 ns;    A <= '1';
        wait for  6 ns;    A <= '0';
        wait for 20 ns;

END PROCESS;

END ARCHITECTURE Behavioural;
```

Figure 6.27: Inertial and transport delays test bench

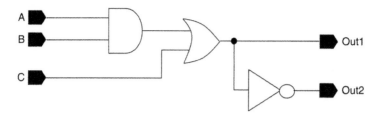

Figure 6.28: Example combinational logic circuit

As the port with a mode OUT cannot be read (i.e., within the architecture), a way to use this signal is to create an internal signal that drives both the port and the inverter gate. An example VHDL code for this design is shown in Figure 6.29.

Here, an internal signal (Out1_Internal) of type STD_LOGIC is created within the architecture body on line 16 before the BEGIN keyword. This internal signal is assigned

```
1     LIBRARY ieee;
2     USE ieee.std_logic_1164.all;
3     USE ieee.std_logic_arith.all;
4     USE ieee.std_logic_unsigned.all;
5
6     ENTITY Signals_1 IS
7         PORT ( A    : IN    STD_LOGIC;
8                B    : IN    STD_LOGIC;
9                C    : IN    STD_LOGIC;
10               Out1 : OUT   STD_LOGIC;
11               Out2 : OUT   STD_LOGIC);
12    END ENTITY Signals_1;
13
14    ARCHITECTURE DataFlow OF Signals_1 IS
15
16    SIGNAL Out1_Internal : STD_LOGIC;
17
18    BEGIN
19
20          Out1_Internal <= (A AND B) OR C;
21
22          Out1 <= Out1_Internal;
23
24          Out2 <= NOT(Out1_Internal);
25
26    END ARCHITECTURE DataFlow;
```

Figure 6.29: VHDL dataflow description for the combinational logic circuit

the output of the AND-OR logic combination. The entity output Out1 is assigned the value of the internal signal, and the entity output Out2 is assigned the inverse logic level (NOT) of the internal signal.

An example test bench for this design is shown in Figure 6.30. This applies all possible input codes, changing values every 10 ns.

```
LIBRARY ieee;
USE ieee.std_logic_1164.all;
USE ieee.std_logic_arith.all;
USE ieee.std_logic_unsigned.all;

ENTITY Test_Signals_1_vhd IS
END ENTITY Test_Signals_1_vhd;

ARCHITECTURE behavioural OF Test_Signals_1_vhd IS

COMPONENT Signals_1
PORT(
                A    : IN std_logic;
                B    : IN std_logic;
                C    : IN std_logic;
                Out1 : OUT std_logic;
                Out2 : OUT std_logic);
END COMPONENT;

SIGNAL A :   STD_LOGIC := '0';
SIGNAL B :   STD_LOGIC := '0';
SIGNAL C :   STD_LOGIC := '0';

SIGNAL Out1 :   STD_LOGIC;
SIGNAL Out2 :   STD_LOGIC;

BEGIN

uut: Signals_1 PORT MAP(
                A    => A,
                B    => B,
                C    => C,
                Out1 => Out1,
                Out2 => Out2);

Test_Bench_Process : PROCESS

BEGIN

        Wait for 0 ns;    A <= '0';  B <= '0';  C <= '0';
        Wait for 10 ns;   A <= '0';  B <= '0';  C <= '1';
        Wait for 10 ns;   A <= '0';  B <= '1';  C <= '0';
        Wait for 10 ns;   A <= '0';  B <= '1';  C <= '1';
        Wait for 10 ns;   A <= '1';  B <= '0';  C <= '0';
        Wait for 10 ns;   A <= '1';  B <= '0';  C <= '1';
        Wait for 10 ns;   A <= '1';  B <= '1';  C <= '0';
        Wait for 10 ns;   A <= '1';  B <= '1';  C <= '1';
        Wait for 10 ns;

END PROCESS;

END ARCHITECTURE behavioural;
```

Figure 6.30: Example combinational logic circuit test bench

6.7.3   Example: Architecture with Internal Variables

Variables are used within processes to compute values. Unlike signals, variables do not have an associated time component. A design can be created using variables rather than signals. For example, consider the two-input AND gate. A behavioral description for the AND gate is shown in Figure 6.31. Here, the description for AND gate behavior is placed within a process.

The process is located between lines 16 and 25. With this design, an internal variable (Tmp) is created within the process itself, and the output of the AND operation on the two inputs (A and B) is held in this variable. The value of the variable is then assigned to the output (Z). The variable assignment symbol (:=) is used rather than the signal assignment symbol (<=).

An example test bench for this design is shown in Figure 3.32.

```
 1    LIBRARY ieee;
 2    USE ieee.std_logic_1164.all;
 3    USE ieee.std_logic_arith.all;
 4    USE ieee.std_logic_unsigned.all;
 5
 6    ENTITY And_Gate_Variables is
 7        PORT ( A    : IN    STD_LOGIC;
 8               B    : IN    STD_LOGIC;
 9               Z    : OUT   STD_LOGIC);
10    END ENTITY And_Gate_Variables;
11
12    ARCHITECTURE Behavioural OF And_Gate_Variables IS
13
14    BEGIN
15
16    AndGate_Process: PROCESS(A, B)
17
18    VARIABLE Tmp: STD_LOGIC;
19
20    BEGIN
21
22                Tmp := (A AND B);
23                Z <= Tmp;
24
25    END PROCESS;
26
27    END ARCHITECTURE Behavioural;
```

Figure 6.31: AND gate with internal variable

```
LIBRARY ieee;
USE ieee.std_logic_1164.all;
USE ieee.std_logic_arith.all;
USE ieee.std_logic_unsigned.all;


ENTITY Test_And_Gate_Variables_vhd IS
END ENTITY Test_And_Gate_Variables_vhd;


ARCHITECTURE Behavioural OF Test_And_Gate_Variables_vhd IS

COMPONENT And_Gate_Variables
PORT(
                A : IN   STD_LOGIC;
                B : IN   STD_LOGIC;
                Z : OUT STD_LOGIC);
END COMPONENT;

SIGNAL A   :   STD_LOGIC := '0';
SIGNAL B   :   STD_LOGIC := '0';

SIGNAL Z   :   STD_LOGIC;


BEGIN

uut: And_Gate_Variables PORT MAP(
                A => A,
                B => B,
                Z => Z);


Test_Bench_Process : PROCESS

BEGIN

      Wait for 0 ns;    A <= '0';   B <= '0';
      Wait for 10 ns;   A <= '0';   B <= '1';
      Wait for 10 ns;   A <= '1';   B <= '0';
      Wait for 10 ns;   A <= '1';   B <= '1';
      Wait for 10 ns;

END PROCESS;

END ARCHITECTURE Behavioural;
```

Figure 6.32: AND gate with internal variable test bench

6.8   Generics

Generics are used where a value required for use within a design may need to be changed whenever the design is used. For example, an AND gate might be defined with an inertial time delay, but among the family of available AND gates, each might have a different delay. It would be possible to create VHDL entity–architecture pairs for each AND gate, where different entity names and delays would be required. It would be possible, however, to create a single AND gate entity–architecture pair and for a delay to be changed whenever the AND gate is to be used. A generic could then be defined within the design entity and used within the design architecture. The value for this generic would be set whenever the AND gate is used.

If an AND gate has an inertial delay of 5 ns, this is written as:

```
C <= (A AND B) AFTER 5 NS;
```

The time delay is replaced using a generic:

```
C <= (A AND B) AFTER delay_time;
```

The generic delay_time is used, and its value is set elsewhere. The use of generics is shown in Figure 6.33. Here, the design contains two two-input AND gates, each with a delay. These are placed within process statements. The first process (Delay1) has a delay of 5 ns explicitly stated within the process. The second process (Delay2) uses a generic (delay_time). This GENERIC is declared in the entity declaration (placed immediately before the PORT declaration). The generic name is delay_time and its type is TIME (for a time delay). Also in the declaration is the default value of 5 ns, although it need not be included if the value for the generic is to be passed to the design when the design is used within a structural style VHDL description.

An example test bench to simulate the design is shown in Figure 6.34. When simulated, the outputs are initially unknown (shown as a U value) until the delay time has passed.

The usefulness of using generics is apparent when a logic gate with a generic delay is used to set the parameter for a logic gate that is used multiple times. Figure 6.35 shows an example three-input AND gate using a generic delay time (delay_time).

```
LIBRARY ieee;
USE ieee.std_logic_1164.all;
USE ieee.std_logic_arith.all;
USE ieee.std_logic_unsigned.all;

ENTITY AND_Gate IS

    GENERIC( delay_time : TIME := 5 ns);

    PORT ( A  :  IN   STD_LOGIC;
           B  :  IN   STD_LOGIC;
           Z1 :  OUT  STD_LOGIC;
           Z2 :  OUT  STD_LOGIC);
END ENTITY AND_Gate;

ARCHITECTURE Behavioural OF AND_gate IS

BEGIN

Delay1: PROCESS(A, B)
BEGIN
             Z1 <= A AND B AFTER 5 NS;
END PROCESS;

Delay2: PROCESS(A, B)
BEGIN
             Z2 <= A AND B AFTER delay_time;
END PROCESS;

END ARCHITECTURE Behavioural;
```

Figure 6.33: AND gate using generic time delay

This has three inputs (A, B, and C), and one output (Z). A default delay time of 5 ns is defined in the generic declaration.

This AND gate is placed within a hierarchical design, and three instances of the AND gate are created. The structural VHDL code for this is shown in Figure 6.36 with three inputs (Ain, Bin, and Cin), and three outputs (Z1, Z2, and Z3).

Instance I1 has a delay of 1 ns, instance I2 has a delay of 5 ns, and instance I3 has a delay of 10 ns. These are all inertial delays that will override the default value of 5 ns.

```
LIBRARY ieee;
USE ieee.std_logic_1164.all;
USE ieee.std_logic_arith.all;
USE ieee.std_logic_unsigned.all;

ENTITY Test_AND_Gate_vhd IS
END Test_AND_Gate_vhd;

ARCHITECTURE Behavioural OF Test_AND_Gate_vhd IS

COMPONENT AND_gate
PORT(
            A  : IN   STD_LOGIC;
            B  : IN   STD_LOGIC;
            Z1 : OUT  STD_LOGIC;
            Z2 : OUT  STD_LOGIC);
END COMPONENT;

SIGNAL A  :  STD_LOGIC := '0';
SIGNAL B  :  STD_LOGIC := '0';

SIGNAL Z1 :  STD_LOGIC;
SIGNAL Z2 :  STD_LOGIC;

BEGIN

uut: AND_gate PORT MAP(
            A  => A,
            B  => B,
            Z1 => Z1,
            Z2 => Z2);

Input_Process : PROCESS

BEGIN

    Wait for 0 ns;   A <= '0';  B <= '0';
    Wait for 10 ns;  A <= '0';  B <= '1';
    Wait for 10 ns;  A <= '1';  B <= '0';
    Wait for 10 ns;  A <= '1';  B <= '1';
    Wait for 10 ns;

END PROCESS;

END ARCHITECTURE Behavioural;
```

Figure 6.34: AND gate using generic time delay test bench

```
LIBRARY ieee;
USE ieee.std_logic_1164.all;
USE ieee.std_logic_arith.all;
USE ieee.std_logic_unsigned_all;

ENTITY Three_Input_AndGate IS

    GENERIC( delay_time : TIME := 5 ns);

    PORT (  A  :  IN   STD_LOGIC;
            B  :  IN   STD_LOGIC;
            C  :  IN   STD_LOGIC;
            Z  :  OUT  STD_LOGIC);
END ENTITY Three_Input_AndGate;

ARCHITECTURE Behavioural OF Three_Input_AndGate IS

BEGIN

Delay_Process: PROCESS(A, B, C)

BEGIN

            Z <= (A AND B AND C) AFTER delay_time;

END PROCESS;

END ARCHITECTURE Behavioural;
```

Figure 6.35: Three-input AND gate using generic time delay

In I1, the top level design, the value for the generic is defined within the GENERIC MAP for the logic gate when it is instantiated. For example,

```
GENERIC MAP(delay_time => 1ns)
```

The delay_time here is set to 1ns. Note the use of the symbol => to set the value.

A test bench for this design is shown in Figure 6.37. Here, the top level design inputs (Ain, Bin, and Cin) are changed every 20 ns. This value was set to allow the effect of the gate delays to be seen in a reasonably short simulation time, while preventing timing problems that might arise when the inputs change too quickly and confuse the interpretation of the simulation results.

```
LIBRARY ieee;
USE ieee.std_logic_1164.all;
USE ieee.std_logic_arith.all;
USE ieee.std_logic_unsigned.all;

ENTITY Top_Design IS
Port ( Ain  : IN   STD_LOGIC;
       Bin  : IN   STD_LOGIC;
       Cin  : IN   STD_LOGIC;
       Z1   : OUT  STD_LOGIC;
       Z2   : OUT  STD_LOGIC;
       Z3   : OUT  STD_LOGIC);
END ENTITY Top_Design;

ARCHITECTURE Structural OF Top_Design IS

COMPONENT Three_Input_AndGate
GENERIC(delay_time : TIME);
PORT(
      A  : IN   STD_LOGIC;
      B  : IN   STD_LOGIC;
      C  : IN   STD_LOGIC;
      Z  : OUT  STD_LOGIC);
END COMPONENT;

BEGIN

I1: Three_Input_AndGate
          GENERIC MAP (delay_time => 1 ns)
          PORT MAP(A => Ain, B => Bin, C => Cin, Z => Z1);

I2: Three_Input_AndGate
          GENERIC MAP (delay_time => 5 ns)
          PORT MAP(A => Ain, B => Bin, C => Cin, Z => Z2);

I3: Three_Input_AndGate
          GENERIC MAP (delay_time => 10 ns)
          PORT MAP(A => Ain, B => Bin, C => Cin, Z => Z3);

END ARCHITECTURE Structural;
```

Figure 6.36: Structural design using AND gates

```
LIBRARY ieee;
USE ieee.std_logic_1164.all;
USE ieee.std_logic_arith.all;
USE ieee.std_logic_unsigned.all;

ENTITY Test_Top_Design_vhd IS
END ENTITY Test_Top_Design_vhd;

ARCHITECTURE Behavioural OF Test_Top_Design_vhd IS

COMPONENT Top_Design
PORT(
        Ain : IN  STD_LOGIC;
        Bin : IN  STD_LOGIC;
        Cin : IN  STD_LOGIC;
        Z1  : OUT STD_LOGIC;
        Z2  : OUT STD_LOGIC;
        Z3  : OUT STD_LOGIC);
END COMPONENT;

SIGNAL Ain :  STD_LOGIC := '0';
SIGNAL Bin :  STD_LOGIC := '0';
SIGNAL Cin :  STD_LOGIC := '0';

SIGNAL Z1 :  STD_LOGIC;
SIGNAL Z2 :  STD_LOGIC;
SIGNAL Z3 :  STD_LOGIC;

BEGIN

uut: Top_Design PORT MAP(
        Ain => Ain,
        Bin => Bin,
        Cin => Cin,
        Z1 => Z1,
        Z2 => Z2,
        Z3 => Z3);

Test_Bench_Process : PROCESS

BEGIN

      Wait for 0 ns;  Ain <= '0';  Bin <= '0';  Cin <= '0';
      Wait for 20 ns; Ain <= '0';  Bin <= '0';  Cin <= '1';
      Wait for 20 ns; Ain <= '0';  Bin <= '1';  Cin <= '0';
      Wait for 20 ns; Ain <= '0';  Bin <= '1';  Cin <= '1';
      Wait for 20 ns; Ain <= '1';  Bin <= '0';  Cin <= '0';
      Wait for 20 ns; Ain <= '1';  Bin <= '0';  Cin <= '1';
      Wait for 20 ns; Ain <= '1';  Bin <= '1';  Cin <= '0';
      Wait for 20 ns; Ain <= '1';  Bin <= '1';  Cin <= '1';
      Wait for 20 ns;

END PROCESS;

END ARCHITECTURE Behavioural;
```

Figure 6.37: Structural design test bench

6.9   Reserved Words

As with all software programming and hardware configuration languages, VHDL includes a number of reserved words. Reserved words cannot be used by the designer for identifiers such as signal names. As VHDL is not case sensitive, the case of the letters within the reserved word is not important. For example, *ELSE* is the same as *Else* is the same as *else*.

Table 6.7 shows the reserved words with special meanings in VHDL.

Table 6.7: Reserved words in VHDL (identifying only the reserved words used in the VHDL code examples)

abs	file	of	then
after	for	open	to
all		or	transport
and	generic	others	type
architecture		out	
array	if		until
	in	package	use
begin	inertial	port	
	inout	process	variable
case	is		
component		rem	wait
configuration	library	report	when
constant	linkage	rol	while
	loop	ror	with
downto			
	mod	select	xnor
else		signal	xor
elsif	nand	sla	
end	next	sll	
entity	nor	sra	
	not	srl	

Note: This is not a comprehensive list of reserved words. There are one hundred reserved words in the standard. Refer to the IEEE Std 1076™-2002 standard document for the full set of reserved words in VHDL.

6.10   Data Types

All objects—ports, signals, variables, and so forth—within VHDL have an associated type. A VHDL type specifies the values that the object may take and determines the operations that can be performed on the objects of that particular type. In this book, types STD_LOGIC and STD_LOGIC_VECTOR are mainly used.

VHDL puts strictly enforced constraints on the data types. Any operation to be performed on the object of a particular type is restricted to the types of operation that are permissible for that type.

VHDL has four classes of data type:

1. scalar

2. composite

3. access

4. file

A scalar type has a value that is a single entity (e.g., a logic 0 or logic 1 on a single signal line), and it can hold only one value at a time. Scalar types include all the simple data types in VHDL such as integer, real, physical, and enumerated. Integer types are integer numbers set between limits ($-$ve to $+$ve). Real types are used to hold floating point numbers. The physical type is a numeric data type that is used to describe physical quantities such as time and voltage. An enumerated type defines the values of the type by listing them within an ordered list. The elements within an enumerated type can be identifiers or character literals.

In many designs, the enumerated types used are BIT and STD_ULOGIC. STD_ULOGIC allows only one value to be put onto a signal. If more than value is put onto a signal, then a conflict occurs.

The IEEE Standard 1164 (in the IEEE.STD_LOGIC_1164 package) defines an enumerated type with nine possible values:

- U, uninitialized

- X, forcing unknown (strong)

- 0, forcing 0

- 1, forcing 1

- Z, high impedance

- W, weak unknown

- L, weak 0

- H, weak 1

- -, don't care

An enumerated standard logic type (STD_ULOGIC) is defined by:

```
type std_ulogic is ('U', 'X', 'O', '1', 'Z', 'W', 'L', 'H', '-');
```

A composite type allows objects to have more than one value. Composite types include arrays and records. Arrays group together elements of the same type as a single object. Elements of the array can be any VHDL data type, including other arrays. In addition, constants can be used to initialize the elements of an array to known values. Records are collections of named elements. These are used to model data structures, which consist of closely related items of potentially different data types.

In many designs, the composite types (vectors) used are BIT_VECTOR and STD_ULOGIC_VECTOR.

An access type is equivalent to pointers in a software programming language.

A file type is used for file access (read and write).

In addition to these four VHDL data types, there are subtypes. Subtypes are used when an object is known only to take on a restricted subset of values from the existing type. STD_ULOGIC has a subtype STD_LOGIC that allows more than one value to be placed on a signal. It has the same set of values as STD_ULOGIC. However, if more than value is put onto a signal, then a resolution function defines the actual state of the signal. The resolution function is named resolved, and the subtype declaration for STD_LOGIC is:

```
subtype std_logic is resolved std_ulogic;
```

The values 0, 1, L, and H are supported by synthesis. The value Z is supported by synthesis for tristate drivers. The value - is supported by synthesis for *Don't care* conditions. The values U, X, and W are not supported by synthesis.

In this text, only STD_LOGIC and STD_LOGIC_VECTOR will be used.

As VHDL is a strongly typed language, operations on an object of any type are restricted. In some cases, it may be necessary to convert the object from one type to another type. This is possible with type conversion. In addition to explicitly undertaking type conversion within the design architecture body, the use of overloaded operators (such as the + operator) allows for these overloaded operators to be used on mixed operand types (such as adding an integer number to a STD_LOGIC_VECTOR).

6.11   Concurrent versus Sequential Statements

VHDL architecture supports design descriptions written as concurrent statements and sequential statements.

- *Concurrent statements* include concurrent signal assignment, concurrent process, and component instantiations. They are written within the body of an architecture and lie outside of a process. The statements conceptually execute concurrently (at the same time), so their order of placement within the VHDL code is not important.

- *Sequential statements* are written within a process statement, function, or procedure. Sequential statements include:
    - *Case* statement
    - *If-then-else* statement
    - *Loop* statement

6.12   Loops and Program Control

As in software programming languages, looping statements and program control statements are also available in VHDL. These are:

1. ***If-then-else***, which takes the form:

```
IF (Reset = '0') THEN
      Q(7 downto 0) <= "00000000";
ELSIF (Clk'EVENT AND Clk = '1') THEN
      Q(7 DOWNTO 0) <= D(7 DOWNTO 0);
ELSE
END IF;
```

2. ***Case-when***, which takes the form:

```
CASE Control IS
      WHEN "00"   => Z <= A;
      WHEN "01"   => Z <= B;
      WHEN OTHERS => Z <= C;
END CASE;
```

3. *When-else*, which takes the form:

```
Z <= A WHEN (Control = "00") ELSE
Z <= B WHEN (Control = "01") ELSE
Z <= C;
```

4. *With-select-when*, which takes the form:

```
WITH Control SELECT
        Z <= A WHEN "00",
        Z <= B WHEN "01",
        Z <= C WHEN OTHERS;
```

5. *While-loop*, which takes the form:

```
While_Loop_Example:
        WHILE (Control > 0) LOOP
                Control := Control - 1;
                Z <= A + B;
        END LOOP While_Loop_Example;
```

6. *For-loop*, which takes the form:

```
For_Loop_Example:
        FOR i IN (1 DOWNTO 0) LOOP
            IF (Control(i) = '0') THEN
                    Z := A + B;
            END IF;
        END LOOP For_Loop_Example;
```

The ability to develop repetitive operations is achieved using both the `for loop` and `while loop` statements. Loop statements have many uses in hardware designs in executing a particular operation a set number of times or executing a particular operation until a certain condition is attained. The basic loop statement identified above can be extended so that they are executed conditionally by incorporating NEXT and/or EXIT statements within the loop itself. For example:

```
1. EXIT loop_label WHEN y = 20;
```

This will exit (terminate) the loop when the condition (`y = 20`) is attained.

2. NEXT loop_label WHEN condition;

This causes the remainder of the loop to be skipped and for the loop to immediately return to the start when a condition is reached.

The loop_label is optional. If the loop_label is included, then the NEXT or EXIT statements apply to that loop. If the loop_label is not included, then the NEXT or EXIT statements will apply to the innermost enclosing loop.

6.13   Coding Styles for VHDL

VHDL design can be written for the following purposes:

- To simulate the design description.

- To synthesize the design description. (This implies that the design has been successfully simulated prior to synthesis as it would be pointless to synthesize a design that does not work.)

- For documentation purposes.

Each of these purposes can be treated as separate or as parts of a whole. When writing VHDL code, the best practice in software design should be adopted:

1. Develop and use a *statement of requirements*. This document is written by the customer at the beginning of a project and can range from a minimal handwritten note to a detailed document that could run into volumes of text.

2. Develop and use a *system specification*. This document is developed from the statement of requirements and describes how the system will function and constraints put on the designer. Such constraints will be *process constraints* on how code development is to be undertaken and/or *product constraints* on the features of the code.

3. *System design* refers to developing the architecture of the code as a collection of blocks to meet the system specification.

4. *Detailed design* refers to defining the algorithms that make up the blocks identified in the system design.

5. *Programming* refers to the stage at which the code for the individual blocks and the overall design is written, and includes:

    a. **Validation and verification**. Validation is a process that ensures that the developed (or developing, if coding is in progress) system matches the user requirements. Verification is a process of checking that the system is a correct reflection of the system specification.

    b. **Integration testing**. This testing refers to the overall design to ensure that the blocks operate together correctly.

How these steps are undertaken depends on the size of the design. For example, a simple design will not require much effort in steps 1 through 4, with most of the effort in step 5. However, for large projects, steps 1 to 4 take a higher priority, and step 5 would require less time.

Whenever any VHDL code is to be written, the designer should consider the layout of the code (its appearance) as well as the correctness of the code. Care should be taken to consider:

- use of a suitable amount of commenting to ensure readability of the code
- use of suitable information concerning the design creation such as:
  - organization name
  - designer name
  - design project name
  - name of the design and intended functionality
  - date of creation and/or last modification
  - code modification history
  - target technology
  - tools used and their versions
- use of spaces and blank lines

An important aspect to consider when writing the VHDL code is its intended use. For example, is the intention to develop a high-level behavioral model for use in simulation studies to verify the operation of the design, but with no intention or attempt to synthesize the design? In this case, the code could be developed using behavioral descriptions that would not be synthesizable but can be quickly developed and simulated. Or is the intention to develop VHDL code that can be simulated but also can be synthesized into logic?

It is possible to develop VHDL code that can be simulated but cannot be synthesized. VHDL code can be either synthesizable or nonsynthesizable. Examples of VHDL code that cannot be synthesized, and so must be avoided if the design is to be synthesized, include:

- The use of the AFTER reserved word to define INERTIAL and TRANSPORT time delays. The AFTER reserved word is useful for simulation (i.e., to define the test stimulus timing in the VHDL test bench), but such time delays have no meaning in actual logic hardware.

- The use of the WAIT reserved word, since this also is associated with time delays.

- The use of file I/O operations for reading from and writing to text files when simulating the operation of the VHDL code design. File I/O operations have no meaning in actual logic hardware.

The use of any initial values for signals and variables, such as these, will be ignored. If values are to be initialized in a circuit, then they must be set using flip-flops with set or reset inputs.

6.14 Combinational Logic Design

6.14.1 Introduction

A logic circuit consists of combinational logic and sequential logic circuit elements. The combinational logic is defined by a Boolean logic expression (refer to Chapter 5 for an introduction to digital logic techniques) made up of the basic logic gates (AND, OR, etc.) whose meanings in VHDL are shown in Table 6.2. The logic operators can be used as they are, or they can be combined into expressions such as

```
Z1 <= (A AND B) OR (C AND D);
Z2 <= NOT(A) AND D;
Z3 <= NOT(A XOR B);
```

Note the use of parentheses to group expressions and to set the order in which the expressions are to be evaluated.

6.14.2   Complex Logic Gates

The complex logic gate shown in Figure 6.38 consists of the functions of three two-input logic gates with the Boolean expression given by:

$$\text{OUTPUT} = \overline{((\overline{A+B}) \cdot C) \cdot D}$$

Such a logic expression could be implemented in logic as three interconnected two-input logic gates or as a single complex logic gate consisting of transistors connected to form the function of the expression.

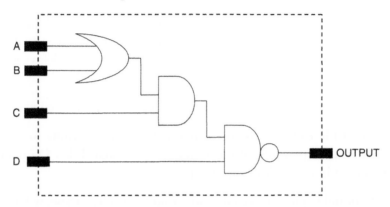

Figure 6.38: Example complex logic gate schematic

6.14.3   One-Bit Half-Adder

An important logic design created from the basic logic gates is the half-adder, shown in Figure 6.39, which has two inputs (A and B) and two outputs (Sum and Carry-Out (Cout)). This cell adds the two binary input numbers and produces sum and carry-out terms.

The truth table for this design is shown in Table 6.8.

The Sum output is only a logic 1 when *either but not both* inputs are logic 1: $\text{Sum} = (\overline{A} \cdot B) + (A \cdot \overline{B})$

Figure 6.39: Half-adder cell

Table 6.8: Half-adder cell truth table

A	B	Sum	Cout
0	0	0	0
0	1	1	0
1	0	1	0
1	1	0	1

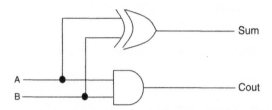

Figure 6.40: Half-adder logic diagram

This is actually the Exclusive-OR function, so:

Sum = $A \oplus B$

The Cout output is logic 1 only when *all two* inputs are logic 1 (i.e., A AND B):

Cout = $A . B$

This can be drawn as a logic diagram as shown in Figure 6.40.

A dataflow VHDL description for the one-bit half-adder that uses the two logic expressions is shown in Figure 6.41. Here, two expressions are placed in the architecture body (one expression for the Sum output on line 17, the second for the Cout output on line 18).

An example VHDL test bench for this design is shown in Figure 6.42.

6.14.4 *Four-to-One Multiplexer*

The multiplexer is a many-input-to-one-output circuit that allows one of many signals to be digitally switched (selected or multiplexed) to a single output under the control of additional control signals. Figure 6.43 shows an example of a four-to-one multiplexer that has four input signals (single bit), any one of which can be selected to become the output. The truth table included in the figure identifies the selected input for each combination of the control inputs C1 and C2.

```
 1    LIBRARY ieee;
 2    USE ieee.std_logic_1164.all;
 3    USE ieee.std_logic_arith.all;
 4    USE ieee.std_logic_unsigned.all;
 5
 6    ENTITY Half_Adder IS
 7        PORT ( A    : IN    STD_LOGIC;
 8               B    : IN    STD_LOGIC;
 9               Sum  : OUT   STD_LOGIC;
10               Cout : OUT   STD_LOGIC);
11    END ENTITY Half_Adder;
12
13    ARCHITECTURE Behavioural OF Half_Adder IS
14
15    BEGIN
16
17        Sum  <= (A XOR B);
18        Cout <= (A AND B);
19
20    END ARCHITECTURE Behavioural;
```

Figure 6.41: VHDL code for a one-bit half-adder

```
 1    LIBRARY ieee;
 2    USE ieee.std_logic_1164.all;
 3    USE ieee.std_logic_arith.all;
 4    USE ieee.std_logic_unsigned.all;
 5
 6    ENTITY Test_Half_Adder_vhd IS
 7    END Test_Half_Adder_vhd;
 8
 9    ARCHITECTURE behavioural OF Test_Half_Adder_vhd IS
10
11    COMPONENT Half_Adder
12    PORT(
13          A    : IN  STD_LOGIC;
14          B    : IN  STD_LOGIC;
15          Sum  : OUT STD_LOGIC;
16          Cout : OUT STD_LOGIC);
17    END COMPONENT;
18
```

Figure 6.42: VHDL test bench for a one-bit half-adder

```
19    SIGNAL A :   STD_LOGIC:= '0';
20    SIGNAL B :   STD_LOGIC:= '0';
21
22    SIGNAL Sum  :   STD_LOGIC;
23    SIGNAL Cout :   STD_LOGIC;
24
25    BEGIN
26
27    uut: Half_Adder PORT MAP(
28          A    => A,
29          B    => B,
30          Sum  => Sum,
31          Cout => Cout);
32
33    Test_Bench_Process : PROCESS
34    BEGIN
35          Wait for 0 ns;    A <= '0';   B <= '0';
36          Wait for 10 ns;   A <= '0';   B <= '1';
37          Wait for 10 ns;   A <= '1';   B <= '0';
38          Wait for 10 ns;   A <= '1';   B <= '1';
39          Wait for 10 ns;
40    END PROCESS;
41
42    END ARCHITECTURE behavioural;
```

Figure 6.42: (Continued)

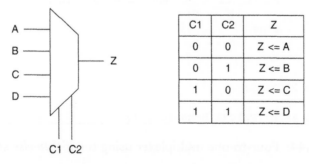

C1	C2	Z
0	0	Z <= A
0	1	Z <= B
1	0	Z <= C
1	1	Z <= D

Figure 6.43: Four-to-one multiplexer

The multiplexer can be created in VHDL using the *If-then-else* statement, as shown in Figure 6.44. Lines 1 to 4 identify the libraries to use. Lines 6 to 14 identify the design entity (Four_To_One_Mux) with four input signals (A, B, C, D), one output signal (Z), and two control input (C1, C2). Lines 16 to 36 identify the design architecture.

```
1    LIBRARY ieee;
2    USE ieee.std_logic_1164.all;
3    USE ieee.std_logic_arith.all;
4    USE ieee.std_logic_unsigned.all;
5
6    ENTITY Four_To_One_Mux is
7    PORT ( A          : IN    STD_LOGIC ;
8           B          : IN    STD_LOGIC;
9           C          : IN    STD_LOGIC;
10          D          : IN    STD_LOGIC;
11          C1         : IN    STD_LOGIC;
12          C2         : IN    STD_LOGIC;
13          Z          : OUT   STD_LOGIC);
14   END ENTITY Four_To_One_Mux;
15
16   ARCHITECTURE Behavioural OF Four_To_One_Mux IS
17
18   BEGIN
19
20   PROCESS (A, B, C, D, C1, C2)
21
22   BEGIN
23
24        If    (C1 = '0' AND C2 = '0') Then
25            Z <= A;
26        ElsIf (C1 = '0' AND C2 = '1') Then
27            Z <= B;
28        ElsIf (C1 = '1' AND C2 = '0') Then
29            Z <= C;
30        Else
31            Z <= D;
32        End If;
33
34   END PROCESS;
35
36   END ARCHITECTURE Behavioural;
```

Figure 6.44: Four-to-one multiplexer using the *If-then-else* statement

The design operation is defined within a single process in lines 20 to 34. This has a sensitivity list with all inputs to enable the process to react to changes in both the signal and control inputs.

An example test bench to simulate the buffer design is shown in Figure 6.45. The inputs change every 10 ns; there is zero time delay in the operation of the design, so this

```
LIBRARY ieee;
USE ieee.std_logic_1164.all;
USE ieee.std_logic_arith.all;
USE ieee.std_logic_unsigned.all;

ENTITY Test_Four_To_One_Mux_vhd IS
END ENTITY Test_Four_To_One_Mux_vhd;

ARCHITECTURE Behavioural OF Test_Four_To_One_Mux_vhd IS

COMPONENT Four_To_One_Mux
PORT(
        A  : IN  STD_LOGIC;
        B  : IN  STD_LOGIC;
        C  : IN  STD_LOGIC;
        D  : IN  STD_LOGIC;
        C1 : IN  STD_LOGIC;
        C2 : IN  STD_LOGIC;
        Z  : OUT STD_LOGIC);
END COMPONENT;

SIGNAL A  :  STD_LOGIC := '0';
SIGNAL B  :  STD_LOGIC := '0';
SIGNAL C  :  STD_LOGIC := '0';
SIGNAL D  :  STD_LOGIC := '0';
SIGNAL C1 :  STD_LOGIC := '0';
SIGNAL C2 :  STD_LOGIC := '0';

SIGNAL Z  :  STD_LOGIC;

BEGIN

uut: Four_To_One_Mux PORT MAP(
        A  =>  A,
        B  =>  B,
        C  =>  C,
        D  =>  D,
        C1 =>  C1,
        C2 =>  C2,
        Z  =>  Z);

Test_Bench_Process : PROCESS

BEGIN

Wait for 0 ns;  A <= '0'; B <= '0'; C <='0'; D <= '0'; C1 <= '0'; C2 <='0';
Wait for 10 ns; A <= '0'; B <= '0'; C <='0'; D <= '1'; C1 <= '0'; C2 <='0';
Wait for 10 ns; A <= '0'; B <= '0'; C <='1'; D <= '0'; C1 <= '0'; C2 <='0';
Wait for 10 ns; A <= '0'; B <= '1'; C <='0'; D <= '0'; C1 <= '0'; C2 <='0';
Wait for 10 ns; A <= '1'; B <= '0'; C <='0'; D <= '0'; C1 <= '0'; C2 <='0';

Wait for 10 ns; A <= '0'; B <= '0'; C <='0'; D <= '0'; C1 <= '0'; C2 <='1';
Wait for 10 ns; A <= '0'; B <= '0'; C <='0'; D <= '1'; C1 <= '0'; C2 <='1';
Wait for 10 ns; A <= '0'; B <= '0'; C <='1'; D <= '0'; C1 <= '0'; C2 <='1';
Wait for 10 ns; A <= '0'; B <= '1'; C <='0'; D <= '0'; C1 <= '0'; C2 <='1';
Wait for 10 ns; A <= '1'; B <= '0'; C <='0'; D <= '0'; C1 <= '0'; C2 <='1';

Wait for 10 ns; A <= '0'; B <= '0'; C <='0'; D <= '0'; C1 <= '1'; C2 <='0';
Wait for 10 ns; A <= '0'; B <= '0'; C <='0'; D <= '1'; C1 <= '1'; C2 <='0';
Wait for 10 ns; A <= '0'; B <= '0'; C <='1'; D <= '0'; C1 <= '1'; C2 <='0';
Wait for 10 ns; A <= '0'; B <= '1'; C <='0'; D <= '0'; C1 <= '1'; C2 <='0';
Wait for 10 ns; A <= '1'; B <= '0'; C <='0'; D <= '0'; C1 <= '1'; C2 <='0';

Wait for 10 ns; A <= '0'; B <= '0'; C <='0'; D <= '0'; C1 <= '1'; C2 <='1';
Wait for 10 ns; A <= '0'; B <= '0'; C <='0'; D <= '1'; C1 <= '1'; C2 <='1';
Wait for 10 ns; A <= '0'; B <= '0'; C <='1'; D <= '0'; C1 <= '1'; C2 <='1';
Wait for 10 ns; A <= '0'; B <= '1'; C <='0'; D <= '0'; C1 <= '1'; C2 <='1';
Wait for 10 ns; A <= '1'; B <= '0'; C <='0'; D <= '0'; C1 <= '1'; C2 <='1';

END PROCESS;

END ARCHITECTURE Behavioural;
```

Figure 6.45: Four-to-one multiplexer test bench for the *If-then-else* statement

short time between input signal changes would not cause any timing problems. The input signal is toggled between logic 0 and 1 for each state of the two control signals.

This design can also be configured using the *Case-when* statement, shown in Figure 6.46, which can be simulated with the test bench identified in Figure 6.47. In this design, however, the multiplexer input select control signal is applied as a

```
1     LIBRARY ieee;
2     USE ieee.std_logic_1164.all;
3     USE ieee.std_logic_arith.all;
4     USE ieee.std_logic_unsigned.all;
5
6     ENTITY Mux_Case_When IS
7     PORT ( A            : IN    STD_LOGIC;
8            B            : IN    STD_LOGIC;
9            C            : IN    STD_LOGIC;
10           D            : IN    STD_LOGIC;
11           Control      : IN    STD_LOGIC_VECTOR(1 downto 0);
12           Z            : OUT   STD_LOGIC);
13    END ENTITY Mux_Case_When;
14
15    ARCHITECTURE Behavioural OF Mux_Case_When IS
16
17    BEGIN
18
19    PROCESS (A, B, C, D, Control)
20
21    BEGIN
22
23    CASE Control IS
24
25              When "00"  =>  Z  <= A;
26              When "01"  =>  Z  <= B;
27              When "10"  =>  Z  <= C;
28              When "11"  =>  Z  <= D;
29
30              When OTHERS => Z  <= A;
31
32    END CASE;
33
34    END PROCESS;
35
36    END ARCHITECTURE Behavioural;
```

Figure 6.46: Four-to-one multiplexer using the *Case-when* statement

```
LIBRARY ieee;
USE ieee.std_logic_1164.all;
USE ieee.std_logic_arith.all;
USE ieee.std_logic_unsigned.all;

ENTITY Test_Mux_Case_When_vhd IS
END ENTITY Test_Mux_Case_When_vhd;

ARCHITECTURE Behavioural OF Test_Mux_Case_When_vhd IS

COMPONENT Mux_Case_When
PORT(
            A         : IN  STD_LOGIC;
            B         : IN  STD_LOGIC;
            C         : IN  STD_LOGIC;
            D         : IN  STD_LOGIC;
            Control   : IN STD_LOGIC_VECTOR(1 downto 0);
            Z         : OUT STD_LOGIC);
END COMPONENT;

SIGNAL A       :  STD_LOGIC := '0';
SIGNAL B       :  STD_LOGIC := '0';
SIGNAL C       :  STD_LOGIC := '0';
SIGNAL D       :  STD_LOGIC := '0';
SIGNAL Control :  STD_LOGIC_VECTOR(1 downto 0) := (others=>'0');

SIGNAL Z :  STD_LOGIC;

BEGIN

uut: Mux_Case_When PORT MAP(
            A => A,
            B => B,
            C => C,
            D => D,
            Control => Control,
            Z => Z);

Test_Bench_Process : PROCESS

BEGIN

Wait for 0 ns;   A <= '0';  B <= '0';  C <='0';  D <= '0'; Control(1 downto 0) <= "00";
Wait for 10 ns;  A <= '0';  B <= '0';  C <='0';  D <= '1'; Control(1 downto 0) <= "00";
Wait for 10 ns;  A <= '0';  B <= '0';  C <='1';  D <= '0'; Control(1 downto 0) <= "00";
Wait for 10 ns;  A <= '0';  B <= '1';  C <='0';  D <= '0'; Control(1 downto 0) <= "00";
Wait for 10 ns;  A <= '1';  B <= '0';  C <='0';  D <= '0'; Control(1 downto 0) <= "00";

Wait for 10 ns;  A <= '0';  B <= '0';  C <='0';  D <= '0'; Control(1 downto 0) <= "01";
Wait for 10 ns;  A <= '0';  B <= '0';  C <='0';  D <= '1'; Control(1 downto 0) <= "01";
Wait for 10 ns;  A <= '0';  B <= '0';  C <='1';  D <= '0'; Control(1 downto 0) <= "01";
Wait for 10 ns;  A <= '0';  B <= '1';  C <='0';  D <= '0'; Control(1 downto 0) <= "01";
Wait for 10 ns;  A <= '1';  B <= '0';  C <='0';  D <= '0'; Control(1 downto 0) <= "01";

Wait for 10 ns;  A <= '0';  B <= '0';  C <='0';  D <= '0'; Control(1 downto 0) <= "10";
Wait for 10 ns;  A <= '0';  B <= '0';  C <='0';  D <= '1'; Control(1 downto 0) <= "10";
Wait for 10 ns;  A <= '0';  B <= '0';  C <='1';  D <= '0'; Control(1 downto 0) <= "10";
Wait for 10 ns;  A <= '0';  B <= '1';  C <='0';  D <= '0'; Control(1 downto 0) <= "10";
Wait for 10 ns;  A <= '1';  B <= '0';  C <='0';  D <= '0'; Control(1 downto 0) <= "10";

Wait for 10 ns;  A <= '0';  B <= '0';  C <='0';  D <= '0'; Control(1 downto 0) <= "11";
Wait for 10 ns;  A <= '0';  B <= '0';  C <='0';  D <= '1'; Control(1 downto 0) <= "11";
Wait for 10 ns;  A <= '0';  B <= '0';  C <='1';  D <= '0'; Control(1 downto 0) <= "11";
Wait for 10 ns;  A <= '0';  B <= '1';  C <='0';  D <= '0'; Control(1 downto 0) <= "11";
Wait for 10 ns;  A <= '1';  B <= '0';  C <='0';  D <= '0'; Control(1 downto 0) <= "11";

END PROCESS;

END ARCHITECTURE Behavioural;
```

Figure 6.47: Four-to-one multiplexer test bench for the _Case-when_ statement

two-bit-wide STD_LOGIC_VECTOR named Control. All four possible inputs for the combinations of logic levels 0 and 1 are defined in the CASE statement (in lines 23 to 32). Line 30 is a *Catch all others* statement that sets the output to input A for all other (if any) possible input combinations! This can be left out (if the results would be the same) or set to a *Don't care* condition with '—' rather than to A.

Another alternative is to configure the multiplexer using the *When-else* statement. This statement is placed in a dataflow description rather than within a behavioral description process, because only sequential statements may be placed within the statement part of a process. An example is shown in Figure 6.48 and can be simulated with the test bench identified in Figure 6.47 (if the reference to Mux_Case_When is replaced with Mux_When_Else).

A final possible configuration uses the *With-select-when* statement. This statement is placed in a dataflow description rather than within a behavioral description process. An example of this is shown in Figure 6.49 and can be simulated with the test bench

```
1    LIBRARY ieee;
2    USE ieee.std_logic_1164.all;
3    USE ieee.std_logic_arith.all;
4    USE ieee.std_logic_unsigned.all;
5
6    ENTITY Mux_When_Else IS
7    PORT ( A              : IN    STD_LOGIC;
8           B              : IN    STD_LOGIC;
9           C              : IN    STD_LOGIC;
10          D              : IN    STD_LOGIC;
11          Control        : IN    STD_LOGIC_VECTOR(1 downto 0);
12          Z              : OUT   STD_LOGIC);
13   END ENTITY Mux_When_Else;
14
15   ARCHITECTURE Dataflow OF Mux_When_Else IS
16
17   BEGIN
18
19        Z   <= A   WHEN (Control(1 downto 0) = "00") ELSE
20               B   WHEN (Control(1 downto 0) = "01") ELSE
21               C   WHEN (Control(1 downto 0) = "10") ELSE
22               D   WHEN (Control(1 downto 0) = "11") ELSE
23               A;
24
25   END ARCHITECTURE Dataflow;
```

Figure 6.48: Four-to-one multiplexer using the *When-else* statement

```
1    LIBRARY ieee;
2    USE ieee.std_logic_1164.all;
3    USE ieee.std_logic_arith.all;
4    USE ieee.std_logic_unsigned.all;
5
6    ENTITY Mux_With_Select_When IS
7    PORT ( A               : IN    STD_LOGIC;
8           B               : IN    STD_LOGIC;
9           C               : IN    STD_LOGIC;
10          D               : IN    STD_LOGIC;
11          Control         : IN    STD_LOGIC_VECTOR(1 downto 0);
12          Z               : OUT   STD_LOGIC);
13   END ENTITY Mux_With_Select_When;
14
15   ARCHITECTURE Dataflow OF Mux_With_Select_When IS
16
17   BEGIN
18
19        WITH Control(1 downto 0) SELECT
20
21                    Z    <=   A      WHEN   "00",
22                              B      WHEN   "01",
23                              C      WHEN   "10",
24                              D      WHEN   "11",
25                              '-'    WHEN   OTHERS;
26
27   END ARCHITECTURE Dataflow;
```

Figure 6.49: Four-to-one multiplexer using the *With-select-when* statement

identified in Figure 6.47 (if the reference to (`Mux_Case_When` is replaced with `Mux_With_Select_When`).

6.14.5 *Thermometer-to-Binary Encoder*

The thermometer-to-binary encoder circuit is used in a number of applications where a circuit produces an output that is a binary representation of a thermometer code input. An example is the flash analogue-to-digital converter (ADC). The basic idea is shown in Table 6.9 with reference to a three-bit binary output. The input code starts with all 0s, and then from the LSB (here X_0), each bit becomes a logic 1 until all the inputs are at a logic 1 level.

This circuit has seven inputs (X_6 to X_0) and three outputs (d_2 to d_0) and can be implemented in VHDL code using the conditional statements. An example using the

Table 6.9: Thermometer code to three-bit binary encoder

X_6	X_5	X_4	X_3	X_2	X_1	X_0	d_2	d_1	d_0
0	0	0	0	0	0	0	0	0	0
0	0	0	0	0	0	1	0	0	1
0	0	0	0	0	1	1	0	1	0
0	0	0	0	1	1	1	0	1	1
0	0	0	1	1	1	1	1	0	0
0	0	1	1	1	1	1	1	0	1
0	1	1	1	1	1	1	1	1	0
1	1	1	1	1	1	1	1	1	1

Case-when statement within a process is shown in Figure 6.50. In this, the inputs and outputs are declared each as a STD_LOGIC_VECTOR.

An example VHDL test bench for this design is shown in Figure 6.51.

6.14.6 Seven-Segment Display Driver

Consider the seven-segment display, a commonly used display device consisting of eight LED segments, each of which can be switched ON or OFF independently. The display is available as either a common cathode (where all the LED cathode connections are connected together) or a common anode (where all the LED anode connections are connected together) device. The display is shown in Figure 6.52. Each segment has a letter identifier (a through f for the character display and dp for the decimal point for number displays).

The idea is to turn the individual LED segments ON or OFF to create letters or numbers. When two or more displays are placed side by side, then messages consisting of words and numbers are created. In digital logic terms, and considering a common cathode display, applying a logic 1 (i.e., high voltage) will turn the LED segment ON. It is common to create numbers 0 to 9 and letters A to F, as shown in Table 6.10. A logic 0 (i.e., low voltage) represents the LED segment OFF.

The display segments ON and OFF for each of the characters in Table 6.10 is shown in Figure 6.53. Where the segment is black, the segment is ON. Where the segment is white, the segment is OFF.

Table 6.10 shows sixteen possible combinations with the characters representing the hexadecimal equivalent of a four-bit binary number. A byte of data (i.e., eight bits)

```
 1    LIBRARY ieee;
 2    USE ieee.std_logic_1164.all;
 3    USE ieee.std_logic_arith.all;
 4    USE ieee.std_logic_unsigned.all;
 5
 6    ENTITY Thermometer_Case_When IS
 7    PORT (  X    : IN   STD_LOGIC_VECTOR(6 downto 0);
 8            d    : OUT  STD_LOGIC_VECTOR(2 downto 0));
 9    END ENTITY Thermometer_Case_When;
10
11    ARCHITECTURE Behavioural OF Thermometer_Case_When IS
12
13    BEGIN
14
15    PROCESS(X)
16
17    BEGIN
18
19    CASE (X) IS
20
21                When "0000000"  =>  d(2 downto 0) <= "000";
22                When "0000001"  =>  d(2 downto 0) <= "001";
23                When "0000011"  =>  d(2 downto 0) <= "010";
24                When "0000111"  =>  d(2 downto 0) <= "011";
25                When "0001111"  =>  d(2 downto 0) <= "100";
26                When "0011111"  =>  d(2 downto 0) <= "101";
27                When "0111111"  =>  d(2 downto 0) <= "110";
28                When "1111111"  =>  d(2 downto 0) <= "111";
29
30                When OTHERS     =>  d(2 downto 0) <= "000";
31
32    END CASE;
33
34    END PROCESS;
35
36    END ARCHITECTURE Behavioural;
```

Figure 6.50: Thermometer code to three-bit binary encoder using the
Case-when **statement**

can be viewed in hexadecimal format on two seven-segment displays as the upper nibble and lower nibble, shown in Figure 6.54.

Circuit implementation must incorporate current-limiting resistors in series with each LED segment, with the value of resistance dependent on the voltage level representing

```
1    LIBRARY ieee;
2    USE ieee.std_logic_1164.all;
3    USE ieee.std_logic_arith.all;
4    USE ieee.std_logic_unsigned.all;
5
6    ENTITY Test_Thermometer_Case_When_vhd IS
7    END ENTITY Test_Thermometer_Case_When_vhd;
8
9    ARCHITECTURE Behavioural OF Test_Thermometer_Case_When_vhd IS
10
11   COMPONENT Thermometer_Case_When
12   PORT(   X    : IN    STD_LOGIC_VECTOR(6 downto 0);
13           d    : OUT   STD_LOGIC_VECTOR(2 downto 0));
14   END COMPONENT;
15
16   SIGNAL  X :  STD_LOGIC_VECTOR(6 downto 0) := (others=>'0');
17
18   SIGNAL  d :  STD_LOGIC_VECTOR(2 downto 0);
19
20   BEGIN
21
22   uut: Thermometer_Case_When PORT MAP(
23           X => X,
24           d => d);
25
26   Test_Bench_Process : PROCESS
27
28   BEGIN
29
30   Wait for 0 ns;  X <= "0000000";
31   Wait for 10 ns; X <= "0000001";
32   Wait for 10 ns; X <= "0000011";
33   Wait for 10 ns; X <= "0000111";
34   Wait for 10 ns; X <= "0001111";
35   Wait for 10 ns; X <= "0011111";
36   Wait for 10 ns; X <= "0111111";
37   Wait for 10 ns; X <= "1111111";
38   Wait for 10 ns;
39
40   END PROCESS;
41
42   END ARCHITECTURE Behavioural;
```

Figure 6.51: Test bench for the thermometer code to a three-bit binary encoder

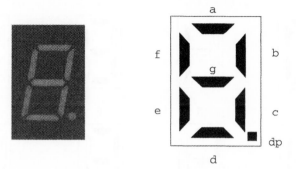

Figure 6.52: Seven-segment display: component (left), segment assignments (right)

Table 6.10: Creating numbers and letters

Character to create	a	b	c	d	e	f	g
0	1	1	1	1	1	1	0
1	0	1	1	0	0	0	0
2	1	1	0	1	1	0	1
3	1	1	1	1	0	0	1
4	0	1	1	0	0	1	1
5	1	0	1	1	0	1	1
6	1	0	1	1	1	1	1
7	1	1	1	0	0	0	0
8	1	1	1	1	1	1	1
9	1	1	1	1	0	1	1
A	1	1	1	0	1	1	1
B	0	0	1	1	1	1	1
C	1	0	0	1	1	1	0
D	0	1	1	1	1	0	1
E	1	0	0	1	1	1	1
F	1	0	0	0	1	1	1

the logic level (e.g., $+3.3$ V $=$ logic 1), the current required by the LED to provide illumination, and the voltage drop across the LED when illuminated. In Figure 6.55, a voltage (V_{IN}) of $+5.0$ V represents a logic high level used to illuminate an LED. The LED requires a current of 20 mA to illuminate, and when operating, a voltage drop of $+2.0$ V is across the LED. A current-limiting resistor drops the difference voltage between V_{IN} and V_D (i.e., $+3.0$ V). From Ohm's Law, then, the value of resistance ($150\,\Omega$) can be calculated.

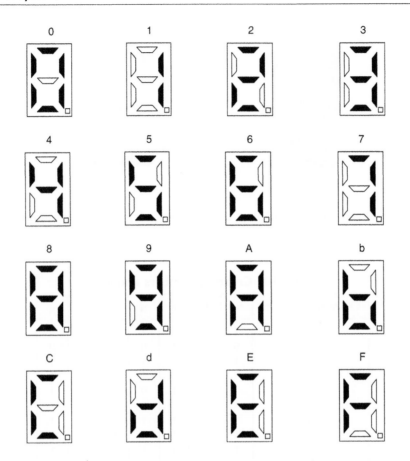

Figure 6.53: Seven-segment display illumination for different characters

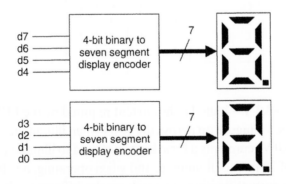

Figure 6.54: Viewing a byte of data on two seven-segment displays

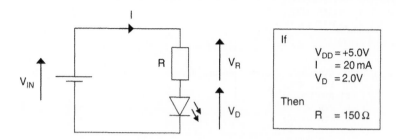

Figure 6.55: LED current-limiting resistor calculation

If a seven-segment display is to display the hexadecimal value of a four-bit input, then a combinational logic circuit is required. The following example designs VHDL code use the following statements:

- *Case-when* statement

- *If-then-else* statement

In this example, the dp (decimal point) segment is ignored (and in the circuit would be connected to 0 V).

Case-when Statement

A design based on a *Case-when* statement is shown in Figure 6.56. Here, the architecture contains a single process with one item in the sensitivity list. Data_In is a four-bit STD_LOGIC_VECTOR and applies the four bits of input data to the entity. The case statement within the process is of the form:

```
CASE Control IS
      WHEN "00"   => Z <= A;
      WHEN "01"   => Z <= B;
      WHEN OTHERS => Z <= C;
END CASE;
```

In this example, the case statement starts with: CASE Data_In IS

The keyword CASE is followed by the expression to use (the input signal Data_In) and the keyword IS.

```
LIBRARY ieee;
USE ieee.std_logic_1164.all;
USE ieee.std_logic_arith.all;
USE ieee.std_logic_unsigned.all;

ENTITY Hex_Converter IS

PORT(
        Data_In : IN  STD_LOGIC_VECTOR(3 downto 0);
        a        : OUT STD_LOGIC;
        b        : OUT STD_LOGIC;
        c        : OUT STD_LOGIC;
        d        : OUT STD_LOGIC;
        e        : OUT STD_LOGIC;
        f        : OUT STD_LOGIC;
        g        : OUT STD_LOGIC);

END ENTITY Hex_Converter;

-----------------------------------------------------------------
-- Hex_Converter Architecture
-----------------------------------------------------------------

ARCHITECTURE Behavioural OF Hex_Converter IS

BEGIN

----------------------------------------
-- Process to perform display encoding

PROCESS(Data_In)

BEGIN

CASE Data_In IS

When "0000" =>    a <= '1'; b <= '1'; c <= '1'; d <= '1'; e <= '1'; f <= '1'; g <= '0';
When "0001" =>    a <= '0'; b <= '1'; c <= '1'; d <= '0'; e <= '0'; f <= '0'; g <= '0';
When "0010" =>    a <= '1'; b <= '1'; c <= '0'; d <= '1'; e <= '1'; f <= '0'; g <= '1';
When "0011" =>    a <= '1'; b <= '1'; c <= '1'; d <= '1'; e <= '0'; f <= '0'; g <= '1';
When "0100" =>    a <= '0'; b <= '1'; c <= '1'; d <= '0'; e <= '0'; f <= '1'; g <= '1';
When "0101" =>    a <= '1'; b <= '0'; c <= '1'; d <= '1'; e <= '0'; f <= '1'; g <= '1';
When "0110" =>    a <= '1'; b <= '0'; c <= '1'; d <= '1'; e <= '1'; f <= '1'; g <= '1';
When "0111" =>    a <= '1'; b <= '1'; c <= '1'; d <= '0'; e <= '0'; f <= '0'; g <= '0';
When "1000" =>    a <= '1'; b <= '1'; c <= '1'; d <= '1'; e <= '1'; f <= '1'; g <= '1';
When "1001" =>    a <= '1'; b <= '1'; c <= '1'; d <= '0'; e <= '0'; f <= '1'; g <= '1';
When "1010" =>    a <= '1'; b <= '1'; c <= '1'; d <= '0'; e <= '1'; f <= '1'; g <= '1';
When "1011" =>    a <= '0'; b <= '0'; c <= '1'; d <= '1'; e <= '1'; f <= '1'; g <= '1';
When "1100" =>    a <= '1'; b <= '0'; c <= '0'; d <= '1'; e <= '1'; f <= '1'; g <= '0';
When "1101" =>    a <= '0'; b <= '1'; c <= '1'; d <= '1'; e <= '1'; f <= '0'; g <= '1';
When "1110" =>    a <= '1'; b <= '0'; c <= '0'; d <= '1'; e <= '1'; f <= '1'; g <= '1';
When "1111" =>    a <= '1'; b <= '0'; c <= '0'; d <= '0'; e <= '1'; f <= '1'; g <= '1';

When OTHERS =>    a <= '0'; b <= '0'; c <= '0'; d <='0'; e <= '0'; f <= '0'; g <= '0';

END CASE;

END PROCESS;

END ARCHITECTURE Behavioural;
```

Figure 6.56: *Case-when* statement example

The case statement ends with the line END CASE;. Within the case statement, each case for the expression is identified and the outputs set for each case. The first case is:

```
When "0000" => a <= '1' b <= '1'; c <= '1';
d <= '1'; e <= '1'; f <= '1'; g <= '0';
```

This states that when Data_In is "0000", then (=>) the outputs will all be set to logic 0 or 1 to create the number 0. Note that everything is placed on one line to save space. A semicolon indicates a new line, so another way of writing this case would be:

```
When "0000" =>

    a <= '1';
    b <= '1';
    c <= '1';
    d <= '1';
    e <= '1';
    f <= '1';
    g <= '0';
```

The designer can use spaces, line indentation, multiple lines, and blank lines to aid readability. It is important to ensure that all possible cases of the input signal values are covered in the case statement. This will ensure that when the design is synthesized, the output of the synthesis will produce a known result. Otherwise, if some states are omitted from the case statement, the synthesis tool will decide what the result of the synthesis should be, and the result might be a circuit that uses more logic than otherwise required. In Figure 6.56, the last line within the case statement is incorporated:

```
When OTHERS => a <= '0'; b <= '0'; c <= '0';
d <='0'; e <= '0'; f <= '0'; g <= '0';
```

This line uses the keyword OTHERS, which states that any cases for the expression that have not been previously defined sets all the outputs to logic 0.

Figure 6.57 shows an example VHDL test bench for the design. This will run through each possible input state for the input signal, changing values every 10 ns.

If-then-else Statement

An alternative to the case statement is the *If-then-else* statement, shown in Figure 6.58. Here, the architecture contains a single process with one item in the sensitivity list. Data_In is a four-bit STD_LOGIC_VECTOR and applies the four

```
LIBRARY ieee;
USE ieee.std_logic_1164.all;
USE ieee.std_logic_arith.all;
USE ieee.std_logic_unsigned.all;

ENTITY Test_Hex_Converter_vhd IS
END ENTITY Test_Hex_Converter_vhd;

ARCHITECTURE Behavioural OF Test_Hex_Converter_vhd IS
COMPONENT Hex_Converter
PORT(
        Data_In  :  IN   STD_LOGIC_VECTOR(3 downto 0);
        a        :  OUT  STD_LOGIC;
        b        :  OUT  STD_LOGIC;
        c        :  OUT  STD_LOGIC;
        d        :  OUT  STD_LOGIC;
        e        :  OUT  STD_LOGIC;
        f        :  OUT  STD_LOGIC;
        g        :  OUT  STD_LOGIC);
END COMPONENT;
SIGNAL Data_In :  std_logic_vector(3 downto 0) := (others=>'0');

SIGNAL a :  std_logic;
SIGNAL b :  std_logic;
SIGNAL c :  std_logic;
SIGNAL d :  std_logic;
SIGNAL e :  std_logic;
SIGNAL f :  std_logic;
SIGNAL g :  std_logic;
BEGIN
uut: Hex_Converter PORT MAP(
            Data_In  =>  Data_In,
            a        =>  a,
            b        =>  b,
            c        =>  c,
            d        =>  d,
            e        =>  e,
            f        =>  f,
            g        =>  g);

Test_Bench_Process : PROCESS

BEGIN
        wait for  0 ns;  Data_In <= "0000";
        wait for 10 ns;  Data_In <= "0001";
        wait for 10 ns;  Data_In <= "0010";
        wait for 10 ns;  Data_In <= "0011";
        wait for 10 ns;  Data_In <= "0100";
        wait for 10 ns;  Data_In <= "0101";
        wait for 10 ns;  Data_In <= "0110";
        wait for 10 ns;  Data_In <= "0111";
        wait for 10 ns;  Data_In <= "1000";
        wait for 10 ns;  Data_In <= "1001";
        wait for 10 ns;  Data_In <= "1010";
        wait for 10 ns;  Data_In <= "1011";
        wait for 10 ns;  Data_In <= "1100";
        wait for 10 ns;  Data_In <= "1101";
        wait for 10 ns;  Data_In <= "1110";
        wait for 10 ns;  Data_In <= "1111";
        wait for 10 ns;

END PROCESS;

END ARCHITECTURE Behavioural;
```

Figure 6.57: *Case-when* statement example test bench

```
LIBRARY ieee;
USE ieee.std_logic_1164.all;
USE ieee.std_logic_arith.all;
USE ieee.std_logic_unsigned.all;

ENTITY Hex_Converter IS

PORT(
        Data_In : IN    STD_LOGIC_VECTOR(3 downto 0);
        a       : OUT   STD_LOGIC;
        b       : OUT   STD_LOGIC;
        c       : OUT   STD_LOGIC;
        d       : OUT   STD_LOGIC;
        e       : OUT   STD_LOGIC;
        f       : OUT   STD_LOGIC;
        g       : OUT   STD_LOGIC);

END ENTITY Hex_Converter;

ARCHITECTURE Behavioural OF Hex_Converter IS

BEGIN

PROCESS(Data_In)

BEGIN

IF (Data_In = "0000") THEN
                a <= '1'; b <= '1'; c <= '1'; d <= '1'; e <= '1'; f <= '1'; g <= '0';
ELSIF (Data_In = "0001") THEN
                a <= '0'; b <= '1'; c <= '1'; d <= '0'; e <= '0'; f <= '0'; g <= '0';
ELSIF (Data_In = "0010") THEN
                a <= '1'; b <= '1'; c <= '0'; d <= '1'; e <= '1'; f <= '0'; g <= '1';
ELSIF (Data_In = "0011") THEN
                a <= '1'; b <= '1'; c <= '1'; d <= '1'; e <= '0'; f <= '0'; g <= '1';
ELSIF (Data_In = "0100") THEN
                a <= '0'; b <= '1'; c <= '1'; d <= '0'; e <= '0'; f <= '1'; g <= '1';
ELSIF (Data_In = "0101") THEN
                a <= '1'; b <= '0'; c <= '1'; d <= '1'; e <= '0'; f <= '1'; g <= '1';
ELSIF (Data_In = "0110") THEN
                a <= '1'; b <= '0'; c <= '1'; d <= '1'; e <= '1'; f <= '1'; g <= '1';
ELSIF (Data_In = "0111") THEN
                a <= '1'; b <= '1'; c <= '1'; d <= '0'; e <= '0'; f <= '0'; g <= '0';
ELSIF (Data_In = "1000") THEN
                a <= '1'; b <= '1'; c <= '1'; d <= '1'; e <= '1'; f <= '1'; g <= '1';
ELSIF (Data_In = "1001") THEN
                a <= '1'; b <= '1'; c <= '1'; d <= '0'; e <= '0'; f <= '1'; g <= '1';
ELSIF (Data_In = "1010") THEN
                a <= '1'; b <= '1'; c <= '1'; d <= '0'; e <= '1'; f <= '1'; g <= '1';
ELSIF (Data_In = "1011") THEN
                a <= '0'; b <= '0'; c <= '1'; d <= '1'; e <= '1'; f <= '1'; g <= '1';
ELSIF (Data_In = "1100") THEN
                a <= '1'; b <= '0'; c <= '0'; d <= '1'; e <= '1'; f <= '1'; g <= '0';
ELSIF (Data_In = "1101") THEN
                a <= '0'; b <= '1'; c <= '1'; d <= '1'; e <= '1'; f <= '0'; g <= '1';
ELSIF (Data_In = "1110") THEN
                a <= '1'; b <= '0'; c <= '0'; d <= '1'; e <= '1'; f <= '1'; g <= '1';
ELSIF (Data_In = "1111") THEN
                a <= '1'; b <= '0'; c <= '0'; d <= '0'; e <= '1'; f <= '1'; g <= '1';
ELSE
                a <= '0'; b <= '0'; c <= '0'; d <= '0'; e <= '0'; f <= '0'; g <= '0';
END IF;

END PROCESS;

END ARCHITECTURE Behavioural;
```

Figure 6.58: *If-then-else* statement example

bits of input data to the entity. The *If-then-else* statement within the process is of the form:

```
IF (Reset = '0&') THEN
      Q(7 downto 0) <= "00000000";
ELSIF (Clk'EVENT AND Clk = '1') THEN
      Q(7 DOWNTO 0) <= D(7 DOWNTO 0);
ELSE
END IF;
```

In this example, the *If-then-else* statement starts with:

```
IF (Data_In = "0000") THEN
```

The keyword IF is followed by the condition (Data_In = "0000") and the keyword THEN.

On the next line, then the outputs for this condition are set:

```
a <= '1'; b <= '1'; c <= '1'; d <='1'; e <= '1';

f <= '1'; g <= '0';
```

The outputs for the other conditions are set using the *Elsif* conditions. The final *Else* condition catches any states that have not been previously defined.

Each condition and its resulting output in this example is created as follows:

```
ELSIF (Data_In = "0001") THEN

a <= '0'; b <= '1'; c <= '1'; d <= '0';

e <= '0'; f <= '0'; g <= '0';
```

This is for space saving purposes. A semicolon indicates a new line, so another way of writing this case would be:

```
ELSIF (Data_In = "0001") THEN
   a <= '0';
   b <= '1';
   c <= '1';
   d <= '0';
   e <= '0';
   f <= '0';
   g <= '0';
```

The test bench for this design is the same as for the *Case-when* statement.

6.14.7   Tristate Buffer

In many computer architectures, multiple devices share a common set of signals—control signals, address lines, and data lines. In a computer architecture where multiple devices share a common set of data lines, these devices can either receive or provide logic levels when the device is enabled (and all other devices are disabled). However, multiple devices could, when enabled, provide logic levels that would typically conflict with the logic levels provided by other devices. To prevent this happening, when a device is disabled, it would not produce a logic level, but would instead be put in a high-impedance state (denoted by the character z). When enabled, the buffer passes the input to the output. When disabled, it blocks the input and the output is seen by the circuit that it is connected to as a high-impedance electrical load. The operation is shown in Figure 6.59. Here, the enable signal may be active high (top, 1 to enable the buffer) or active low (bottom, 0 to enable the buffer).

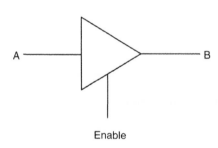

Enable	A	B
0	0	Z
0	1	Z
1	0	0
1	1	1

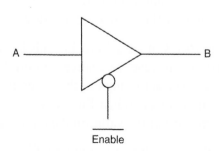

$\overline{\text{Enable}}$	A	B
0	0	0
0	1	1
1	0	Z
1	1	Z

Figure 6.59: Tristate buffer symbol

```
 1    LIBRARY ieee;
 2    USE ieee.std_logic_1164.all;
 3    USE ieee.std_logic_arith.all;
 4    USE ieee.std_logic_unsigned.all;
 5
 6    ENTITY One_Bit_Buffer is
 7    PORT ( Signal_In  : IN   STD_LOGIC;
 8           Enable     : IN   STD_LOGIC;
 9           Signal_Out : OUT  STD_LOGIC);
10    END ENTITY One_Bit_Buffer;
11
12    ARCHITECTURE Behavioural OF One_Bit_Buffer IS
13
14    BEGIN
15
16    PROCESS (Signal_In, Enable)
17
18    BEGIN
19          If (Enable = '1') Then
20                Signal_Out <= Signal_In;
21          Else
22                Signal_Out <= 'Z';
23          End If;
24    END PROCESS;
25
26    END ARCHITECTURE Behavioural;
```

Figure 6.60: One-bit tristate buffer

The tristate buffer can be created in VHDL using the *If-then-else* statement, as shown in Figure 6.60. Lines 1 to 4 identify the libraries and packages to use. Lines 6 to 10 identify the design entity (One_Bit_Buffer) with a signal input (Signal_In) and an enable control input (Enable). Lines 12 to 26 identify the design architecture.

The design operation is defined within a single process in lines 16 to 24. This has a sensitivity list with both inputs to enable the process to react to changes in both the signal and control inputs. The tristate buffer is set to be active high so that when Enable is a 1, the input signal value is passed to the output signal. When Enable is a 0, the output is held in a high impedance state (Z) irrespective of the value on the input signal.

An example test bench to simulate the buffer design is shown in Figure 6.61. Here, the inputs change every 10 ns; there is zero time delay in the operation of the design,

```
LIBRARY ieee;
USE ieee.std_logic_1164.all;
USE ieee.std_logic_arith.all;
USE ieee.std_logic_unsigned.all;

ENTITY Test_One_Bit_Buffer_vhd IS
END ENTITY Test_One_Bit_Buffer_vhd;

ARCHITECTURE Behavioural OF Test_One_Bit_Buffer_vhd IS

COMPONENT One_Bit_Buffer
PORT(
            Signal_In   : IN    STD_LOGIC;
            Enable      : IN    STD_LOGIC;
            Signal_Out  : OUT   STD_LOGIC);
END COMPONENT;

SIGNAL Signal_In :  STD_LOGIC := '0';
SIGNAL Enable    :  STD_LOGIC := '0';

SIGNAL Signal_Out :  STD_LOGIC;

BEGIN

UUT:  One_Bit_Buffer PORT MAP(
            Signal_In   =>  Signal_In,
            Enable      =>  Enable,
            Signal_Out  =>  Signal_Out);

Test bench_Process : PROCESS

BEGIN

      wait for 0 ns;   Signal_In <= '0';  Enable <= '0';
      wait for 10 ns;  Signal_In <= '1';  Enable <= '0';
      wait for 10 ns;  Signal_In <= '0';  Enable <= '1';
      wait for 10 ns;  Signal_In <= '1';  Enable <= '1';
      wait for 10 ns;

END PROCESS;

END ARCHITECTURE Behavioural;
```

Figure 6.61: One-bit tristate buffer test bench

so this short time between input signal changes would not cause any timing problems. The input signal is toggled between logic 0 and 1 for each state of the enable signal.

The one-bit tristate buffer description in VHDL can be readily modified to produce the multibit tristate buffer commonly used in computer architectures. For example, if a device is to be connected to an eight-bit-wide data bus, the one-bit tristate buffer description in VHDL can be readily modified to allow for this. Figure 6.62 shows a code example where both input and output signals are eight-bit-wide STD_LOGIC_VECTORS.

An example test bench for this design is shown in Figure 6.63.

```
LIBRARY ieee;
USE ieee.std_logic_1164.all;
USE ieee.std_logic_arith.all;
USE ieee.std_logic_unsigned.all;

ENTITY Eight_Bit_Buffer is
PORT ( Signal_In  : IN   STD_LOGIC_VECTOR(7 downto 0);
       Enable     : IN   STD_LOGIC;
       Signal_Out : OUT  STD_LOGIC_VECTOR(7 downto 0));
END ENTITY Eight_Bit_Buffer;

ARCHITECTURE Behavioural OF Eight_Bit_Buffer IS

BEGIN

PROCESS (Signal_In, Enable)

BEGIN

     If (Enable = '1') Then
          Signal_Out(7 downto 0) <= Signal_In(7 downto 0);
     Else
          Signal_Out <= "ZZZZZZZZ";
     End If;

END PROCESS;

END ARCHITECTURE Behavioural;
```

Figure 6.62: Eight-bit tristate buffer using the *If-then-else* statement

```
LIBRARY ieee;
USE ieee.std_logic_1164.all;
USE ieee.std_logic_arith.all;
USE ieee.std_logic_unsigned.all;


ENTITY Test_Eight_Bit_Buffer_vhd IS
END ENTITY Test_Eight_Bit_Buffer_vhd;


ARCHITECTURE behavioural OF Test_Eight_Bit_Buffer_vhd IS

COMPONENT Eight_Bit_Buffer
PORT(
            Signal_In   : IN    STD_LOGIC_VECTOR(7 downto 0);
            Enable      : IN    STD_LOGIC;
            Signal_Out  : OUT   STD_LOGIC_VECTOR(7 downto 0));
END COMPONENT;


SIGNAL Signal_In :  STD_LOGIC_VECTOR(7 downto 0) := "00000000";
SIGNAL Enable    :  STD_LOGIC := '0';

SIGNAL Signal_Out :  STD_LOGIC_VECTOR(7 downto 0);

BEGIN

UUT:  Eight_Bit_Buffer PORT MAP(
            Signal_In(7 downto 0)    =>  Signal_In(7 downto 0),
            Enable                   =>  Enable,
            Signal_Out(7 downto 0)   =>  Signal_Out(7 downto 0));

Test bench_Process : PROCESS

BEGIN

      wait for 0 ns;    Signal_In <= "00000000";   Enable <= '0';
      wait for 10 ns;   Signal_In <= "11111111";   Enable <= '0';
      wait for 10 ns;   Signal_In <= "00000000";   Enable <= '1';
      wait for 10 ns;   Signal_In <= "11111111";   Enable <= '1';
      wait for 10 ns;

END PROCESS;

END ARCHITECTURE behavioural;
```

Figure 6.63: Eight-bit tristate buffer test bench using the *If-then-else* statement

```
LIBRARY ieee;
USE ieee.std_logic_1164.all;
USE ieee.std_logic_arith.all;
USE ieee.std_logic_unsigned.all;

-------------------------------------------------------------
-- One_Bit_Buffer Entity
-------------------------------------------------------------

ENTITY Eight_Bit_Buffer is
PORT ( Signal_In  : IN   STD_LOGIC_VECTOR(7 downto 0);
       Enable     : IN   STD_LOGIC;
       Signal_Out : OUT  STD_LOGIC_VECTOR(7 downto 0));
END ENTITY Eight_Bit_Buffer;

-------------------------------------------------------------
-- One_Bit_Buffer Architecture
-------------------------------------------------------------

ARCHITECTURE DataFlow OF Eight_Bit_Buffer IS

BEGIN

Signal_Out(7 downto 0) <= Signal_In(7 downto 0) When Enable = '1' Else "ZZZZZZZZ";

END ARCHITECTURE DataFlow;
```

Figure 6.64: Eight-bit tristate buffer using the *When-else* statement

The tristate buffer can also be created using the *When-else* statement, as shown in Figure 6.64. In this design, a dataflow description is used.

6.15   Sequential Logic Design

6.15.1   *Introduction*

Sequential logic circuits are based on combinational logic circuit elements (AND, OR, etc.) working alongside sequential circuit elements (latches and flip-flops that will be grouped together to form registers). A generic sequential logic circuit is shown in Figure 6.65. Here the circuit inputs to the circuit are applied to the combinational logic, and the circuit outputs are derived from this combinational logic block. The sequential logic circuit elements store an output from the combinational logic, and this is fed back to the combinational logic to form the present state of the circuit. The output from the combinational logic forming the inputs to the sequential logic circuit elements in turn forms the next state of the circuit. The circuit changes from the present state to the next state on a clock control input. Commonly the D latch and

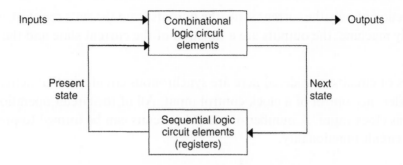

Figure 6.65: Generic sequential logic circuit (counter or state machine)

D-type flip-flop are used, and these sequential circuit elements will be used in this text (rather than other forms of latch and flip-flop such as the S-R, toggle, and J-K flip-flops).

Such sequential logic circuit designs create counters and state machines. The state machines are based on either the Moore machine or Mealy machine, as shown in Figure 6.66.

The diagrams shown in Figure 6.66 are a modification of the basic structure identified in Figure 6.65 by separating the combinational logic block into two blocks, one to create the next state logic (inputs to the state register, an array of flip-flops, that store the state of the circuit) and the output logic. In the Moore machine, the outputs

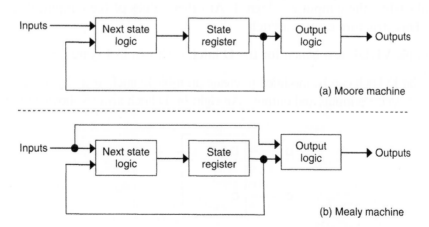

Figure 6.66: Moore and Mealy state machines

are a function of only the current state (the outputs from the state register), whereas in the Mealy machine, the outputs are a function of the current state and the current inputs.

The types of circuits considered here are synchronous circuits in that activity will occur under the control of a clock control input. All of the circuit operation will be tied to this clock input. A number of possible circuits can be formed to produce the required circuit functionality.

6.15.2   Latches and Flip-Flops

The two sequential logic circuit elements used are the latch and the flip-flop. The operation of these circuit elements, discussed in Chapter 5, can be modeled in VHDL, thereby allowing the ability to model, simulate, and synthesize counters and state machines. In this text, the D latch and the D-type flip-flop will be considered.

The basic D latch circuit symbol, shown in Figure 6.67, has two inputs—the data input (D, value to store) and the control input (C)—and one output (Q).

In the D latch, when the CLK input is at a logic 1, the Q output is assigned the value of the D input. When the CLK input is a logic 0, the Q output holds its current value even when the D input changes. In VHDL, this can be written as shown in Figure 6.68 (where the latch control input C is renamed CLK within the VHDL code).

This uses the *If-then-else* statement on lines 20 to 22 to identify an operation to undertake when the C input is a logic 1. At other values of the C input, there is no action. This structure creates a latch.

An example VHDL test bench for the D latch is shown in Figure 6.69.

The one-bit D latch can be modified to create an n-bit D latch array by using STD_LOGIC_VECTOR inputs and outputs. An eight-bit D latch array is shown in Figure 6.70.

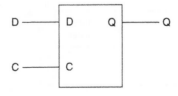

Figure 6.67: D latch circuit symbol

```
1    LIBRARY IEEE;
2    USE IEEE.STD_LOGIC_1164.ALL;
3    USE IEEE.STD_LOGIC_ARITH.ALL;
4    USE IEEE.STD_LOGIC_UNSIGNED.ALL;
5
6    ENTITY D_Latch is
7        PORT ( D    : IN    STD_LOGIC;
8               CLK  : IN    STD_LOGIC;
9               Q    : OUT   STD_LOGIC);
10   END ENTITY D_Latch;
11
12   ARCHITECTURE Behavioural OF D_Latch IS
13
14   BEGIN
15
16   PROCESS(CLK, D)
17
18   BEGIN
19
20           If (CLK = '1') THEN
21                   Q <= D;
22           END IF;
23
24   END PROCESS;
25
26   END ARCHITECTURE Behavioural;
```

Figure 6.68: VHDL code for the D latch

The basic D-type flip-flop circuit symbol, shown in Figure 6.71, has two inputs—the data input (D, value to store) and the clock input (CLK)—and one output (Q).

In the D-type flip-flop, when the CLK input changes from a 0 to a 1 (positive edge triggered) or from a 1 to a 0 (negative edge triggered), the Q output is assigned the value of the D input. When the CLK input is steady at a logic 0 or a 1, the Q output holds its current value even when the D input changes. In VHDL, this can be written as shown in Figure 6.72. This code example is a modification of the basic D latch and is for a positive edge triggered flip-flop.

This uses the *If-then-else* statement on lines 20 to 22 to identify an operation to undertake when the CLK input changes. This is a positive edge triggered flip-flop in

```
1    LIBRARY ieee;
2    USE ieee.std_logic_1164.all;
3    USE ieee.std_logic_arith.all;
4    USE ieee.std_logic_unsigned.all;
5
6    ENTITY Test_D_Latch_vhd IS
7    END Test_D_Latch_vhd;
8
9    ARCHITECTURE Behavioural OF Test_D_Latch_vhd IS
10
11   COMPONENT D_Latch
12        PORT(
13               D   : IN  STD_LOGIC;
14               CLK : IN  STD_LOGIC;
15               Q   : OUT STD_LOGIC);
16   END COMPONENT;
17
18   SIGNAL D   : STD_LOGIC := '0';
19   SIGNAL CLK : STD_LOGIC := '0';
20
21   SIGNAL Q : STD_LOGIC;
22
23   BEGIN
24
25   uut: D_Latch PORT MAP(
26               D    => D,
27               CLK  => CLK,
28               Q    => Q);
29
30   CLK_Process : PROCESS
31
32   BEGIN
33
34               Wait for 0 ns;   CLK <= '0';
35               Wait for 20 ns;  CLK <= '1';
36               Wait for 20 ns;  CLK <= '0';
37
38   END PROCESS;
39
40   D_Process : PROCESS
41
42   BEGIN
43
44               Wait for 0 ns;   D <= '0';
45               Wait for 60 ns;  D <= '1';
46               Wait for 22 ns;  D <= '0';
47               Wait for 2 ns;   D <= '1';
48               Wait for 2 ns;   D <= '0';
49
50         Wait for 16 ns;
51
52   END PROCESS;
53
54   END ARCHITECTURE Behavioural;
```

Figure 6.69: VHDL test bench for the D latch

```
1    LIBRARY IEEE;
2    USE IEEE.STD_LOGIC_1164.ALL;
3    USE IEEE.STD_LOGIC_ARITH.ALL;
4    USE IEEE.STD_LOGIC_UNSIGNED.ALL;
5
6    ENTITY D_Latch_Array is
7        PORT ( D    : IN    STD_LOGIC_VECTOR(7 downto 0);
8               CLK  : IN    STD_LOGIC;
9               Q    : OUT   STD_LOGIC_VECTOR(7 downto 0));
10   END ENTITY D_Latch_Array;
11
12   ARCHITECTURE Behavioural OF D_Latch_Array IS
13
14   BEGIN
15
16   PROCESS(CLK, D)
17
18   BEGIN
19
20               If (CLK = '1') THEN
21                   Q(7 downto 0) <= D(7 downto 0);
22               END IF;
23
24   END PROCESS;
25
26   END ARCHITECTURE Behavioural;
```

Figure 6.70: VHDL code for an eight-bit D latch array

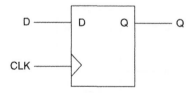

Figure 6.71: D-type flip-flop

that the clock control statement identifies that the clock is on an edge (change) and is a logic 1 (the final value for the clock):

```
IF (CLK'Event AND CLK = '1') THEN
```

It is common, however, for the flip-flop to have a reset or set input to initialize the output Q to either logic 0 (reset) or logic 1 (set). This reset/set input can be either

```
1    LIBRARY IEEE;
2    USE IEEE.STD_LOGIC_1164.ALL;
3    USE IEEE.STD_LOGIC_ARITH.ALL;
4    USE IEEE.STD_LOGIC_UNSIGNED.ALL;
5
6    ENTITY D_Type_Flip_Flop is
7        PORT ( D    : IN    STD_LOGIC;
8               CLK  : IN    STD_LOGIC;
9               Q    : OUT   STD_LOGIC);
10   END ENTITY D_Type_Flip_Flop;
11
12   ARCHITECTURE Behavioural OF D_Type_Flip_Flop IS
13
14   BEGIN
15
16   PROCESS(CLK, D)
17
18   BEGIN
19
20               IF (CLK'EVENT AND CLK = '1') THEN
21                     Q <= D;
22               END IF;
23
24   END PROCESS;
25
26   END ARCHITECTURE Behavioural;
```

Figure 6.72: VHDL code for the D-type flip-flop

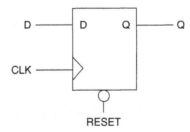

Figure 6.73: D-type flip-flop with active low reset

asynchronous (independent of the clock) or synchronous (the reset/set occurs on a clock edge), and active high (active when a logic 1) or active low (active when a logic 0). The circuit symbol for the D-type flip-flop with active low reset is shown in Figure 6.73.

This idea for this design is shown in Figure 6.74. Here, eight bits of data are stored in a register consisting of eight flip-flops, and the flip-flops are asynchronous active low reset.

To include the reset capability, the action to undertake when the reset input (RESET) is a logic 0 is included. The activity takes place between lines 21 and 29.

```
1     LIBRARY IEEE;
2     USE IEEE.STD_LOGIC_1164.ALL;
3     USE IEEE.STD_LOGIC_ARITH.ALL;
4     USE IEEE.STD_LOGIC_UNSIGNED.ALL;
5
6     ENTITY D_Type_Register is
7         PORT ( D        : IN    STD_LOGIC_VECTOR(7 downto 0);
8                CLK      : IN    STD_LOGIC;
9                RESET    : IN    STD_LOGIC;
10               Q        : OUT   STD_LOGIC_VECTOR(7 downto 0));
11    END ENTITY D_Type_Register;
12
13    ARCHITECTURE Behavioural OF D_Type_Register IS
14
15    BEGIN
16
17    PROCESS(CLK, D, RESET)
18
19    BEGIN
20
21            IF (RESET = '0') THEN
22
23                    Q(7 downto 0) <= "00000000";
24
25            ELSIF (CLK'EVENT AND CLK = '1') THEN
26
27                    Q(7 downto 0) <= D(7 downto 0);
28
29            END IF;
30
31    END PROCESS;
32
33    END ARCHITECTURE Behavioural;
```

Figure 6.74: VHDL code for the D-type flip-flop register with active low asynchronous reset

6.15.3 Counter Design

A counter is a circuit that passes through a set sequence of states on the change (edge) of a clock signal, and the only inputs to the circuit are a clock and a reset or set. The simplest counter is a straight binary up-counter whose output (Q outputs from the flip-flops used in the counter) is a straight binary count. The D-type flip-flop register design can be modified to produce this operation. If all the flip-flops can be reset, then the initial count value after a reset has occurred is 0_{10}. The next count value will be 1_{10}, followed by 2_{10}, and so on. The output from a four-bit counter is shown in Table 6.11. The Q outputs from the counter, as shown in Figure 6.75, are Q3, Q2, Q1, and Q0.

Table 6.11: Four-bit counter output (reset to state 0)

State (count value)	Q3	Q2	Q1	Q0
0	0	0	0	0
1	0	0	0	1
2	0	0	1	0
3	0	0	1	1
4	0	1	0	0
5	0	1	0	1
6	0	1	1	0
7	0	1	1	1
8	1	0	0	0
9	1	0	0	1
10	1	0	1	0
11	1	0	1	1
12	1	1	0	0
13	1	1	0	1
14	1	1	1	0
15	1	1	1	1

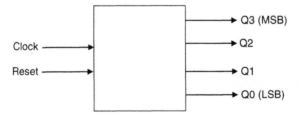

Figure 6.75: Four-bit binary counter

```
1    library IEEE;
2    use IEEE.STD_LOGIC_1164.ALL;
3    use IEEE.STD_LOGIC_ARITH.ALL;
4    use IEEE.STD_LOGIC_UNSIGNED.ALL;
5
6    ENTITY Four_Bit_Counter is
7        PORT ( Clock : IN    STD_LOGIC;
8               Reset : IN    STD_LOGIC;
9               Count : OUT   STD_LOGIC_VECTOR (3 downto 0));
10   END ENTITY Four_Bit_Counter;
11
12   ARCHITECTURE Behavioural of Four_Bit_Counter is
13
14   SIGNAL Count_Int : STD_LOGIC_VECTOR(3 downto 0);
15
16   BEGIN
17
18   PROCESS(Clock, Reset)
19
20   BEGIN
21
22        IF (Reset = '0') THEN
23
24              Count_Int(3 downto 0) <= "0000";
25
26        ELSIF (Clock'Event AND Clock = '1') THEN
27
28              Count_Int(3 downto 0) <= Count_Int(3 downto 0) + 1;
29
30        END IF;
31
32   END PROCESS;
33
34   Count(3 downto 0) <= Count_Int(3 downto 0);
35
36   END ARCHITECTURE Behavioural;
```

Figure 6.76: VHDL code for the four-bit binary counter

The VHDL code for this counter is shown in Figure 6.76. Here, the counter activity within the architecture is specified between lines 18 and 34. When the Reset input is a logic 0, the internal signal Count_Int (a four-bit-wide STD_LOGIC_VECTOR) is set to 0_{10}. This is an asynchronous active low reset. At all other values of Reset, the value for Count_Int increments by 1 on the edge of the Clock signal:

```
ELSIF (Clock'Event AND Clock = '1') THEN

    Count_Int(3 downto 0) <= Count_Int(3 downto 0) + 1;
```

When the value for Count_Int reaches 15_{10} (1111_2), the next count value will wrap around back to 0_{10}. Outside the process, the STD_LOGIC_VECTOR output Count is assigned the value of Count_Int. This is required so that the port signal of mode OUT cannot be read inside from within the entity.

An example VHDL test bench for this design is shown in Figure 6.77. The two inputs Clock and Reset are created within their own process statements.

The output from this counter was taken from the flip-flop Q outputs. However, there are three possible ways to create the output:

1. by taking the flip-flop Q outputs directly

2. by passing the flip-flop Q outputs through combinational logic, then the outputs of the combinational logic become the counter outputs to form a nonregistered output

3. by passing the flip-flop Q outputs through combinational logic, then applying these to the inputs of one or more flip-flops to form a registered output, and the outputs of the flip-flops become the counter outputs.

These possible arrangements are shown in Figure 6.78.

A straight binary count forms the states of this counter. For n flip-flops, there are 2^n possible count states. However, the number of count states could be reduced. For example, if a counter is required to count from 0_{10} to 4_{10} and back to 0_{10}, this is achieved with three flip-flops (forming a possible maximum of eight count states). The counter must detect that from count state 4, on the next clock edge, the counter will jump to count state 0 rather than automatically moving on to count state 5.

Although this design uses a binary count sequence, any count sequence that can be coded in VHDL can be used. The one-hot encoding is an example of this. It uses n flip-flops to represent n states. Table 6.12 shows this arrangement for a four-bit counter, along with the binary count state equivalent.

In the one-hot encoding scheme, to change from one state to the next, only two flip-flop outputs change: the first from a 1 to a 0, and the second from a 0 to a 1. The advantage to this scheme is that the combinational logic to create the next state value is less than required for other encoding schemes. However, it comes at the expense of

```
1    LIBRARY ieee;
2    USE ieee.std_logic_1164.all;
3    USE ieee.std_logic_arith.all;
4    USE ieee.std_logic_unsigned.all;
5
6    ENTITY Test_Four_Bit_Counter_vhd IS
7    END Test_Four_Bit_Counter_vhd;
8
9    ARCHITECTURE Behavioural OF Test_Four_Bit_Counter_vhd IS
10
11   COMPONENT Four_Bit_Counter
12   PORT(
13                Clock : IN  std_logic;
14                Reset : IN  std_logic;
15                Count : OUT std_logic_vector(3 downto 0));
16   END COMPONENT;
17
18   SIGNAL Clock :  std_logic := '0';
19   SIGNAL Reset :  std_logic := '0';
20
21   SIGNAL Count :  std_logic_vector(3 downto 0);
22
23   BEGIN
24
25   uut: Four_Bit_Counter PORT MAP(
26                Clock => Clock,
27                Reset => Reset,
28                Count => Count);
29
30   Reset_Process: PROCESS
31
32   BEGIN
33
34       Wait For 0 ns;    Reset <= '0';
35       Wait For 160 ns;  Reset <= '1';
36       Wait;
37
38   END PROCESS;
39
40   Clock_Process: PROCESS
41
42   BEGIN
43
44       Wait For 0 ns;   Clock <= '0';
45       Wait For 20 ns;  Clock <= '1';
46       Wait For 20 ns;  Clock <= '0';
47
48   END PROCESS;
49
50   END ARCHITECTURE Behavioural;
```

Figure 6.77: VHDL test bench for the four-bit binary counter

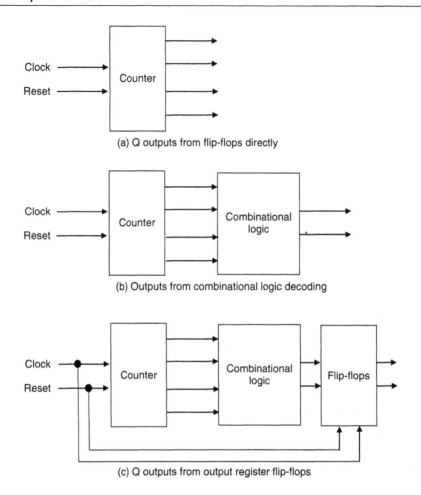

(a) Q outputs from flip-flops directly

(b) Outputs from combinational logic decoding

(c) Q outputs from output register flip-flops

Figure 6.78: Decoding the counter output

dealing with the increase in the number of possible states (although only a small number of the possible states would be used in normal operation). The possibility for the counter to move into one of the unused states must be considered.

6.15.4 State Machine Design

The state machine is a circuit that reacts to one or more inputs that direct it to move into one of a number of possible states, depending on the value of the current state and the value of the current input. State machines are based on either the Moore or Mealy machines. The state transition diagram is drawn to represent state machine

Table 6.12: One-hot encoding

State	One-hot code	Binary code
0	0000000000000001	0000
1	0000000000000010	0001
2	0000000000000100	0010
3	0000000000001000	0011
4	0000000000010000	0100
5	0000000000100000	0101
6	0000000001000000	0110
7	0000000010000000	0111
8	0000000100000000	1000
9	0000001000000000	1001
10	0000010000000000	1010
11	0000100000000000	1011
12	0001000000000000	1100
13	0010000000000000	1101
14	0100000000000000	1110
15	1000000000000000	1111

operation. This aid is invaluable, particularly when writing VHDL code based descriptions of the state machine to be able to visualize its operation and relate this to the VHDL code.

A state machine can be modeled in VHDL as a structural description, dataflow description, or a behavioral description. In this text, the behavioral description is considered. The structure of the behavioral description is based here on two processes within the architecture of the design:

1. The first process describes the transition from the current state to the next state.

2. The second process describes the output values for the current state and describes the next state.

The behavioral description uses the *If-then-else* and *Case-when* statements to achieve the required state machine behavior. Two case study designs are presented to show how this can be achieved: 1001 sequence detector and UART receiver.

Example 1: 1001 Sequence Detector

Consider a state machine, such as that shown in Figure 6.79, that is to detect the sequence 1001 on a data input, then produce a logic 1 output when the sequence has

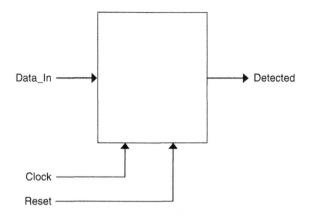

Figure 6.79: 1001 sequence detector

been detected. The state machine has three inputs—one `Data_In` to be monitored for the sequence and two control inputs, `Clock` and `Reset`—and one output, `Detected`. Such a state machine could be used in a digital combinational lock circuit.

The state transition diagram for this design is shown in Figure 6.80.

This state machine defines five states: state 0, state 1, state 2, state 3, and state 4. The circuit is initially reset to state 0 (a design decision) and monitors the `Data_In`

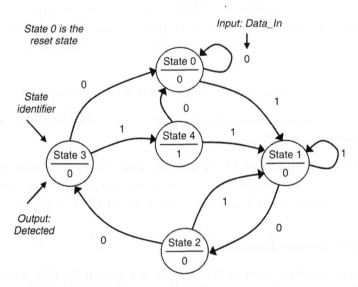

Figure 6.80: 1001 sequence detector state transition diagram (Moore machine)

input. The Detected output is set to a logic 0. State machine changes are summarized below:

- **At state 0**: When Data_In input remains at logic 0, the state remains in state 0. When a logic 1 is detected (the start of a possible 1001 sequence), the state changes to state 1. All state transitions will occur here on the positive edge of the Clock input.

- **At state 1**: When a logic 1 is detected on the Data_In input, the state remains in state 1 because the sequence has been broken. However, if a logic 0 is detected (10 of the sequence), the state changes to state 2.

- **At state 2**: When a logic 1 is detected on the Data_In input, the state changes back to state 1 because the sequence has been broken, but the logic 1 input could be the start of a sequence. However, if a logic 0 is detected (100 of the sequence), the state changes to state 3.

- **At state 3**: When a logic 0 is detected on the Data_In input, the state changes back to state 0 because the sequence has been broken. However, if a logic 1 is detected (1001 of the sequence), the state changes to state 4 and sets the Detected output to a logic 1.

- **At state 4**: When a logic 0 is detected on the Data_In input, the state changes back to state 0. However, if a logic 1 is detected (the logic 1 input could be the start of a sequence), the state changes to state 1. Whether the state is 0 or 1, the Detected output is reset to a logic 0.

In VHDL, this sequence detector can be described using a behavioral description with two processes. The first process describes the transition from the current state to the next state. The second process describes the output values for the current state and describes the next state. This is shown in Figure 6.81.

The first process has the same structure as a counter, except now the signals are defined in the architecture as a defined type, State_Type, and with names Current_State and Next_State (shown in lines 18 and 20). The five possible states (State0, State1, State2, State3, and State4) relate to the names of the states given in the state diagram.

The second process identifies the values of the outputs for each possible state and identifies the value of the next state (Next_State) for the current state (Current_State) and the data input (Data_In).

```
1     LIBRARY ieee;
2     USE ieee.std_logic_1164.all;
3     USE ieee.std_logic_arith.all;
4     USE ieee.std_logic_unsigned.all;
5
6     ENTITY Sequence_Detector is
7         PORT ( Data_In  : IN    STD_LOGIC;
8                Clock    : IN    STD_LOGIC;
9                Reset    : IN    STD_LOGIC;
10               Q2       : OUT   STD_LOGIC;
11               Q1       : OUT   STD_LOGIC;
12               Q0       : OUT   STD_LOGIC;
13               Detected : OUT   STD_LOGIC);
14    END ENTITY Sequence_Detector;
15
16    ARCHITECTURE Behavioural OF Sequence_Detector IS
17
18    TYPE State_Type IS (State0, State1, State2, State3, State4);
19
20    SIGNAL Present_State, Next_State : State_Type;
21
22    BEGIN
23
24    PROCESS(Clock, Reset)
25    BEGIN
26
27          IF (Reset = '0') THEN
28
29                      Present_State <= State0;
30
31          ELSIF (Clock'Event AND Clock = '1') THEN
32
33                      Present_State <= Next_State;
34
35          END IF;
36
37    END PROCESS;
38
39    PROCESS(Present_State, Data_In)
30    BEGIN
41
42          CASE Present_State IS
43
44
45          WHEN State0 =>  Detected <= '0';  Q2 <= '0';  Q1 <= '0';  Q0 <= '0';
46
47                      IF (Data_In = '0') THEN
48                                    Next_State <= State0;
49                      ELSE
50                                    Next_State <= State1;
51                        END IF;
```

Figure 6.81: VHDL code for the 1001 sequence detector

```
52
53
54              WHEN State1 =>  Detected <= '0';  Q2 <= '0';  Q1 <= '0';  Q0 <= '1';
55
56                      IF (Data_In = '0') THEN
57                                      Next_State <= State2;
58                      ELSE
59                                      Next_State <= State1;
60                      END IF;
61
62
63          WHEN State2 =>  Detected <= '0'; Q2 <= '0';  Q1 <= '1';  Q0 <= '0';
64
65                      IF (Data_In = '0') THEN
66                                      Next_State <= State3;
67                      ELSE
68                                      Next_State <= State1;
69                      END IF;
70
71
72          WHEN State3 =>  Detected <= '0';  Q2 <= '0';  Q1 <= '1';  Q0 <= '1';
73
74                      IF (Data_In = '0') THEN
75                                      Next_State <= State0;
76                      ELSE
77                                      Next_State <= State4;
78                      END IF;
79
80
81          WHEN State4 =>  Detected <= '1'; Q2 <= '1';  Q1 <= '0';  Q0 <= '0';
82
83                      IF (Data_In = '0') THEN
84                                  Next_State <= State0;
85                      ELSE
86                                  Next_State <= State1;
87                      END IF;
88
89
90          END CASE;
91
92      END PROCESS;
93
94      END ARCHITECTURE Behavioural;
```

Figure 6.81: (Continued)

This design can be simulated and synthesized into logic. However, the one thing that is not defined is the state encoding: for example, the transition from one state to the next could be a binary count, one-hot encoding, or another form of encoding. In this case, either the synthesis tool is configured to select automatically a suitable form of encoding, or the user can direct the synthesis tool to use a particular form of encoding.

An example VHDL test bench for this design is shown in Figure 6.82.

Example 2: UART Receiver

The UART (universal asynchronous receiver transmitter) is used in RS-232 serial communications to receive and transmit serial data. It consists of a receiver circuit and a transmitter circuit.

The UART receiver circuit receives a serial input (Rx) and requires the use of Clock and Reset control inputs. The clock frequency is sixteen times the baud rate of the serial data transmission. The circuit, shown in Figure 6.83, produces an eight-bit parallel output (Data_Rx(7:0)) and provides a status output (DR) that becomes a logic 1 when a byte of data has been received.

The timing waveform for the serial data format on the Rx input is shown in Figure 6.84. Although serial data transmission protocol has a number of possible scenarios, the following sequence of bits received will be considered here:

1. one start bit (logic 0)

2. eight data bits are transmitted (LSB first, MSB last)

3. one stop bit (logic 1)

4. no parity checking

When the receiver circuit is waiting for data, the Rx input is a logic 1. A start (of serial data transmission by an external circuit) bit is indicated when the Rx input becomes a logic 0. The eight data bits are then transmitted, and the data transmission completes with a stop bit (logic 1).

The operation of the UART receiver can be developed in VHDL using a structural description, a dataflow description, or a behavioral description. A behavioral description for this circuit in VHDL is shown in Figure 6.85. A state machine monitors the Rx input and remains in its initial state while the Rx input remains at a logic 1. This sets up two processes. The first process is a binary counter controlled (to the next count state) by the current count state and the value on the Rx input.

```
1    LIBRARY ieee;
2    USE ieee.std_logic_1164.all;
3    USE ieee.std_logic_arith.all;
4    USE ieee.std_logic_unsigned.all;
5
6
7    ENTITY Test_Sequence_Detector_vhd IS
8    END Test_Sequence_Detector_vhd;
9
10
11   ARCHITECTURE Behavioural OF Test_Sequence_Detector_vhd IS
12
13   COMPONENT Sequence_Detector
14   PORT(
15               Data_In   : IN    STD_LOGIC;
16               Clock     : IN    STD_LOGIC;
17               Reset     : IN    STD_LOGIC;
18               Q2        : OUT   STD_LOGIC;
19               Q1        : OUT   STD_LOGIC;
20               Q0        : OUT   STD_LOGIC;
21               Detected  : OUT   STD_LOGIC);
22   END COMPONENT;
23
24   SIGNAL Data_In  :   STD_LOGIC := '0';
25   SIGNAL Clock    :   STD_LOGIC := '0';
26   SIGNAL Reset    :   STD_LOGIC := '0';
27
28   SIGNAL Q2       :   STD_LOGIC;
29   SIGNAL Q1       :   STD_LOGIC;
30   SIGNAL Q0       :   STD_LOGIC;
31   SIGNAL Detected :   STD_LOGIC;
32
33   BEGIN
34
35
36   uut: Sequence_Detector PORT MAP(
37               Data_In   => Data_In,
38               Clock     => Clock,
39               Reset     => Reset,
40               Q2        => Q2,
41               Q1        => Q1,
42               Q0        => Q0,
43               Detected  => Detected);
```

Figure 6.82: VHDL test bench for the 1001 sequence detector

```
44
45   Reset_Process : PROCESS
46
47   BEGIN
48
49       Wait for 0 ns;   Reset <= '0';
50       Wait for 5 ns;   Reset <= '1';
51       Wait;
52
53   END PROCESS;
54
55   Clock_Process : PROCESS
56
57   BEGIN
58
59       Wait for 0 ns;   Clock <= '0';
60       Wait for 10 ns;  Clock <= '1';
61       Wait for 10 ns;  Clock <= '0';
62
63   END PROCESS;
64
65   Data_In_Process : PROCESS
66
67   BEGIN
68
69       Wait for 0 ns;    Data_In <= '0';
70
71       Wait for 80 ns;   Data_In <= '1';
72       Wait for 20 ns;   Data_In <= '0';
73       Wait for 20 ns;   Data_In <= '0';
74       Wait for 20 ns;   Data_In <= '1';
75
76       Wait for 20 ns;   Data_In <= '0';
77
78       Wait for 80 ns;   Data_In <= '1';
79       Wait for 20 ns;   Data_In <= '0';
80       Wait for 20 ns;   Data_In <= '0';
81       Wait for 20 ns;   Data_In <= '0';
82
83       Wait for 20 ns;   Data_In <= '0';
84
85   END PROCESS;
86
87   END ARCHITECTURE Behavioural;
```

Figure 6.82: (Continued)

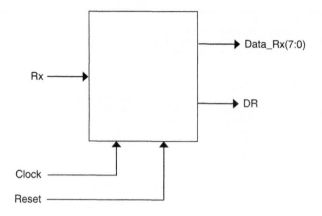

Figure 6.83: UART receiver circuit

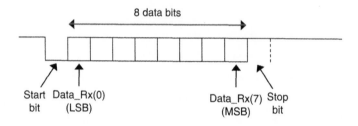

Figure 6.84: Serial data format

The second process uses the current count state and undertakes an action on the clock rising edge on specific count states only. The question is, is this actually the best approach to take, and what will happen in the other count states?

When Rx becomes a 0, then the following sequence of events happens:

1. The internal counter counts for 24 clock pulses and then stores data bit Data_Rx(0).

2. The internal counter counts for 16 clock pulses and then stores data bit Data_Rx(1).

3. The internal counter counts for 16 clock pulses and then stores data bit Data_Rx(2).

```
 1      LIBRARY IEEE;
 2      USE IEEE.STD_LOGIC_1164.ALL;
 3      USE IEEE.STD_LOGIC_ARITH.ALL;
 4      USE IEEE.STD_LOGIC_UNSIGNED.ALL;
 5
 6
 7      ENTITY Uart_Receiver is
 8          PORT ( Rx        : IN    STD_LOGIC;
 9                 Clock     : IN    STD_LOGIC;
10                 Reset     : IN    STD_LOGIC;
11                 Data_Rx   : OUT   STD_LOGIC_VECTOR(7 downto 0);
12                 DR        : OUT   STD_LOGIC);
13      END ENTITY Uart_Receiver;
14
15
16      ARCHITECTURE Behavioural of Uart_Receiver is
17
18      SIGNAL Count     : STD_LOGIC_VECTOR(7 downto 0);
19      SIGNAL Data_Int  : STD_LOGIC_VECTOR(7 downto 0);
20
21      BEGIN
22
23      PROCESS (Clock, Reset)
24
25      BEGIN
26
27              IF (Reset='0') then
28
29                      Count <= "00000000";
30
31              ELSIF (Clock'Event and Clock = '1') then
32
33              IF (Rx='1' AND (Count = "00000000" or Count = "10101011")) THEN
34                              Count <= "00000000";
35              ELSE
36                              Count <= Count + 1;
37              END IF;
38
39              END IF;
40
41      END PROCESS;
42
43      PROCESS (Clock, Reset, Count)
44
45      BEGIN
46
47              IF (Reset='0') THEN
48
49                      Data_Int(7 downto 0) <= "00000000";
50                      Data_Rx(7 downto 0)  <= "00000000";
51                      DR <= '0';
52
53              ELSIF (Clock'Event and Clock = '1') THEN
54
55                      IF (COUNT = "00000000") THEN
56                              DR <= '0';
57                      END IF;
58
```

Figure 6.85: VHDL code for a UART receiver

```
59                         IF (COUNT = "00000001") THEN
60                              DR <= '0';
61                         END IF;
62
63                         IF (COUNT = "00011000") THEN
64                              Data_Int(0) <= Rx;
65                         END IF;
66
67                         IF (COUNT = "00101000") THEN
68                              Data_Int(1) <= Rx;
69                         END IF;
70
71                         IF (COUNT = "00111000") THEN
72                              Data_Int(2) <= Rx;
73                         END IF;
74
75                         IF (COUNT = "01001000") THEN
76                              Data_Int(3) <= Rx;
77                         END IF;
78
79                         IF (COUNT = "01011000") THEN
80                              Data_Int(4) <= Rx;
81                         END IF;
82
83                         IF (COUNT = "01101000") THEN
84                              Data_Int(5) <= Rx;
85                         END IF;
86
87                         IF (COUNT = "01111000") THEN
88                              Data_Int(6) <= Rx;
89                         END IF;
90
91                         IF (COUNT = "10001000") THEN
92                              Data_Int(7) <= Rx;
93                         END IF;
94
95                         IF (COUNT = "10011000") THEN
96                              Data_Rx(7 downto 0) <= Data_Int(7 downto 0);
97                         END IF;
98
99                         IF (COUNT = "10101000") THEN
100                             DR <= '1';
101                        END IF;
102
103                        IF (COUNT = "10101010") THEN
104                             DR <= '0';
105                        END IF;
106
107               END IF;
108
109     END PROCESS;
110
111     END ARCHITECTURE Behavioural;
```

Figure 6.85: (Continued)

4. The internal counter counts for 16 clock pulses and then stores data bit Data_Rx(3).

5. The internal counter counts for 16 clock pulses and then stores data bit Data_Rx(4).

6. The internal counter counts for 16 clock pulses and then stores data bit Data_Rx(5).

7. The internal counter counts for 16 clock pulses and then stores data bit Data_Rx(6).

8. The internal counter counts for 16 clock pulses and then stores data bit Data_Rx(7).

9. The internal counter counts for 16 clock pulses and then outputs the byte of data.

10. The internal counter counts for 16 clock pulses and then sets DR to logic 1.

11. The internal counter counts for 4 clock pulses and then sets DR to logic 0.

12. The counter returns to its initial state and waits for Rx to become logic 0, indicating the receipt of the next byte.

The choice of which count states to act on ensures that the action will be taken in the middle of the received bit (start, data, or stop) so that a correct value is read from Rx.

The received data is initially stored inside the design using the Data_Int signal, which allows the byte to be available at the output of the design at one single time. This is a simple circuit model that decodes a counter output and produces a registered output to store the received data byte. There are of course possible improvements to this description.

An example VHDL test bench for this design is shown in Figure 6.86. Two points should be noted about the code identified in Figure 6.85:

- The *If-then-else* statements would be formatted better as *Elsif* statements.

- The *Case-when* statement could be used instead.

```
1     LIBRARY ieee;
2     USE ieee.std_logic_1164.all;
3     USE ieee.std_logic_arith.all;
4     USE ieee.std_logic_unsigned.all;
5
6     ENTITY Test_Uart_Receiver_vhd IS
7     END Test_Uart_Receiver_vhd;
8
9     ARCHITECTURE Behavioural OF Test_Uart_Receiver_vhd IS
10
11    COMPONENT Uart_Receiver
12    PORT(
13                    Rx       : IN std_logic;
14                    Clock    : IN std_logic;
15                    Reset    : IN std_logic;
16                    Data_Rx  : OUT std_logic_vector(7 downto 0);
17                    DR       : OUT std_logic);
18    END COMPONENT;
19
20    SIGNAL Rx      :  std_logic := '0';
21    SIGNAL Clock   :  std_logic := '0';
22    SIGNAL Reset   :  std_logic := '0';
23
24    SIGNAL Data_Rx  :  std_logic_vector(7 downto 0);
25    SIGNAL DR       :  std_logic;
26
27    BEGIN
28
29    uut: Uart_Receiver PORT MAP(
30                    Rx        => Rx,
31                    Clock     => Clock,
32                    Reset     => Reset,
33                    Data_Rx   => Data_Rx,
34                    DR        => DR);
35
36    Reset_Process: PROCESS
37
38    BEGIN
39
30          Wait for 0 ns;  Reset <= '0';
41          Wait for 5 ns;  Reset <= '1';
42          Wait;
43
44    END PROCESS;
45
46    Clock_Process: PROCESS
47
```

Figure 6.86: VHDL test bench for the UART receiver

```
48      BEGIN
49
50          Wait for 0 ns;    Clock <= '0';
51          Wait for 10 ns;   Clock <= '1';
52          Wait for 10 ns;   Clock <= '0';
53
54      END PROCESS;
55
56      Rx_Process: PROCESS
57
58      BEGIN
59
60          Wait for 0 ns;     Rx <= '1';
61          Wait for 100 ns;   Rx <= '0';
62          Wait for 320 ns;   Rx <= '1';
63
64          Wait for 5000 ns;
65
66          Wait for 0 ns;     Rx <= '1';
67          Wait for 100 ns;   Rx <= '0';
68          Wait for 320 ns;   Rx <= '1';
69          Wait for 320 ns;   Rx <= '0';
70          Wait for 320 ns;   Rx <= '1';
71
72          Wait for 5000 ns;
73
74      END PROCESS;
75
76      END ARCHITECTURE behavioural;
```

Figure 6.86: (Continued)

6.16 Memories

6.16.1 Introduction

Semiconductor memories can be found in many electronic and microelectronic applications such as the personal computer (PC) and are required to store data and program code that can be accessed and/or modified. These circuits are typically found in microprocessor (μP), microcontroller (μC), and digital signal processor (DSP) based systems.

In general, memory can be used in these three types of data and program code storage:

1. *Permanent storage* for values that can be read only within the application and can be changed (if at all) only by removing the memory from the application and reprogramming or replacing it.

2. *Semi-permanent storage* for values that are normally read only within the application (as with permanent storage). However, stored values can be modified by reprogramming while the memory remains in the circuit.

3. *Temporary storage* for values needed only for temporary (immediate) use and requiring fast access or modification (such as program code within a computer system that can be removed when no longer needed).

The two types of memory are read-only memory (ROM) and random access memory (RAM), which is sometimes referred to as read-write memory (RWM).

The memory is usually a fixed block of circuitry that the designer is required to interface correctly with an existing circuit. In this case, a VHDL simulation model for the memory is provided by the memory supplier. In other cases, a memory model might be written by the circuit designer and, if correctly structured, synthesized. However, care must be taken with the resulting memory operation and circuit size. Where certain field programmable gate arrays (FPGA) incorporate memory blocks, a synthesis tool recognizes this and synthesizes the VHDL code to utilize the memories within the FPGA.

6.16.2   Random Access Memory

RAM can be modeled in a number of ways in VHDL. In the example RAM model [4] in Figure 6.87, the address, data, and control signals are shown. Each of 16 addresses holds eight bits of data. Data is written to the memory when the CE (chip enable) and the WE (write enable) signals are active low, and data is read from the memory when the CE and the OE (output enable) signals are active low. In models of this type, care is needed to identify what will happen in the circuit if unexpected control signals are applied on CE, WE, and OE. Line 26 in the code sets the RAM output to high

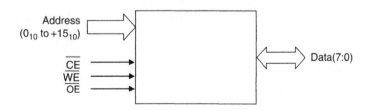

Figure 6.87: 16 address × 8 data bit RAM

impedance under other conditions. An example VHDL code implementation for this design is shown in Figure 6.88.

In this example, the input address signal is an `integer` type, and the data is a bidirectional (`INOUT`) standard logic vector.

An example VHDL test bench for this design is shown in Figure 6.89. As data is written to and read from the RAM model, the applied stimulus is set to high

```
1    LIBRARY ieee;
2    USE ieee.std_logic_1164.all;
3    USE ieee.std_logic_arith.all;
4    USE ieee.std_logic_unsigned.all;
5
6    ENTITY RAM_Model is
7        PORT ( Address  : IN      Integer range 0 to 15;
8               CE       : IN      STD_LOGIC;
9               WE       : IN      STD_LOGIC;
10              OE       : IN      STD_LOGIC;
11              Data     : INOUT STD_LOGIC_VECTOR(7 downto 0));
12   END ENTITY RAM_Model;
13
14   ARCHITECTURE Behavioural OF RAM_Model IS
15
16   BEGIN
17
18   PROCESS(Address, CE, WE, OE) IS
19
20   TYPE Ram_Array IS ARRAY (0 to 15) OF STD_LOGIC_VECTOR(7 downto 0);
21
22   VARIABLE Mem: Ram_Array;
23
24   BEGIN
25
26   Data(7 downto 0) <= (others => 'Z');
27
28       IF (CE = '0') THEN
29           IF (WE = '0') THEN
30               Mem(Address) := Data(7 downto 0);
31           ELSIF (OE = '0') THEN
32               Data(7 downto 0) <= Mem(Address);
33           END IF;
34       END IF;
35
36   END PROCESS;
37
38   END ARCHITECTURE Behavioural;
```

Figure 6.88: 16 × 8 RAM

```
1       LIBRARY ieee;
2       USE ieee.std_logic_1164.all;
3       USE ieee.std_logic_arith.all;
4       USE ieee.std_logic_unsigned.all;
5
6       ENTITY Test_RAM_Model_vhd IS
7       END Test_RAM_Model_vhd;
8
9       ARCHITECTURE behavioural OF Test_RAM_Model_vhd IS
10
11      COMPONENT RAM_Model
12      PORT(
13              Address  : IN       Integer range 0 to 15;
14              CE       : IN       STD_LOGIC;
15              WE       : IN       STD_LOGIC;
16              OE       : IN       STD_LOGIC;
17              Data     : INOUT    STD_LOGIC_VECTOR(7 downto 0));
18      END COMPONENT;
19
20      SIGNAL CE      :  STD_LOGIC := '0';
21      SIGNAL WE      :  STD_LOGIC := '0';
22      SIGNAL OE      :  STD_LOGIC := '0';
23      SIGNAL Address :  Integer range 0 to 15;
24
25      SIGNAL Data :  STD_LOGIC_VECTOR(7 downto 0);
26
27      BEGIN
28
29      uut: RAM_Model PORT MAP(
30                      Address  =>  Address,
31                      CE       =>  CE,
32                      WE       =>  WE,
33                      OE       =>  OE,
34                      Data     =>  Data);
35
36      Test_Bench_Process : PROCESS
37
38      BEGIN
39
40      wait for 0 ns;   Address <= 0;  Data <= "ZZZZZZZZ";
41                       CE <= '1';     WE <= '1';   OE <= '1';
42
43      wait for 10 ns;  Address <= 0;  Data <= "10000001";
44      wait for 10 ns;  CE <= '0';     WE <= '1';   OE <= '1';
45      wait for 10 ns;  CE <= '0';     WE <= '0';   OE <= '1';
46      wait for 10 ns;  CE <= '1';     WE <= '1';   OE <= '1';   Data <= "ZZZZZZZZ";
47
48      wait for 10 ns;  Address <= 0;  Data <= "ZZZZZZZZ";
49      wait for 10 ns;  CE <= '0';     WE <= '1';   OE <= '1';
50      wait for 10 ns;  CE <= '0';     WE <= '1';   OE <= '0';
51      wait for 10 ns;  CE <= '1';     WE <= '1';   OE <= '1';
52
53      wait for 10 ns;
54
55      END PROCESS;
56
57      END ARCHITECTURE Behavioural;
```

Figure 6.89: VHDL test bench for the 16 × 8 RAM

impedance (z) when data is to be read from the memory, and set to the logic levels required to store when data is written to the memory. In this test bench, a value of 129_{10} is written to the memory address 0 and then read back.

6.16.3   Read-Only Memory

ROM can be modeled in a number of ways in VHDL. In the simplest terms, the ROM can be modeled as a combinational logic circuit, with the address providing the input that creates the data output. This is achieved using statements such as *If-then-else* and *Case-when*. Another way of considering the ROM is by creating a look-up table. Both these models in their basic form, shown in Figure 6.90, do not consider the control signals (ROM enable and read signals) that would normally exist with the address and data signals.

A simple ROM design can be generated by creating an array and filling it with the data values. In the example shown in Figure 6.91, an array type (Rom_Array) is defined within the architecture with a size (number of elements) equal to the number of address locations in the memory. This has sixteen address locations and eight data bits per address.

The array is filled with values in the CONSTANT data object. The Data output is assigned the value held within selected element of the Rom_Array. The Address input selects the array element to access. The Address here is an INTEGER data type with values from 0_{10} to $+15_{10}$. An example VHDL test bench for this design is shown in Figure 6.92.

Figure 6.90: Simple read-only memory model

```
1    LIBRARY IEEE;
2    USE IEEE.STD_LOGIC_1164.ALL;
3    USE IEEE.STD_LOGIC_ARITH.ALL;
4    USE IEEE.STD_LOGIC_UNSIGNED.ALL;
5
6    ENTITY ROM is
7        Port ( Address :   IN     INTEGER Range 0 to 15;
8               Data    :   OUT    STD_LOGIC_VECTOR(7 downto 0));
9    END ENTITY ROM;
10
11   ARCHITECTURE Behavioural of ROM is
12
13   TYPE Rom_Array IS Array (0 to 15) of STD_LOGIC_VECTOR(7 downto 0);
14
15   CONSTANT ROM: Rom_Array := (
16                                 "11000000",
17                                 "00010011",
18                                 "00100000",
19                                 "00110000",
20                                 "01000000",
21                                 "01010000",
22                                 "01100000",
23                                 "01110000",
24                                 "10000000",
25                                 "10011000",
26                                 "10100000",
27                                 "10110000",
28                                 "11000000",
29                                 "11010000",
30                                 "11100011",
31                                 "11111111");
32
33   BEGIN
34
35         Data <= Rom(Address);
36
37   END ARCHITECTURE Behavioural;
```

Figure 6.91: 16 address × 8 data bit ROM

This design could be modified either by adding control signals and a tristate output (the output being high impedance when the ROM is not selected), or by considering the use of STD_LOGIC_VECTOR data input for the address rather than an INTEGER data input.

```
LIBRARY ieee;
USE ieee.std_logic_1164.all;
USE ieee.std_logic_arith.all;
USE ieee.std_logic_unsigned.all;

ENTITY Test_ROM IS
END ENTITY Test_ROM;

ARCHITECTURE Behavioural OF Test_ROM IS

COMPONENT ROM
PORT(   Address :  IN    INTEGER Range 0 to 15;
        Data    :  OUT   STD_LOGIC_VECTOR(7 downto 0));
END COMPONENT;

SIGNAL Address :  Integer range 0 to 15;
SIGNAL Data    :  STD_LOGIC_VECTOR (7 downto 0);

BEGIN

uut: ROM PORT MAP(
                   Address => Address,
                   Data    => Data);

Test_Stimulus : PROCESS

BEGIN

        wait for 0 ns;   Address <= 0;
        wait for 10 ns;  Address <= 1;
        wait for 10 ns;  Address <= 2;
        wait for 10 ns;  Address <= 3;
        wait for 10 ns;  Address <= 4;
        wait for 10 ns;  Address <= 5;
        wait for 10 ns;  Address <= 6;
        wait for 10 ns;  Address <= 7;
        wait for 10 ns;  Address <= 8;
        wait for 10 ns;  Address <= 9;
        wait for 10 ns;  Address <= 10;
        wait for 10 ns;  Address <= 11;
        wait for 10 ns;  Address <= 12;
        wait for 10 ns;  Address <= 13;
        wait for 10 ns;  Address <= 14;
        wait for 10 ns;  Address <= 15;
        wait for 10 ns;

END PROCESS;

END ARCHITECTURE Behavioural;
```

Figure 6.92: 16 × 8 ROM test bench

6.17   Unsigned versus Signed Arithmetic

6.17.1   Introduction

In many applications, unsigned arithmetic is sufficient to implement the required functionality using only positive numbers. However, for applications such as digital signal processing, digital filtering, and digital control where positive and negative numbers must be handled, signed arithmetic is required. VHDL can manage this in several ways.

In the examples presented so far, the designs use STD_LOGIC and STD_LOGIC_VECTOR inputs and outputs. The arithmetic operations have used unsigned arithmetic (using the STD_LOGIC_ARITH and STD_LOGIC_UNSIGNED packages). Alternately, the STD_LOGIC_VECTOR signals could have been converted to UNSIGNED, the arithmetic operation performed, and finally the UNSIGNED result converted back to STD_LOGIC_VECTOR. Signed arithmetic operations are accomplished by either using the STD_LOGIC_ARITH and STD_LOGIC_SIGNED packages, or by firstly converting the input from STD_LOGIC_VECTOR to SIGNED, performing the arithmetic operations, and finally converting the SIGNED result back to STD_LOGIC_VECTOR. This idea is shown in Figure 6.93.

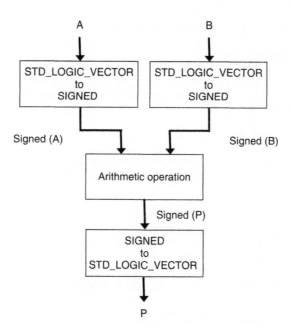

Figure 6.93: Internal signal conversion

Two STD_LOGIC_VECTOR inputs (A and B) are internally converted to signed values, an operation performed on these internal signals, and the result converted back to a STD_LOGIC_VECTOR for output (P). This requires the creation of three internal signals.

This idea is shown in the following two arithmetic operation examples using an eight-bit data $8 + 8$ adder and an 8×8 multiplier.

6.17.2 Adder Example

In this design, two eight-bit input numbers are added together to produce a nine-bit output. An unsigned addition is shown in Figure 6.94. Internal to the unsigned adder, the wordlength is increased to nine bits, and the MSB of each signal is set to logic 0 to prevent overflow of the signal.

```
1    LIBRARY IEEE;
2    USE IEEE.STD_LOGIC_1164.ALL;
3    USE IEEE.STD_LOGIC_ARITH.ALL;
4    USE IEEE.STD_LOGIC_UNSIGNED.ALL;
5
6    ENTITY Adder1 IS
7        PORT ( A : IN    STD_LOGIC_VECTOR (7 downto 0);
8               B : IN    STD_LOGIC_VECTOR (7 downto 0);
9               P : OUT   STD_LOGIC_VECTOR (8 downto 0));
10   END ENTITY Adder1;
11
12   ARCHITECTURE DataFlow of Adder1 is
13
14   SIGNAL A_Int : STD_LOGIC_VECTOR(8 downto 0);
15   SIGNAL B_Int : STD_LOGIC_VECTOR(8 downto 0);
16
17   BEGIN
18
19         A_Int(8) <= '0';
20         A_Int(7 downto 0) <= A(7 downto 0);
21         B_Int(8) <= '0';
22         B_Int(7 downto 0) <= B(7 downto 0);
23         P(8 downto 0) <= A_Int(8 downto 0) + B_Int(8 downto 0);
24
25   END ARCHITECTURE DataFlow;
```

Figure 6.94: Unsigned addition

```
1    LIBRARY IEEE;
2    USE IEEE.STD_LOGIC_1164.ALL;
3    USE IEEE.STD_LOGIC_ARITH.ALL;
4    -- USE IEEE.STD_LOGIC_UNSIGNED.ALL;
5
6    ENTITY Adder2 IS
7        PORT ( A : IN    STD_LOGIC_VECTOR (7 downto 0);
8               B : IN    STD_LOGIC_VECTOR (7 downto 0);
9               P : OUT   STD_LOGIC_VECTOR (8 downto 0));
10   END ENTITY Adder2;
11
12   ARCHITECTURE DataFlow of Adder2 is
13
14   SIGNAL Signed_A : SIGNED(8 downto 0);
15   SIGNAL Signed_B : SIGNED(8 downto 0);
16   SIGNAL Signed_P : SIGNED(8 downto 0);
17
18   BEGIN
19
20           Signed_A(8) <= Signed_A(7);
21           Signed_A(7 downto 0)   <= Signed(A(7 downto 0));
22           Signed_B(8) <= Signed_B(7);
23           Signed_B(7 downto 0)   <= Signed(B(7 downto 0));
25
26           Signed_P(8 downto 0) <= Signed_A(8 downto 0) + Signed_B(8 downto 0)
27           P <= STD_LOGIC_VECTOR(Signed_P(8 downto 0));
28
29   END ARCHITECTURE DataFlow;
```

Figure 6.95: Signed addition

The unsigned adder can be modified to become a signed adder (using 2s complement arithmetic) as shown in Figure 6.95. Here, the wordlength is increased to nine bits, and the MSB of each signal is set to the value of bit 7.

An example VHDL test bench for both unsigned and signed adders is shown in Figure 6.96.

6.17.3 Multiplier Example

The multiplier is an important design in most DSP applications. The multiplier accepts two inputs and multiplies them to produce a result. The multiplication in on two binary numbers that can be either unsigned or signed; signed arithmetic commonly uses 2s complement. Unsigned and signed multiplication require different circuits. When the multiplier accepts two numbers that can vary, the circuit is created in logic using either basic logic gates (AND, OR, etc.) with a particular algorithm, or in some programmable logic devices (such as the Xilinx® Spartan™-3 FPGA), the

```
 1    LIBRARY ieee;
 2    USE ieee.std_logic_1164.all;
 3    USE ieee.std_logic_arith.all;
 4    USE ieee.std_logic_unsigned.all;
 5
 6    ENTITY Test_Adder1_vhd IS
 7    END Test_Adder1_vhd;
 8
 9    ARCHITECTURE Behavioural OF Test_Adder1_vhd IS
10
11    COMPONENT Adder1
12    PORT(
13                    A : IN std_logic_vector(7 downto 0);
14                    B : IN std_logic_vector(7 downto 0);
15                    P : OUT std_logic_vector(7 downto 0));
16    END COMPONENT;
17
18    SIGNAL A :   std_logic_vector(7 downto 0) := (others=>'0');
19    SIGNAL B :   std_logic_vector(7 downto 0) := (others=>'0');
20
21    SIGNAL P :   std_logic_vector(7 downto 0);
22
23    BEGIN
24
25    uut: Adder1 PORT MAP(
26                    A => A,
27                    B => B,
28                    P => P);
29
30    Test_Bench_Process : PROCESS
31
32    BEGIN
33
34    WAIT for 0 ns;    A <= "00000000";    B <= "00000000";
35    WAIT for 10 ns;   A <= "00000001";    B <= "00000000";
36    WAIT for 10 ns;   A <= "00000001";    B <= "00000001";
37    WAIT for 10 ns;   A <= "11111111";    B <= "00000000";
38    WAIT for 10 ns;   A <= "11111111";    B <= "00000001";
39    WAIT for 10 ns;   A <= "11111101";    B <= "11111101";
40    WAIT for 10 ns;   A <= "11111111";    B <= "11111111";
41    WAIT for 10 ns;
42
43    END PROCESS;
44
45    END ARCHITECTURE Behavioural;
```

Figure 6.96: Addition test bench

```
1    LIBRARY IEEE;
2    USE IEEE.STD_LOGIC_1164.ALL;
3    USE IEEE.STD_LOGIC_ARITH.ALL;
4    USE IEEE.STD_LOGIC_UNSIGNED.ALL;
5
6
7    ENTITY Unsigned_Multiplier is
8        PORT ( A : IN  STD_LOGIC_VECTOR (7 downto 0);
9               B : IN  STD_LOGIC_VECTOR (7 downto 0);
10              P : OUT STD_LOGIC_VECTOR (15 downto 0));
11   END ENTITY Unsigned_Multiplier;
12
13
14   ARCHITECTURE DataFlow of Unsigned_Multiplier is
15
16   BEGIN
17
18       P <= A * B;
19
20   END ARCHITECTURE DataFlow;
```

Figure 6.97: Eight-bit unsigned multiplication

design will use one or more of the built-in hardware multipliers. When a hardware multiplier is available, the synthesis tool will utilize the multiplier if possible.

Consider a multiplier operating on two unsigned eight-bit numbers. An example VHDL code for this is shown in Figure 6.97. Here, two eight-bit STD_LOGIC_VECTORS are multiplied to produce a 16-bit result.

This is an example of a dataflow description, and the multiplication is undertaken on line 18:

```
P <= A * B;
```

As A, B, and P are STD_LOGIC_VECTORS, unsigned arithmetic is undertaken by default. With unsigned binary arithmetic, there is no sign bit (the MSB is a value rather than a sign), so the multiplier input signal range is 0_{10} to $+255_{10}$ and the output signal range is 0_{10} to $+65025_{10}$.

An example VHDL test bench for this design is shown in Figure 6.98. This applies the binary values with a decimal equivalent, as shown in Table 6.13, changing values once every 10 ns.

```
LIBRARY IEEE;
USE IEEE.STD_LOGIC_1164.ALL;
USE IEEE.STD_LOGIC_ARITH.ALL;
USE IEEE.STD_LOGIC_UNSIGNED.ALL;

ENTITY Test_Unsigned_Multiplier_vhd IS
END ENTITY Test_Unsigned_Multiplier_vhd;

ARCHITECTURE Behavioural OF Test_Unsigned_Multiplier_vhd IS

COMPONENT Unsigned_Multiplier
PORT(
                A : IN  STD_LOGIC_VECTOR(7 downto 0);
                B : IN  STD_LOGIC_VECTOR(7 downto 0);
                P : OUT STD_LOGIC_VECTOR(15 downto 0));
END COMPONENT;

SIGNAL A :  STD_LOGIC_VECTOR(7 downto 0) := (others=>'0');
SIGNAL B :  STD_LOGIC_VECTOR(7 downto 0) := (others=>'0');

SIGNAL P :  STD_LOGIC_VECTOR(15 downto 0);

BEGIN

uut: Unsigned_Multiplier PORT MAP(
                A => A,
                B => B,
                P => P);

Test_Bench_Process : PROCESS

BEGIN

        wait for  0 ns;   A <= "00000000";  B <= "00000000";
        wait for 10 ns;   A <= "00000001";  B <= "00000001";
        wait for 10 ns;   A <= "10000000";  B <= "10000000";
        wait for 10 ns;   A <= "00000010";  B <= "00000010";
        wait for 10 ns;   A <= "11111111";  B <= "00000001";
        wait for 10 ns;   A <= "11111111";  B <= "11111111";
        wait for 10 ns;

END PROCESS;

END ARCHITECTURE Behavioural;
```

Figure 6.98: Eight-bit unsigned multiplication test bench

**Table 6.13: Unsigned multiplier input and output
values from test bench**

A	B	P
0	0	0
1	1	1
2	2	4
128	128	16384
255	1	255
255	255	65205

Depending on the type of device chosen, the results of the synthesis step will differ. For example, for a Xilinx® Coolrunner™-II CPLD, the result of synthesis as viewed in the ISE™ tools as a technology schematic, shown in Figure 6.99. The synthesis tool understands the multiplier operator, and the resulting circuit is made up of the basic logic gates available in the device. However, for a Spartan™-3 device that incorporates 18×18 signed hardware multiplier blocks, a very different technology schematic results, as shown in Figure 6.100.

The unsigned multiplier can be modified to produce a signed adder (using 2s complement arithmetic) as shown in Figure 6.101. Here, the output wordlength is again increased to 16 bits.

Applying the same test bench inputs as for the unsigned multiplier gives the results shown in Table 6.14. The decimal values shown here are the decimal equivalent of 2s complement binary numbers.

The schematic views for the synthesised VHDL code identifies the differences in the resulting hardware implementation for the CPLD and FPGA devices.

6.18   Testing the Design: The VHDL Test Bench

In the VHDL code examples presented in this chapter, each design has been accompanied by a VHDL test bench to simulate the design. Simulation is an essential part of the design process to verify the correct operation of the VHDL code prior to and after synthesis. No design should be implemented in its target technology unless it has been verified through simulation.

The code within the test bench is the same code that would be used within a design entity and architecture. The main difference is that the test bench need not be synthesized and so can use behavioral descriptions that are not necessarily synthesizable.

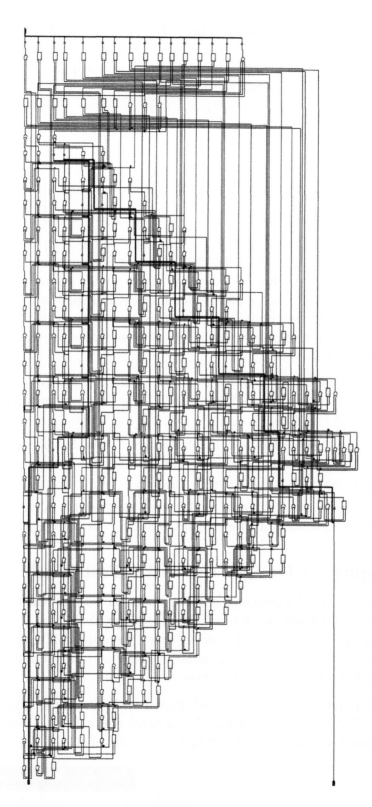

Figure 6.99: Eight-bit unsigned multiplier: synthesis results using Coolrunner™-II CPLD

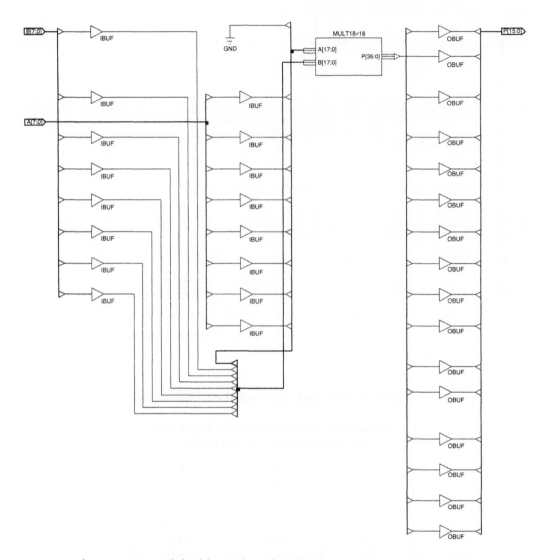

Figure 6.100: Eight-bit unsigned multiplier: synthesis results using Spartan™-3 FPGA

The test bench is a VHDL design unit that creates a test stimulus that is connected to an instance of the design to be tested. Unlike the design entity, the test bench does not have any inputs or outputs (Figure 6.102).

```
1    LIBRARY IEEE;
2    USE IEEE.STD_LOGIC_1164.ALL;
3    USE IEEE.STD_LOGIC_ARITH.ALL;
4    -- USE IEEE.STD_LOGIC_UNSIGNED.ALL;
5
6
7    ENTITY Signed_Multiplier is
8        PORT ( A : IN  STD_LOGIC_VECTOR (7 downto 0);
9               B : IN  STD_LOGIC_VECTOR (7 downto 0);
10              P : OUT STD_LOGIC_VECTOR (15 downto 0));
11   END ENTITY Signed_Multiplier;
12
13
14
15   ARCHITECTURE DataFlow of Signed_Multiplier is
16
17   SIGNAL   A_Signed: SIGNED(7 downto 0);
18   SIGNAL   B_Signed: SIGNED(7 downto 0);
19   SIGNAL   P_Signed: SIGNED(15 downto 0);
20
21   BEGIN
22
23           A_Signed(7 downto 0)  <=  SIGNED(A(7 downto 0));
24           B_Signed(7 downto 0)  <=  SIGNED(B(7 downto 0));
25
26           P_Signed(15 downto 0) <= A_Signed(7 downto 0) * B_Signed(7 downto 0);
27
28           P(15 downto 0) <= STD_LOGIC_VECTOR(P_Signed(15 downto 0));
29
30   END ARCHITECTURE DataFlow;
```

Figure 6.101: Eight-bit signed multiplication

Table 6.14: Signed multiplier input and output values from test bench

A	B	P
0	0	0
1	1	1
2	2	4
128	128	16384
255	1	65535
255	255	1

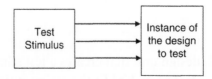

Figure 6.102: VHDL test bench

An example VHDL test bench is shown in Figure 6.103. This is a test bench for a two-input AND gate with inputs A and B and an output Z. The architecture has three main parts:

1. *Top part* identifies the reference libraries and packages to use within the test bench (lines 1 and 2).

2. *Middle part* identifies the test bench entity (lines 5 and 6).

3. *Bottom part* identifies the test bench architecture (lines 9 to 43).

The entity does not have any port declarations because there are no inputs to or outputs from the test bench. The structure of the VHDL test bench architecture is as follows:

1. The component declaration declares the design entity to test. This has the same format as the design entity, except now the keyword ENTITY is replaced with the keyword COMPONENT. This is seen on lines 11 to 17.

2. The input signals to the instance of the design to test are declared. These are SIGNALs on lines 19 and 20.

3. The output signal from the instance of the design to test is declared. This is a SIGNAL on line 22.

4. The architecture BEGIN keyword is on line 24.

5. The design to test is placed as an instance in a structural VHDL description. This is seen on lines 26 to 29.

6. A process is created for the test stimulus. Although only one process is shown here, two or more processes are common for signal generators operating in parallel (as would happen in a real hardware test set-up). The process uses WAIT FOR statements to create time delays between changes in the stimulus. Here, the inputs A and B are applied and changed every 10 ns. When the last statement in the process is acted on, it will loop back to the top of the process. The process here will continually loop until the simulation comes to an end. If the WAIT FOR statement is used (without a time) by itself (i.e., WAIT;), then the process comes to a halt.

The simulation results from this test bench would typically be viewed using a waveform viewer supported by the particular simulation tool. This is only one manner in which to use the test bench.

```
1    LIBRARY ieee;
2    USE ieee.std_logic_1164.ALL;
3
4
5    ENTITY test_And_Gate_vhd IS
6    END ENTITY test_And_Gate_vhd;
7
8
9    ARCHITECTURE Behavioural OF test_And_Gate_vhd IS
10
11   COMPONENT And_Gate
12          PORT(
13                  A : IN std_logic;
14                  B : IN std_logic;
15                  Z : OUT std_logic
16                  );
17   END COMPONENT;
18
19   SIGNAL A :   std_logic := '0';
20   SIGNAL B :   std_logic := '0';
21
22   SIGNAL Z :   std_logic;
23
24   BEGIN
25
26   uut: And_Gate PORT MAP(
27                  A => A,
28                  B => B,
29                  Z => Z);
30
31   Input_Process : PROCESS
32
33   BEGIN
34
35          WAIT FOR 0 ns;   A <= '0'; B <= '0';
36          WAIT FOR 10 ns;  A <= '1'; B <= '0';
37          WAIT FOR 10 ns;  A <= '0'; B <= '1';
38          WAIT FOR 10 ns;  A <= '1'; B <= '1';
39          WAIT FOR 10 ns;
40
41   END PROCESS;
42
43   END ARCHITECTURE Behavioural;
```

Figure 6.103: Two-input AND gate test bench

In general there are three main test bench styles:

1. **Self controlled test bench**, which itself creates the test stimulus and checks the response of the design to test for correct operation. There are no external connections to and from the test bench.

2. **File driven test bench**, in which the test stimulus is held within a data file, and the test bench will read the contents and apply them to the design to be tested. The response of the design is written to a results data file. The user will interact with the stimulus data file and results data file to analyze the results of the simulation run.

3. **Command oriented test bench**, the operation of which is similar to the file driven test bench, except that a command interpreter is placed between the files and the design to be tested. For the test stimulus, commands are stored in the file to be read and converted to test vectors within VHDL. The response of the design is also passed through a command interpreter to create the data to store. With this style, it is possible to reduce the amount of data to read from, and write to files.

6.19 File I/O for Test Bench Development

In scenarios such as test bench development and use in simulation studies of circuit designs, it is often useful to read from and write to text files. Files to be read would typically contain the test stimulus to apply to the circuit under test. Files to be written to would then contain the results of the simulation study. The text held within these files can be formatted to provide for both machine-readable and human readable file contents. In order to access file I/O operations, then the TEXTIO package is used. This package contains the declarations and subprograms to support formatted I/O operations on text files. In the library declaration section in the VHDL test bench file, then the following text would be added:

```
USE STD.TEXTIO.ALL;
```

A file would initially be opened for read or write access and procedures (and functions) would exist for reading formatted text from and writing formatted text to a text file. The IEEE standard document for the VHDL language provides the full description of the procedures and their associated syntax.

Table 6.15: File I/O procedures and functions used in this text. Adapted from IEEE Std. 1076-2002. Copyright 2002, by IEEE. All rights reserved.

Procedure or Function	Example
Procedure: FILE_OPEN	FILE_OPEN(Stimulus_File, "C:\Circuit1_ Stimulus.txt", READ_MODE);
Procedure: READLINE	READLINE(Stimulus_File, Input_Pattern);
Procedure: READ	READ(Input_Pattern, CHAR, Read_OK);
Procedure: WRITELINE	WRITELINE(Results_File, Results_Pattern);
Procedure: WRITE	WRITE(Results_Pattern, "----------------------------");
Function: ENDFILE	WHILE (NOT ENDFILE(Stimulus_File)) LOOP
Procedure: FILE_CLOSE	FILE_CLOSE(Stimulus_File);

The procedures and functions to be considered here are identified in Table 6.15. Here, there are procedures to open a file (for read or write), to read a line of text from a file (opened for reading), to write a line of text to a file (opened for writing), to read values from a line that has been read in from a text file, to write values to a line that is to be written to a text file. Finally, a function is used to identify that an *end of file* has been reached (for file read operations).

The standard types for read and write operations are:

- BIT,

- BIT_VECTOR,

- BOOLEAN,

- CHARACTER,

- INTEGER,

- REAL,

- STRING,

- TIME.

The READLINE and WRITELINE procedures are used to read and write entire lines of a file of type TEXT:

1. READLINE is used to read the next line from a file and returns the value to the defined parameter,

2. WRITELINE is used to write the current line designated by the defined parameter to the file.

The READ and WRITE procedures are used to read values from a line that has been read in using the READLINE procedure, or write values to a line that is to be written using the WRITELINE procedure. It is common to use *whitespace characters* to format the text held in a text file in order to aid readability. However, care must be taken as to the inclusion of *whitespace characters* and the meaning (both to the developer of the code and how it is interpreted by the particular software package used (for simulation, etc.). The *whitespace characters* are:

1. Space   SP

2. Non-breaking space   NBSP

3. horizontal tabulation character   HT

As an example of how file I/O access can be achieved, then Figure 6.104 provides the VHDL code for a dataflow description of a combinational logic circuit with three inputs (A, B and C), and one output (Z).

This is to be simulated by applying all eight possible input codes in a binary count sequence at a rate of 10 ns. The stimulus is to be held in a separate text file with a name:

```
C:\Circuit1_Stimulus.txt
```

This is to hold the text identified in Figure 6.105. This contains lines of comments (commencing with a "–") and the stimulus (logic '0' and '1' values) placed in a line format of:

```
<Horizontal tabulation character> <A> <Space> <B> <Space> <C>
```

When this file is read by the VHDL test bench, then there will be the need to firstly open the file for reading, secondly identify and discard comment lines, thirdly identify stimulus lines and extract the relevant values from these lines, and fourthly close the file when the end of file has been reached.

When the stimulus values have been extracted from the input lines, then they will be applied to the circuit under test. A second file will be opened and the input stimulus along with the output response will be written to this results file with a name:

```
C:\Circuit1_Results.txt
```

```
1    --------------------------------------------------------------
2    -- Combinational Logic Circuit Design: Dataflow Description
3    --------------------------------------------------------------
4
5    --------------------------------------------------------------
6    -- Libraries and packages to use
7    --------------------------------------------------------------
8
9    LIBRARY ieee;
10   USE ieee.std_logic_1164.all;
11   USE ieee.std_logic_arith.all;
12   USE ieee.std_logic_unsigned.all;
13
14   --------------------------------------------------------------
15   -- Top Entity
16   --------------------------------------------------------------
17
18   ENTITY Circuit1 IS
19
20       PORT (    A : IN    STD_LOGIC;
21                 B : IN    STD_LOGIC;
22                 C : IN    STD_LOGIC;
23                 Z : OUT   STD_LOGIC);
24
25   END ENTITY Circuit1;
26
27   --------------------------------------------------------------
28   -- Top Architecture
29   --------------------------------------------------------------
30
31   ARCHITECTURE Behavioural OF Circuit1 IS
32
33   BEGIN
34
35        Z  <=  (A AND B) OR NOT((NOT(A OR B)) AND (A OR C));
36
37   END ARCHITECTURE Behavioural;
38
39   --------------------------------------------------------------
40   -- End of File
41   --------------------------------------------------------------
```

Figure 6.104: Combinational logic circuit description

An example test bench for this design simulation study is shown in Figure 6.106.

The test bench structure is the same as previously used. However now, within the single process created to apply the test stimulus to the circuit under test is additional code to create file access operations. Everything is local to the process, and the two files (Stimulus_File and Results_File) and variables to use are then declared within the process (i.e. local to the process itself).

On commencing the process, the first stage is to open the two files, the Stimulus_File in READ_MODE and the Results_File in WRITE_MODE.

```
 1   -----------------------------------
 2   -- Circuit1 test stimulus input file
 3   -----------------------------------
 4   -- The stimulus format is:
 5   -- A B C
 6   -----------------------------------
 7            0 0 0
 8            0 0 1
 9            0 1 0
10            0 1 1
11            1 0 0
12            1 0 1
13            1 1 0
14            1 1 1
15   -----------------------------------
16   -- End of stimulus file
17   -----------------------------------
```

Figure 6.105: Combinational logic circuit test stimulus file

In the next step, header comments are written to the `Results_File`.

The process then enters a loop that repeats until the end of the `Stimulus_File` is reached:

    WHILE (NOT ENDFILE(Stimulus_File)) LOOP

The first line of the `Stimulus_File` is read and the contents placed in the variable `Input_Pattern`.

The first character of the line is read to detect whether it is a *horizontal tabulation character* or not. If not, the loop starts again and the remainder of the loop is not performed. If the first character is a *horizontal tabulation character*, then the contents of the line are used noting that there is a *white space character* between the input values.

The values to apply to the circuit under test are then assigned the values of the variables.

A time delay of 10 ns occurs and then the results are written to the `Results_File`.

The loop repeats until the end of the `Stimulus_File` is reached, at which point footer comments are written to the `Results_File`.

Both files are closed and the process then halts. If the process repeats, as would happen if the simulation time were to be longer than the time taken for the process to run once, the process would repeat an the `Results_File` contents would be overwritten.

The test bench in Figure 6.106 assumes that the values to apply to the circuit under test (STD_LOGIC) can be assigned the values of the variables (CHARACTERS). If this cannot happen directly, then the values of the variables can alternatively be created as

```
1     ------------------------------------------------------------
2     -- Test bench for Circuit1
3     ------------------------------------------------------------
4
5     ------------------------------------------------------------
6     -- Libraries and packages to use
7     ------------------------------------------------------------
8
9     LIBRARY ieee;
10    USE ieee.std_logic_1164.ALL;
11    USE ieee.std_logic_unsigned.ALL;
12    USE ieee.numeric_std.ALL;
13
14    USE std.textio.ALL;
15
16    ------------------------------------------------------------
17    -- Test bench Entity
18    ------------------------------------------------------------
19
20    ENTITY Test_Circuit1_vhd IS
21    END Test_Circuit1_vhd;
22
23    ARCHITECTURE Behavioural OF Test_Circuit1_vhd IS
24
25    --------------------------------------------------
26    -- Component Declaration for the Unit Under Test (UUT)
27    --------------------------------------------------
28
29    COMPONENT Circuit1
30    PORT(
31                    A : IN   STD_LOGIC;
32                    B : IN   STD_LOGIC;
33                    C : IN   STD_LOGIC;
34                    Z : OUT  STD_LOGIC);
35    END COMPONENT;
36
37    --------------------------------------------------
38    -- Inputs
39    --------------------------------------------------
40
41    SIGNAL A :  STD_LOGIC := '0';
42    SIGNAL B :  STD_LOGIC := '0';
43    SIGNAL C :  STD_LOGIC := '0';
44
45    --------------------------------------------------
46    -- Outputs
47    --------------------------------------------------
48
49    SIGNAL Z :  STD_LOGIC;
50
51    --------------------------------------------------
52
53    BEGIN
54
55    --------------------------------------------------
56    -- Instantiate the Unit Under Test (UUT)
57    --------------------------------------------------
58
59    uut: Circuit1 PORT MAP(
60                    A  =>  A,
61                    B  =>  B,
62                    C  =>  C,
63                    Z  =>  Z);
64
```

Figure 6.106: Combinational logic circuit test bench (1)

```
65    ----------------------------------------------------------
66    -- Read from stimulus file and apply to circuit process
67    ----------------------------------------------------------
68
69    Process_1 : PROCESS
70
71    FILE       Stimulus_File      : TEXT;
72    FILE       Results_File       : TEXT;
73
74    VARIABLE   Input_Pattern      : LINE;
75    VARIABLE   Results_Pattern    : LINE;
76    VARIABLE   Read_OK            : BOOLEAN;
77    VARIABLE   Char               : CHARACTER;
78
79    VARIABLE   A_In, B_In, C_In   : STD_LOGIC;
80
81    ------------------------------------------
82
83    BEGIN
84
85    ------------------------------------------
86    -- Open files for READ and WRITE
87    ------------------------------------------
88
89         FILE_OPEN(Stimulus_File, "C:\Circuit1_Stimulus.txt", READ_MODE);
90         FILE_OPEN(Results_File,  "C:\Circuit1_Results.txt",  WRITE_MODE);
91
92    ------------------------------------------
93    -- Write header text to results file
94    ------------------------------------------
95
96         WRITE(Results_Pattern, "----------------------------");
97         WRITELINE(Results_File, Results_Pattern);
98         WRITE(Results_Pattern, "ABC Z");
99         WRITELINE(Results_File, Results_Pattern);
100        WRITE(Results_Pattern, "----------------------------");
101        WRITELINE(Results_File, Results_Pattern);
102
103   ------------------------------------------
104   -- Loop read from file and apply
105   -- stimulus until end of file
106   ------------------------------------------
107
108        WHILE (NOT ENDFILE(Stimulus_File)) LOOP
109
110   ------------------------------------------
111   -- Read line from 'Stimulus_File' into
112   -- variable 'Input_Pattern'
113   ------------------------------------------
114
115        READLINE(Stimulus_File, Input_Pattern);
116
117   ------------------------------------------
118   -- Read first character from
119   -- 'Input_Pattern'
120   ------------------------------------------
121
122        READ(Input_Pattern, CHAR, Read_OK);
123
124   ------------------------------------------
125   -- If line is not good or the first
126   -- character is not a TAB, then
127   -- skip remainder of loop is not good
128   ------------------------------------------
129
130        IF((NOT Read_OK) OR (CHAR /=HT)) THEN NEXT;
131        END IF;
132
133   ------------------------------------------
134   -- Read first stimulus bit
135   -- Read second stimulus bit
136   -- Read third stimulus bit
137   ------------------------------------------
138
139        READ(Input_Pattern, A_In);
140        READ(Input_Pattern, CHAR);
```

Figure 6.106: (Continued)

```
141          READ(Input_Pattern, B_In);
142          READ(Input_Pattern, CHAR);
143          READ(Input_Pattern, C_In);
144
145          A   <=   A_In;
146          B   <=   B_In;
147          C   <=   C_In;
148
149    ------------------------------------
150    -- Wait for time before applying next
151    -- test stimulus
152    ------------------------------------
153
154          WAIT FOR 10 ns;
155
156    ------------------------------------
157    -- Write stimulus and output to output
158    -- file
159    ------------------------------------
160
161          WRITE(Results_Pattern, A);
162          WRITE(Results_Pattern, B);
163          WRITE(Results_Pattern, C);
164          WRITE(Results_Pattern, " ");
165          WRITE(Results_Pattern, Z)
166
167          WRITELINE(Results_File, Results_Pattern);
168
169    ------------------------------------
170    -- End of Loop
171    ------------------------------------
172
173          END LOOP;
174
175    ------------------------------------
176    -- Write footer text to results file
177    ------------------------------------
178
179          WRITE(Results_Pattern, "-----------------------------");
180          WRITELINE(Results_File, Results_Pattern);
181          WRITE(Results_Pattern, "-- Test completed");
182          WRITELINE(Results_File, Results_Pattern);
183          WRITE(Results_Pattern, "-----------------------------");
185          WRITELINE(Results_File, Results_Pattern);
185
186    ------------------------------------
187    -- Close the OPENed files
188    ------------------------------------
189
190          FILE_CLOSE(Stimulus_File);
191          FILE_CLOSE(Results_File);
192
193    ------------------------------------
194    -- Stop process or it will repeat if
195    -- simulation time longer than a
196    -- single pass of the input and will
197    -- overwrite results file
198    ------------------------------------
199
200          WAIT;
201
202    ------------------------------------
203    END PROCESS;
204
205    ----------------------------------------------------
206
207    END ARCHITECTURE Behavioural;
208
209    ----------------------------------------------------------------
210    -- End of File
211    ----------------------------------------------------------------
```

Figure 6.106: (Continued)

```
 1    ------------------------------------------------------------------
 2    -- Test bench for Circuit1
 3    ------------------------------------------------------------------
 4
 5    ------------------------------------------------------------------
 6    -- Libraries and packages to use
 7    ------------------------------------------------------------------
 8
 9    LIBRARY ieee;
10    USE ieee.std_logic_1164.ALL;
11    USE ieee.std_logic_unsigned.ALL;
12    USE ieee.numeric_std.ALL;
13
14    USE std.textio.ALL;
15
16    ------------------------------------------------------------------
17    -- Test bench Entity
18    ------------------------------------------------------------------
19
20    ENTITY Test_Circuit1_vhd IS
21    END Test_Circuit1_vhd;
22
23    ARCHITECTURE Behavioural OF Test_Circuit1_vhd IS
24
25    ----------------------------------------------------------
26    -- Component Declaration for the Unit Under Test (UUT)
27    ----------------------------------------------------------
28
29    COMPONENT Circuit1
30    PORT(
31                    A : IN   STD_LOGIC;
32                    B : IN   STD_LOGIC;
33                    C : IN   STD_LOGIC;
34                    Z : OUT  STD_LOGIC);
35    END COMPONENT;
36
37    --------------------------------------------------------
38    -- Inputs
39    --------------------------------------------------------
40
41    SIGNAL A :  STD_LOGIC := '0';
42    SIGNAL B :  STD_LOGIC := '0';
43    SIGNAL C :  STD_LOGIC := '0';
44
45    --------------------------------------------------------
46    -- Outputs
47    --------------------------------------------------------
48
49    SIGNAL Z :  STD_LOGIC;
50
51    --------------------------------------------------------
52
53    BEGIN
54
55    --------------------------------------------------------
56    -- Instantiate the Unit Under Test (UUT)
57    --------------------------------------------------------
58
59    uut: Circuit1 PORT MAP(
60                    A  =>  A,
61                    B  =>  B,
62                    C  =>  C,
63                    Z  =>  Z);
64
```

Figure 6.107: Combinational logic circuit test bench (2)

```
65    -------------------------------------------------------------
66    -- Read from stimulus file and apply to circuit process
67    -------------------------------------------------------------
68
69    Process_1 : PROCESS
70
71    FILE       Stimulus_File     : TEXT;
72    FILE       Results_File      : TEXT;
73
74    VARIABLE   Input_Pattern     :  LINE;
75    VARIABLE   Results_Pattern   :  LINE;
76    VARIABLE   Read_OK           :  BOOLEAN;
77    VARIABLE   Char              :  CHARACTER;
78
79    VARIABLE   A_In, B_In, C_In  :  BIT;
80    Variable   Z_Out             :  BIT;
81
82    ------------------------------------
83
84    BEGIN
85
86    ------------------------------------
87    -- Open files for READ and WRITE
88    ------------------------------------
89
90         FILE_OPEN(Stimulus_File, "C:\Circuit1_Stimulus.txt", READ_MODE);
91         FILE_OPEN(Results_File,  "C:\Circuit1_Results.txt",  WRITE_MODE);
92
93    ------------------------------------
94    -- Write header text to results file
95    ------------------------------------
96
97         WRITE(Results_Pattern, "-----------------------------");
98         WRITELINE(Results_File, Results_Pattern);
99         WRITE(Results_Pattern, "ABC Z");
100        WRITELINE(Results_File, Results_Pattern);
101        WRITE(Results_Pattern, "-----------------------------");
102        WRITELINE(Results_File, Results_Pattern);
103
104   ------------------------------------
105   -- Loop read from file and apply
106   -- stimulus until end of file
107   ------------------------------------
108
109        WHILE (NOT ENDFILE(Stimulus_File)) LOOP
110
111   ------------------------------------
112   -- Read line from 'Stimulus_File' into
113   -- variable 'Input_Pattern'
114   ------------------------------------
115
116        READLINE(Stimulus_File, Input_Pattern);
117
118   ------------------------------------
119   -- Read first character from
120   -- 'Input_Pattern'
121   ------------------------------------
122
123        READ(Input_Pattern, CHAR, Read_OK);
124
125   ------------------------------------
126   -- If line is not good or the first
127   -- character is not a TAB, then
128   -- skip remainder of loop is not good
129   ------------------------------------
130
131        IF((NOT Read_OK) OR (CHAR /=HT)) THEN NEXT;
132        END IF;
133
134   ------------------------------------
135   -- Read first stimulus bit
136   -- Read second stimulus bit
137   -- Read third stimulus bit
138   ------------------------------------
139
140        READ(Input_Pattern, A_In);
```

Figure 6.107: (Continued)

```
141         READ(Input_Pattern, CHAR);
142         READ(Input_Pattern, B_In);
143         READ(Input_Pattern, CHAR);
144         READ(Input_Pattern, C_In);
145
146   ------------------------------------
147   -- Apply test stimulus
148   -- Initially convert inputs (A, B, C)
149   -- as BIT to STD_LOGIC - only consider
150   -- logic '0' or logic '1'
151   ------------------------------------
152
153         IF    (A_In = '1') THEN  A <= '1';
154         ELSE  A <= '0';
155         END IF;
156
157         IF    (B_In = '1') THEN  B <= '1';
158         ELSE  B <= '0';
159         END IF;
160
161         IF    (C_In = '1') THEN  C <= '1';
162         ELSE  C <= '0';
163         END IF;
164
165   ------------------------------------
166   -- Wait for time before applying next
167   -- test stimulus
168   ------------------------------------
169
170         WAIT FOR 10 ns;
171
172   ------------------------------------
173   -- Convert 'Z' STD_LOGIC to 'Z_Out'
174   -- BIT - only consider logic '0' and
175   -- logic '1' and unknown 'X'
176   ------------------------------------
177
178         IF    (Z = '1') THEN  Z_Out := '1';
179         ELSE  Z_Out := '0';
180         END IF;
181
182   ------------------------------------
183   -- Write stimulus and output to output
185   -- file
185   ------------------------------------
186
187         WRITE(Results_Pattern, A_In);
188         WRITE(Results_Pattern, B_In);
189         WRITE(Results_Pattern, C_In);
190         WRITE(Results_Pattern, " ");
191         WRITE(Results_Pattern, Z_Out);
192
193         WRITELINE(Results_File, Results_Pattern);
194
195   ------------------------------------
196   -- End of Loop
197   ------------------------------------
198
199         END LOOP;
200
201   ------------------------------------
202   -- Write footer text to results file
203   ------------------------------------
204
205         WRITE(Results_Pattern, "-----------------------------");
206         WRITELINE(Results_File, Results_Pattern);
207         WRITE(Results_Pattern, "-- Test completed");
208         WRITELINE(Results_File, Results_Pattern);
209         WRITE(Results_Pattern, "-----------------------------");
210         WRITELINE(Results_File, Results_Pattern);
211
212   ------------------------------------
213   -- Close the OPENed files
214   ------------------------------------
```

Figure 6.107: (Continued)

```
215
216              FILE_CLOSE(Stimulus_File);
217              FILE_CLOSE(Results_File);
218
219       ------------------------------------
220       -- Stop process or it will repeat if
221       -- simulation time longer than a
222       -- single pass of the input and will
223       -- overwrite results file
224       ------------------------------------
225
226              WAIT;
227
228       ------------------------------------
229
230       END PROCESS;
231
232       ----------------------------------------------
233
234       END ARCHITECTURE Behavioural;
235
236       ------------------------------------------------------
237       -- End of File
238       ------------------------------------------------------
```

Figure 6.107: (Continued)

BIT types and then the values to apply to the circuit under test (STD_LOGIC) can be assigned the values of the variables (BITS) using a suitable type conversion operation. An example of how this could be achieved is shown in Figure 6.107. Here, the difference between this test bench and the previous test bench is the type of the variables declared and the addition of local routines to convert from BIT to STD_LOGIC and from STD_LOGIC to BIT. Care should however be taken here as the limitation of this is that the values are only considered as logical '0' and '1' values.

The contents of the results file for the simulation of the circuit using the test bench in Figure 6.107 is shown in Figure 6.108.

```
1        ---------------------------
2        ABC Z
3        ---------------------------
4        000 1
5        001 0
6        010 1
7        011 1
8        100 1
9        101 1
10       110 1
11       111 1
12       ---------------------------
13
14       ---------------------------
```

Figure 6.108: Test bench simulation results file contents

References

[1]   MacMillen, D., et al., "An Industrial View of Electronic Design Automation," *IEEE Transactions on Computer Aided Design of Integrated Circuits and Systems*, Vol. 19, No. 12, December 2000, pp. 1428–1448.

[2]   Salcic, Z., and Smailagic, A., *Digital Systems Design and Prototyping Using Field Programmable Logic*, Kluwer Academic Publishers, 1998, ISBN 0-7923-9935-8.

[3]   Maxfield, C., *The Design Warrior's Guide to FPGAs*, Newnes, 2004, ISBN 0-7506-7604-3.

[4]   Zwolinski, M., *Digital System Design with VHDL*, Pearson Education Limited, 2000, England, ISBN 0-201-36063-2.

[5]   Skahill, K., *VHDL for Programmable Logic*, Addison-Wesley, 1996, ISBN 0-201-89573-0.

[6]   Navabi, Z., *VHDL Analysis and Modeling of Digital Systems*, McGraw-Hill International Editions, 1993, ISBN 0-07-112732-1.

[7]   The Institute of Electrical and Electronics Engineers, IEEE Standard 1364-2001, Verilog® Hardware Description Language, IEEE, USA, http://www.ieee.org

[8]   The Institute of Electrical and Electronics Engineers, IEEE Standard 1076-2002, VHDL Language Reference Manual, IEEE, USA, http://www.ieee.org

[9]   The Institute of Electrical and Electronics Engineers, http://www.ieee.org

[10] Open Verilog International, OVI.

Student Exercises

1.1 What are the advantages of using VHDL in the design of digital circuits and systems for PLDs and digital ASICs?

1.2 What are the disadvantages of using VHDL in the design of digital circuits and systems for PLDs and digital ASICs?

1.3 Why could a digital circuit design written in VHDL be simulatable but not necessarily synthesizable?

1.4 What are the advantages of developing designs for PLDs in an HDL rather than using schematic capture design entry methods?

1.5 Why would a designer choose to use VHDL rather than Verilog®-HDL in the design of a digital logic circuit?

1.6 What are the main differences between a soft-core and a hard-core intellectual property (IP) block? For the main PLD vendors, identify the types of processors that they support as IP blocks and whether they are provided as soft cores or hard cores.

1.7 How could a 2s complement digital hardware multiplier be designed and implemented within a Xilinx® Coolrunner™-II CPLD? Compare this implementation to that in a Xilinx® Sparatan™-3 FPGA?

1.8 What does EDIF stand for? Identify and discuss an example of an EDIF description for:
   a. A combinational logic circuit containing a maximum of five logic gates
   b. A sequential logic circuit using D-type flip-flops containing a maximum of four flip-flops.

1.9 Develop the VHDL code for an eight-bit up-counter (straight binary count) with a synchronous load and an asynchronous reset. Data is loaded when a load input signal is a 1; otherwise the counter acts to increment on the positive edge of the clock input. All eight bits are to be loaded in parallel.

1.10 Modify the design in Exercise 1.9 to implement the counter using Gray code rather than binary.

1.11 Modify the design in Exercise 1.9 to implement the counter using one-hot coding rather than binary.

1.12 Develop the VHDL code for a four-to-one multiplexer where each input to the MUX is eight bits wide. Simulate the design on the CPLD and check that the design operates as expected. Modify the code to use generics so that an n-bit wide input four-to-one multiplexer can be created.

1.13 Using the CPLD development board (refer to Appendix F—see the last paragraph of the Preface for instructions regarding how to access this online content), develop the VHDL code to implement a checker pattern to test the operation of the SRAM IC on the development board.

1.14 Using the CPLD design tools, enter the design for an 8×8 Baugh Wooley hardware multiplier design as a schematic.

    a. Simulate the design on the CPLD and check that the design operates as expected.

    b. Implement the design on the CPLD and check that the design operates as expected.

1.15 Using the CPLD development board and the seven-segment display board (refer to Appendix F), develop the VHDL code to implement a digital clock. When the CPLD is reset, the clock output returns to zero. The display is to read hours (two characters), minutes (two characters), and seconds (two characters).

    a. Simulate the design on the CPLD and check that the design operates as expected.

    b. Implement the design on the CPLD and check that the design operates as expected.

1.16 Using the CPLD development board and the seven-segment display board (refer to Appendix F), develop the VHDL code to write a message to the six displays. The message will use all six characters, although a character may be left blank. Each seven-segment display will be programmed to turn on and off at the following frequencies: 1 Hz, 10 Hz, 100 Hz, and 1 kHz.

    a. Simulate the design on the CPLD and check that the design operates as expected.

    b. Implement the design on the CPLD and check that the design operates as expected.

1.17 Modify the design in Exercise 1.16 to allow a user to send the message from a PC via a UART receiver (9600 baud rate, no parity checking). Each character is to be sent in turn as a byte of data.

1.18 Using the CPLD development board and the seven-segment display board (refer to Appendix F—see the last paragraph of the Preface for instructions regarding how to access this online content), develop the VHDL code to write the VHDL code to implement a ten-bit binary counter. The seven-segment display board is to display each count state as an integer number (0_{10} to $+1023_{10}$).

    a. Simulate the design on the CPLD and check that the design operates as expected.

b Implement the design on the CPLD and check that the design operates as expected.

1.19 Modify the design in Exercise 1.18 to allow the binary output to represent signed binary (i.e., -512_{10} to $+511_{10}$).

1.20 Modify the design in Exercise 1.16 to allow the binary output to represent the counter output as hexadecimal numbers.

1.21 Using the CPLD development board and the LCD and hex keypad board (refer to Appendix F—see the last paragraph of the Preface for instructions regarding how to access this online content), develop the VHDL code to implement a two-line message board. The message is to change every second, and there are to be five two-line messages. Choose suitable messages to display.

a. Simulate the design on the CPLD and check that the design operates as expected.

b. Implement the design on the CPLD and check that the design operates as expected.

1.22 Modify the design in Exercise 1.21 to allow the user to enter a number between 1 and 5 and, if entered from the hex keypad, the appropriate message will be displayed.

1.23 Using the CPLD development board and the LCD and hex keypad board (refer to Appendix F—see the last paragraph of the Preface for instructions regarding how to access this online content), develop the VHDL code to implement a four-number combinational lock. The lock code is to be hard-wired into the design. If a user successfully enters the four numbers in the correct sequence, a message is to be displayed on the LCD display. If any number is incorrectly entered, a message is displayed after all four numbers have been entered.

a. Simulate the design on the CPLD and check that the design operates as expected.

b. Implement the design on the CPLD and check that the design operates as expected.

1.24 Modify the design in Exercise 1.23 to allow the user a maximum of four attempts at entering a number before being locked out of the system. The system must then be manually reset to start again.

1.25 How is file I/O dealt with in VHDL?

1.26 Develop a UART receiver design that uses the *Case-when* statement.

1.27 Develop a UART transmitter design that uses the *If-then-else* statement.

1.28 Develop a UART transmitter design that uses the *Case-when* statement.

CHAPTER 7

Introduction to Digital Signal Processing

7.1   Introduction

The processing of analogue electrical signals and digital data from one form to another is fundamental to many electronic circuits and systems. Both analogue (voltage and current) signals and digital (logic value) data can be processed by many types of circuits, and the task of finding the right design is a sometimes confusing but normal part of the design process. It depends on identifying the benefits and limitations of the possible implementations to select the most appropriate solution for the particular scenario. Initial concerns are:

- Is the input analogue or digital?

- Is the output analogue or digital?

- Will signal processing use analogue or digital techniques?

This idea is shown in Figure 7.1, where signal processing uses either an analogue signal processor (ASP) or a digital signal processor (DSP). If an analogue signal is to be processed or output as digital data, then the analogue signal must be converted to digital using the analogue-to-digital converter (ADC). The operation of this circuit is discussed in Chapter 8. If a digital signal is to be processed or output as an analogue signal, then the digital data will be converted to analogue using the digital-to-analogue converter (DAC). The operation of this circuit is also discussed in Chapter 8.

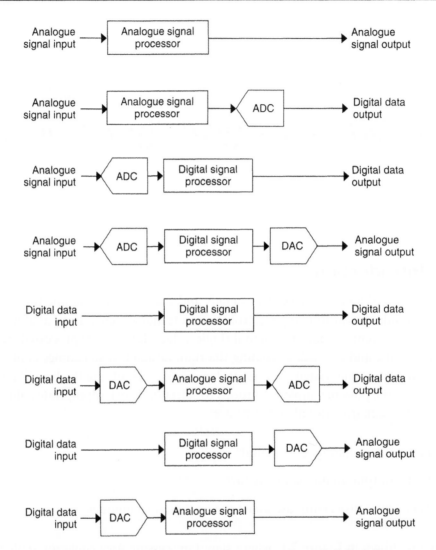

Figure 7.1: Processing of analogue signals and digital data

ASP and DSP each has its own advantages and disadvantages:

Analogue implementation:

Advantages:

- high bandwidth (from DC up to high signal frequencies)
- high resolution

- ease of design
- good approach for simpler design solutions

Disadvantages:

- component value change occurs with component aging
- component value change occurs with temperature variations
- behavior variance between manufactured circuits due to component tolerances
- difficult to change circuit operation

Digital implementation:

Advantages:

- programmable and configurable solution (either programmed in software on a processor or configured in hardware on a CPLD/FPGA)
- operation insensitive to temperature variations
- precise behavior (no behavior variance due to varying component tolerances)
- can implement algorithms that cannot be implemented in analogue
- ease of upgrading and modifying the design

Disadvantages:

- implementation issues due to issues related to numerical calculations
- requires high-performance digital processing
- design complexity
- higher cost

Increasingly, digital implementations are the preferred choice because of their advantages over analogue and because of the ability to implement advanced algorithms that are only possible in the digital domain. In many cases where there are analogue signals and also a requirement for analogue circuitry, the analogue circuitry is kept to a minimum, and the majority of the work performed by the circuit uses digital techniques. The two main areas for digital signal processing considered in this text are digital filters [1–4] and digital control algorithms [5–7].

These can be implemented both in software on the microprocessor (μP), microcontroller (μC), or the digital signal processor (DSP) and in hardware on the complex programmable logic device (CPLD) or field programmable gate array (FPGA). The basis for all possible implementation approaches is a circuit design that will accept samples of digitized analogue signals or direct digital data, perform an algorithm that uses the current sampled value and previous sampled values, and output the digital data directly or in analogue form. The algorithm to be implemented is typically developed using the Z-transform. This algorithm is an equation (or set of equations) that defines a current output in terms of the sums and differences of a current input sample and previous input samples, along with weighting factors. However, to achieve a working implementation of the algorithm, a number of key steps are required:

- analysis of the signal to filter or system to control

- creation of the design specification

- design of the algorithm to fulfill the design requirements

- simulation of the operation of the algorithm

- analysis of the stability of the resulting system

- implementation of the algorithm in the final system

- testing of the final system

It is not the purpose of this text to provide a comprehensive introduction to the Z-transform, but rather to highlight its key points and how the algorithm can be implemented in hardware within a CPLD or FPGA.

Whether digital filtering or digital control is required, a typical system for undertaking DSP tasks is shown in Figure 7.2. Here, the digital system accepts an analogue input and outputs an analogue response. This is undertaken on one or more inputs and creates one or more outputs. In the view shown in Figure 7.2,

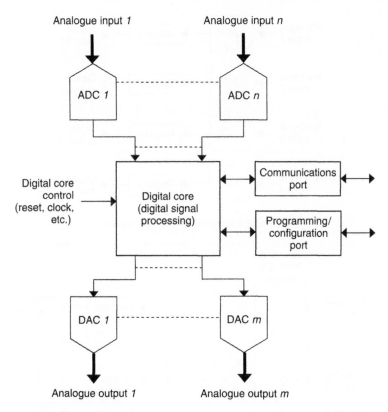

Figure 7.2: Generic digital signal processing arrangement (with analogue I/O)

a DSP core contains the algorithm to implement in addition to a control unit that creates the necessary control signals for ADC control, DAC control, communications port control, and the correct operation of the algorithm. Also shown is a programming/configuration port used to upload a software program (processor-based system) or a hardware configuration (FPGA- or CPLD-based system).

An alternative to using multiple ADCs to sample the analogue input is to use a single ADC, then switch the different analogue inputs to the ADC in turn. The system that utilizes individual ADCs for each analogue input has the capability to sample all analogue inputs in parallel. A system that uses a single switched ADC must sample each input in series (one after another). A parallel ADC arrangement provides for a short sampling period (compared to the serial arrangement, whose signal sampling period

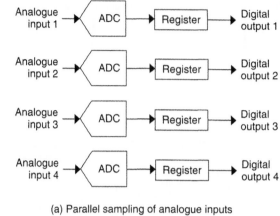

(a) Parallel sampling of analogue inputs

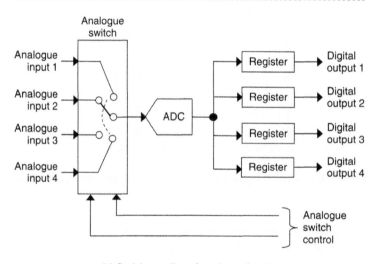

(b) Serial sampling of analogue inputs

Figure 7.3: Parallel or serial sampling of an analogue input

equals the time taken to sample one analogue input multiplied by the number of inputs). However, the need for a parallel or serial arrangement depends on the system requirements and the signal sampling period required. Figure 7.3 shows this idea for a system with four analogue inputs and each digital output is stored in a register.

The choice of ADC architecture determines the number of control pins required by the ADCs and DACs and the conversion time (A/D and D/A). The choice of digital code (e.g., unsigned straight binary or 2s complement signed binary) influences the amount of digital signal encoding and decoding required within the digital core.

It should now be noted that integral to the design of these circuits but not shown here are anti-aliasing filters at the system input (analogue input) to remove any high-frequency signals that would cause aliasing problems with the sampled data.

Example 1: Single-Input, Single-Output DSP Top-Level Description

The basic design architecture shown in Figure 7.2 can be coded in VHDL for a particular design requirement. Consider a custom digital signal processor design that is to sample a single analogue input via an eight-bit ADC, undertake a particular digital signal processing algorithm, and produce an analogue output via an eight-bit DAC. The digital design is to be implemented in hardware using a CPLD or FPGA. The timing of the digital design is to be controlled by a digital input master clock and an active low asynchronous reset. The basic architecture for this design is shown in Figure 7.4. Here, the DSP core:

- uses the AD7575 eight-bit LC^2MOS successive approximation ADC [8]

- uses the AD7524 eight-bit buffered multiplying DAC [9]

- incorporates a simple UART (universal asynchronous receiver transmitter) for communications between the DSP core and an external digital system, using only the *Tx* (transmit) and *Rx* (receive) serial data connections

The digital core contains the algorithm to implement, the necessary control unit that will create the ADC and DAC control signals, and the UART control and data signals. The data to pass to the UART transmitter and the data (or commands) to be received from the UART receiver are specified in the design requirements. The UART has a DR (data received) output used to inform the control unit that a byte has been received from the external digital system and a Transmit input that is used to instruct the UART to transmit a byte of data.

The set-up is shown in Figure 7.5. Here, a CPLD implements the digital actions and interfaces directly with the ADC and DAC. All devices are considered to operate on the same power supply voltage (e.g., +3.3 V) and use the same I/O standards. A suitable clock frequency must be chosen to ensure that all operations can be

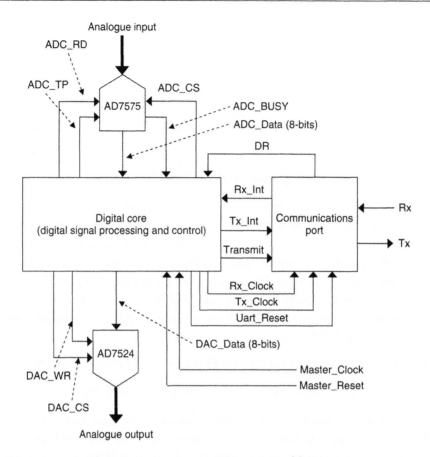

Figure 7.4: Custom DSP core architecture

undertaken within the CPLD (or FPGA) in the required time. The CPLD interfaces with an external system (here a PC) via the RS-232C interface. To enable this, the voltage levels created and accepted by the CPLD must be level-shifted to those required by the RS-232C standard.

The top-level design for the digital circuitry to be configured into the CPLD (or FPGA) can be coded in VHDL. The VHDL structural code (the name of the top-level design here is *top*) is shown in Figure 7.6. Here, the core within the CPLD or FPGA contains two main functional blocks: the first contains the digital core (Dsp_Core), and the second contains the UART (Uart).

The I/O pins for the design are detailed in Table 7.1.

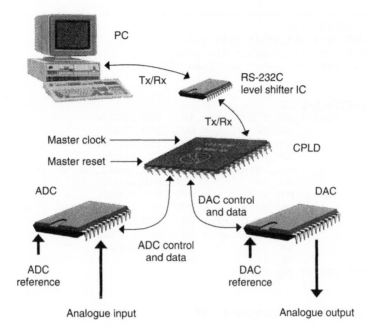

Figure 7.5: System set-up

The internal signals used within the design are detailed in Table 7.2.

The basic operation of the digital system is shown in the flow chart in Figure 7.7. At the start of the circuit operation, the circuit is in a reset state. It then follows a repetitive sequence—sample the analogue input, run the digital algorithm, and update the analogue output—that continues until the circuit is reset back to the reset state.

An example DSP core structure for this design is shown in Figure 7.8. The algorithm, control unit, and I/O register functions are placed in separate blocks. The VHDL code for this structure is shown in Figure 7.9, where the control unit is designed to create four control signals (algorithm control (3:0)) to control the movement and storage of data through the algorithm block. There will be as many control signals as required for the particular algorithm.

An example UART structure for this design is shown in Figure 7.10. The receiver and transmitter functions are placed in separate blocks. The VHDL code for this structure is shown in Figure 7.11.

```
1     LIBRARY ieee;
2     USE ieee.std_logic_1164.all;
3     USE ieee.std_logic_arith.all;
4     USE ieee.std_logic_unsigned.all;
5
6
7     ENTITY Top IS
8         PORT ( ADC_BUSY     : IN    STD_LOGIC;
9                ADC_TP       : OUT   STD_LOGIC;
10               ADC_RD       : OUT   STD_LOGIC;
11               ADC_CS       : OUT   STD_LOGIC;
12               ADC_Data     : IN    STD_LOGIC_VECTOR (7 downto 0);
13               DAC_WR       : OUT   STD_LOGIC;
14               DAC_CS       : OUT   STD_LOGIC;
15               DAC_Data     : OUT   STD_LOGIC_VECTOR (7 downto 0);
16               Master_Clock : IN    STD_LOGIC;
17               Master_Reset : IN    STD_LOGIC;
18               Rx           : IN    STD_LOGIC;
19               Tx           : OUT   STD_LOGIC);
20    END ENTITY Top;
21
22
23    ARCHITECTURE Structural OF Top IS
24
25    SIGNAL Tx_Int     :  STD_LOGIC_VECTOR (7 downto 0);
26    SIGNAL Tx_Clock   :  STD_LOGIC;
27    SIGNAL Rx_Int     :  STD_LOGIC_VECTOR (7 downto 0);
28    SIGNAL Rx_Clock   :  STD_LOGIC;
29    SIGNAL Uart_Reset :  STD_LOGIC;
30    SIGNAL DR         :  STD_LOGIC;
31    SIGNAL Transmit   :  STD_LOGIC;
32
33
34    COMPONENT Dsp_Core IS
35        PORT ( ADC_BUSY     : IN    STD_LOGIC;
36               ADC_TP       : OUT   STD_LOGIC;
37               ADC_RD       : OUT   STD_LOGIC;
38               ADC_CS       : OUT   STD_LOGIC;
39               ADC_Data     : IN    STD_LOGIC_VECTOR (7 downto 0);
40               DAC_WR       : OUT   STD_LOGIC;
41               DAC_CS       : OUT   STD_LOGIC;
42               DAC_Data     : OUT   STD_LOGIC_VECTOR (7 downto 0);
43               Master_Clock : IN    STD_LOGIC;
44               Master_Reset : IN    STD_LOGIC;
45               Rx           : IN    STD_LOGIC_VECTOR (7 downto 0);
46               Rx_Clock     : OUT   STD_LOGIC;
47               Tx           : OUT   STD_LOGIC_VECTOR (7 downto 0);
48               Tx_Clock     : OUT   STD_LOGIC;
49               Uart_Reset   : OUT   STD_LOGIC;
```

Figure 7.6: Top-level structural VHDL code

```
50          DR              : IN   STD_LOGIC;
51          Transmit        : OUT  STD_LOGIC);
52    END COMPONENT Dsp_Core;
53
54    COMPONENT Uart IS
55        PORT (  Uart_Reset  : IN   STD_LOGIC;
56                Rx_Clock    : IN   STD_LOGIC;
57                Tx_Clock    : IN   STD_LOGIC;
58                Rx_Int      : OUT  STD_LOGIC_VECTOR (7 downto 0);
59                Tx_Int      : IN   STD_LOGIC_VECTOR (7 downto 0);
60                Rx          : IN   STD_LOGIC;
61                Tx          : OUT  STD_LOGIC;
62                DR          : OUT  STD_LOGIC;
63                Transmit    : IN   STD_LOGIC);
64    END COMPONENT Uart;
65
66
67    BEGIN
68
69    I1 : Dsp_Core
70        PORT MAP( ADC_BUSY        =>  ADC_BUSY,
71                  ADC_TP          =>  ADC_TP,
72                  ADC_RD          =>  ADC_RD,
73                  ADC_CS          =>  ADC_CS,
74                  ADC_Data        =>  ADC_Data,
75                  DAC_WR          =>  DAC_WR,
76                  DAC_CS          =>  DAC_CS,
77                  DAC_Data        =>  DAC_Data,
78                  Master_Clock    =>  Master_Clock,
79                  Master_Reset    =>  Master_Reset,
80                  Rx              =>  Rx_Int,
81                  Rx_Clock        =>  Rx_Clock,
82                  Tx              =>  Tx_Int,
83                  Tx_Clock        =>  Tx_Clock,
84                  Uart_Reset      =>  Uart_Reset,
85                  DR              =>  DR,
86                  Transmit        =>  Transmit);
87
88    I2 : Uart
89        PORT MAP( Uart_Reset      =>  Uart_Reset,
90                  Rx_Clock        =>  Rx_Clock,
91                  Tx_Clock        =>  Tx_Clock,
92                  Rx_Int          =>  Rx_Int,
93                  Tx_Int          =>  Tx_Int,
94                  Rx              =>  Rx,
95                  Tx              =>  Tx,
96                  DR              =>  DR,
97                  Transmit        =>  Transmit);
98
99    END ARCHITECTURE Structural;
```

Figure 7.6: (Continued)

Table 7.1: Example I/O pins

Pin name	Direction	Purpose
ADC_BUSY	Input	ADC converts analogue input to digital
ADC_TP	Output	Connect to logic 1 in application (test use only)
ADC_RD	Output	ADC read (active low)
ADC_CS	Output	ADC chip select (active low)
ADC_Data	Input	8-bit data from ADC
DAC_WR	Output	DAC write (active low)
DAC_CS	Output	DAC chip select (active low)
DAC_Data	Output	8-bit data to DAC
Master_Clock	Input	Clock input
Master_Reset	Input	Reset control input (active low asynchronous reset)
Rx	Input	Serial data input to UART
Tx	Output	Serial data output from UART

Table 7.2: Example internal signals

Signal name	Purpose
Tx_Int	8-bit data (byte) to send out via the UART
Tx_Clock	UART transmitter clock (x16 baud rate)
Rx_Int	8-bit data (byte) received from the UART
Rx_Clock	UART receiver clock (x16 baud rate)
Uart_Reset	Reset control input (active low asynchronous reset)
DR	Byte of data received on UART Rx input
Transmit	Control signal to initiate the transmission of a byte of data on the UART Tx output

Example 2: Switched Analogue Input

Consider now a circuit that accepts two analogue inputs and produces a single analogue output. The basic architecture for this design is shown in Figure 7.12 where the DSP core:

- uses the AD7575 eight-bit LC^2MOS successive approximation ADC [8]

- uses the AD7524 eight-bit buffered multiplying DAC [9]

- incorporates a simple UART for communications between the DSP core and an external digital system, with just the Tx (transmit) and Rx (receive) serial data connections used

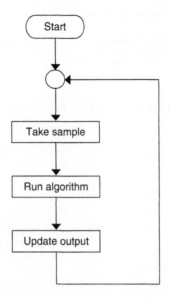

Figure 7.7: Overview of core operation (flow chart)

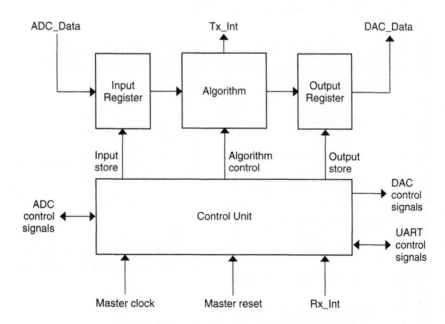

Note: All blocks have a common master reset input.

Figure 7.8: Example DSP core structure

```
 1    LIBRARY ieee;
 2    USE ieee.std_logic_1164.all;
 3    USE ieee.std_logic_arith.all;
 4    USE ieee.std_logic_unsigned.all;
 5
 6
 7    ENTITY Dsp_Core IS
 8        PORT ( ADC_BUSY      :   IN    STD_LOGIC;
 9               ADC_TP        :   OUT   STD_LOGIC;
10               ADC_RD        :   OUT   STD_LOGIC;
11               ADC_CS        :   OUT   STD_LOGIC;
12               ADC_Data      :   IN    STD_LOGIC_VECTOR (7 downto 0);
13               DAC_WR        :   OUT   STD_LOGIC;
14               DAC_CS        :   OUT   STD_LOGIC;
15               DAC_Data      :   OUT   STD_LOGIC_VECTOR (7 downto 0);
16               Master_Clock  :   IN    STD_LOGIC;
17               Master_Reset  :   IN    STD_LOGIC;
18               Rx            :   IN    STD_LOGIC_VECTOR (7 downto 0);
19               Rx_Clock      :   OUT   STD_LOGIC;
20               Tx            :   OUT   STD_LOGIC_VECTOR (7 downto 0);
21               Tx_Clock      :   OUT   STD_LOGIC;
22               Uart_Reset    :   OUT   STD_LOGIC;
23               DR            :   IN    STD_LOGIC;
24               Transmit      :   OUT   STD_LOGIC);
25    END ENTITY Dsp_Core;
26
27
28    ARCHITECTURE Structural OF Dsp_Core IS
29
30
31    SIGNAL     ADC_Data_Int        : STD_LOGIC_VECTOR (7 downto 0);
32    SIGNAL     DAC_Data_Int        : STD_LOGIC_VECTOR (7 downto 0);
33    SIGNAL     Algorithm_Control   : STD_LOGIC_VECTOR (3 downto 0);
34    SIGNAL     Input_Store         : STD_LOGIC;
35    SIGNAL     Output_Store        : STD_LOGIC;
36
37
38    COMPONENT Algorithm IS
39        PORT ( ADC_Data_In       : IN   STD_LOGIC_VECTOR(7 downto 0);
40               Reset             : IN   STD_LOGIC;
41               Algorithm_Control : IN   STD_LOGIC_VECTOR(3 downto 0);
42               Tx                : OUT  STD_LOGIC_VECTOR(7 downto 0);
43               DAC_Data_Out      : OUT  STD_LOGIC_VECTOR(7 downto 0));
44    END COMPONENT  Algorithm;
45
46
47    COMPONENT Register_8_Bit IS
48        PORT ( Store     : IN  STD_LOGIC;
49               Reset     : IN  STD_LOGIC;
50               Data_In   : IN  STD_LOGIC_VECTOR(7 downto 0);
51               Data_Out  : OUT STD_LOGIC_VECTOR(7 downto 0));
52    END COMPONENT Register_8_Bit;
53
54
55    COMPONENT Control_Unit IS
56        PORT ( Master_Clock  : IN   STD_LOGIC;
57               Master_Reset  : IN   STD_LOGIC;
```

Figure 7.9: Example DSP core structure VHDL code

```
58              Rx                    : IN    STD_LOGIC_VECTOR(7 downto 0);
59              Uart_Reset            : OUT   STD_LOGIC;
60              Rx_Clock              : OUT   STD_LOGIC;
61              Tx_Clock              : OUT   STD_LOGIC;
62              Transmit              : OUT   STD_LOGIC;
63              DR                    : IN    STD_LOGIC;
64              ADC_BUSY              : IN    STD_LOGIC;
65              ADC_TP                : OUT   STD_LOGIC;
66              ADC_RD                : OUT   STD_LOGIC;
67              ADC_CS                : OUT   STD_LOGIC;
68              DAC_WR                : OUT   STD_LOGIC;
69              DAC_CS                : OUT   STD_LOGIC;
70              Input_Store           : OUT   STD_LOGIC;
71              Output_Store          : OUT   STD_LOGIC);
72     END COMPONENT Control_Unit;
73
74
75     BEGIN
76
77
78     I_Algorithm : Algorithm
79          PORT MAP ( ADC_Data_In        =>   ADC_Data_Int,
80                     Reset              =>   Master_Reset,
81                     Algorithm_Control  =>   Algorithm_Control,
82                     Tx                 =>   Tx,
83                     DAC_Data_Out       =>   DAC_Data_Int);
84
85
86     I_ControlUnit : Control_Unit
87          PORT MAP ( Master_Clock       =>   Master_Clock,
88                     Master_Reset       =>   Master_Reset,
89                     Rx                 =>   Rx,
90                     Uart_Reset         =>   Uart_Reset,
91                     Rx_Clock           =>   Rx_Clock,
92                     Tx_Clock           =>   Tx_Clock,
93                     Transmit           =>   Transmit,
94                     DR                 =>   DR,
95                     ADC_BUSY           =>   ADC_BUSY,
96                     ADC_TP             =>   ADC_TP,
97                     ADC_RD             =>   ADC_RD,
98                     ADC_CS             =>   ADC_CS,
99                     DAC_WR             =>   DAC_WR,
100                    DAC_CS             =>   DAC_CS,
101                    Input_Store        =>   Input_Store,
102                    Output_Store       =>   Output_Store);
103
104
105    Input_Register : Register_8_Bit
106         PORT MAP ( Store              =>   Input_Store,
107                    Reset              =>   Master_Reset,
108                    Data_In            =>   DAC_Data_Int,
109                    Data_Out           =>   DAC_Data);
110
111    Outut_Register : Register_8_Bit
112         PORT MAP ( Store              =>   Output_Store,
113                    Reset              =>   Master_Reset,
114                    Data_In            =>   DAC_Data_Int,
115                    Data_Out           =>   DAC_Data);
116
117
118    END ARCHITECTURE Structural;
```

Figure 7.9: (Continued)

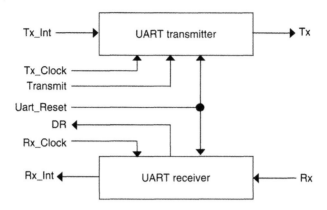

Figure 7.10: Example UART structure

```
1     LIBRARY ieee;
2     USE ieee.std_logic_1164.all;
3     USE ieee.std_logic_arith.all;
4     USE ieee.std_logic_unsigned.all;
5
6
7     ENTITY Uart IS
8         PORT (  Uart_Reset   : IN    STD_LOGIC;
9                 Rx_Clock     : IN    STD_LOGIC;
10                Tx_Clock     : IN    STD_LOGIC;
11                Rx_Int       : OUT   STD_LOGIC_VECTOR (7 downto 0);
12                Tx_Int       : IN    STD_LOGIC_VECTOR (7 downto 0);
13                Rx           : IN    STD_LOGIC;
14                Tx           : OUT   STD_LOGIC;
15                DR           : OUT   STD_LOGIC;
16                Transmit     : IN    STD_LOGIC);
17    END ENTITY Uart;
18
19
20    ARCHITECTURE Structural OF Uart IS
21
22
23    COMPONENT Transmitter IS
24        PORT ( Tx_Clock    : IN    STD_LOGIC;
25               Reset       : IN    STD_LOGIC;
```

Figure 7.11: Example UART structure VHDL code

```
26                    Transmit    : IN    STD_LOGIC;
27                    Tx_Int      : IN    STD_LOGIC_VECTOR(7 downto 0);
28                    Tx          : OUT   STD_LOGIC);
29      END COMPONENT Transmitter;
30
31
32      COMPONENT Receiver IS
33          PORT ( Rx_Clock    : IN    STD_LOGIC;
34                 Reset        : IN    STD_LOGIC;
35                 Rx           : IN    STD_LOGIC;
36                 DR           : OUT   STD_LOGIC;
37                 Rx_Int       : OUT   STD_LOGIC_VECTOR(7 downto 0));
38      END COMPONENT Receiver;
39
40
41      BEGIN
42
43      I1: Transmitter
44              PORT MAP ( Tx_Clock =>   Tx_Clock,
45                         Reset    =>   Uart_Reset,
46                         Transmit =>   Transmit,
47                         Tx_Int   =>   Tx_Int,
48                         Tx       =>   Tx);
49
50      I2 : Receiver
51          PORT MAP ( Rx_Clock =>   Rx_Clock,
52                     Reset    =>   Uart_Reset,
53                     Rx       =>   Rx,
54                     DR       =>   DR,
55                     Rx_Int   =>   Rx_Int);
56
57      END ARCHITECTURE Structural;
```

Figure 7.11: (Continued)

The design is basically the same as that described in Example 1, plus an additional output (Input_Select) from the control unit that selects the analogue input using the analogue switch such that:

- When Input_Select = 0, then analogue input 1 is selected.

- When Input_Select = 1, then analogue input 2 is selected.

The basic operation of the digital system is shown in the flowchart in Figure 7.13. At the start of the circuit operation, the circuit is in a reset state. It then

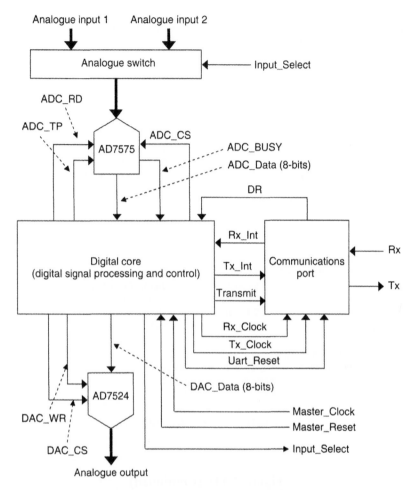

Figure 7.12: Custom DSP core architecture

follows a repetitive sequence—sample both analogue inputs, run the digital algorithm, and update the analogue output—until the circuit is reset back to the reset state.

The top-level design for the digital circuitry to be configured into the CPLD (or FPGA) can be coded in VHDL. The VHDL structural code (the name of the top-level

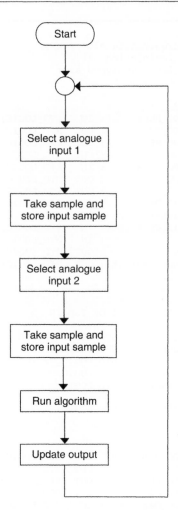

Figure 7.13: Overview of core operation (flowchart)

design here is *Top*) is shown in Figure 7.14. Here, the core within the CPLD or FPGA contains two main functional blocks: the first contains the digital core (Dsp_Core), and the second contains the UART (Uart).

The I/O pins for the design are detailed in Table 7.3.

```
1     LIBRARY ieee;
2     USE ieee.std_logic_1164.all;
3     USE ieee.std_logic_arith.all;
4     USE ieee.std_logic_unsigned.all;
5
6
7     ENTITY Top IS
8         PORT ( ADC_BUSY      : IN    STD_LOGIC;
9                ADC_TP        : OUT   STD_LOGIC;
10               ADC_RD        : OUT   STD_LOGIC;
11               ADC_CS        : OUT   STD_LOGIC;
12               ADC_Data      : IN    STD_LOGIC_VECTOR (7 downto 0);
13               DAC_WR        : OUT   STD_LOGIC;
14               DAC_CS        : OUT   STD_LOGIC;
15               DAC_Data      : OUT   STD_LOGIC_VECTOR (7 downto 0);
16               Master_Clock  : IN    STD_LOGIC;
17               Master_Reset  : IN    STD_LOGIC;
18               Rx            : IN    STD_LOGIC;
19               Tx            : OUT   STD_LOGIC;
20               Input_Select  : OUT   STD_LOGIC);
21    END ENTITY Top;
22
23
24    ARCHITECTURE Structural OF Top IS
25
26    SIGNAL Tx_Int      :   STD_LOGIC_VECTOR (7 downto 0);
27    SIGNAL Tx_Clock    :   STD_LOGIC;
28    SIGNAL Rx_Int      :   STD_LOGIC_VECTOR (7 downto 0);
29    SIGNAL Rx_Clock    :   STD_LOGIC;
30    SIGNAL Uart_Reset  :   STD_LOGIC;
31    SIGNAL DR          :   STD_LOGIC;
32    SIGNAL Transmit    :   STD_LOGIC;
33
34
35    COMPONENT Dsp_Core IS
36        PORT ( ADC_BUSY      : IN    STD_LOGIC;
37               ADC_TP        : OUT   STD_LOGIC;
38               ADC_RD        : OUT   STD_LOGIC;
39               ADC_CS        : OUT   STD_LOGIC;
40               ADC_Data      : IN    STD_LOGIC_VECTOR (7 downto 0);
41               DAC_WR        : OUT   STD_LOGIC;
42               DAC_CS        : OUT   STD_LOGIC;
43               DAC_Data      : OUT   STD_LOGIC_VECTOR (7 downto 0);
44               Master_Clock  : IN    STD_LOGIC;
45               Master_Reset  : IN    STD_LOGIC;
46               Rx            : IN    STD_LOGIC_VECTOR (7 downto 0);
47               Rx_Clock      : OUT   STD_LOGIC;
48               Tx            : OUT   STD_LOGIC_VECTOR (7 downto 0);
49               Tx_Clock      : OUT   STD_LOGIC;
50               Uart_Reset    : OUT   STD_LOGIC;
51               DR            : IN    STD_LOGIC;
52               Transmit      : OUT   STD_LOGIC;
53               Input_Select  : OUT   STD_LOGIC);
```

Figure 7.14: Top-level structural VHDL code

```
54      END COMPONENT Dsp_Core;
55
56      COMPONENT Uart IS
57          PORT (  Uart_Reset    : IN    STD_LOGIC;
58                  Rx_Clock      : IN    STD_LOGIC;
59                  Tx_Clock      : IN    STD_LOGIC;
60                  Rx_Int        : OUT   STD_LOGIC_VECTOR (7 downto 0);
61                  Tx_Int        : IN    STD_LOGIC_VECTOR (7 downto 0);
62                  Rx            : IN    STD_LOGIC;
63                  Tx            : OUT   STD_LOGIC;
64                  DR            : OUT   STD_LOGIC;
65                  Transmit      : IN    STD_LOGIC);
66      END COMPONENT Uart;
67
68
69      BEGIN
70
71      I1 : Dsp_Core
72          PORT MAP( ADC_BUSY          =>  ADC_BUSY,
73                    ADC_TP            =>  ADC_TP,
74                    ADC_RD            =>  ADC_RD,
75                    ADC_CS            =>  ADC_CS,
76                    ADC_Data          =>  ADC_Data,
77                    DAC_WR            =>  DAC_WR,
78                    DAC_CS            =>  DAC_CS,
79                    DAC_Data          =>  DAC_Data,
80                    Master_Clock      =>  Master_Clock,
81                    Master_Reset      =>  Master_Reset,
82                    Rx                =>  Rx_Int,
83                    Rx_Clock          =>  Rx_Clock,
84                    Tx                =>  Tx_Int,
85                    Tx_Clock          =>  Tx_Clock,
86                    Uart_Reset        =>  Uart_Reset,
87                    DR                =>  DR,
88                    Transmit          =>  Transmit,
89                    Input_Select      =>  Input_Select);
90
91      I2 : Uart
92          PORT MAP( Uart_Reset        =>  Uart_Reset,
93                    Rx_Clock          =>  Rx_Clock,
94                    Tx_Clock          =>  Tx_Clock,
95                    Rx_Int            =>  Rx_Int,
96                    Tx_Int            =>  Tx_Int,
97                    Rx                =>  Rx,
98                    Tx                =>  Tx,
99                    DR                =>  DR,
100                   Transmit          =>  Transmit);
101
102     END ARCHITECTURE Structural;
```

Figure 7.14: (Continued)

Table 7.3: Example I/O pins

Pin name	Direction	Purpose
ADC_BUSY	Input	ADC converts analogue input to digital
ADC_TP	Output	Connect to logic 1 in application (test use only)
ADC_RD	Output	ADC read (active low)
ADC_CS	Output	ADC chip select (active low)
ADC_Data	Input	8-bit data from ADC
DAC_WR	Output	DAC write (active low)
DAC_CS	Output	DAC chip select (active low)
DAC_Data	Output	8-bit data to DAC
Master_Clock	Input	Clock input
Master_Reset	Input	Reset control input (active low asynchronous reset)
Rx	Input	Serial data input to UART
Tx	Output	Serial data output from UART
Input_Select	Output	Analogue switch control (0 = analogue input 1 selected, 1 = analogue input 2 selected)

7.2   Z-Transform

The Z-transform is used in the design and analysis of sampled data systems to describe the properties of a sampled data signal and/or a system. It is used in all aspects of digital signal processing as a way to:

- describe the properties of a sampled data signal and/or a system

- transform a continuous time system described using Laplace transforms into a discrete time equivalent

- mathematically analyze the signal and/or system

- view a sampled data signal and/or a system graphically as a block diagram

The Laplace transform is used in continuous time systems to describe a transfer function (the system input-output relationship) with a set of poles and zeros. A continuous time transfer function of a system is represented by the equation:

$$\frac{Y(s)}{X(s)} = G(s) = \frac{N(s)}{D(s)}$$

where:

Y(s) is the output signal from the system

X(s) is the input signal to the system

G(s) is the system transfer function

N(s) is the numerator of the equation

D(s) is the denominator of the equation

This equation is then expanded to become:

$$\frac{Y(s)}{X(s)} = \frac{b_0 + b_1 s + b_2 s^2 + ... + b_m . s^m}{a_0 + a_1 s + a_2 s^2 + ... + a_n . s^n}$$

The poles of the characteristic equation can be found by solving the denominator for:

$$D(s) = 0$$

The zeros of the characteristic equation can be found by solving the denominator for:

$$N(s) = 0$$

Analysis of the poles and zeros determines the performance of the system in both the time and frequency domains. These poles and zeros are complex numbers composed of real (Re(s)) and imaginary (Im(s)) parts. For a system to be stable, the poles of the system must lie to the left of the imaginary axis on the graph of the real and imaginary parts (the Argand diagram), as shown in Figure 7.15. Any pole to the right of the axis indicates an unstable system. A pole that appears on the imaginary axis corresponds to a marginally stable system. The available analysis techniques are described in many DSP, digital filter design, and digital control texts, so they will not be covered further in this text.

The Z-transform is used in discrete time systems to create a discrete time transfer function of the system with a set of poles and zeros. It is a formal transformation for

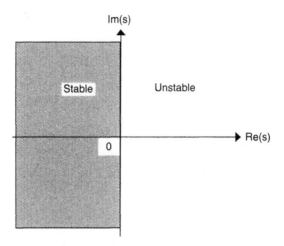

**Figure 7.15: Argand diagram to analyze the stability
of a continuous-time system**

discrete time signals (signals described in terms of their samples) to a new complex
variable called z. For a discrete time signal x(n), then:

$$x(n) = x(0), x(1), x(2), \ldots, \text{etc.}$$

Parentheses indicate the signal sample number. The Z-transform for this is written as
an infinite power series in terms of the complex variable z as:

$$Z\{x(n)\} = x(0) + x(1)z^{-1} + x(2)z^{-2} + \ldots$$

This could be also written as:

$$Z\{x(n)\} = X(z) = \sum x(n)z^{-1}$$

The pulse transfer function of a system is now defined as the Z-transform of the
output divided by the Z-transform of the input and is written as:

$$G(z) = \frac{Z\{y(n)\}}{Z\{x(n)\}} = \frac{Y(z)}{X(z)}$$

where:

- Y(z), is the output signal from the system

- X(z), is the input signal to the system

- G(z), is the pulse transfer function

for a general discrete time transfer function written as:

$$G(z) = \frac{Y(z)}{X(z)} = \frac{N(z)}{D(z)}$$

where:

- Y(z), is the output signal from the system

- X(z), is the input signal to the system

- G(z), is the system transfer function

- N(z), is the numerator of the general discrete time transfer function

- D(z), is the denominator of the general discrete time transfer function

This is then expanded to become:

$$\frac{Y(z)}{X(z)} = \frac{b_0 + b_1 z + b_2 z^2 + \ldots + b_m.z^m}{a_0 + a_1 z + a_2 z^2 + \ldots + a_n.z^n}$$

The poles of the characteristic equation can be found by solving the denominator for:

$$D(z) = 0$$

The zeros of the characteristic equation can be found by solving the denominator for:

$$N(z) = 0$$

Analysis of the poles and zeros determines the performance of the system in both the time and frequency domains. These poles and zeros are complex numbers

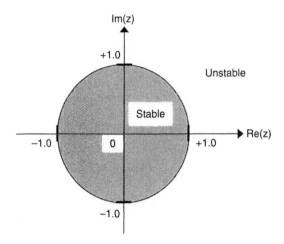

Figure 7.16: Argand diagram showing the unit circle to analyze the stability of a discrete-time system

composed of real (Re(z)) and imaginary (Im(z)) parts. For a system to be stable, the poles of the system must lie within the unit circle on the graph of the real and imaginary parts (the Argand diagram), as shown in Figure 7.16. Any pole outside the unit circle indicates an unstable system. A pole that appears on the unit circle corresponds to a marginally stable system. The available analysis techniques are described in many DSP, digital filter design, and digital control texts, so they will not be covered further in this text.

Comparing systems defined using the Laplace transform and the Z-transform, a continuous time system with a pole at s will have the same dynamic characteristics as a discrete time system with a pole at:

$$z = e^{sT}$$

Here, T is the sampling period of the signal sampling. This allows a discrete-time system to be designed initially as a continuous-time system, then to be translated to a discrete-time implementation. The discrete-time implementation uses signal samples (the current sample and delayed [previous] samples).

However, care must be taken in the implementation of the discrete-time system to account for implementation limitations and for the effect of frequency warping, which occurs when an analogue prototype system is translated to a discrete-time implementation. These aspects are discussed in the next section, on digital control.

The effect of delaying a signal by n samples is to multiply its Z-transform by z^{-n}. This effect is used to implement a discrete-time transfer function either in software or in hardware by sampling and delaying signals. A delay by one sample (Z^{-1}) is shown in Figure 7.17,

where:

$$(\text{Data Output}(z)) = (\text{Data input}(z))z^{-1}$$

Here, D-type flip-flops with asynchronous active low resets store the input data. The Store input is the clock input to each of the flip-flops (all flip-flops are considered to have a common clock input) provides the control for the storage of the data input.

A delay element design used to store a value and delay by one sample is a register. An eight-bit data delay element design in VHDL is shown in Figure 7.18.

Figure 7.19 provides an example VHDL test bench for the delay element.

The individual delay elements can be cascaded to provide a delay-by-m output where m is an integer number that identifies how many clock control signals are required before the input signal becomes an output.

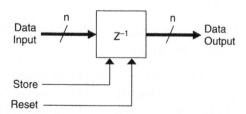

Figure 7.17: Delay element (n-bit register)

```
1     LIBRARY ieee;
2     USE ieee.std_logic_1164.all;
3     USE ieee.std_logic_arith.all;
4     USE ieee.std_logic_unsigned.all;
5
6     ENTITY Delay IS
7         PORT ( Data_In       : IN    STD_LOGIC_VECTOR(7 downto 0);
8                Store         : IN    STD_LOGIC;
9                Reset         : IN    STD_LOGIC;
10               Data_Out      : OUT   STD_LOGIC_VECTOR(7 downto 0));
11    END ENTITY Delay;
12
13    ARCHITECTURE Behavioural OF Delay IS
14
15    BEGIN
16
17    Store_Process: PROCESS(Store, Data_In, Reset)
18
19    BEGIN
20
21          IF (Reset = '0') THEN
22
23                Data_Out(7 downto 0) <= "00000000";
24
25          ELSIF (Store'EVENT AND Store = '1') THEN
26
27                Data_Out(7 downto 0) <= Data_In(7 downto 0);
28
29          END IF;
30
31    END PROCESS Store_Process;
32
33    END ARCHITECTURE Behavioural;
```

Figure 7.18: Delay element (eight-bit register)

Example 3: Delay-by-3 Circuit

To illustrate the delay-by-m circuit, consider a delay-by-3 circuit using three delay elements as shown in Figure 7.20, where:

$$
\begin{aligned}
(\text{No\_Delay}(z)) &= (\text{Data input}(z)) \\
(\text{Delay\_By\_One}(z)) &= (\text{Data input}(z)z^{-1} \\
(\text{Delay\_By\_Two}(z)) &= (\text{Data input}(z))z^{-2} \\
(\text{Delay\_By\_Three}(z)) &= (\text{Data input}(z))z^{-3}
\end{aligned}
$$

```vhdl
1    LIBRARY ieee;
2    USE ieee.std_logic_1164.all;
3    USE ieee.std_logic_arith.all;
4    USE ieee.std_logic_unsigned.all;
5
6
7    ENTITY Test_Delay_vhd IS
8    END Test_Delay_vhd;
9
10
11   ARCHITECTURE Behavioural OF Test_Delay_vhd IS
12
13   COMPONENT Delay
14   PORT(
15               Data_In  : IN  STD_LOGIC_VECTOR(7 downto 0);
16               Store    : IN  STD_LOGIC;
17               Reset    : IN  STD_LOGIC;
18               Data_Out : OUT STD_LOGIC_VECTOR(7 downto 0));
19   END COMPONENT;
20
21   SIGNAL Store    :   STD_LOGIC := '0';
22   SIGNAL Reset    :   STD_LOGIC := '0';
23   SIGNAL Data_In  :   STD_LOGIC_VECTOR(7 downto 0) := (others=>'0');
24
25   SIGNAL Data_Out :   STD_LOGIC_VECTOR(7 downto 0);
26
27   BEGIN
28
29   uut: Delay PORT MAP(
30               Data_In  =>  Data_In,
31               Store    =>  Store,
32               Reset    =>  Reset,
33               Data_Out =>  Data_Out);
34
35
36   Reset_Process : PROCESS
37
38   BEGIN
39
40               Wait for 0 ns;  Reset <= '0';
41               Wait for 5 ns;  Reset <= '1';
42               Wait;
43
44   END PROCESS Reset_Process;
45
46
47   Store_Process : PROCESS
48
49   BEGIN
50
51               Wait for 0 ns;   Store  <= '0';
52               Wait for 10 ns;  Store  <= '1';
53               Wait for 10 ns;  Store  <= '0';
54
55   END PROCESS Store_Process;
56
57
58   DataIn_Process : PROCESS
59
60   BEGIN
61
62               Wait for 0 ns;   Data_In  <= "00000000";
63               Wait for 60 ns;  Data_In  <= "11111111";
64               Wait for 20 ns;  Data_In  <= "00000000";
65               Wait for 20 ns;  Data_In  <= "11111111";
66               Wait for 20 ns;  Data_In  <= "00000000";
67
68               Wait for 20 ns;
69
70   END PROCESS DataIn_Process;
71
72   END ARCHITECTURE Behavioural;
```

Figure 7.19: VHDL test bench for delay element

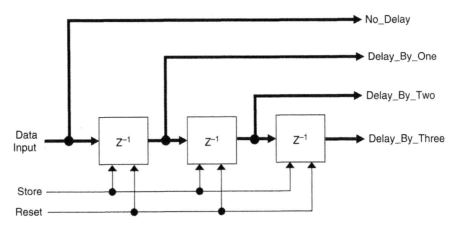

Figure 7.20: Delay-by-3 circuit schematic

Here, the input data and each of the delay element outputs is also available for monitoring signal progression through the circuit.

Such a circuit could be coded in VHDL using a dataflow, behavioral, or structural description. Figure 7.21 shows a behavioral description for this design using two processes. The first process is created to store the input signal in three eight-bit registers, the outputs of which are internal signals. The second process takes the internal signals and provides these as outputs. In the structure illustrated here, the internal signals can be read by another process within the design if this delay-by-3 circuit is modified within a larger design.

Figure 7.22 provides an example VHDL test bench for the delay-by-3 behavioral description.

Using the delay element shown in Figure 7.18, then a structural VHDL description for the delay-by-3 circuit can be created. An example of this is shown in Figure 7.23.

In this design, the outputs from the delay elements are now buffered using an eight-bit buffer (Buffer_Cell). The VHDL code for this buffer design is shown in Figure 7.24.

```vhdl
1    LIBRARY ieee;
2    USE ieee.std_logic_1164.all;
3    USE ieee.std_logic_arith.all;
4    USE ieee.std_logic_unsigned.all;
5
6
7    ENTITY Delay_By_3_Behavioural IS
8         PORT (   Data_In          : IN    STD_LOGIC_VECTOR(7 downto 0);
9                  Store            : IN    STD_LOGIC;
10                 Reset            : IN    STD_LOGIC;
11                 No_Delay         : OUT   STD_LOGIC_VECTOR(7 downto 0);
12                 Delay_By_One     : OUT   STD_LOGIC_VECTOR(7 downto 0);
13                 Delay_By_Two     : OUT   STD_LOGIC_VECTOR(7 downto 0);
14                 Delay_By_Three   : OUT   STD_LOGIC_VECTOR(7 downto 0));
15   END ENTITY Delay_By_3_Behavioural;
16
17
18   ARCHITECTURE Behavioural OF Delay_By_3_Behavioural IS
19
20   SIGNAL Internal_1 : STD_LOGIC_VECTOR(7 downto 0);
21   SIGNAL Internal_2 : STD_LOGIC_VECTOR(7 downto 0);
22   SIGNAL Internal_3 : STD_LOGIC_VECTOR(7 downto 0);
23
24   BEGIN
25
26   Store_Process : PROCESS(Store, Data_In, Internal_1, Internal_2, Internal_3, Reset)
27
28   BEGIN
29
30         IF (Reset = '0') THEN
31
32                 Internal_1 (7 downto 0)   <=  "00000000";
33                 Internal_2 (7 downto 0)   <=  "00000000";
34                 Internal_3 (7 downto 0)   <=  "00000000";
35
36         ELSIF (Store'EVENT AND Store = '1') THEN
37
38                 Internal_1(7 downto 0)   <=  Data_In(7 downto 0);
39                 Internal_2(7 downto 0)   <=  Internal_1(7 downto 0);
40                 Internal_3(7 downto 0)   <=  Internal_2(7 downto 0);
41
42         END IF;
43
44   END PROCESS Store_Process;
45
46
47   Update_Outputs: PROCESS(Data_In, Internal_1, Internal_2, Internal_3)
48
49   BEGIN
50
51   No_Delay(7 downto 0)          <=  Data_In(7 downto 0);
52   Delay_By_One(7 downto 0)      <=  Internal_1(7 downto 0);
53   Delay_By_Two(7 downto 0)      <=  Internal_2(7 downto 0);
54   Delay_By_Three(7 downto 0)    <=  Internal_3(7 downto 0);
55
56   END PROCESS Update_Outputs;
57
58
59   END ARCHITECTURE Behavioural;
```

Figure 7.21: Delay-by-3 circuit behavioral VHDL description

```
1     LIBRARY ieee;
2     USE ieee.std_logic_1164.all;
3     USE ieee.std_logic_arith.all;
4     USE ieee.std_logic_unsigned.all;
5
6
7     ENTITY Test_Delay_By_3_Behavioural_vhd IS
8     END Test_Delay_By_3_Behavioural_vhd;
9
10
11    ARCHITECTURE Behavioural OF Test_Delay_By_3_Behavioural_vhd IS
12
13    COMPONENT Delay_By_3_Behavioural
14      PORT (  Data_In          : IN    STD_LOGIC_VECTOR(7 downto 0);
15              Store            : IN    STD_LOGIC;
16              Reset            : IN    STD_LOGIC;
17              No_Delay         : OUT   STD_LOGIC_VECTOR(7 downto 0);
18              Delay_By_One     : OUT   STD_LOGIC_VECTOR(7 downto 0);
19              Delay_By_Two     : OUT   STD_LOGIC_VECTOR(7 downto 0);
20              Delay_By_Three   : OUT   STD_LOGIC_VECTOR(7 downto 0));
21    END COMPONENT;
22
23    SIGNAL Store   : STD_LOGIC:= '0';
24    SIGNAL Reset   : STD_LOGIC := '0';
25    SIGNAL Data_In : STD_LOGIC_VECTOR(7 downto 0) := (others=>'0');
26
27    SIGNAL  Data_Out       :  STD_LOGIC_VECTOR(7 downto 0);
28    SIGNAL  No_Delay       :  STD_LOGIC_VECTOR(7 downto 0);
29    SIGNAL  Delay_By_One   :  STD_LOGIC_VECTOR(7 downto 0);
30    SIGNAL  Delay_By_Two   :  STD_LOGIC_VECTOR(7 downto 0);
31    SIGNAL  Delay_By_Three :  STD_LOGIC_VECTOR(7 downto 0);
32
33
34    BEGIN
35
36    uut: Delay_By_3_Behavioural PORT MAP(
37                    Data_In          =>   Data_In,
38                    Store            =>   Store,
39                    Reset            =>   Reset,
40                    No_Delay         =>   No_Delay,
41                    Delay_By_One     =>   Delay_By_One,
42                    Delay_By_Two     =>   Delay_By_Two,
43                    Delay_By_Three   =>   Delay_By_Three);
44
45
```

Figure 7.22: VHDL test bench for delay-by-3 circuit behavioral VHDL description

```
46    Reset_Process : PROCESS
47
48    BEGIN
49
50                    Wait for 0 ns;   Reset <= '0';
51                    Wait for 5 ns;   Reset <= '1';
52                    Wait;
53
54    END PROCESS Reset_Process;
55
56
57    Store_Process : PROCESS
58
59    BEGIN
60
61                    Wait for 0 ns;   Store  <= '0';
62                    Wait for 10 ns;  Store  <= '1';
63                    Wait for 10 ns;  Store  <= '0';
64
65    END PROCESS Store_Process;
66
67
68    DataIn_Process : PROCESS
69
70    BEGIN
71
72                    Wait for 0 ns;   Data_In  <= "00000000";
73                    Wait for 60 ns;  Data_In  <= "11111111";
74                    Wait for 20 ns;  Data_In  <= "00000000";
75                    Wait for 20 ns;  Data_In  <= "11111111";
76                    Wait for 20 ns;  Data_In  <= "00000000";
77
78        Wait for 20 ns;
79
80    END PROCESS DataIn_Process;
81
82
83    END ARCHITECTURE Behavioural;
```

Figure 7.22: (Continued)

```
1    LIBRARY ieee;
2    USE ieee.std_logic_1164.all;
3    USE ieee.std_logic_arith.all;
4    USE ieee.std_logic_unsigned.all;
5
6    ENTITY Delay_By_3_Structural IS
7            PORT ( Data_In          : IN    STD_LOGIC_VECTOR(7 downto 0);
8                   Store            : IN    STD_LOGIC;
9                   Reset            : IN    STD_LOGIC;
10                  No_Delay         : OUT   STD_LOGIC_VECTOR(7 downto 0);
11                  Delay_By_One     : OUT   STD_LOGIC_VECTOR(7 downto 0);
12                  Delay_By_Two     : OUT   STD_LOGIC_VECTOR(7 downto 0);
13                  Delay_By_Three   : OUT   STD_LOGIC_VECTOR(7 downto 0));
14   END ENTITY Delay_By_3_Structural;
15
16   ARCHITECTURE Structural OF Delay_By_3_Structural IS
17
18   SIGNAL Internal_1 : STD_LOGIC_VECTOR(7 downto 0);
19   SIGNAL Internal_2 : STD_LOGIC_VECTOR(7 downto 0);
20   SIGNAL Internal_3 : STD_LOGIC_VECTOR(7 downto 0);
21
22   COMPONENT Delay IS
23       PORT ( Data_In      : IN    STD_LOGIC_VECTOR(7 downto 0);
24              Store        : IN    STD_LOGIC;
25              Reset        : IN    STD_LOGIC;
26              Data_Out     : OUT   STD_LOGIC_VECTOR(7 downto 0));
27   END COMPONENT Delay;
28
29   COMPONENT Buffer_Cell IS
30       PORT ( Data_In      : IN    STD_LOGIC_VECTOR(7 downto 0);
31              Data_Out     : OUT   STD_LOGIC_VECTOR(7 downto 0));
32   END COMPONENT Buffer_Cell;
33
34   BEGIN
35   I_Delay1 : Delay
36           PORT MAP( Data_In  => Data_In,
37                     Store    => Store,
38                     Reset    => Reset,
39                     Data_Out => Internal_1);
40
41   I_Delay2 : Delay
42           PORT MAP( Data_In  => Internal_1,
43                     Store    => Store,
44                     Reset    => Reset,
45                     Data_Out => Internal_2);
46
47   I_Delay3 : Delay
48           PORT MAP( Data_In  => Internal_2,
49                     Store    => Store,
50                     Reset    => Reset,
51                     Data_Out => Internal_3);
52
53   I_Buffer1 : Buffer_Cell
54           PORT MAP( Data_In  => Data_In,
55                     Data_Out => No_Delay);
56
57   I_Buffer2 : Buffer_Cell
58           PORT MAP( Data_In  => Internal_1,
59                     Data_Out => Delay_By_One);
60
61   I_Buffer3 : Buffer_Cell
62           PORT MAP( Data_In  => Internal_2,
63                     Data_Out => Delay_By_Two);
64
65   I_Buffer4 : Buffer_Cell
66           PORT MAP( Data_In  => Internal_3,
67                     Data_Out => Delay_By_Three);
68
69   END ARCHITECTURE Structural;
```

Figure 7.23: Delay-by-3 circuit structural VHDL description

```
1    LIBRARY ieee;
2    USE ieee.std_logic_1164.all;
3    USE ieee.std_logic_arith.all;
4    USE ieee.std_logic_unsigned.all;
5
6
7    ENTITY Buffer_Cell IS
8        Port ( Data_In  : IN    STD_LOGIC_VECTOR (7 downto 0);
9               Data_Out : OUT   STD_LOGIC_VECTOR (7 downto 0));
10   END ENTITY Buffer_Cell;
11
12
13   ARCHITECTURE Behavioural OF Buffer_Cell IS
14
15   BEGIN
16
17   Buffer_Process: PROCESS(Data_In)
18
19   BEGIN
20
21           Data_Out(7 downto 0) <= Data_In(7 downto 0);
22
23   END PROCESS Buffer_Process;
24
25   END ARCHITECTURE Behavioural;
```

Figure 7.24: Eight-bit buffer VHDL description

7.3   Digital Control

A control system is composed of two subsystems, a plant and a controller. The plant is the object controlled by the controller. The plant and controller can be either analogue or digital, although digital control algorithms have become more popular because they can be quickly and cost-effectively implemented. In many cases, digital algorithms are implemented using a software program running on a suitable processor within a PC or processor-based embedded system, so the implementer need not have the skills and/or tools to design controllers in hardware on FPGAs and CPLDs. The fundamental algorithm design is however the same, whether the implementation is in hardware or software, and a hardware implementation using an FPGA or CPLD might in some situations be the preferred option. A custom digital controller in hardware has several benefits over processor-based implementation:

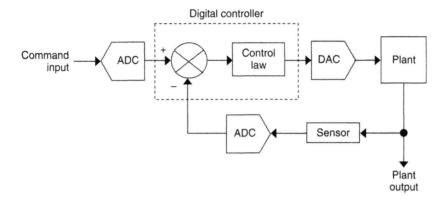

Figure 7.25: Basic computer-based control system

- Custom hardware can be optimized for the application.

- Any processor features not required in the application are not included in the design.

- A software program to run on the target hardware need not be developed.

As an example, Figure 7.25 shows a basic computer-based control system with two analogue inputs and an analogue output. The user sets the required plant output by applying a suitable command input signal. The controller responds to the command input and creates a plant control signal based on the difference between the command input and a feedback signal from the plant. The control law chosen determines how the controller and plant respond to the command input.

In Figure 7.25, then:

- This is an automatic control system in that once a user has set the command input, the system will automatically perform to the requirements of the command input (i.e., it will automatically set the plant to the value set by the command input).

- Using a digital controller, this is also referred to as direct digital control (DDC).

- The first analogue input is a DC voltage (here rather than a current), which sets the value required for the plant (the output load to be controlled). In a motor speed control system, for example, the DC voltage represents the required motor speed. This is the command input. Increasing the command input in a positive direction increases the motor shaft speed in one direction of motor shaft rotation. Increasing the command input in a negative direction increases the motor shaft speed in the opposite direction of motor shaft rotation. A command input of zero indicates a the motor shaft speed of zero.

- The second analogue input is a feedback voltage whose value indicates the value attained by the plant. In a motor speed control system, for example, the DC feedback voltage represents the actual motor shaft speed.

- The analogue output is a signal that is applied to the plant. In a motor speed control system, for example, this is the voltage applied to the motor terminals.

This is an example of a closed-loop control system in that the feedback signal applied to the controller is subtracted from the command input to form an error signal. This error signal is applied to the control law (the algorithm to act on the current sampled input and previous sampled inputs). In general, there can be one or more inputs and one or more outputs. The plant is a continuous time plant, and the inputs to and output from the digital controller are analogue signals.

In general, this leads to the following nine possible arrangements:

1. The control system is either an open-loop system (no feedback) or a closed-loop system (feedback).

2. The command input can be either analogue or digital.

3. The feedback can be either analogue or digital.

4. The controller output can be either analogue or digital.

5. There can be one or more command inputs.

6. There can be one or more feedback signals.

7. There can be one or more plant control signals (outputs from the controller).

8. The controller can implement one or more control algorithms.

9. The digital control algorithm can be designed directly in digital, or it can be created by first creating an analogue prototype, then converting the analogue control law to a digital control law.

The digital controller (or filter) is designed to undertake the required operations using a particular circuit architecture. This architecture is chosen to enable the required operations in the required time using the minimal amount of circuitry (or size of software program) and effectively using the available resources provided by the target technology. The architecture might use a predefined standard computer architecture or a custom architecture. A custom architecture either is based on a processor architecture, or it implements the algorithm exactly as represented by the control law or filter equation.

Standard computer architecture is based on either the Von Neumann or Harvard computer architecture, shown in Figure 7.26. In Von Neumann architecture, the data and instructions share memory and buses, meaning that both cannot be read at the same time. In some applications, this sequential access of data and instructions limits the speed of operation. The Harvard architecture separates the data and instructions storage and buses, thereby providing higher speed of operation than a Von Neumann computer architecture but at the price of increased design complexity.

The processor used within the computer architecture is based on CISC (complex instruction set computer) or RISC (reduced instruction set computer) architecture. The CISC is designed to complete a task in as few lines of processor

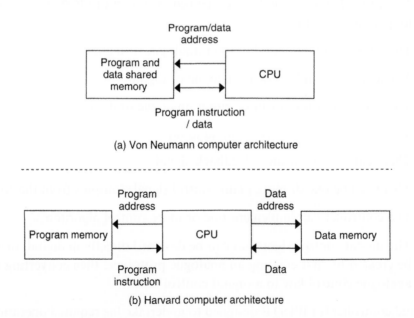

Figure 7.26: Von Neumann and Harvard computer architectures

assembly code as possible, which it achieves by incorporating hardware into the processor that can understand and execute a sequence of operations. The RISC architecture, on the other hand, uses a set of simple instructions that are executed quickly; to perform a complex operation, those simple instructions are combined to form the overall complex operation. Although the RISC approach requires more lines of processor assembly code, it enables smaller and faster processors to be designed. RISC processors are incorporated into many embedded systems.

In a digital control or digital filtering application, a number of operations that need to be performed are common to all applications, and the choice of which operations to incorporate and in which order depends on the application. Table 7.4 identifies the types of operation required.

The overflow prevention operation is required to prevent a value from exceeding its positive and negative limits for correct operation. For example, a four-bit, 2s complement signed number has a range from -8_{10} (1000_2) to $+7_{10}$ (0111_2). If the number is at a value of $+7_{10}$ (0111_2) and one is added to it, the resulting binary code

Table 7.4: Basic operations for digital control and digital filtering

Type of operation	Description
Arithmetic	Perform the basic operations of addition, subtraction, multiplication, and division.
Value store	Store a value in a register for use at a later time.
Wordlength increase/decrease	Increase/Decrease the wordlength of a value to account for the value increasing/decreasing as an arithmetic operation is performed on it.
Overflow prevention	Prevent a value from exceeding a predefined limit (both positive and negative values).
Value truncation	Limit the wordlength of a value by truncation.
Value rounding	Limit the wordlength of a value by rounding.
Conversion	Convert values from one form to another (e.g., unsigned binary to 2s complement signed binary and vice versa).
Sample input control	Control the sampling of the analogue signal(s) to use as the input(s) to the digital controller or filter.
Update output control	Control the output of the analogue signal(s) result(s) as the output(s) from the digital controller or filter.
External communications	Communicate with external systems.

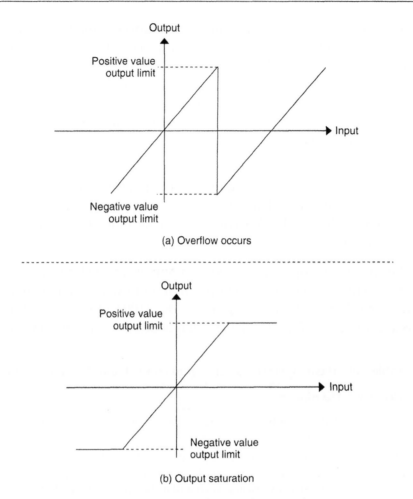

(a) Overflow occurs

(b) Output saturation

Figure 7.27: Overflow and saturation

would be 1000_2. This is -8_{10} in the number system, even though the number should be $+8_{10}$. This effect is referred to as overflow and must be prevented, either by designing circuitry to detect the possibility of overflow and preventing it, or by ensuring that the situation would never occur in the normal operation of the design. Figure 7.27 shows the effect of saturation on an adder that adds 2s complement numbers where (a) there is no overflow prevention, and (b) the output of the adder is designed to saturate rather than overflow. The detection circuitry and saturation can be coded in VHDL. An example schematic for such a circuit is shown in Figure 7.28. Here, the 2s

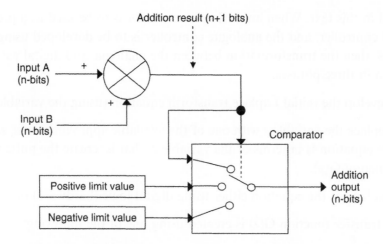

Figure 7.28: 2s complement adder with overflow prevention

complement adder receives two n-bit words and performs an n + 1 addition. The result of this addition is then compared to value limits (positive and negative), and depending on the result of the comparison, the circuit will produce one of three outputs:

1. the result of the addition (n-bits of the n + 1 bit number)

2. the positive limit value (n-bits)

3. the negative limit value (n-bits)

This occurs in a situation where the result of an n-bit arithmetic operation remains n-bits in size. However, in a custom architecture, the potential exists for the range of values to increase or decrease in wordlength as it passes through the arithmetic operations. The designer has this choice.

The choice of wordlength and the truncation or rounding of values as they pass through a digital filter or digital control algorithm affects the result; specifically, how closely the digital result in the implementation represents the result of the calculation if truncation or rounding had not occurred. Additionally, the examples considered in this text apply to fixed-point arithmetic. Designs can also accommodate floating point arithmetic.

The digital control algorithm can be designed using any of a number of possible methods. In many cases, the proportional plus integral plus derivative (PID) controller is used, and the implementation of this algorithm in digital will be considered in this text. When an analogue controller is to be used as a prototype for the digital controller, and the analogue controller is to be developed using Laplace transforms, then the transformation between the analogue and digital will be undertaken in three phrases:

1. Develop the initial Laplace transform equation (using the variable s).

2. Replace the variable s with one of the available approximations, so that now the equation is in terms of the variable z; that is, create the pulse transfer function G(z).

3. Implement the equation either using digital logic (hardware) or in software.

The pulse transfer function G(z) is created using one of the following:

- Forward difference or Euler's method:

$$s = \frac{z - 1}{T}$$

- Backward difference method:

$$s = \frac{z - 1}{zT}$$

- Tustin's approximation (also referred to as the bilinear transform):

$$s = \frac{2}{T} \cdot \frac{z - 1}{z + 1}$$

Here, T is the signal sampling period. These methods are readily applied by hand to transform from s to z.

Example 4: Proportional (P) Control

Consider a digital controller that is to perform proportional control. The controller will accept two inputs, the command input and feedback signals, and will output a single controller output, the controller effort signal. The two inputs are initially subtracted and multiplied by a gain value (the proportional gain Kp is set here to +7). This gain value is held in a ROM. The arrangement for this controller is shown in

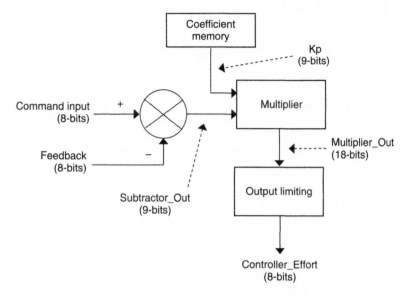

Figure 7.29: Digital proportional gain

Figure 7.29, and here, the internal wordlength increases as the values pass through the arithmetic operations, but finally will be limited to eight bits at the controller output (the inputs are also eight bits). The multiplication in this example is undertaken using a digital multiplier. Figure 7.30 provides the VHDL code for the structure of this design. In this implementation, each block will be coded as a unique entity-architecture pair, although this might not necessarily be the best solution. The design here is purely combinational logic and as such includes no clock or reset inputs.

Figure 7.31 shows the schematic for the synthesized VHDL code using the Xilinx® ISE™ tools. When a digital multiplier is required and the coefficient is fixed, then an alternative to using a digital multiplier is to use a shift-and-add operation. For example, multiplying a value by 2 is a shift-left operation by one bit—simple and easy to do in digital logic and avoids the need for a large digital multiplier.

Example 5: Discrete-Time Integrator

In many situations, then integral action is added to the proportional action in order to achieve the required response from the plant. The integral action can be represented

```
1     LIBRARY ieee;
2     USE ieee.std_logic_1164.all;
3     USE ieee.std_logic_arith.all;
4     USE ieee.std_logic_unsigned.all;
5
6
7     ENTITY Proportional_Gain IS
8         PORT ( Command_Input       : IN   STD_LOGIC_VECTOR (7 downto 0);
9                Feedback            : IN   STD_LOGIC_VECTOR (7 downto 0);
10               Controller_Effort   : OUT  STD_LOGIC_VECTOR (7 downto 0));
11    END ENTITY Proportional_Gain;
12
13
14    ARCHITECTURE Structural OF Proportional_Gain IS
15
16
17    SIGNAL  Subtractor_Out  :  STD_LOGIC_VECTOR(8 downto 0);
18    SIGNAL  Kp              :  STD_LOGIC_VECTOR(8 downto 0);
19    SIGNAL  Multiplier_Out  :  STD_LOGIC_VECTOR(17 downto 0);
20
21
22    COMPONENT Subtractor IS
23        PORT ( Data_In_1    : IN   STD_LOGIC_VECTOR (7 downto 0);
24               Data_In_2    : IN   STD_LOGIC_VECTOR (7 downto 0);
25               Data_Out     : OUT  STD_LOGIC_VECTOR (8 downto 0));
26    END COMPONENT Subtractor;
27
28
29    COMPONENT Coefficient_Memory IS
30        PORT ( Data_Out   : OUT STD_LOGIC_VECTOR (8 downto 0));
31    END COMPONENT Coefficient_Memory;
32
33
34    COMPONENT Multiplier IS
35        PORT ( Data_In     : IN   STD_LOGIC_VECTOR (8 downto 0);
36               Coefficient : IN   STD_LOGIC_VECTOR (8 downto 0);
37               Data_Out    : OUT  STD_LOGIC_VECTOR (17 downto 0));
38    END COMPONENT Multiplier;
39
40
41    COMPONENT Output_Limit IS
42        PORT ( Data_In    : IN   STD_LOGIC_VECTOR (17 downto 0);
43               Data_Out   : OUT  STD_LOGIC_VECTOR (7 downto 0));
44    END COMPONENT Output_Limit;
45
46
47    BEGIN
48
49    I1 : Subtractor
50        PORT MAP ( Data_In_1   =>  Command_Input,
51                   Data_In_2   =>  Feedback,
52                   Data_Out    =>  Subtractor_Out);
53
54    I2 : Coefficient_Memory
55        PORT MAP ( Data_Out    =>  Kp);
56
57    I3 : Multiplier
58        PORT MAP ( Data_In     =>  Subtractor_Out,
59                   Coefficient =>  Kp,
60                   Data_Out    =>  Multiplier_Out);
61
62    I4 : Output_Limit
63        PORT MAP ( Data_In     =>  Multiplier_Out,
64                   Data_Out    =>  Controller_Effort);
65
66
67    END ARCHITECTURE Structural;
```

Figure 7.30: Digital proportional gain VHDL structure code

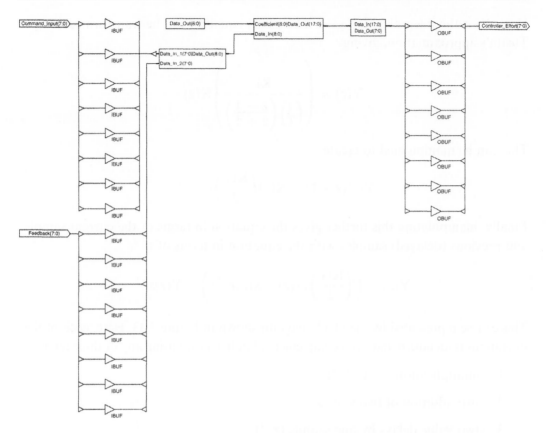

Figure 7.31: Digital proportional gain schematic for the synthesized VHDL code

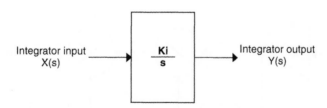

Figure 7.32: Integral action (Laplace transform)

using Z-transforms. Taking an integral action represented initially using a Laplace transform as shown in Figure 7.32, this can be translated to a Z-transform by one of a number of transforms.

Here, Ki is the integral action gain. The (Ki/s) equation can be transformed using Tustin's approximation, giving:

$$Y(z) = \left(\frac{Ki}{\left(\frac{2}{T}\right)\left(\frac{z-1}{z+1}\right)} \right) X(z)$$

This can be manipulated to create:

$$Y(z)(z-1) = X(z)\left(\frac{KiT}{2}\right)(z+1)$$

Finally, manipulating this further gives the equation in terms of the current sample and previous (delayed) samples with the equation in terms of z^{-n}:

$$Y(z) = \left(\left(\frac{KiT}{2}\right)\left(x(z) + x(z)z^{-1}\right) \right) + Y(z)z^{-1}$$

This can be represented by the block diagram shown in Figure 7.33. Here, each of the operations is identified and can be implemented in hardware using any of three methods:

1. multiplication by (KiT/2)

2. two addition of two values

3. two value delays by one sample (z^{-1})

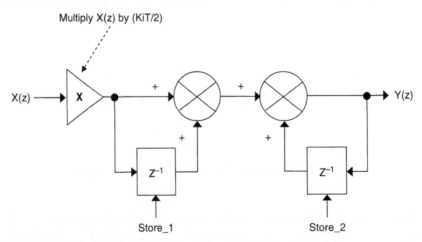

Figure 7.33: Discrete-time integral action

This uses the same basic building blocks as previous examples and can be implemented in VHDL as a structural description (using discrete designs for each of the functional blocks) or as a behavioral or dataflow description.

The multiplication is positioned before the first addition operation. However, the multiplication can be placed after the addition, and if necessary, values can be scaled within the design to address the potential problem of ever-increasing wordlengths due to the range of values that could be encountered in the design.

A modification to the integrator design shown in Figure 7.33 would include an antiwindup circuit. Integrator windup can occur when an input is of a size and polarity that, over time, causes the integrator output to become larger and larger. It can take a substantial amount of time for the integrator output to reduce when the input signal reverses polarity. Additionally, as the values within the integrator become larger, the potential for overflow occurs, which must be taken into account in the design of the circuit.

Example 6: Discrete-Time Differentiator

In addition to the proportional and integral actions, derivative action (a differentiator) can be added to achieve the required response from the plant. The derivative action can be represented using Z-transforms. A derivative action represented initially using a Laplace transform, as shown in Figure 7.34, can be translated to a Z-transform by any of a number of transforms.

Here, Kd is the derivative action gain. The (Kd s) equation can be transformed using Tustin's approximation. This then gives:

$$Y(z) = (Kd)\left(\left(\frac{2}{T}\right)\left(\frac{z-1}{z+1}\right)\right)X(z)$$

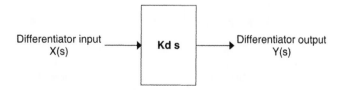

Figure 7.34: Differential action (Laplace transform)

This can be manipulated to create:

$$Y(z)(z+1) = X(z)\left(\frac{2Kd}{T}\right)(z-1)$$

Finally, manipulating further gives the equation in terms of the current sample and previous (delayed) samples with the equation in terms of z^{-n}:

$$Y(z) = \left(\left(\frac{2Kd}{T}\right)\left(x(z) - x(z)z^{-1}\right)\right) - Y(z)z^{-1}$$

This can be represented by the block diagram shown in Figure 7.35. Here, each of the operations is identified and this can be implemented in hardware using any of three methods:

1. multiplication by (2Kd/T)

2. two subtraction of two values

3. two value delays by one sample (z^{-1})

This uses the same basic building blocks as previous examples and can be implemented in VHDL as a structural description (using discrete designs for each of the functional blocks) or as a behavioral or dataflow description.

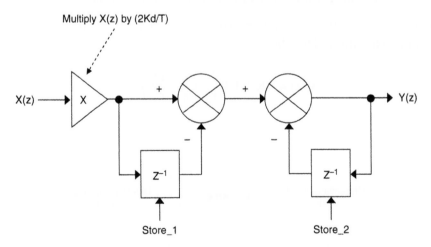

Figure 7.35: Discrete-time derivative action

The multiplication is positioned before the first subtraction operation. However, the multiplication could be placed after the subtraction, and if necessary, values can be scaled within the design to address the potential problem of ever-increasing wordlengths due to the range of values that could be encountered in the design.

Although this structure is similar to the discrete-time integrator, it would not suffer from windup because the feedback signal to the second subtractor is subtracted from the internal signal rather than added.

Example 7: PID Controller

The proportional, integral, and derivative control actions can be brought together to create a PID controller. Figure 7.36 shows an example of how this can be created. As the design increases in complexity, the need for more additions/subtractions and multiplications/divisions increases. This highlights the need to develop an architecture that uses hardware efficiently and can operate within the time constraints of the design. The arithmetic operations to be undertaken either can be designed to be either separate actions (each action requiring its own dedicated hardware) or can be shared (each addition, subtraction, multiplication, or division has a single common block, as is typical in the architecture of an arithmetic and logic unit, ALU). Hence, design speed of operation can be considered against the size of the hardware circuit required for a given architecture.

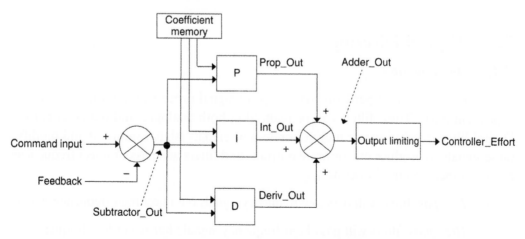

Figure 7.36: Digital PID controller

With the design shown in Figure 7.36, the actions can be implemented such that one of the following two scenarios exists:

1. Each action identified in the block diagram can be created using its own dedicated hardware.

2. Resources can be shared. Table 7.5 provides an example flow of actions for an implementation using shared resources.

Table 7.5: Shared resources for the digital PID controller

Action number	Action description
1	Subtract the feedback input from the command input
2	Store result (Subtractor_Out)
3	Read Subtractor_Out and apply to proportional action
4	Store result (Prop_Out)
5	Read Subtractor_Out and apply to integral action
6	Store result (Int_Out)
7	Read Subtractor_Out and apply to derivative action
8	Store result (Deriv_Out)
9	Read Prop_Out, Int_Out, and Deriv_Out
10	Add Prop_Out, Int_Out, and Deriv_Out
11	Store result (Adder_Out)
12	Read Adder_Out
13	Apply output limiting
14	Store result (Controller_Effort)

7.4 Digital Filtering

7.4.1 Introduction

A filter is a circuit that performs some type of signal processing on a frequency-dependent basis. These filters can be realized in both analogue and digital circuits. Digital filters receive one or more discrete time signals (signal samples) and modify these signals to produce one or more outputs, and filters will pass or reject frequencies based on their required operation:

1. *Low-pass* filters will pass low-frequency signals but reject high-frequency signals.

2. *High-pass* filters will pass high-frequency signals but reject low-frequency signals.

3. *Band-pass* filters will pass a band of signal frequencies but will reject frequencies lower than or higher than the pass range.

4. *Band-reject* or *notch* filters will reject a band of signal frequencies but will pass frequencies lower than or higher than the pass range.

The idealized response for each of the filters is shown in Figure 7.37. On each plot, the X-axis is the frequency (f), and the Y-axis is the magnitude (|H|) of the filter output signal at a particular frequency. The response of an actual filter will deviate from this idealized response. Additionally, although only the filter signal output magnitude is shown, both the signal magnitude and phase response would need to be considered.

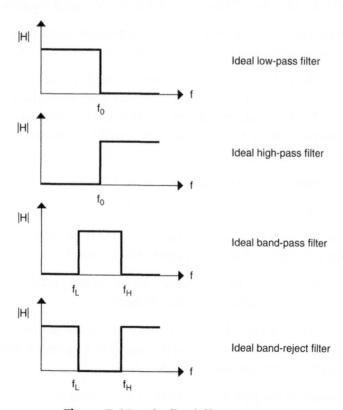

Figure 7.37: Idealized filter response

There are four types of filter design [10]:

1. Bessel filter

2. Butterworth filter

3. Chebyshev filter

4. elliptic filter

The ideal filter response is also referred to as a brick-wall response due to its shape. In low-pass and high-pass filters, the cut-off frequency is f_0. For the band-pass filter, two cut-off frequencies exist, lower (f_L) and upper (f_H), and signals are passed between them. The center frequency is in the center of the pass-band. The frequency range between the lower and upper cut-off frequencies is the bandwidth of the filter. The band-reject filter is the complement of the band-pass filter. To the four responses identified in Figure 7.37 is added a fifth, the all-pass filter. With this, all signal frequencies are passed.

Analogue filters are either passive filters (containing resistors, capacitors, and inductors) or active filters (using active devices such as a transistor or operational amplifier).

Digital filters use DSP techniques on either software- or hardware-based systems. The general structure for a digital filter, shown in Figure 7.38, is similar to the digital controller previously discussed, but the architecture here is presented in a slightly different arrangement. Only one analogue input signal is to be sampled, and the output is a single analogue signal.

The following components are identified in Figure 7.38:

- The *digital filter core* contains three main blocks:

  - The *digital filter algorithm* is responsible for implementing the algorithm operations (add, subtract, multiply, divide, store).

  - The *filter coefficient memory* stores the coefficients used by the digital filter algorithm for multiplications and divisions.

  - The *control unit* provides the necessary timing for actions to occur (ADC input sampling, DAC output updating, filter coefficient memory access, and digital filter algorithm operation).

- The *communications port* allows the filter to communicate with an external digital system.

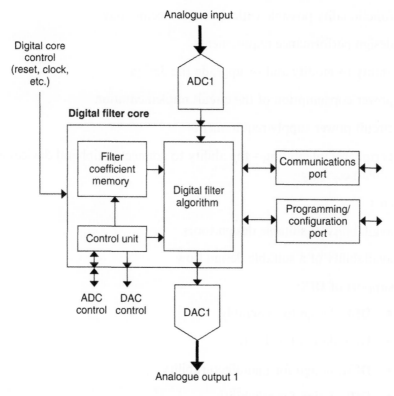

Figure 7.38: General digital filter architecture (with analogue I/O)

- The *programming/configuration port* uploads a software program (in a processor-based system) or a hardware configuration (in an FPGA- or CPLD-based system).

The complete system is controlled by external control signals such as a clock and reset.

This architecture can be modified to provide for different scenarios and specific implementation requirements. In general, the choice of architecture must consider a range of design and implementation issues that include:

1. whether to use standard processor type architecture or to develop a custom architecture

2. available hardware resources

3. functionality possible with the target technology

4. design performance requirements

5. ability to modify and/or upgrade the design

6. power consumption of the circuit implementation

7. circuit power supply requirements

8. peripheral integration—the ability to connect peripheral devices as and when necessary

9. cost

10. availability of suitable design tools

11. availability of a suitable design flow

12. support of **DfX**:

    *   **DfA**, design for assembly

    *   **DfD**, design for debug

    *   **DfM**, design for manufacturability

    *   **DfR**, design for reliability

    *   **DfT**, design for testability

    *   **DfY**, design for yield

Example 8: Digital Filter Structure

Consider the filter architecture shown in Figure 7.38. This can be coded for in VHDL as a structural description. Consider the case where the digital filter algorithm requires four control signals and eight fixed 16-bit coefficients. The filter coefficients are stored in a ROM within the design and are set when the CPLD is configured. A structural description for each of the main blocks within the digital filter core is shown in Figure 7.39.

The filter coefficient memory has three address lines and sixteen data lines. There are no memory control signals, so when an address is applied to the memory, the data stored in that address is applied to the digital filter algorithm.

The ADC used is the AD7575, and the DAC used is the AD7524.

```vhdl
1    LIBRARY ieee;
2    USE ieee.std_logic_1164.all;
3    USE ieee.std_logic_arith.all;
4    USE ieee.std_logic_unsigned.all;
5
6
7    ENTITY Filter_Core IS
8        PORT ( Master_Clock   :  IN   STD_LOGIC;
9               Master_Reset   :  IN   STD_LOGIC;
10              ADC_Data_In    :  IN   STD_LOGIC_VECTOR(7 downto 0);
11              ADC_BUSY       :  IN   STD_LOGIC;
12              ADC_TP         :  OUT  STD_LOGIC;
13              ADC_RD         :  OUT  STD_LOGIC;
14              ADC_CS         :  OUT  STD_LOGIC;
15              DAC_Data_Out   :  OUT  STD_LOGIC_VECTOR(7 downto 0);
16              DAC_WR         :  OUT  STD_LOGIC;
17              DAC_CS         :  OUT  STD_LOGIC);
18   END ENTITY Filter_Core;
19
20
21   ARCHITECTURE Structural OF Filter_Core IS
22
23
24   SIGNAL Coefficient_Internal     : STD_LOGIC_VECTOR(15 downto 0);
25   SIGNAL Control_Internal         : STD_LOGIC_VECTOR(3 downto 0);
26   SIGNAL Memory_Address_Internal  : STD_LOGIC_VECTOR(2 downto 0);
27
28
29   COMPONENT Algorithm IS
30       PORT ( Filter_In      :  IN   STD_LOGIC_VECTOR (7 downto 0);
31              Reset          :  IN   STD_LOGIC;
32              Coefficient    :  IN   STD_LOGIC_VECTOR (15 downto 0);
33              Filter_Control :  IN   STD_LOGIC_VECTOR (3 downto 0);
34              Filter_Out     :  OUT  STD_LOGIC_VECTOR (7 downto 0));
35   END COMPONENT Algorithm;
36
37
38   COMPONENT Coefficient_Memory IS
39       PORT ( Address        :  IN   STD_LOGIC_VECTOR (2 downto 0);
40              Data           :  OUT  STD_LOGIC_VECTOR (15 downto 0));
41   END COMPONENT Coefficient_Memory;
42
43
44   COMPONENT Control_Unit IS
45       PORT ( Master_Clock   :  IN   STD_LOGIC;
46              Master_Reset   :  IN   STD_LOGIC;
47              Filter_Control :  OUT  STD_LOGIC_VECTOR (3 downto 0);
48              Memory_Address :  OUT  STD_LOGIC_VECTOR (2 downto 0);
49              ADC_BUSY       :  IN   STD_LOGIC;
50              ADC_TP         :  OUT  STD_LOGIC;
51              ADC_RD         :  OUT  STD_LOGIC;
52              ADC_CS         :  OUT  STD_LOGIC;
53              DAC_WR         :  OUT  STD_LOGIC;
54              DAC_CS         :  OUT  STD_LOGIC);
55   END COMPONENT Control_Unit;
```

Figure 7.39: Digital filter core example

```
56
57
58       BEGIN
59
60       I1 : Algorithm
61           PORT MAP( Filter_In        =>   ADC_Data_In,
62                     Reset            =>   Master_Reset,
63                     Coefficient      =>   Coefficient_Internal,
64                     Filter_Control   =>   Control_Internal,
65                     Filter_Out       =>   DAC_Data_Out);
66
67       I2 : Coefficient_Memory
68           PORT MAP( Address          =>   Memory_Address_Internal,
69                     Data             =>   Coefficient_Internal);
70
71       I3 : Control_Unit
72           PORT MAP( Master_Clock     =>   Master_Clock,
73                     Master_Reset     =>   Master_Reset,
74                     Filter_Control   =>   Control_Internal,
75                     Memory_Address   =>   Memory_Address_Internal,
76                     ADC_BUSY         =>   ADC_BUSY,
77                     ADC_TP           =>   ADC_TP,
78                     ADC_RD           =>   ADC_RD,
79                     ADC_CS           =>   ADC_CS,
80                     DAC_WR           =>   DAC_WR,
81                     DAC_CS           =>   DAC_CS);
82
83       END ARCHITECTURE Structural;
```

Figure 7.39: (Continued)

The control unit identifies the control signals for the memory, algorithm, ADC, and DAC, and does not include any control signals for the communications interface.

Figure 7.40 shows the schematic for the synthesized VHDL code using the Xilinx® ISE™ tools.

VHDL entity-architecture pairs can then be created to complete the design by adding the required detail to the algorithm, memory, and control unit blocks.

Example 9: Multiply by Two

Although a digital implementation could be created to solve a given problem, it is not always suitable. Consider the need to amplify an analogue voltage by two. This could be implemented in analogue or digital, and Figure 7.41 shows a possible implementation of both. The analogue circuit uses a noninverting operational amplifier (op-amp). The digital circuit is rather more complex.

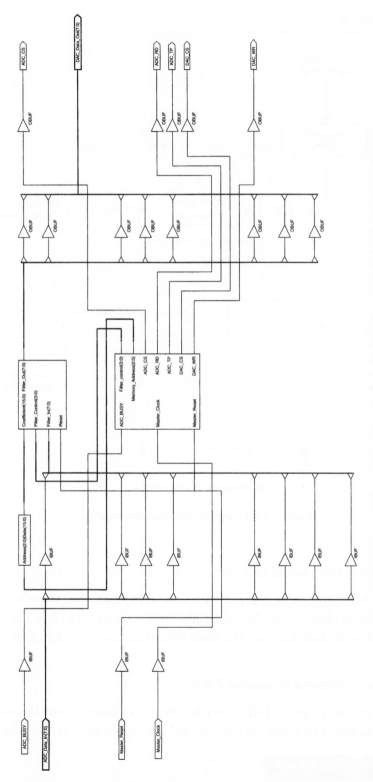

Figure 7.40: Digital filter example schematic for the synthesized VHDL code

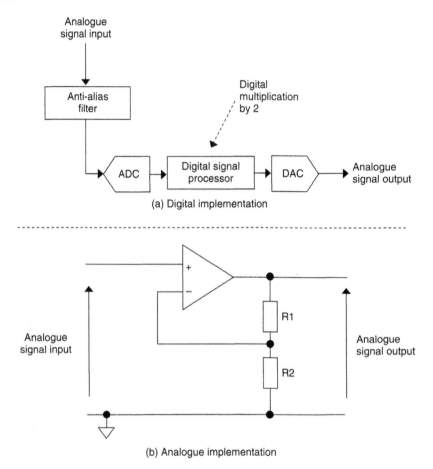

Figure 7.41: Amplifier implementation

Which implementation would be better?

Filters are of two types: infinite impulse response (IIR) and finite impulse response (FIR). The type of filter chosen determines the architecture of the filter and what values are to be used in the calculations. The basic filter structures are identified below.

7.4.2   Infinite Impulse Response Filters

The infinite impulse response (IIR) filter is a recursive filter in that the output from the filter is computed by using the current and previous inputs and previous outputs.

Because the filter uses previous values of the output, there is feedback of the output in the filter structure. The design of the IIR filter is based on identifying the pulse transfer function G(z) that satisfies the requirements of the filter specification. This can be undertaken either by developing an analogue prototype and then transforming it to the pulse transfer function, or by designing directly in digital. Figure 7.42 shows typical IIR filter architecture.

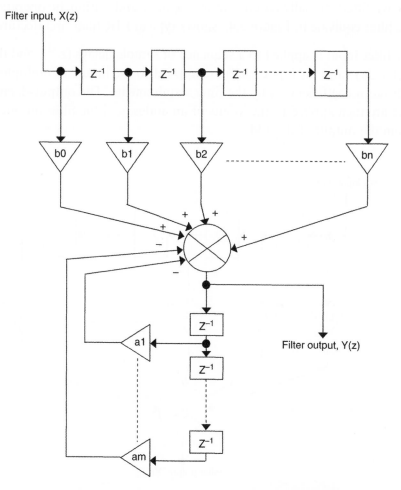

Figure 7.42: Typical architecture of an IIR filter

7.4.3 Finite Impulse Response Filters

The finite impulse response (FIR) filter is a nonrecursive filter in that the output from the filter is computed by using the current and previous inputs. It does not use previous values of the output, so there is no feedback in the filter structure. The design of the FIR filter is based on identifying the pulse transfer function G(z) that satisfies the requirements of the filter specification. This can be undertaken either by developing an analogue prototype and then transforming this to the pulse transfer function, or by designing directly in digital. A nonrecursive filter is always stable, and the amplitude and phase characteristics can be arbitrarily specified. However, a nonrecursive filter generally requires more memory and arithmetic operations than a recursive filter equivalent. Figure 7.43 shows typical FIR filter architecture.

Here, the filter input is applied to a sequence of sample delays (z^{-1}), and the outputs from each delay (and the input itself) are applied to the inputs of multipliers. Each multiplier has a coefficient set by the filter requirements. The outputs from each multiplier are then applied to the inputs of an adder, and the filter output is then taken from the output of the adder.

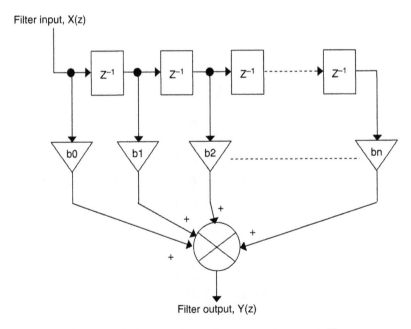

Figure 7.43: Typical architecture of an FIR filter

References

[1] Terrell, T. J., *Introduction to Digital Filters*, The MacMillan Press Ltd., 1980, ISBN 0-333-24671-3.

[2] Kamen, E. W., and Heck, B. S., *Fundamentals of Signals and Systems Using the Web and MATLAB®*, Pearson Education Ltd., 2007, ISBN 0-13-168737-9.

[3] Ifeachor, E. C., and Jervis, B. W., *Digital Signal Processing: A Practical Approach*, Pearson Education Ltd., 2002, ISBN 0-201-59619-9.

[4] Meade, M. L., and Dillon, C. R., *Signals and Systems Models and Behaviour*, Chapman & Hall, 1991, ISBN 0-412-40110-x.

[5] Hanselman, D., and Littlefield, B., *Mastering MATLAB® 6—A Comprehensive Tutorial and Reference*, Prentice Hall Inc., 2001, ISBN 0-13-019468-9.

[6] Golten, J., and Verwer, A., *Control System Design and Simulation*, McGraw-Hill, 1991, ISBN 0-07-707412-2.

[7] Astrom, K. J., and Wittenmark, B., *Computer-Controlled Systems Theory and Design*, Second Edition, Prentice Hall International, 1990, ISBN 0-13-172784-2.

[8] Analog Devices Inc., *AD7575 LC²MOS Successive Approximation ADC* datasheet.

[9] Analog Devices Inc., *AD7524 CMOS 8-Bit Buffered Multiplying DAC* datasheet.

[10] Schaumann, R., and Van Valkenburg, M., *Design and Analog Filters*, Oxford University Press, 2001, ISBN 0-19-511877-4.

Student Exercises

7.1   Develop the VHDL code for a design that will perform the following three functions:

- Sample an analogue signal from a 12-bit ADC. (Choose an ADC and obtain the required control signals from the device data sheet.)
- Multiply the signal by 0.76 with an error of no more than 5 percent.
- Output the result to a 12-bit DAC. (Choose an DAC and obtain the required control signals from the device data sheet.)

7.2   Modify the design in Exercise 7.1 so that the sample is multiplied by a value set from a PC via a simple UART receiver (using the integer value of the byte received from the UART).

7.3   From analysis of the data sheets from available ADCs, create the VHDL code that will control the sampling from the following ADCs:

- 8-bit
- 10-bit
- 12-bit
- 14-bit
- 16-bit
- 18-bit

7.4   From analysis of the data sheets from available DACs, create the VHDL code that will control the output of data to the following DACs:

- 8-bit
- 10-bit
- 12-bit
- 14-bit
- 16-bit
- 18-bit

7.5   Develop the VHDL code for a PID controller where each of the actions in the controller is defined in its own entity-architecture pair. The coefficients for each of the control actions are to be stored in a ROM. What assumptions are made in the implementation?

7.6   Develop the VHDL code for a PID controller where the additions/subtractions and multiplications/divisions are shared by all of the control actions. What assumptions are made in the implementation?

CHAPTER 8

Interfacing Digital Logic to the Real World: A/D Conversion, D/A Conversion, and Power Electronics

8.1 Introduction

Developing a digital algorithm to be implemented in hardware or software is the key task for many designers. If they develop the required digital logic or software program to run on an existing electronic circuit, the requirements for interfacing the algorithm to an external system—commonly referred to as the real world—for typical applications such as electronic circuit test, control, and instrumentation will have already been established. The necessary digital control and data signals to access the external system electronics will then need to be developed. In many other applications, however, the designer must develop and implement the digital algorithm in hardware or software, as well as the circuitry for interfacing the algorithm to an external system.

An example of interfacing a digital processor for electronic circuit test applications is in the testing of semiconductor devices during device production [1–3].

Aside: Consider the discussions in this section as an example of how electronic circuits and systems can be formed in order to create something useful. Consider in particular how the system is created, the different functions it is required to undertake, the variations on the basic idea of the semiconductor tester, and the need to carefully consider future as well as current requirements of the system. This discussion should be read along with Chapter 9, Testing the Electronic System.

The production testing of integrated circuits (ICs) is undertaken both at the wafer level (prior to die packaging, see Figure 8.1) and on the final packaged devices. The wafer contains multiple copies of a single design (or multiple designs), as well as special drop-in circuits used to measure specific parameters of the fabrication process.

An example set-up for testing a packaged device is shown in Figure 8.2. A typical set-up has the tester using a PC or workstation running a Windows®-, UNIX™-,

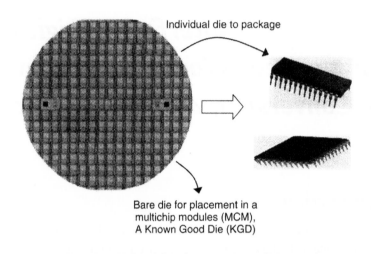

Individual die to package

Bare die for placement in a
multichip modules (MCM),
A Known Good Die (KGD)

(a) Fabricated silicon wafer   (b) Packaged ICs

Figure 8.1: Wafer to packaged IC

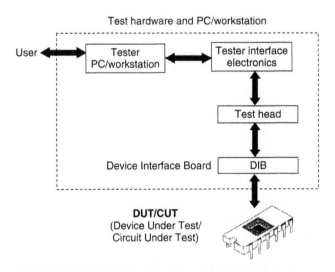

Test hardware and PC/workstation

Figure 8.2: Semiconductor device test set-up

or Linux®-based operating system that runs the software test routines, user I/O, and results analysis programs. A hardware interface (serial or parallel) from the PC or workstation connects to the tester interface electronics via a suitable cable arrangement. The tester interface electronics provides for analogue and digital output (test stimulus) and analogue and digital input (results capture). Some testers include additional processors for results analysis purposes, in particular the use of digital signal processors for analysis routines such as fast Fourier transform (FFT).

A cable arrangement connects the tester interface electronics to the test head. The test head is a mechanical arrangement with additional electronics built into it. The mechanical movement capability allows it to be moved around (and away from the tester PC or workstation) to facilitate fixing additional equipment such as hot air blowers for burn-in device testing. Burn-in testing applies a higher operating temperature and higher applied signal and power voltage levels to the device than would normally be encountered in the final application. By stressing the device electrically and thermally, any manufacturing defects will develop and fail faster than they would in normal operation. This test is undertaken for reliability reasons.

A device interface board (DIB), sometimes called a load board, is a custom-designed printed circuit board (PCB) that provides local electrical connections to the pins of the device and sometimes local (i.e., in close proximity to the device) interfacing electronic circuits.

In general, such test equipment used may be categorized as:

- **Dedicated test equipment**, which is specially designed to measure specific parameters for a device and dedicated to a particular device or small set of devices.

- **General purpose testers**, which are used to test a range of devices, where the devices may have vastly different operational parameters. This type of tester is temporarily customized to a particular IC via a software test program and a hardware DIB.

In production test, the need is for the required tests to be undertaken in a suitably comprehensive manner for the particular product and product application area, but at the lowest cost possible. This means using the most cost-effective equipment and minimizing the test time per device. The semiconductor testers used require both hardware and software subsystems to set up and control the tester and test program execution. During the production test stage, automatic test equipment (ATE) is used to reduce the test time by automating as much of the test process as practical. The test equipment used will be designed for the testing of particular types of device:

- **Digital**, including dedicated digital logic, memory, microprocessors, and programmable logic

- **Analogue**, including operational amplifiers, filters, and amplifiers

- **Mixed-Signal**, including analogue-to-digital converters (ADC), digital-to-analogue converters (DAC), phase-locked loops (PLLs), and analogue switches

Each type of test equipment has specific input/output (I/O) capabilities, suitable for the types of devices it tests. In general, testers are categorized as:

- *Digital* testers are optimized for digital circuits and systems with typically a large number of high-speed digital I/O pins and a limited analogue capability.

- *Memory* testers are optimized for testing of memory devices.

- *Analogue* testers are optimized for analogue circuits with high-performance analogue I/O current and voltage pins and high-performance data acquisition, but with a limited digital capability.

- *Mixed-Signal* testers provide a good level of both digital and analogue I/O capabilities, but may not necessarily reach the performance levels attained by digital or analogue testers.

- *System on a Chip* (SoC) testers provide specific support for complex (mainly digital) ICs that are considered to be complete electronic systems within the packaged device.

- *Design for Testability* (DfT) testers provide specific support for devices (mainly digital) that contain structures available from the major ATE vendors.

In addition to the basic arrangement shown in Figure 8.2, it is often necessary to utilize signal generators and results capture devices external to the basic tester itself. This happens where the tester does not incorporate the required electronic test and measurement equipment required for a particular test or set of tests. This is shown in Figure 8.3.

The test set-up is modified here by typically interfacing external equipment with both the tester computer and the DIB. Therefore, the computer needs suitable I/O ports that are compatible with the I/O ports of the test equipment such as RS-232, USB

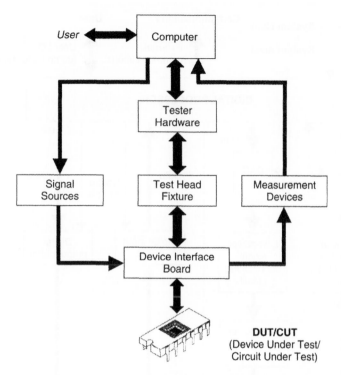

Figure 8.3: Semiconductor device hardware arrangement

(universal serial bus), and GPIB (general purpose interface bus). The software test program must have suitable software drivers (programs to access the I/O) available on the computer. The DIB must be designed so that the required signals on the DIB can be connected to the signal sources and measurement devices via appropriate connectors and required interface standards.

This example is only one of many possible examples of a test set-up. In applications such as test, control, and instrumentation, the typical arrangement is to have a digital processor at the core of the design and suitable analogue and digital I/O devices connecting to suitable connections on the digital processing circuit. A general arrangement for this is shown in Figure 8.4.

Here, the digital processor is chosen for the target application based on a set of criterion (also discussed in Chapter 2):

- Software programmed processor: microprocessor (μP), microcontroller (μC), or digital signal processor (DSP).

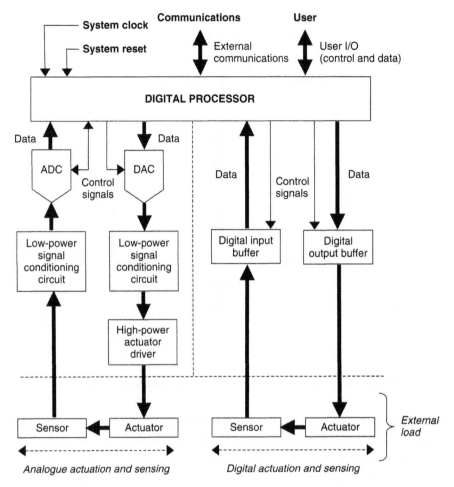

Figure 8.4: General digital processor and interfacing hardware

- Hardware configured programmable logic device: simple programmable logic device (SPLD), complex programmable logic device (CPLD), or field programmable gate array (FPGA).

The processor is required interface with the following components, systems, or subsystems:

- *Control signals* typically require an external signal clock for internal timing and a reset signal for both power-on reset when the power is first applied and user manual reset during normal operation

- *User I/O* typically is required to input data from the user (via switches, keypad, or keyboard) and output data to the user (via lights, light-emitting diode [LED] displays, and liquid crystal displays [LCD]).

- *Communications* to an external digital system uses wired, optical fiber, or wireless means, and follows either a standard communications protocol or a custom protocol for the particular application.

- *Digital I/O* is used to create digital control and data signals to drive an external actuator (or circuit) and to read in from a digital sensor (or circuit). Digital I/O signals are buffered by additional logic buffers (simple logic buffers or tristate buffers) for three reasons: (i) to protect the processor from the external actuator (or circuit) and digital sensor (or circuit), (ii) to provide logic level translation (the changing of the voltage levels that represent the logic 0 and 1 levels), and (iii) to provide electrical or optical buffering required to enable correct and robust interfacing to the external circuitry.

- *Analogue I/O* is used to create analogue signals to drive an external actuator (or circuit) and to read in from an analogue sensor (or circuit).

Because the digital processor provides digital signals, in order to connect the digital to the analogue world, both DAC and ADC devices are required. The DAC receives a digital input signal and produces an analogue output (either a voltage or a current); the ADC receives an analogue input (either a voltage or a current) and produces a digital output. The analogue output from the DAC is typically applied to a low-power analogue signal conditioning circuit that translates the DAC output signal to a signal level (voltage and current), which in turn is either directly required by the external actuator (or circuit) or acts as the input to a high-power actuator driver circuit such as a power amplifier to drive an electric motor. The low-power signal conditioning circuitry is normally based on the operational amplifier (op-amp).

8.2   Digital-to-Analogue Conversion

8.2.1   Introduction

The DAC (also called a D/A converter) is an electronic circuit that provides a link between the digital and the analogue domains [4]. The device accepts a digital word (group of digital bits) and outputs an analogue voltage or current. This analogue value can be either a single-ended signal, which is a single-node connection referenced

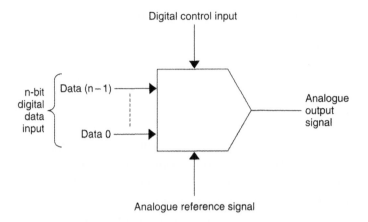

Figure 8.5: Basic DAC arrangement

to the common node (ground or 0 V), or a differential signal, which is a two-node connection where the difference between the two nodes is the output signal.

The basic arrangement for the DAC is shown in Figure 8.5. Here, the output is a singled-ended signal, and all voltages are referenced to a common node.

The number of input data bits (n) is usually between 8 and 24, which produces a number of possible codes that the DAC can accept, shown in Table 8.1. Each individual code produces a unique analogue output level: number of codes $= 2^n$.

The number of bits to use depends on the particular application. For example, 8 to 12 bits of digital input are common for control applications, but 16 to 18 bits are typical of test and instrumentation purposes, and 22 to 24 for audio applications. As the

Table 8.1: DAC input codes for number of input data bits

Number of bits (n)	Number of codes (2^n)
8	256
10	1024
12	4096
14	16,384
16	65,536
18	262,144
20	1,048,576
22	4,194,304
24	16,777,216

number of bits increases, it becomes more important to ensure that the effects of the manufacturing process variations are reduced or removed. The main methods used to account for process variations are:

- **Trimming**. Passive components (resistors and capacitors) can be trimmed after fabrication. Laser trimming of resistors and the switching of components in or out of a circuit are used to trim component values.

- **Dynamic element matching**. Process variations are averaged out over time by the dynamic switching of mismatched components.

- **Self-calibration**. Additional circuitry is included in the design, and an external control signal causes the DAC to run through a self-calibration algorithm automatically. It is important to understand what the self-calibration routine does and how this affects the performance of the data converter.

In addition to the digital data inputs, the device will also accept digital control input signals. These are used to control the operation of the DAC to enable the device to be selected, usually with a chip select (CS) signal that will be either active high (CS) or active low (/CS), and a data write (WR) signal, also either active high (WR) or active low (/WR). The timing of these signals is important; violating it will cause incorrect operation of the DAC. Timing data is identified on the device datasheet. In addition to these two basic control signals, some DAC architectures also require additional control signals such as a master clock signal (for control of internal sequential logic circuits within the DAC).

Another input to the DAC is the analogue reference signal. This is a DC voltage or current that sets the output voltage or current range of the DAC. This may be a unique input to the DAC, and the user will be able to set the value of the analogue reference signal to fit his or her own circuit design requirements; or the reference signal may be connected to the power supply voltage and not adjustable by the user. The datasheet for the particular DAC provides the required information on the creation and use of this analogue reference signal.

The final inputs to the DAC that are not shown in Figure 8.5 are the power supply connections to provide power to both the digital and analogue circuitry within the device. The digital and analogue circuitry may be powered from a single power supply input, or they may have their own unique power supply pins (V_{DD} and V_{SS}). Additionally, the power supply may be a single fixed value or may be adjustable within limits by the user.

The key characteristics to consider for DAC choice and use are:

1. **Output range (voltage or current)**. The DAC produces an analogue output with a specific range from a minimum value (minimum input code) to a maximum value (maximum input code) either as a unipolar output (output of a single polarity, positive or negative) or as a bipolar output (an output ranging from a negative value to a positive value). The output range determines the amount of signal conditioning circuitry to be placed at the DAC output before the signal can be used.

2. **Parallel or serial load**. With high-resolution (with the data converter resolution considered here as the number of digital data inputs applied to the DAC) data converters, the number of digital I/Os significantly increases. Traditional DACs used parallel data transfer—that is, the applied digital data is available at a single time, with all the data bits connected to an external system in parallel. However, for discrete packaged ICs, each data bit requires a device pin, which leads to physically large packages and in turn increases package cost and introduces package level nonideal circuit operation. An alternative to the parallel data load of the DAC is to serially clock in (load) the data one bit at a time. This requires the use of a serial communications protocol (in the simplest terms, the data bits are applied in a set order) and a clock signal to load the serial data. Internal to the data converter, this serial data is converted to parallel and applied to the data converter circuitry. Although this requires an additional clock signal, more complex external digital processing for the serial data transfer, and additional time to load the data converter with the correct number of bits, it also results in a fewer package pins (for smaller and cheaper packages) and a smaller footprint on the final PCB application.

3. **Power supply voltage**. Early data converters were designed to operate on a power supply voltage range in excess of $+5\,V$ DC. In many cases, the DAC could be used in a range typically from $+5\,V$ to $+12\,V$ DC (unipolar) or in bipolar operation. However, the more recent trend toward lower power supply voltages, driven by user requirements for portable, low-power circuit operation, has seen a reduction in the power supply voltage to between $+5\,V$ and $+3\,V$ DC, and potentially lower. The need to run the electronics on the available battery voltage levels of $+1.5\,V$ and $+3\,V$ DC is driving the need for low-voltage (and also low-power) electronics.

4. **Voltage or current output**. The output signal can be either a voltage or a current, depending on the data converter architecture. For a current output, external circuitry (op-amp) is used to convert the current to a voltage.

5. **Signal-ended or differential output**. The DAC can provide either a single-ended output with a single signal that is referenced to the common point (0 V) in the circuit, or it can provide a differential output, providing a signal that is measured as a difference between the two outputs.

6. **DAC architecture**. A range of DAC architectures are available [5, 6], including:

   - resistor string DAC

   - binary weighted resistor DAC

   - binary weighted current DAC

   - R-2R ladder DAC

   - segmented resistor string DAC

   - current steering DAC

   - sigma-delta ($\Sigma\Delta$) DAC

7. **Device packaging**. Devices are available in different package types (through-hole or surface mount, refer to Appendix C, Integrated Circuit Package Types—see the last paragraph of the Preface for instructions regarding how to access this online content) and in package case materials (plastic or ceramic). Figure 8.6 shows an example DAC and the types of pin connections required in a plastic DIL package type.

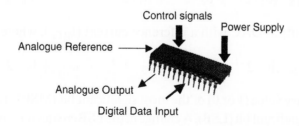

Figure 8.6: DAC in a DIL package

8. **Operating temperature range** is the range in temperature that the IC can handle without damage during component use and depends on the application. The IC will be one of the following types:

- commercial: 0°C to +70°C

- industrial: −40°C to +85°C

- military: −55°C to +125°C

9. **Performance**. The performance of the DAC is considered in one of three categories: static (DC) parameters, transfer curve parameters, and dynamic parameters. These parameters are guaranteed for a particular device, as defined in the device datasheet, by production testing and/or by end-user testing. The performance also includes the time required to convert the digital input before the analogue output appears at the DAC output. The speed of conversion depends on the DAC architecture, so the architecture determines the types of applications in which the converter can be used.

8.2.2   DAC Characteristics

The DAC analogue output varies from a set minimum to a set maximum value that is a function of the reference signal and the value of the digital input code. For the DAC, the full-scale voltage (V_{FS}) or full-scale current (I_{FS}) sets the limit of operation. It is common for the DAC input to be an unsigned binary value, although (particularly for bipolar operation) the digital input might also be provided in signed binary (2s complement) representation or in BCD (binary coded decimal) representation.

The output of the DAC can be written mathematically. For voltage output DAC with a reference voltage (V_{REF}), where V_{FS} is set by V_{REF}:

$$V_{out} = V_{FS} \cdot (b_1 \cdot 2^{-1} + b_2 \cdot 2^{-2} + b_3 \cdot 2^{-3} + \cdots\cdots + b_n \cdot 2^{-n}) + V_{OS}$$

And for current output DAC with a reference current (I_{REF}), where I_{FS} is set by I_{REF}:

$$I_{out} = I_{FS} \cdot (b_1 \cdot 2^{-1} + b_2 \cdot 2^{-2} + b_3 \cdot 2^{-3} + \cdots\cdots + b_n \cdot 2^{-n}) + I_{OS}$$

Here, b_1 is the binary value (1 or 0) of the most significant bit (MSB) and b_n is the binary value of the least significant bit (LSB). A change in the LSB creates the smallest single change in the output signal. It is common for the digital input code to be an unsigned binary count, starting at 0_{10} and incrementing in unit steps. However, with the use of suitable digital signal encoding, any digital code (e.g., 2s complement signed binary) could be used. A change in the MSB creates the largest single change in the output signal. The above equations include an offset voltage (V_{OS}) or offset current (I_{OS}) if the output signal is not zero for an input code of 0_{10}. Therefore, the output of the DAC can be either unipolar (either a positive or negative value only) or bipolar (both positive and negative values can be generated).

The resolution of the converter is given above as the number of digital bits at the converter input. This is one way to define the resolution of the converter and is independent of the analogue reference signal value. This is how the resolution of the converter will be referred to in this text.

A second way to define the resolution of the converter is to identify the minimum output voltage (or current) change, which occurs with a change of 1 LSB in the digital input code. For a voltage output converter, the voltage change (V_{LSB}) when the input changes by 1 LSB is the LSB step size and is given by:

$$V_{LSB} = (2^{-n} \cdot V_{FS}) \text{volts}$$

The resolution of the converter in terms of the voltage change (V_{LSB}) is the value of the LSB step size with units of volts per bit.

$$\text{Resolution} = (V_{LSB}) \text{volts/bit}$$

For example, for a unipolar eight-bit DAC with a full-scale voltage of 5.0 V, $V_{LSB} = 19.53$ mV. For a 16-bit DAC with a full-scale voltage of 5 V, then $V_{LSB} = 76.29 \mu$V. Therefore the resolution figure given here is the value of V_{LSB} and is dependent on the value of the analogue reference signal value. For a voltage output DAC, this value can also be presented as a percentage of the full-scale voltage:

$$\text{Resolution} (\%) = (V_{LSB}/V_{FS}) \times 100\%$$

The characteristics of the DAC must be considered as either ideal or real. An ideal DAC identifies the operation of the DAC when all values are set to their designed (or ideal) values. However, manufacturing tolerances of the DAC circuitry causes real DAC operation to deviate from the ideal. In this case, the DAC maximum deviation will be defined in the DAC datasheet and will be guaranteed by the manufacturer. To understand the operation of the DAC, begin by considering the ideal DAC, then identify how a real DAC could deviate from this.

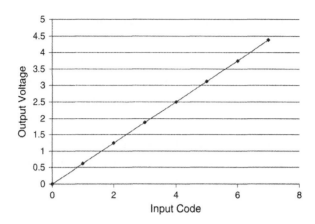

Figure 8.7: Ideal three-bit DAC transfer curve

Consider a three-bit DAC (for simplicity) that has an input code ranging from 000_2 (0_{10}) to 111_2 (7_{10}) and that produces an output voltage. The ideal DAC input code−output voltage relationship, called the transfer curve, is shown in Figure 8.7. The full-scale voltage is +5.0 V. This is a unipolar DAC with an output voltage of 0 V to +5.0 V. In this range, a change in 1 LSB of the input code creates a step change in the output voltage. For each 1 LSB step change, the output voltage changes by the same amount. This voltage level change is given by:

$$V_{LSB} = \left(2^{-n} \cdot V_{FS}\right)$$
$$V_{LSB} = \left(2^{-3} \times 5.0\right)$$
$$V_{LSB} = (0.125 \times 5.0) = 0.625\,V$$

The output voltage range 0 to V_{FS} is separated into eight (2^n) equal levels. In Figure 8.7, each input code produces a unique output voltage. When these points are drawn on the transfer curve graph, they will be joined with a straight line.

The input-code-to-output-voltage values are shown in Table 8.2. The first thing to notice is that with this DAC, the output voltage rises only to a maximum of 1 LSB less than the full-scale voltage. This is common of many DACs and of the types considered here. However, some DACs are designed so that the maximum output voltage reaches the full-scale voltage.

An actual DAC output voltage would deviate from the ideal for particular or all input codes. Consider the DAC test results shown in Table 8.3. The table shows the ideal and actual results taken by measuring the output of a theoretical three-bit DAC.

Table 8.2: Ideal three-bit DAC input-code-to-output-voltage mapping (1 LSB = 0.625 V)

Input code (binary)	Input code (decimal equivalent)	Output voltage (V)
000	0	0.0
001	1	0.625
010	2	1.25
011	3	1.875
100	4	2.5
101	5	3.125
110	6	3.75
111	7	4.375

Table 8.3: Three-bit DAC example

Input code (binary)	Input code (decimal equivalent)	Ideal output voltage (V)	Actual output voltage (V)
000	0	0.0	0.0500
001	1	0.625	0.3125
010	2	1.25	1.2500
011	3	1.875	1.8750
100	4	2.5	2.7000
101	5	3.125	3.0000
110	6	3.75	4.0000
111	7	4.35	4.3400

The performance in this particular theoretical DAC is poor. The results for the ideal and actual DACs are plotted in Figure 8.8.

The general shape of the graph of the real DAC results is similar to the ideal, but the deviation is a measure of the quality of the real DAC. From this transfer curve, then the transfer curve parameters can be identified.

As the resolution (number of bits) of the converter increases and the operating voltage range of the DAC decreases, the LSB step size (volts or current) will decrease. The effect of this is that the analogue signal levels become the same order of value as the noise generated in the circuit, and the inevitable manufacturing process variations have a more significant impact, leading to problems with design and ultimate use of these converters. Unwanted circuit effects not seen with the lower-resolution data converters are seen with the higher-resolution data converters.

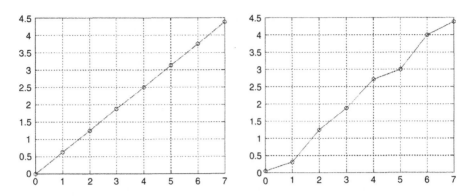

Figure 8.8: Plot of DAC results: ideal (left) and actual (right)

The characteristics of the DAC are categorized into three types of parameters—static (DC) parameters, transfer curve parameters, or dynamic parameters—whose details are specified in the tables that follow.

The static (DC) parameters are identified in Table 8.4:

Table 8.4: Static (DC) parameters

Parameter number	Parameter name	Parameter description
1	Code specific parameters	A measure of the output signal value for specific input codes. Typically the minimum, midpoint, and maximum input codes are of interest.
2	Full-scale range (V_{FSR})	The difference between the maximum and minimum output analogue signal values.
3	DC gain error	A measure of the deviation of the slope of the straight-line approximation of the actual converter output from the ideal converter straight-line output.
4	Offset error	Offset of the actual converter output. The offset error may be taken at the lowest input code, the mid-code, or the highest input code.
5	LSB step size	A measure of the average step size between input codes (quoted in volts per bit).
6	DC PSS	DC power supply sensitivity, a measure of the sensitivity of the DAC circuitry to variations in the power supply voltage.

The transfer curve parameters are identified in Table 8.5:

Table 8.5: Transfer curve parameters

Parameter number	Parameter name	Parameter description
1	Absolute error	The difference between the ideal DAC output curve and actual DAC output curve for each input code.
2	Monotonicity	When the input code increments by 1 bit, there should be an increment in the output signal. This situation occurs when the DAC is monotonic. In a nonmonotonic DAC, an increase in the input code results in a decrease in the output signal for certain code transitions.
3	Integral nonlinearity (INL)	The deviation of the actual converter data point from the point on the straight-line approximation. The ideal converter, end-points, or best-fit straight-line approximation can be used. Where the ideal converter is used, this value will be the same as the absolute error. This is normally quoted in LSBs.
4	Differential nonlinearity (DNL)	Where a binary input code change of 1 bit occurs, the output should change by 1 LSB. The DNL is the difference between each output step size of the converter and an ideal step size of 1 LSB. For a given input code, the output step size is taken between the current input code and the previous code. This is normally quoted in LSBs.

The dynamic parameters are identified in Table 8.6.

In addition to the previously identified parameters, the DAC signal frequency response is usually analyzed. By sampling the DAC output signal and undertaking a FFT (fast Fourier transform) on the digitized samples, the frequency components of the DAC output signal can be identified. This can be used to determine the correct or incorrect operation of the DAC. Where samples of an analogue signal are taken, DSP techniques can perform analysis of the signal in both the time and frequency domains. The FFT is undertaken on sampled signals to identify the frequency components of a complex signal in terms of the signal magnitude per root mean squared (RMS) value and phase at different frequencies. The FFT is an efficient implementation of the discrete Fourier transform (DFT), which uses samples of a signal taken at a chosen sampling frequency (f_S).

By analyzing the FFT plot, it is possible to identify a number of converter characteristics:

- Signal-to-noise ratio (SNR) is the ratio of the signal power (in the fundamental frequency when a single-frequency sine wave is applied to the

Table 8.6: DAC dynamic parameters

Parameter number	Parameter name	Parameter description
1	Conversion time (settling time)	When the input changes, the output change tends to take awhile to settle. The conversion time is the time needed for the output to settle within a specified error band after the input code has been set.
2	Overshoot and undershoot	As the output change settles, it tends to overshoot (go past and become greater than) and undershoot (go back to be less than) the final value.
3	Rise and fall times	Time needed for the output to rise or fall from 10% to 90% between the initial and final values.
4	DAC-to-DAC skew	Timing mismatch between DACs to be used in matched groups.
5	Glitch energy	Specification common to high-frequency DACs.
6	Clock and data feedthrough	A measure of the cross-talk of the digital signals to the analogue output signal.

DAC) and the noise power over the frequency band of interest. For an ideal converter, the SNR (in decibels, dB) is given by:

$$SNR_{dB} = 6.02N + 1.76$$

where SNR_{dB} is the SNR quoted in dB, and N is the resolution of the converter (number of bits). For an eight-bit DAC, the SNR is 49.92 dB. For a 16-bit DAC, the SNR is 98.08 dB. This equation links the SNR of the converter to the number of bits.

- Spurious free dynamic range (SFDR) is the difference (in dB) between the fundamental frequency (for a single-frequency sine wave input) and the largest ray (of all other frequencies identified on the FFT plot). It is the usable dynamic range of the converter before noise effects become noticeable.

- Total harmonic distortion (THD) is the ratio of the sum of the power in the signal harmonics to the power in the fundamental signal (in dB).

- Signal-to-noise plus total harmonic distortion (S/(N + THD)) is the plot of the actual SNR curve versus input signal magnitude.

- Signal-to-noise and distortion (SINAD) is a combination of SNR and THD.

- Effective Number of Bits (ENOB) is a measure of how close the actual converter is to the theoretical model.

8.2.3 Types of DAC

A number of architectures have been developed for the DAC, each providing its own unique operating characteristic and limitations. Any given DAC, regardless of architecture, falls into one of two categories:

- Nyquist rate DAC

- oversampling DAC

With the Nyquist rate DAC, the input signal bandwidth is equal to the Nyquist frequency. The Nyquist frequency is half the DAC update frequency (which is the Nyquist rate).

With the oversampling DAC, the converter update frequency is much greater than the Nyquist rate. Typical values of oversampling ratio (OSR) are 64, 128, and 256:

$$OSR = (f_S/2 \cdot f_N)$$

where f_S is the update frequency of the DAC and f_N is the Nyquist frequency.

For an ADC, the Nyquist frequency is half the minimum sampling frequency of the ADC required to reconstruct fully the original signal from the sampled signal. This means that no information has been lost in the sampling process. The Nyquist rate is when the sampling frequency is twice the bandwidth of the signal being sampled. When a signal that has been sampled and converted to digital at the Nyquist rate, then to reconstruct the analogue signal with a DAC, the DAC update frequency must also be at the Nyquist rate.

The main available DAC architectures are:

- resistor string DAC

- binary weighted resistor DAC

- binary weighted current DAC

- R-2R ladder DAC

- segmented resistor string DAC

- current steering DAC

- sigma-delta ($\Sigma\Delta$) DAC

The resistor string DAC uses a resistor string and transistor switches to select a voltage. The basic idea is shown in the circuit diagram in Figure 8.9, where the output of the switch array in the DAC is applied to the input of an op-amp unity gain buffer. This buffer, with its high input impedance and low output impedance, prevents the circuit to be connected to the DAC output from electrically loading the resistor string. With this design for an n-bit DAC, 2^n resistors are needed. In silicon, resistors require a substantial amount of area on the die and cannot be fabricated with a low tolerance (i.e., they cannot be fabricated to accurate values) because of process variations. Therefore, this type of DAC is not suited for high-resolution data converters. Digital control logic determines which switches must be closed based on the input code applied to the DAC.

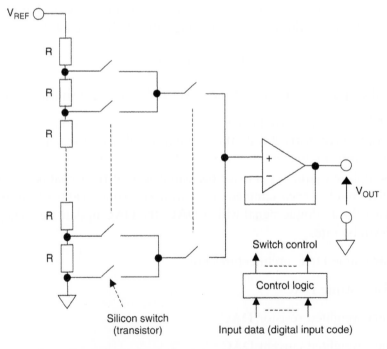

Figure 8.9: Resistor string DAC

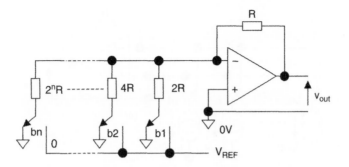

Figure 8.10: Binary weighted resistor DAC

The binary weighted resistor DAC uses resistors with binary weighted values (n, 2n, 4n, 8n, etc.) that must be switched between the circuit common voltage and a DC reference voltage (V_{REF}). The basic idea is shown in the circuit diagram in Figure 8.10, where the common node is 0 V. In this implementation, a logic 1 digital input control to the switch connects the switch to the common node, and a logic 0 connects the switch to the V_{REF} node. This is essentially a summing amplifier with the currents flowing through the binary weighted resistors and combining at the inverting input node. The feedback resistor between the op-amp output and inverting node can be altered to provide voltage gain. Because the op-amp is configured to be a summing amplifier, which by its operation provides an inverted output, if the reference voltage is a positive value, then the output voltage is a negative value and vice-versa. In this arrangement, the LSB is controlled by switch *bn* and the MSB is controlled by switch *b1*.

A variant on this converter is the binary weighted capacitor DAC. This uses switched capacitor techniques rather than resistors to implement the circuit resistor function.

The binary weighted current DAC uses the same basic architecture as the binary weighted resistor DAC, except now the resistors have been replaced with constant current sources. The basic idea is shown in the circuit diagram in Figure 8.11. This produces a variable input (current) to a current-to-voltage converter (I-V) converter, which is an op-amp with a feedback resistor. The output voltage from the op-amp is proportional to the current flowing through the resistor R. In this implementation, a logic 1 digital input control to the switch connects the switch to the op-amp inverting node, and a logic 0 connects the switch to the op-amp noninverting node. The constant current source is created using a transistor current mirror. In this arrangement, the LSB is controlled by switch *bn* and the MSB is controlled by switch *b1*.

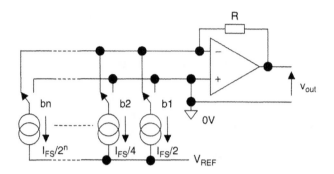

Figure 8.11: Binary weighted current DAC

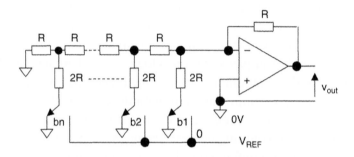

Figure 8.12: R-2R ladder DAC

The R-2R ladder DAC uses a combination of resistors in a R-2R arrangement to produce the current input to an op-amp based I-V converter. The basic idea is shown in the circuit diagram in Figure 8.12. This produces a variable output voltage dependent on the switch positions. In this implementation, a logic 1 digital input control to the switch connects the switch to the common node, and a logic 0 connects the switch to the V_{REF} node. In this arrangement, the LSB is controlled by switch *bn* and the MSB is controlled by switch *b1*.

The segmented resistor string DAC is a modification of the resistor string DAC. This consists of two resistor strings, a coarse string as in the resistor string DAC and a fine resistor string that subdivides the voltage step created by the coarse string.

The current steering DAC uses current mirrors and switches to switch a current around the circuit. This produces an output current that can be converted to a voltage using an I-V converter.

The sigma-delta ($\Sigma\Delta$) DAC is an example of an oversampling DAC. It uses oversampling and interpolation to produce the requirements of an n-bit DAC. The design consists

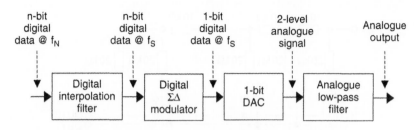

Figure 8.13: One-bit sigma-delta (ΣΔ) DAC

of a digital interpolation filter, a digital modulator, and usually a one-bit DAC to produce the n-bit DAC. The output of the ΣΔ DAC is a bitstream pattern that then must be low-pass filtered. A ΣΔ DAC is similar to a ΣΔ ADC, except that the noise shaping in the DAC is performed using a digital modulator rather than an analogue modulator. The basic architecture of a one-bit ΣΔ DAC is shown in Figure 8.13.

The input to the digital interpolation filter is at the Nyquist rate (F_N), which is then increased to produce a digital signal at the update frequency (f_S). This is applied to a digital ΣΔ modulator, which produces a one-bit digital output at the update frequency (f_S). The one-bit digital output is then converted to the analogue signal levels by a one-bit DAC.

8.2.4   DAC Control Example

An FPGA or CPLD could be used to control the DAC to create an analogue output signal. Consider the AD7524 DAC (from Analog Devices Inc.) [7]. This is an eight-bit multiplying DAC consisting of a thin film R-2R ladder network and n-channel (nMOS transistor) current switches in a single device. The functional diagram for this DAC is shown in Figure 8.14. The eight digital inputs control the position of switches S1 to S8. This is turn directs the current flow from the reference voltage (V_{REF}) to either OUT1 or OUT2 node. The operation of the DAC is controlled by two control inputs /CS (Chip Select, active low signal) and /WR (Write, active low signal). The switches are used to select the appropriate node to connect the end of the appropriate 20 kΩ resistor to when the particular data logic level is either 0 or 1.

The DAC has two modes of operation:

1. **Hold mode**, which holds the value of the last digital data input on DB7-DB0 at a time prior to either /WR or /CS control inputs, assuming a high (logic 1) state. The digital data input is latched, and the analogue output remains static.

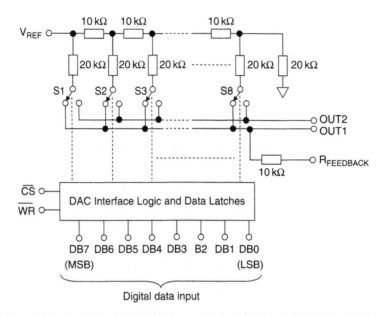

Figure 8.14: AD7524 eight-bit DAC functional diagram. Figure adapted from the AD7524 datasheet. Copyright © Analog Devices, Inc. Used with permission.

2. **Write mode**, in which, when both /WR and /CS are low, the analogue output responds to the digital data input on DB7-DB0 and acts like a nonlatched DAC.

The timing diagram for the write mode is shown in Figure 8.15. To write a value to the DAC, initially the /CS and /WR control signals are high and the data can change. The /CS control signal goes low first, then the /WR signal goes low. The data input

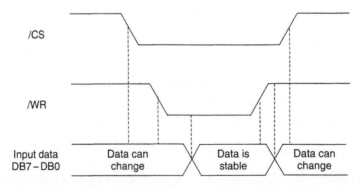

Figure 8.15: Write mode timing diagram. Figure adapted from the AD7524 datasheet. Copyright © Analog Devices, Inc. Used with permission.

must then become stable (i.e., does not change). When the /WR signal goes high, the data is latched into the DAC. The /CS then goes high and the DAC is then ready to latch a new data value.

With this architecture, the DAC produces two output currents that are externally converted to a single output voltage using an op-amp (as an I-V converter). A typical arrangement is shown in Figure 8.16. Here, both outputs are connected to the inputs of a suitable op-amp. (The datasheet provides a choice of suitable op-amps to use with this particular DAC.) The output OUT2 is also connected to the common (0 V) node.

A CPLD could be configured to provide the data and control signals for the DAC. The block diagram shown in Figure 8.17 has a register that stores the data to apply to the DAC and a counter with decoded outputs to provide the DAC control signals and an internal register Update (data store) signal. The operation of the control signals is synchronized to the input clock signal (with a frequency suited to the update frequency of the DAC so that the timing requirements of the DAC are not violated).

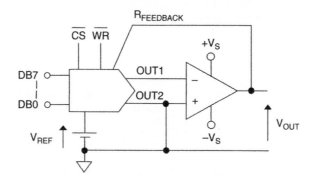

Figure 8.16: Typical arrangement for single-ended output voltage creation. Figure adapted from the AD7524 datasheet. Copyright © Analog Devices, Inc. Used with permission.

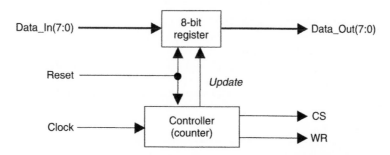

Figure 8.17: Example digital controller for the AD7524 DAC

Table 8.7: Counter decoding

Counter state	Update	CS	WR
0	0	1	1
1	1	1	1
2	0	0	1
3	0	0	0
4	0	0	1

The clock controls a counter with the states decoded to produce the control signals, as shown in Table 8.7. The counter resets (asynchronous, active low) to state 0. When the counter is in state 4, the next state will be state 0.

The VHDL code to achieve this is shown in Figure 8.18. This description uses processes to create the register, counter, and counter output decoding operations.

```
LIBRARY ieee;
USE ieee.std_logic_1164.all;
USE ieee.std_logic_arith.all;
USE ieee.std_logic_unsigned.all;

ENTITY AD7524_Controller IS
    PORT ( Clock      : IN   STD_LOGIC;
           Reset      : IN   STD_LOGIC;
           Data_In    : IN   STD_LOGIC_VECTOR (7 downto 0);
           Data_Out   : OUT  STD_LOGIC_VECTOR (7 downto 0);
           CS         : OUT  STD_LOGIC;
           WR         : OUT  STD_LOGIC);
END ENTITY AD7524_Controller;

ARCHITECTURE Behavioural OF AD7524_Controller IS

SIGNAL Count  : STD_LOGIC_VECTOR(2 downto 0);
SIGNAL Update : STD_LOGIC;

BEGIN

PROCESS (Update, Reset, Data_In)
BEGIN

     If (Reset = '0') Then
                  Data_Out(7 downto 0) <= "00000000";
     ElsIf (Update'event and Update = '1') Then
                  Data_Out(7 downto 0) <= Data_In(7 downto 0);
     End If;

END PROCESS;
```

Figure 8.18: VHDL code for DAC controller

```
PROCESS (Clock, Reset)
BEGIN

        If (Reset = '0') Then

                        Count(2 downto 0) <= "000";

        ElsIf (Clock'event and Clock = '1') Then

                        If (Count = "100") Then
                                        Count <= "000";
                Else
                                Count <= Count + 1;
                End If;

        End If;

END PROCESS;

PROCESS (Count)
BEGIN

        If (Count = "000") Then
                Update <='0';  CS <= '1';  WR <= '1';
        ElsIf (Count = "001") Then
                Update <='1';  CS <= '1';  WR <= '1';
        ElsIf (Count = "010") Then
                Update <='0';  CS <= '0';  WR <= '1';
        ElsIf (Count = "011") Then
                Update <='0';  CS <= '0';  WR <= '0';
        ElsIf (Count = "100") Then
                Update <='0';  CS <= '0';  WR <= '1';
        Else
                Update <='0';  CS <= '1';  WR <= '1';
        End If;

END PROCESS;

END ARCHITECTURE Behavioural;
```

Figure 8.18: (Continued)

Figure 8.19 shows an example VHDL test bench for simulating the design. This creates a clock signal and a reset signal. The clock signal has a period of 20 ns, although in reality this would be too short a time for the DAC to react. However, here with no circuit implementation (gate and interconnect) and DAC timing

```
LIBRARY ieee;
USE ieee.std_logic_1164.all;
USE ieee.std_logic_arith.all;
USE ieee.std_logic_unsigned.all;

ENTITY Test_AD7524_Controller_vhd IS
END Test_AD7524_Controller_vhd;

ARCHITECTURE Behavioural OF Test_AD7524_Controller_vhd IS

COMPONENT AD7524_Controller
PORT(
              Clock     : IN  STD_LOGIC;
              Reset     : IN  STD_LOGIC;
              Data_In   : IN  STD_LOGIC_VECTOR (7 downto 0);
              Data_Out  : OUT STD_LOGIC_VECTOR (7 downto 0);
              CS        : OUT STD_LOGIC;
              WR        : OUT STD_LOGIC);
END COMPONENT;

SIGNAL Clock    : STD_LOGIC := '0';
SIGNAL Reset    : STD_LOGIC := '0';
SIGNAL Data_In  : STD_LOGIC_VECTOR (7 downto 0) := (others=>'0');

SIGNAL Data_Out : STD_LOGIC_VECTOR(7 downto 0);
SIGNAL CS       : STD_LOGIC;
SIGNAL WR       : STD_LOGIC;

BEGIN

uut: AD7524_Controller PORT MAP(
              Clock    => Clock,
              Reset    => Reset,
              Data_In  => Data_In,
              Data_Out => Data_Out,
              CS       => CS,
              WR       => WR);

Reset_Process : PROCESS

BEGIN
       Wait for 0 ns;  Reset <= '0';
       Wait for 5 ns;  Reset <= '1';
       Wait;
END PROCESS;

Clock_Process : PROCESS

BEGIN
       Wait for 0 ns;   Clock <= '0';
       Wait for 10 ns;  Clock <= '1';
       Wait for 10 ns;  Clock <= '0';
END PROCESS;

END ARCHITECTURE Behavioural;
```

Figure 8.19: VHDL test bench for DAC controller

considerations taken into account, the basic functionality of the VHDL code design can be simulated.

The last point to note is that the AD7524 was designed to operate on a $+5\,V$ power supply for TTL logic compatibility. As such, the choice of FPGA or CPLD requires either a device with the same power supply and logic voltage levels or use of a level-shifting circuit for matching the logic voltage levels.

8.3   Analogue-to-Digital Conversion

8.3.1   Introduction

The ADC, or A/D converter, is an electronic circuit that provides a link between the analogue and digital domains. The device accepts an analogue input signal (voltage or current) and produces a digital output. This analogue value can also be either a single-ended signal (single-node connection referenced to the common node, a ground or $0\,V$) or a differential signal (two-node connection where the difference between the two nodes is the output signal).

The basic arrangement for the ADC is shown in Figure 8.20. Here, the input is a singled-ended signal. All voltages are referenced to a common node.

The ADC may also include input circuitry to store an analogue input (a sample-and-hold circuit) to store a single-input signal value at a particular point in time while the input continuously changes. The number of output data bits (n) is usually between

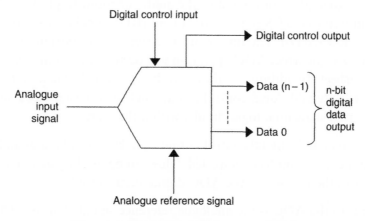

Figure 8.20:   Basic ADC arrangement

Table 8.8: ADC output codes for number of output data bits

Number of bits (n)	Number of codes (2^n)
8	256
10	1024
12	4096
14	16,384
16	65,536
18	262,144
20	1,048,576
22	4,194,304
24	16,777,216

8 and 24 bits, which produces a number of possible codes that the ADC can produce, as shown in Table 8.8.

The number of bits to use will depend on the particular application. For example, 8 to 12 bits of digital input are common for control applications, 16 to 24 bits for test and instrumentation purposes, and 22 to 24 bits for audio applications. As the number of bits increases, be careful to ensure that the manufacturing process variations do not cause problems.

The device will also accept digital control input signals. These are used to control ACD operation to enable the device to be selected, usually with a chip select (CS) signal that is either active high (CS) or active low (/CS). This function may also act as a control signal to start a conversion of the analogue input to produce the digital output. In addition to the CS signal, a data read (RD) signal is also either active high (RD) or active low (/RD)). The timing of these signals is important and must not be violated, otherwise incorrect ADC operation will occur. This timing data is located in the device datasheet. In addition to these two basic control signals, some ADC architectures require additional control signals such as a master clock signal for control of internal sequential logic circuits within the ADC.

Some ADCs also include digital control output signals. Typical is a digital output that indicates that a conversion has completed. This can be used by an external digital circuit to control the reading of the ADC digital data output.

The final input to the ADC is the analogue reference signal. This is a DC voltage or current that sets the input voltage or current range of the ADC. This might be a unique input to the ADC, in which case the user can set the value of the analogue reference signal to fit his or her own circuit design requirements, or the reference signal will be connected to the power supply voltage and not adjustable by the user. The datasheet for the particular ADC will provide the required information on the creation and use of this analogue reference signal.

Other inputs to the ADC that are not shown on the symbol are the power supply connections. The power supply provides the power for both the digital and analogue circuitry within the device. The digital and analogue circuitry may be powered from a single power supply input, or may have their own unique power supply pins (V_{DD} and V_{SS}). Additionally, the power supply may be a single fixed value or may be adjustable within limits by the user.

The key characteristics to consider for DAC choice and use are:

- input range (voltage or current)

- parallel or serial data read

- power supply voltage

- voltage or current input

- signal ended or differential input

- ADC architecture

- device packaging

- operating temperature range

- performance

These requirements are the same as for the DAC, except now the input rather than the output is an analogue signal. The performance also includes the time required to convert the analogue input before the digital output becomes a valid representation of the analogue input. The speed of conversion depends on the ADC architecture, so the architecture determines the types of applications appropriate for the converter.

As with the DAC, the ADC can be provided for in a range of packages. For example, Figure 8.21 shows an example ADC and the types of pin connections required in a plastic DIL package type.

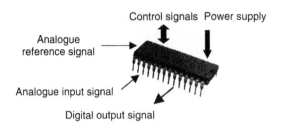

Control signals Power supply

Analogue
reference signal

Analogue input signal

Digital output signal

Figure 8.21: ADC in a DIL package

8.3.2   ADC Characteristics

The ADC analogue input (voltage or current) can vary from a set minimum value to a set maximum value to provide a valid digital output representation of the analogue input. Any analogue input that exceeds these limits (both positive and negative inputs) can damage the ADC as well as saturate the output at a minimum or maximum digital output value. It is common for the ADC output to be an unsigned binary value, although (particularly for bipolar operation) the digital output might also be provided in signed binary (2s complement) representation.

The digital output is a discrete level signal with a value that represents a range of analogue input signal levels. As such, there will be a quantization of the analogue input signal. The ADC creates a quantization error that results from the conversion of the infinitely variable analogue input signal to a discrete level output signal. This quantization error will be important to the choice of the ADC resolution (number of bits). The higher the resolution of the ADC for a given input signal range, the smaller the quantization error as the number of possible output codes for the given input signal range increases. This effect is sometimes referred to as a many-to-one mapping.

The conversion process can be considered with the generalized form shown in Figure 8.22. In this model, two main operations are identified, the sampling operation and the quantization operation.

The analogue input signal is sampled using an ideal sampling block at a sampling rate (sampling frequency) of f_S Hz. The process converts a continuous time signal into a discrete time signal. The sampled signal is then fed to a quantization block that produces the digital output $x(n)$ where n indicates the sample number, as well as process produces quantization noise.

Both ideal and real characteristics of the ADC must be considered. An ideal ADC identifies the operation of the ADC when all values are set to their designed (or ideal)

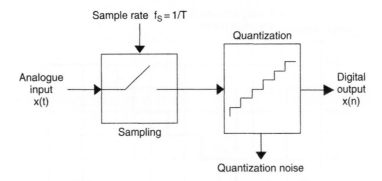

Figure 8.22: Generalized A/D conversion

values. However, due to manufacturing tolerances of the ADC circuitry, real ADC operation deviates from the ideal. In this case, the ADC maximum deviation is defined in the ADC data sheet and guaranteed by the manufacturer. To understand the operation of the ADC, it is common to begin by considering the ideal ADC and then identify how a real ADC could deviate from this.

Consider a three-bit ADC (for simplicity) with an input voltage that ranges from 0 V to +5.0 V and an unsigned binary output code. This is a unipolar ADC whose input voltage ranges from 0 V to the full-scale voltage (V_{FS}). The output code ranges from 000_2 (0_{10}) to 111_2 (7_{10}). The ideal ADC transfer curve, the input voltage–output code relationship, is shown in Figure 8.23. In this view, the input signal conversion range (from the minimum input voltage value 0 V to the maximum input voltage value V_{FS}) is divided into 2^n, where n is the ADC resolution (number of output bits) equal segments, and the point at which the output code moves from one value to the next value falls in the middle of each segment (except for the end points). For a three-bit ADC, the voltage range is split into eight equal segments. For an eight-bit ADC, the voltage range is split into 256 equal segments. A change in 1 LSB of the input voltage creates a step change in the output code of 1 bit. For each 1 LSB step change in the input voltage, the voltage level range is given by:

$$V_{LSB} = (2^{-n} \cdot V_{FS})$$
$$V_{LSB} = (2^{-3} \times 5.0)$$
$$V_{LSB} = (0.125 \times 5.0) = 0.625 \text{ V}$$

The width of a segment is 1 LSB. The point at which the output code changes from one value to the next is the code transition point. When the output code is 000_2, the code changes to 001_2 in the middle of the segment, and the width of this

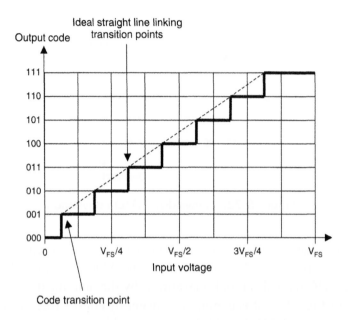

Figure 8.23: ADC transfer curve

code is ½ LSB. When the output code is 110_2, the code changes to 111_2 in the middle of the segment, and the width of the final code is 1½ LSBs. For an ideal ADC, the corner points at the code transition points can be joined with a straight line. Nonideal converters have characteristics that deviate from this straight line.

The input voltage transition point to output code values are shown in Table 8.9.

Table 8.9: Ideal three-bit ADC input voltage to output code mapping
(1 LSB = 0.625 V)

Code transition point	Input voltage at code transition point (V)	Output code (binary)	Output code (decimal equivalent)
Minimum input voltage	0.0	000	0
1st code transition point	0.3125	001	1
2nd code transition point	0.9375	010	2
3rd code transition point	1.5625	011	3
4th code transition point	2.1875	100	4
5th code transition point	2.8175	101	5
6th code transition point	3.4375	110	6
7th code transition point	4.0625	111	7
Full-scale voltage	5.0	111	7

As the resolution (number of bits) of the converter increases and the operating voltage range of the ADC decreases, the LSB step size (voltage or current) will decrease. The effect of this is that the analogue signal levels become the same order of value as the noise generated in the circuit, and the inevitable manufacturing process variations have a more significant impact, leading to problems with design and ultimate use of these converters. Unwanted circuit effects not seen with the lower-resolution data converters are then seen with the higher-resolution data converters.

The characteristics of the ADC are categorized into three types of parameters—static (DC) parameters, transfer curve parameters, or dynamic parameters—whose details are specified in the tables that follow.

The static (DC) and transfer curve parameters are closely related and are considered here together in Table 8.10.

Table 8.10: Static (DC) and transfer curve parameters

Parameter number	Parameter name	Parameter description
1	DC gain error	A measure of the deviation of the slope of the straight-line approximation of the actual converter output from the ideal converter straight-line output. The best-fit straight-line approximation is used for the actual converter straight-line approximation.
2	Offset	The deviation of the first code transition point from the expected. The ideal converter or best-fit straight-line approximation is used. This is normally quoted in LSBs.
3	Integral nonlinearity (INL)	A measure of the deviation of the actual converter code transition point from the straight-line approximation for each code. The best-fit straight-line approximation is used. This is normally quoted in LSBs.
4	Differential nonlinearity (DNL)	The difference between the width (range of input signal) between converter output code changes and an ideal step size of 1 LSB. For a given input code, the output step size is taken between the current input code and the previous code. This is normally quoted in LSBs.
5	Monotonicity	The output code should increase with an increase in input signal; this is a monotonic ADC. A nonmonotonic ADC has an output code that decreases (at particular codes) as the input signal increases.
6	Missing codes	The converter output (digital) should generate 2^n codes where n is the resolution of the converter. Problems may occur within the converter where certain codes are not generated.

Table 8.11: Dynamic parameters

Parameter number	Parameter name	Parameter description
1	Conversion time	There must be a guaranteed maximum conversion time (time from start of conversion to conversion completed).
2	Recovery time	Some ADCs require a minimum time after a conversion has been completed before the next conversion may start.
3	Sampling frequency	Testing of ADC at maximum sampling frequency and ensuring that no errors occur.
4	Aperture jitter	Variations in the sampling period cause an error in the digitized value. Aperture jitter will add noise to the digitized signal.
5	Sparkling	Results from digital timing race conditions. The ADC occasionally produces an output with a larger than expected offset error.

The static and transfer curve tests do not look at the dynamic operation of the ADC and the effects of signal changes and frequency related effects. The dynamic parameters, identified in Table 8.11, describe these effects.

8.3.3 Types of ADC

A number of architectures have been developed for the ADC, and each of which provides its own unique operating characteristic and limitations. Any given ADC, regardless of architecture, falls into one of two categories:

- Nyquist rate ADC
- oversampling ADC

With the Nyquist rate ADC, the input signal bandwidth is equal to the Nyquist frequency. The Nyquist frequency is ½ the ADC sampling frequency (the ADC sampling frequency being the Nyquist rate).

With the oversampling ADC, the converter sampling frequency is much greater than the Nyquist rate. Typical values of oversampling ratio (OSR) are 64, 128, and 256:

$$OSR = (f_S/2 \cdot f_N)$$

where f_S is the sampling frequency of the ADC and f_N is the Nyquist frequency.

For an ADC, the Nyquist frequency is half the minimum sampling frequency of the ADC required to reconstruct fully the original signal from the sampled signal. This means that no information has been lost in the sampling process. The Nyquist rate is when the sampling frequency is twice the bandwidth of the signal to sample.

The main available ADC architectures are:

- successive approximation ADC

- integrating ADC

- ramp ADC

- flash ADC

- sigma-delta ($\Sigma\Delta$) ADC

The successive approximation ADC, shown in Figure 8.24, is commonly found in the implementation of 8- to 16-bit ADCs. This is for a voltage input ADC, and the architecture for this n-bit ADC uses the following main building blocks:

1. **Sample-and-hold circuit**. A control signal to this circuit allows the circuit to sample an input voltage at an instant in time and store this signal. The voltage is stored as a charge on a capacitor.

2. **Comparator**. This compares the stored input voltage to a feedback voltage created by an n-bit DAC in the feedback loop. The comparator is a logic level, 0 or 1, dependent on whether the sampled voltage is less than or greater than the DAC output voltage.

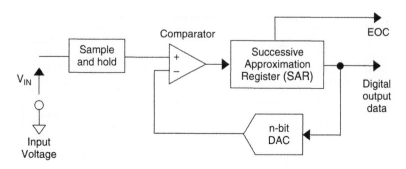

Figure 8.24: Successive approximation ADC architecture

3. **N-bit DAC**. This uses the output of a digital logic block, the successive approximation register, which produces the digital output from the ADC.

4. **Successive approximation register** (SAR). The SAR increments or decrements the digital output data in a binary search pattern depending on the value of the comparator output.

This ADC requires additional control input to initiate or start a conversion, along with clock and reset inputs (to clock and reset the circuitry within the SAR). It will typically produce an EOC (End of Conversion) signal to indicate the end of conversion has occurred and that the digital output data is valid to read.

The integrating ADC applies the input voltage to an analogue integrator circuit, as shown in Figure 8.25, and applies the output of the integrator to the input of a comparator. The integrator input is actually switched between the input signal for a set time, then to a reference voltage ($-V_{REF}$) for a set time. By comparing the output of the integrator to $0\,V$, the comparator produces an output that is either a logic 0 or 1, depending on the result of the comparison. This comparator output produces a control input to a control logic block that continues to count until the integrator output falls to zero. The result of the counter is the ADC output and is stored in an output register.

The ramp ADC is a simple ADC with what might initially seem like a similar architecture to the successive approximation ADC, except now the DAC input code within the ADC is a simple ramp produced by a binary counter. The basic architecture is shown in Figure 8.26. The counter clock is ANDed with the main clock

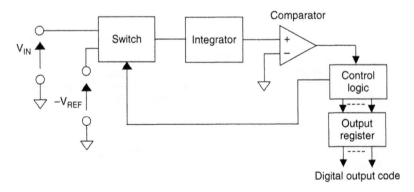

Figure 8.25: Integrating ADC

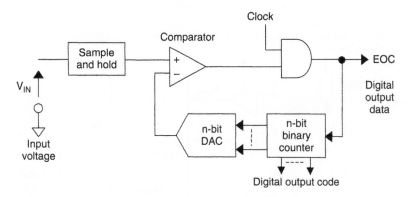

Figure 8.26: Ramp ADC

input to control the count. As long as the input voltage is less than the DAC-generated voltage, the comparator output is a 1 and the counter increments, thereby incrementing the DAC output voltage. As soon as the DAC-generated voltage exceeds the input voltage, the comparator output becomes a 0 and the counter stops counting. The AND gate output can also be used to indicate an EOC.

The flash ADC is a fast converter in which the analogue input is applied to a resistor string and comparator array [8]. The basic converter architecture is shown in Figure 8.27 for an n-bit ADC. The input signal (V_{IN}) is applied to one input of a comparator, with the other input derived from taps from a resistor string. Each comparator output changes at a different input voltage level set by the reference voltage (V_{REF}). The output from the comparator array is a thermometer code. The comparator output (digital) is passed through a combinational logic block that creates the binary output signal. Although this ADC is fast, suited for high-frequency applications such as radar, it needs a large number of resistors and comparators: for an n-bit ADC, 2^n resistors and 2^n-1 comparators are required. Resistor tolerances must also be taken into account (i.e., fabrication process variations), so flash ADCs are restricted to lower-resolution converters. The comparator array creates a thermometer code output, which is then encoded using combinational logic to produce a binary output. This is shown in Table 8.12 for a three-bit ADC.

The sigma-delta ($\Sigma\Delta$) ADC is an example of an oversampling ADC [9]. It uses oversampling and decimation to produce the requirements of an n-bit ADC. The design consists of an analogue $\Sigma\Delta$ modulator followed by a digital decimator. The $\Sigma\Delta$

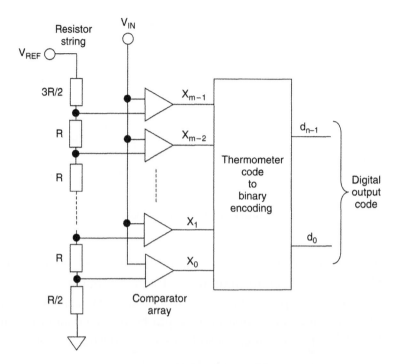

Figure 8.27: Flash ADC architecture

Table 8.12: Three-bit flash ADC thermometer code to binary output

X_6	X_5	X_4	X_3	X_2	X_1	X_0	d_2	d_1	d_0
0	0	0	0	0	0	0	0	0	0
0	0	0	0	0	0	1	0	0	1
0	0	0	0	0	1	1	0	1	0
0	0	0	0	1	1	1	0	1	1
0	0	0	1	1	1	1	1	0	0
0	0	1	1	1	1	1	1	0	1
0	1	1	1	1	1	1	1	1	0
1	1	1	1	1	1	1	1	1	1

modulator usually has a one-bit (two-level) output and samples the analogue signal at the sampling frequency (f_S). The basic arrangement is shown in Figure 8.28. The analogue input to the ADC is sampled before application to the $\Sigma\Delta$ modulator. Here, the $\Sigma\Delta$ modulator produces a one-bit digital output, and the decimator consists of

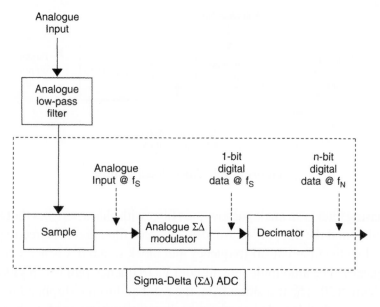

Figure 8.28: Sigma-delta ($\Sigma\Delta$) ADC architecture

a digital low-pass filter followed by sample rate reduction down to the Nyquist rate. It is common also to pass the analogue input through a low-pass filter to avoid aliasing effects.

8.3.4 Aliasing

When a signal is sampled by an ADC, the minimum sampling frequency (f_S) required is twice that of the maximum frequency in the signal (f_O). If the signal frequency increases above this frequency (i.e., $f_O > f_S/2$), then aliasing of the signal can occur [10]. A sampling frequency of, for example, 48 kHz would restrict the signal frequency range from 0 Hz (DC) to 24 kHz. The effect of not sampling at a high enough frequency is that a high-frequency input signal appears in the sampled signal as a low-frequency signal. This problem results from undersampling the signal, when the sampling frequency is less than twice the maximum signal frequency.

To avoid aliasing, an anti-aliasing filter is used at the analogue input of the ADC. This arrangement is shown in Figure 8.29. Here, the anti-aliasing filter is an analogue low-pass filter (usually an op-amp based filter) with a suitable cut-off frequency and filter order (first order, second order, third order, etc.) to ensure that any frequencies above half of the ADC sampling frequency are suitably attenuated.

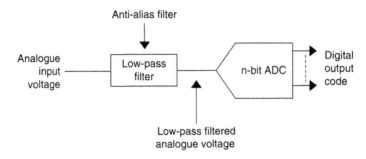

Figure 8.29: Anti-aliasing filter

The anti-aliasing filter is an analogue low-pass filter with characteristics that are based on the type of filter and the filter order. An ideal low-pass filter is considered to pass a signal frequency below the filter cut-off frequency and block signal frequencies above the cut-off frequency. The amplitude (magnitude) Vs frequency plot for this ideal filter is shown in Figure 8.30. The transition from the passband to the stopband is at a single frequency, commonly referred to as a brick wall response because of its shape. However, both the output signal magnitude and phase will be of interest to the filter designer.

The amplitude axis is presented in dB with the low-frequency response at 0 dB. This represents the output signal amplitude being the same amplitude at the input signal. A practical low-pass filter response does not have the ideal response with a transition at a single frequency. There will be a transition region, as shown in Figure 8.31.

The filter design must provide the necessary attenuation of the signal in the stopband with a suitable width transition region, and will be one of the following filter types [11]:

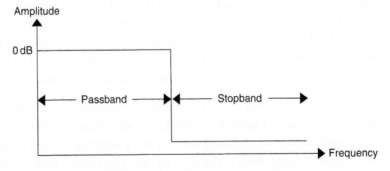

Figure 8.30: Ideal low-pass filter

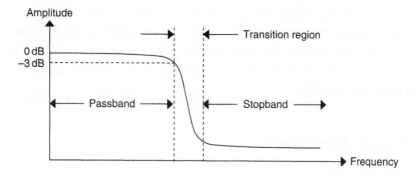

Figure 8.31: Practical low-pass filter

- Bessel filter

- Butterworth filter

- Chebyshev filter

- elliptic filter

The choice depends on the filter design requirements and the shape of the filter response (in both the passband and stopband), as well as ensuring that the resulting size of filter (typically implemented using op-amps, resistors, and capacitors) is not too large and costly. An example of a filter commonly used for anti-aliasing is the second-order Sallen & Key filter. This uses one op-amp, four resistors, and two capacitors, as shown in Figure 8.32. In this circuit, the passband gain of the filter is set

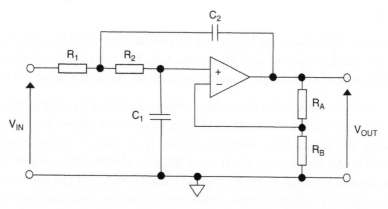

Figure 8.32: Second-order Sallen & Key filter

by resistors R_A and R_B. When this gain is set to 1.586 (+4 dB), the filter produces a second-order Butterworth response. However, it is possible to set the gain to other values, and if the resistors are omitted and the output is connected directly to the op-amp inverting input, then the op-amp is configured as a unity gain buffer and the passband gain is 1.0 (0 dB).

8.4   Power Electronics

8.4.1   Introduction

Devices such as FPGAs and CPLDs operate at low voltage levels (5 V and lower) and at low current levels (μA to mA). These are considered low-power devices as they are not designed to operate at high-voltage and/or high-current levels. However, in many situations, low-power devices must control large electrical loads such as DC motors, AC motors, stepper motors, and audio amplifiers. High-power electronics [12–14] will need to interface the loads to the low-power electronics. The high-power circuit components considered here are the:

- diode
- power transistor
- thyristor
- gate turn-off thyristor
- asymmetric thyristor
- triac

Each component type has a particular set of characteristics and use within an electronic system. Low-power and high-power devices are interfaced either by direct electrical connection or through an opto-isolator, as show in Figure 8.33.

The opto-isolator connects the electronics using an optical link rather than an electrical link, so it electrically isolates the electronics but allows for signals to be transferred by the optical link. This is particularly useful for situations where high voltage levels in the high-power electronics must be electrically isolated from the low-power electronics, and where high-power electronics and electrical load create a substantial amount of electrical noise that could interfere with the operation of the low-power electronics.

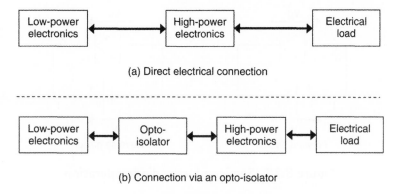

(a) Direct electrical connection

(b) Connection via an opto-isolator

Figure 8.33: Connecting the electronics

8.4.2 Diodes

The diode is a two-layer, two-terminal semiconductor device. When n-type and p-type semiconductor material are joined together, this forms a PN junction, which is referred to as a diode. The semiconductor diode operates to allow current to flow through it in one direction but not the other. The basic structure and circuit symbol for the semiconductor diode are shown in Figure 8.34. The two terminals are named the anode (A) and the cathode (K).

Conventional current flows through the diode from the anode to the cathode (the electrons flow from the cathode to the anode). The current carriers in p-type semiconductors are the holes, whereas the current carriers in n-type semiconductors are the electrons. Normal diffusion at the junction of the two materials will cause

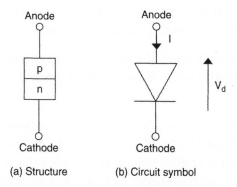

(a) Structure   (b) Circuit symbol

Figure 8.34: Semiconductor diode

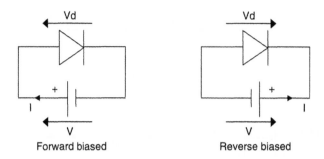

Figure 8.35: Semiconductor diode operation

some of the holes to drift into the n-type material and some of the electrons to drift into the p-type material. This creates a small charge across the junction that repels any further diffusion of holes and electrons. The charged region at the junction is referred to as the depletion region or barrier region. The operation of the diode is considered when either the diode is forward biased or reverse biased, as shown in Figure 8.35. Here, a voltage (V) is applied, and the current (I) can be measured.

Typical applications for the semiconductor diode include AC signal rectification in power supplies, peak detector circuits, signal level clamping (to prevent a signal voltage level from exceeding a safe level, called circuit input protection), telecommunications, and inductive circuit back EMF catch circuits (to remove large voltages created by a rapidly changing current in an inductor).

When the diode is forward biased, this has the effect of reducing the depletion region. If the diode is sufficiently biased (by a suitably high value of V), then current (I) starts to flow. If however, the diode is reverse biased, this has the effect of increasing the depletion region and this prevents the flow of current.

An ideal diode conducts only when the diode is forward biased, and then the voltage drop across the diode (Vd) is zero. When the ideal diode is reverse biased, no current flows.

In a real diode, when the diode is forward biased, there is a finite voltage drop (Vd) across the diode: approximately 0.6 V for a silicon diode and approximately 0.4 V for a germanium diode. With any applied voltage below this value, there will be no current flow. When a real diode is reverse biased, there will be a small but finite leakage current. The current-voltage (I-V) characteristic of the silicon diode is shown in Figure 8.36.

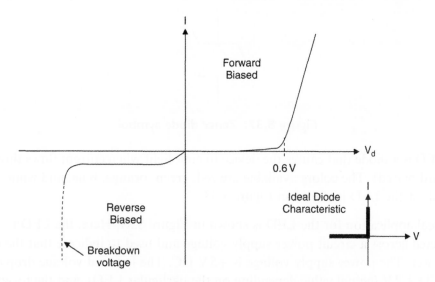

Figure 8.36: Semiconductor diode characteristic (forward biased and reverse biased scales not equal)

When forward biased, the diode equation is given by:

$$I = Is \cdot \left(e^{q \cdot V / K \cdot T} - 1\right)$$

where I is the current flowing in the diode, Is is the saturation or leakage current (typically in the order of 10^{-14} A), V is the voltage across the diode (i.e., V_d), q is the charge on an electron, k is Boltzman's constant, and T is the absolute temperature (in degrees Kelvin). For a circuit operating around 20°C, k.T/q is usually taken to be 25m V.

Variations on the semiconductor diode commonly found in electronic circuits are the Zener diode, the light-emitting diode (LED), and the photodiode.

If the reverse bias voltage exceeds a maximum value, the breakdown voltage, the diode will conduct current and an excessive current flow can destroy the device. This is called avalanche breakdown. A second form of breakdown, tunneling (or Zener) breakdown, can also occur.

The Zener diode has a controlled reverse breakdown voltage. Tunneling or Zener breakdown occurs when the control voltage is exceeded. The symbol for the Zener diode is shown in Figure 8.37. The Zener diode is used in applications such as power supplies and voltage reference circuits.

Figure 8.37: Zener diode symbol

The LED is a diode that causes the device to emit light when current flows through it (forward biased). The colors available are red, green, orange, blue, and white. The symbol for the LED is shown in Figure 8.38.

A typical application for the LED is shown in Figure 8.39. Here, the LED is connected across a circuit power supply voltage and used to indicate that the circuit has power. The power supply voltage is +5 V DC. The forward voltage drop across the LED is 2 V (actual value depending on the particular LED), and the forward current for standard LEDs is 20 mA (actual value depending on the particular LED). To connect the LED to the +5 V source, the current flowing through the diode must be limited by a suitable value resistor.

The photodiode can be used to measure light intensity as it produces a current dependent on the amount of light falling onto the pn junction.

Figure 8.38: LED symbol

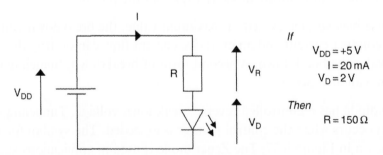

Figure 8.39: LED operation

8.4.3 Power Transistors

The transistor is a three-layer, three-terminal device. It can either be a bipolar junction transistor (BJT) or a metal oxide semiconductor field effect transistor (MOSFET). Transistors are generally categorized by the manufacturer according to their intended application area:

- *General purpose, small-signal transistors* are designed for low- to medium-power (under 1 W) operation or for switching applications.

- *Power transistors* are designed for handling large currents and/or large voltages.

- *RF* (radio frequency) transistors are designed for high-frequency operation such as communications applications.

The BJT is either an NPN or a PNP transistor, shown in Figure 8.40, with three terminals, the base, collector, and emitter. The BJT is sometimes thought of as two diodes connected in series to produce the n-p-n or p-n-p structure.

The flow of a base current (I_B) allows a larger collector current (I_C) to flow. The emitter current is the sum of the base and collector currents. The BJT acts as a current amplifier, although in many cases, this current is passed through a resistor to produce a voltage. By connecting the BJT with resistors (and capacitors), the resulting circuits can provide both current and voltage amplification.

The MOSFET is either an nMOS or pMOS transistor, shown in Figure 8.41, with three terminals, the gate, drain, and source. Some MOSFETs also have a fourth connection, the bulk or substrate, but with a three-terminal device, the bulk is internally connected to the source of the transistor.

The application of a voltage between the gate and source (V_{GS}) of the MOS transistor (a voltage greater than the threshold voltage for the transistor) allows a drain current (I_D) to flow. The gate input to the transistor is capacitive, and only a small gate current (a leakage current in a nonideal capacitor) flows in a device. (In simple analysis, this gate current is assumed to be zero for an ideal capacitor.) The MOSFET uses an input voltage to control an output current. In many cases, this current is passed through a resistor to produce a voltage. By connecting the MOSFET with resistors (and capacitors), the resulting circuits can provide voltage and current output.

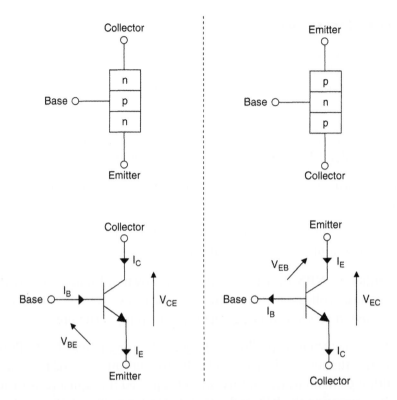

**Figure 8.40: BJT: structure (top) and circuit symbol (bottom),
NPN transistor (left) and PNP transistor (right)**

Both BJTs and MOSFETs can be used to produce either amplifier or analogue filter circuits (linear applications) or switching applications (nonlinear applications). Example applications for power transistors include:

- DC motor control
- AC motor control
- stepper motor control
- audio amplifiers (the output stage of the amplifier driving the speakers)
- switched-mode power supplies

For the power transistor, the safe operating region (SOAR) specifies the safe limits of operation for the transistor in terms of the operating voltages and currents for continuous

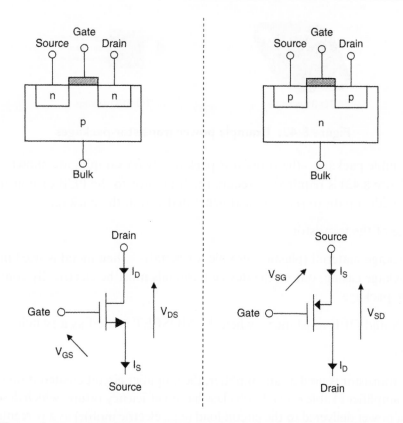

Figure 8.41: MOSFET: structure (top) and circuit symbol (bottom), nMOS transistor (left) and pMOS transistor (right)

operation (continuous current and voltage levels) as well as for levels that exceed the continuous operation region for a limited time period. When used as a switch (particularly applicable for motor control), the switching ON and OFF times also must be considered to ensure correct operation of the circuit in which the transistor is used. If the circuit attempts to switch the transistor ON and OFF too fast, the transistor cannot react fast enough and the result will be incorrect circuit operation.

The choice of which power transistor to use depends on a number of factors:

- availability of a transistor capable of operating to the required voltage, current, and temperature levels

- maximum transistor power dissipation

TO-3 package   TO-218 package

Figure 8.42: Example power transistor packages

- suitable package—the transistor package (two examples are shown in Figure 8.42) is required to secure the transistor to the PCB or housing and to provide a path to remove heat generated within the package

- size of the transistor

- package material (plastic, ceramic, or metal)—when metal is used in the package casing, one of the device terminals must be electrically connected to the package

- ON and OFF resistance—when the MOSFET is used as a switch

- cost

When the transistor is used as an amplifier, the amplifier circuit created is one of five classes of amplifier (Table 8.13). Each class has an efficiency rating, which describes the amount of power delivered to the circuit load (e.g., electric motor) as a percentage of the power delivered to the amplifier. An efficiency of 100 percent means that the amplifier does not dissipate any power (as heat), but an efficiency of 100 percent is not attainable.

Table 8.13: Amplifier classes

Amplifier class	Description
Class A	The transistor conducts during the complete period of the input signal. The efficiency is low, with a maximum of 25%.
Class B	The transistor conducts during one half of the input signal period. The efficiency is higher, with a maximum of about 78%.
Class AB	The amplifier operation is somewhere between Class A and Class B.
Class C	The transistor conducts for less than half the period of the input signal. The efficiency approaches 100%, but produces large distortion of the input signal.
Class D	The transistor is used as a switch (ON or OFF) and produces an amplifier with good efficiency. These are often called switching amplifiers or switch-mode amplifiers.

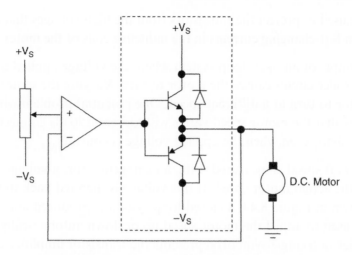

Figure 8.43: Open-loop motor speed control

Power transistors may be used in motor control to provide motor speed, position, or torque control. An example transistor amplifier circuit to provide speed control for a DC electric motor is shown in Figure 8.43:

- The circuit is operated from a dual-rail power supply where $+V_S$ is positive power supply voltage and $-V_S$ is negative power supply voltage.

- A user sets the position of the potentiometer to produce a voltage that represents a required motor speed.

- The potentiometer output is buffered using an op-amp.

- The output of the op-amp drives a class B amplifier.

- The class B amplifier drives the DC motor.

The class B amplifier uses one NPN and one PNP transistor. When the input voltage (output voltage from the op-amp) is positive (with respect to the common node), the NPN transistor conducts. Current flows from the positive power supply to the common node through the motor, and the motor turns in one direction. When the input voltage (output voltage from the op-amp) is negative (with respect to the common node), the PNP transistor conducts. Current flows from the common node to the negative power supply through the motor, and the motor turns in the other direction. The two reverse-biased diodes are connected across the transistor collector-emitter nodes and are used to protect the transistors from the high voltages that could be produced from fast-changing currents in the inductive coils of the motor.

This is an example of an open-loop system where the voltage applied to the motor from the controller circuit causes the motor to turn. Varying the motor voltage will cause the motor to turn at a different speed. One potential problem with this arrangement is that the motor speed varies with different loads connected to the motor output shaft, even when the applied voltage is constant.

If the motor shaft speed is measured using a tachogenerator, a voltage is generated according to the actual motor speed. If this voltage is then fed back to the controller circuit, as shown in Figure 8.44, a closed-loop system is produced and this feedback signal can be used to adjust the motor speed up or down automatically. Here, the power amplifier (a triangle symbol) represents the transistor amplifier circuit. The user input sets the required speed, and the controller circuit automatically adjusts the motor speed to the correct value. The dynamics of the resulting control system depend on the motor dynamics and the control algorithm used.

The control system shown in Figure 8.44 can be realized by developing a digital control circuit with analogue input and output. The basic arrangement is shown in Figure 8.45. Here, the CPLD implements a digital control algorithm such as proportional plus integral (PI) control. The motor speed is set by the user with an analogue voltage. The polarity of the command input determines the direction of motor shaft rotation, and a magnitude determines the speed of rotation of the motor shaft.

The digital output from the controller provides the data input to an n-bit DAC. The output voltage from the DAC is applied through an op-amp based signal conditioning circuit, and this provides the input to a class B amplifier. The op-amp based signal

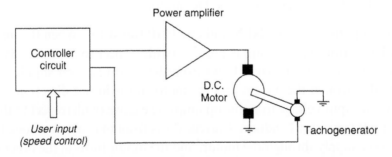

Figure 8.44: Closed-loop motor speed control

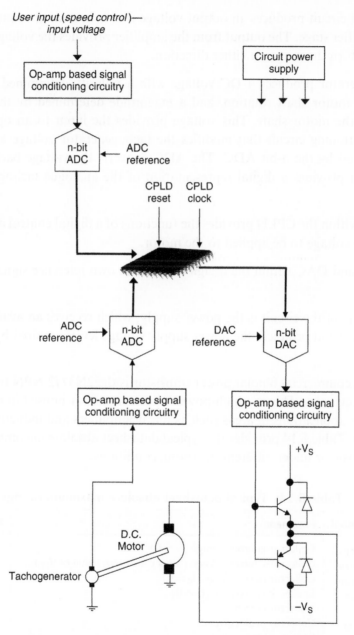

Figure 8.45: CPLD control of a DC motor example

conditioning circuit produces an output voltage that is in a range required by the power amplifier stage. The output from the amplifier provides the voltage and current required to turn the motor in either direction.

A tachogenerator produces a DC voltage with a polarity determined by the direction of motor shaft rotation, and a magnitude determined by the speed of rotation of the motor shaft. This voltage provides the input to an op-amp based signal conditioning circuit that modifies the tachogenerator voltage levels to the levels required by the n-bit ADC. The ADC converts the voltage back to a digital value, which provides a digital representation of the analogue tachogenerator voltage.

The circuit within the CPLD provides the functions of a digital control algorithm that controls the voltage to be applied to the motor.

Each ADC and DAC within the design requires its own reference signal (typically a voltage).

The final part of the circuit is the power supply, which receives an available power supply voltage and produces the power supply voltage levels required by each part of the design.

An example commercial bipolar power transistor is the 2N3772 NPN transistor from ST Microelectronics. This is a high-power silicon transistor housed in a TO-3 metal case and has applications in areas such as linear amplifiers and inductive switching applications. Table 8.14 provides the typical datasheet absolute maximum ratings for a power transistor under different operating conditions.

Table 8.14: Typical datasheet absolute maximum ratings

Symbol	Parameter	Units
V_{CE0}	Collector-emitter voltage ($I_E = 0$)	V
V_{CEV}	Collector-emitter voltage (for set non-zero value of V_{BE})	V
V_{CB0}	Collector-base voltage ($I_B = 0$)	V
V_{EB0}	Emitter-base voltage ($I_C = 0$)	V
I_C	Collector current	A
I_{CM}	Collector peak current	A
I_B	Base current	A
I_{BM}	Base peak current	A
P_{tot}	Total power dissipation at set temperature conditions (T_C)	W
T_{stg}	Storage temperature	°C

An example commercial MOS power transistor is the STF2NK60Z N-Channel transistor from ST Microelectronics. This is a high-power silicon transistor available in the following packages: TO-92, TO-220, IPAK, and TO-220FP. Internal to the transistor are protection Zener diodes. Applications include low-power battery chargers, switched-mode power supplies, and control of fluorescent lamps.

8.4.4   Thyristors

The thyristor is a four-layer, three-terminal semiconductor device used to control the flow of current. It consists of three p-n junctions, as shown in Figure 8.46, and three terminals named anode, cathode, and gate. Uses of the thyristor include overvoltage (crowbar) protection of electronic circuits, motor control, domestic aids (such as electrical kitchen aids), and voltage regulation circuits.

When turned OFF, no current (I) flows from the anode to the cathode. The thyristor can be turned ON, or put into a conducting state, by injecting a current into the p-type layer connected to the gate. When switched ON, it will continue to conduct current (from the anode to the cathode) as long as the conducting current remains above a holding current level. This is independent of the gate current.

Figure 8.47 shows a thyristor controlling the current flowing through a resistor. A sine wave input voltage is applied as the signal to control, and current will flow when the thyristor is in a conducting state and the conducting current remains above the holding current level for the thyristor. For commercial devices, the datasheet provides

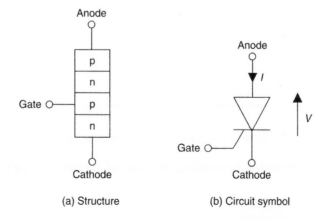

(a) Structure   (b) Circuit symbol

Figure 8.46: Thyristor structure and circuit symbol

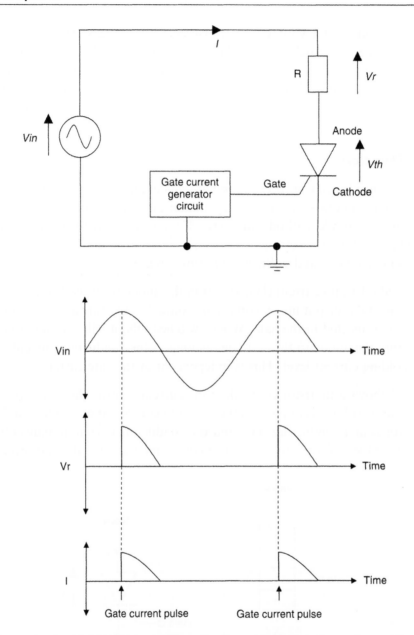

Figure 8.47: Thyristor controlling the flow of current through a resistor

this information. A gate current generator circuit generates the necessary signals to control the operation of the thyristor. Typically, the circuit generates pulses at the appropriate point on the input signal sine wave, in this example turning on the thyristor at the peak of the input signal voltage. Current (I) flows as long as this current is above the holding current level. If the load is inductive (as in electric motors), a phase difference between the voltage and current will need to be considered. The current will only flow from the anode to the cathode, so an AC signal must be rectified. With this action, the thyristor is also referred to as a silicon controlled rectifier (SCR).

The thyristor characteristics are viewed on one of two graphs:

1. **Thyristor characteristic with zero gate current**, Figure 8.48 shows the device voltage (anode-to-cathode voltage) Vs current (current flow into the anode) characteristic when the gate is not operated. Initially, with the thyristor turned OFF, there is no conducting current and only a small forward leakage current will flow. As the voltage across the thyristor is increased, only the small forward leakage current will flow until the voltage reaches a value at which the current can increase to a value (the latching current) at which the thyristor will itself turn ON. The voltage across the thyristor drops to a forward voltage drop level. The thyristor will continue to conduct (independent of the gate current) as long as the forward current remains above a holding current level. When the

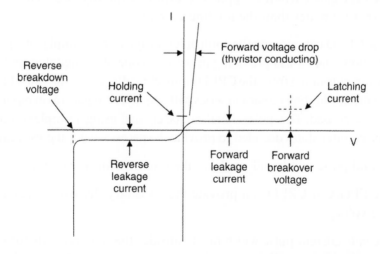

Figure 8.48: Thyristor characteristic with zero gate current

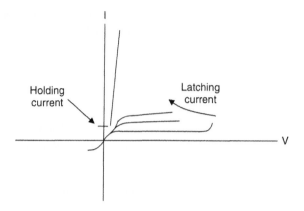

Figure 8.49: Thyristor switching characteristic

thyristor is OFF and a reverse voltage is applied across the anode and cathode, there will be a small reverse leakage current until the applied voltage reaches a magnitude that causes a reverse breakdown (the reverse breakdown voltage) to occur. At this point, the current flow can dramatically increase and, if not limited, can cause device breakdown. These voltage and current levels must be taken into account during the circuit design to prevent unwanted circuit operation and potential circuit failure.

2. **Thyristor switching characteristic**, Figure 8.49 shows the device characteristic when the gate current is applied to turn ON the thyristor. Here, the latching current is greater than the holding current.

An FPGA or CPLD can provide the control for a thyristor. A simple set-up, shown in Figure 8.50, shows the CPLD providing pulses from one of its digital outputs. Here, the circuit shows an output pin from the CPLD connecting directly to the gate of the thyristor. However, a current-limiting resistor in series with the thyristor gate (as in bipolar transistor circuits) may be needed. This pulse signal can be created using a simple counter, with the output of the counter states decoded to provide the necessary 0-1-0 pulse sequence.

The circuit and pulse width will need to take into account for factors:

1. The FPGA or CPLD can provide the necessary thyristor gate current and gate voltage.

2. The gate current pulse width must consider the requirements for the thyristor turn-on and turn-off times, and the frequency of the AC signal to control.

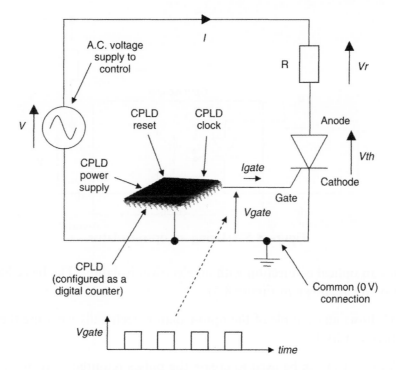

Figure 8.50: CPLD control of a thyristor

3. The point in time during the AC voltage cycle at which to create the gate pulse signal. To create an accurately timed pulse (synchronized to the AC signal), then the AC signal must be monitored, and the point on the signal cycle to create the pulse is determined by the value of the monitored signal. A comparator and DC reference voltage (the signal voltage at which to create the pulse), with the comparator output as an input to the CPLD (and hence a suitable digital state machine within the CPLD), provides this timing.

4. Suitable care must be taken to isolate any low-voltage and high-voltage circuits.

A way in which to isolate electrically any low-voltage and high-voltage circuits is with an opto-isolator. This is a device that provides an optical connection between two circuits, but an electrical isolation. The opto-isolator consists of an LED and a phototransistor on a single package. An externally applied input signal turns the LED ON or OFF on the input circuit. When the LED is ON, the light generated falls onto a phototransistor, switching it ON when illuminated and OFF when not illuminated.

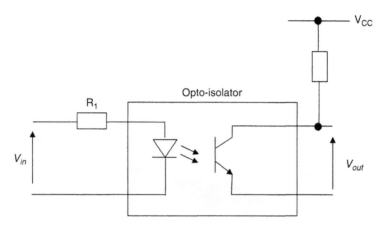

Figure 8.51: Using an opto-isolator

This creates an optical connection with an electrical isolation. The basic idea for the opto-isolator is shown in Figure 8.51.

Figure 8.52 shows an example of the opto-isolator electrically isolating the CPLD from the thyristor itself.

An FPGA or CPLD can be used to create the pulses required to turn on the thyristor. Consider the situation where a 50 Hz sine wave voltage is to be controlled for the circuit shown in Figure 8.50. Here, the pulse is controlled to be incremented in 1 ms steps, derived from a 1 kHz clock (the clock period is 1 ms). If this 1 kHz clock is derived from a higher clock frequency, then a counter can be developed to

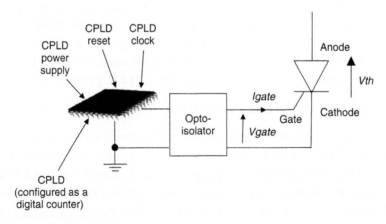

Figure 8.52: Opto-isolation circuit example

produce a clock divider circuit. A simple way to derive the pulse is to create a counter and decode the states of the counter output to produce the pulse signal. The pulse must repeat on every cycle of the sine wave, so the counter must repeat every 20 clock cycles (representing a time period of 20 ms, 1/50 Hz). The pulse is created (i.e., would be a logic 1) on the positive half cycle of the sine wave. No information is given about how the circuit would detect where the time is on the sine wave cycle, so it is assumed that when the sine wave is on the crossover point (i.e., zero) going from a negative value to a positive value (see Figure 8.53), the counter will be in its initial state (state 0).

An example VHDL code design for this arrangement can be seen with reference to the block diagram shown in Figure 8.54. This shows a pictorial representation of the VHDL code (shown in Figure 8.55) and also a counter design with decoded outputs that is controlled from a 50 MHz master clock and an active low asynchronous reset. This VHDL code design is implemented within four processes: The first process creates a 50,000-count counter using the 50 MHz input clock. The second process

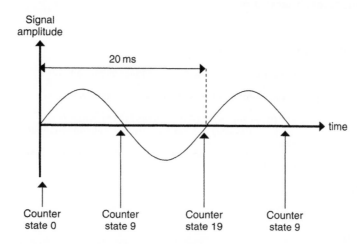

Figure 8.53: Sine wave cycle position to counter state mapping

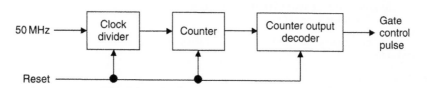

Figure 8.54: Digital design to create thyristor gate pulse

```
LIBRARY ieee;
USE ieee.std_logic_1164.all;
USE ieee.std_logic_arith.all;
USE ieee.std_logic_unsigned.all;


ENTITY Pulse_Generator is
    PORT ( Master_Clock    : IN   STD_LOGIC;
           Master_Reset    : IN   STD_LOGIC;
           Gate_Control    : OUT  STD_LOGIC);
END ENTITY Pulse_Generator;


ARCHITECTURE Behavioural OF Pulse_Generator IS

SIGNAL  Divider : STD_LOGIC_VECTOR(15 downto 0);
SIGNAL  Int_Clock : STD_LOGIC;
SIGNAL  Count    : STD_LOGIC_VECTOR(4 downto 0);


BEGIN


PROCESS(Master_Clock, Master_Reset)

BEGIN

        If (Master_Reset = '0') Then

                        Divider(15 downto 0) <= "0000000000000000";

        ElsIf (Master_Clock'event and Master_Clock = '1') Then

                        If (Divider = "1100001101001111") Then
                                Divider(15 downto 0) <= "0000000000000000";
                        Else
                                Divider(15 downto 0) <= Divider(15 downto 0) + 1;
                        End If;

        End If;
END PROCESS;

PROCESS(Divider)

BEGIN

        If (Divider = "1100001101001111") Then
                Int_Clock <= '1';
        Else
                Int_Clock <= '0';
        End If;

END PROCESS;
```

Figure 8.55: Thyristor gate control pulse generator

```
PROCESS(Int_Clock, Master_Reset)

BEGIN

        If (Master_Reset = '0') Then

                        Count(4 downto 0) <= "00000";

        ElsIf (Int_Clock'event and Int_Clock = '1') Then

                        If (Count = "10011") Then
                                        Count(4 downto 0) <= "00000";
                        Else
                                        Count(4 downto 0) <= Count(4 downto 0) + 1;
                        End If;

        End If;

END PROCESS;




PROCESS(Count)

BEGIN

        If (Count = "00001") Then
                Gate_Control <= '1';
        Else
                Gate_Control <= '0';
        End If;

END PROCESS;

END ARCHITECTURE Behavioural;
```

Figure 8.55: (Continued)

creates an internal 1 kHz clock by decoding the output from the first process. The third process creates a 20-state counter, and the fourth process decodes this counter output to produce the thyristor gate control signal.

An example VHDL test bench for this design is shown in Figure 8.56.

The point of the input signal on which to trigger the thyristor gate pulse can be detected with a circuit like that shown in Figure 8.57. Here, a comparator is used to detect when the input signal exceeds a set DC reference voltage (V_{REF}).

In this arrangement, the two resistors (R_1 and R_2) are used to reduce the value of the sine wave input voltage (V_{IN}) to a safe level that can be used by the comparator without causing damage to the comparator itself.

```
LIBRARY ieee;
USE ieee.std_logic_1164.all;
USE ieee.std_logic_arith.all;
USE ieee.std_logic_unsigned.all;



ENTITY Test_Pulse_Generator_vhd IS
END Test_Pulse_Generator_vhd;



ARCHITECTURE Behavioural OF Test_Pulse_Generator_vhd IS



COMPONENT Pulse_Generator
PORT(
              Master_Clock : IN    STD_LOGIC;
              Master_Reset : IN    STD_LOGIC;
              Gate_Control : OUT   STD_LOGIC);
END COMPONENT;



SIGNAL Master_Clock :   STD_LOGIC := '0';
SIGNAL Master_Reset :   STD_LOGIC := '0';

SIGNAL Gate_Control :   STD_LOGIC;



BEGIN

uut: Pulse_Generator PORT MAP(
              Master_Clock => Master_Clock,
              Master_Reset => Master_Reset,
              Gate_Control => Gate_Control);



Master_Reset_Process : PROCESS

BEGIN

      Wait for 0 ns; Master_Reset <= '0';
      Wait for 5 ns; Master_Reset <= '1';
      Wait;

END PROCESS;



Master_Clock_Process : PROCESS
```

Figure 8.56: Thyristor gate control pulse generator test bench

```
BEGIN

        Wait for 0 ns;  Master_Clock <= '0';
        Wait for 10 ns; Master_Clock <= '1';
        Wait for 10 ns; Master_Clock <= '0';

END PROCESS;

END ARCHITECTURE Behavioural;
```

Figure 8.56: (Continued)

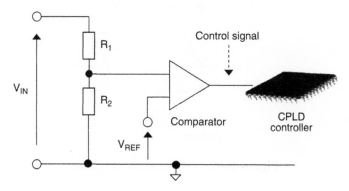

Figure 8.57: Detecting a value of the input sine wave

8.4.5 Gate Turn-Off Thyristors

The gate turn-off (GTO) thyristor is a variant of the basic thyristor that enables the forward-conducting current to be switched ON by the application of a positive gate current and switched OFF by the application of a negative gate current. The symbol for the GTO is shown in Figure 8.58.

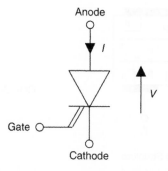

Figure 8.58: GTO symbol

8.4.6   Asymmetric Thyristors

The asymmetric thyristor is a combined thyristor and reverse diode in parallel in a single device, shown in Figure 8.59. The forward current flow is controlled by the gate current; the device always conducts current in the reverse direction.

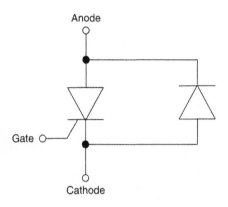

Figure 8.59: Asymmetric thyristor

8.4.7   Triacs

The triac is a multilayer device that has the same electrical operation as two thyristors connected in parallel but in reverse direction, as shown in Figure 8.60. This allows current to flow in both directions. The three terminals are named T_1, T_2, and Gate.

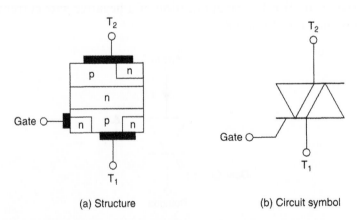

(a) Structure   (b) Circuit symbol

Figure 8.60: Triac structure and circuit symbol

Either a positive or a negative gate current will turn ON the triac. Applications for the triac include office equipment (such as photocopiers), motor control, switched-mode power supplies, light dimmers, and domestic equipment such as hair dryers, televisions, and refrigerators.

The triac is operated according to the polarity of the gate current pulse (positive or negative) and the polarity of the voltage at terminal T_2 with respect to terminal T_1. This produces four possible quadrants (I through IV) of operation, as shown in Figure 8.61.

A variation on the triac is the diac. This is a triac that does not have a gate connection, as shown in Figure 8.62. The diac is designed to break down at low voltage levels in both the forward and reverse directions.

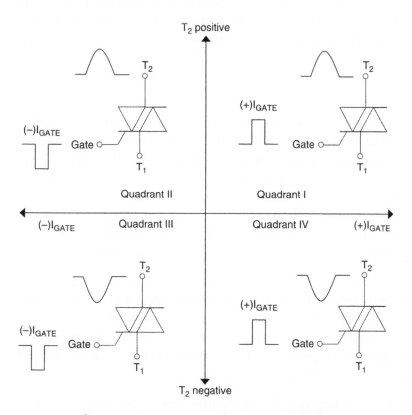

Figure 8.61: Triac, quadrants of operation

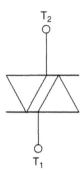

Figure 8.62: Diac circuit symbol

8.5   Heat Dissipation and Heatsinks

When operating power devices, the current flowing through the device will cause the device to dissipate power as heat. The result is that the device temperature will rise above the ambient temperature, and this temperature rise must be controlled. If the temperature rises above a maximum limit, the device could be damaged and fail. The power dissipated as heat must be transferred by one of the mechanisms of heat transfer:

- *Conduction* occurs within a body of material and when two bodies of material are in physical contact with each other. For example, if a rod of metal is heated at one end, then the temperature of the other end will rise as heat is conducted through the metal. Conduction occurs within a body when its different parts are at different temperatures, the direction of heat flow always being from a higher temperature to a lower temperature. Most metals are good conductors of both electricity and heat. In conduction, energy transfer is by molecular vibration and electron motion within the material.

- *Convection* involves the movement of mass from the point to another, such as the movement of hot water around the pipes in a home central heating system. When the material is forced to move by a pump or blower (see Figure 8.63), this is referred to as forced convection. When material moves in response to differences in density caused by thermal expansion of the material, this is referred to as natural or free convection.

- *Radiation* is heat transfer by electromagnetic radiation, such as visible light, infrared radiation, and ultraviolet radiation.

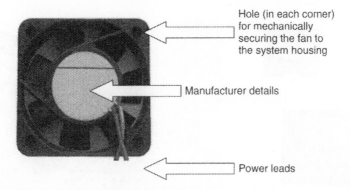

Hole (in each corner) for mechanically securing the fan to the system housing

Manufacturer details

Power leads

Figure 8.63: Example fan for cooling electronic circuits (small fan operating on a +12 V DC power supply)

In low-power applications, the heat generated by a device might be low, so the temperature rise would be small and well within the limits set by the device manufacturer (as defined in the device data sheet). The device package allows the heat to dissipate by natural convection away from the device to the ambient environment. However, for high-power applications, the heat generated can be significantly unable to dissipate fast enough by natural convection. In this case, a heatsink is attached to the device to draw the heat from the device, then release it through natural convection to the ambient environment. The choice of heatsink shape, size, and material depends on the device casing (the shape of the device package) and the operating temperatures involved. In some circumstances, forced convection is also required, such as the fans in a PC.

Example heatsinks are shown in Figure 8.64. The left side shows three different shapes for mounting ICs and transistors; the bottom heatsink shows a transistor mounted onto the heatsink. The right side shows a +12 V voltage regulator IC (a 7812 [+12 V] regulator IC in a TO-18 package) mounted onto the heatsink, which is in turn mounted onto a PCB. A capacitor is sited next to the IC (to the right). In this example, the IC is secured directly to the heatsink, and no electrical insulation is used to insulate the IC from the metal of the heatsink. A screw and nut arrangement secures the components to the PCB.

To insulate the components electrically from the heatsink while maintaining a good level of thermal conduction, an insulating sheet of material (a thermal pad) is placed between the base of the IC and the heatsink, and a plastic insulator covers the screw as it passes through the IC base. This arrangement is shown in Figure 8.65.

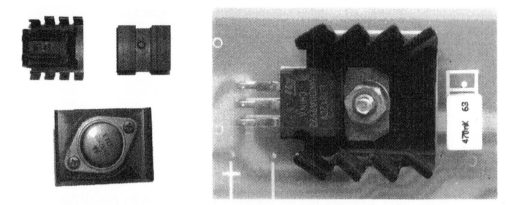

Figure 8.64: Example heatsinks

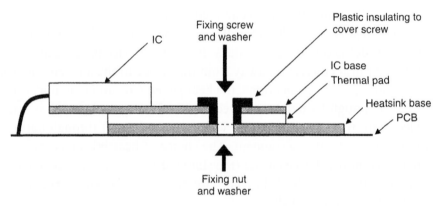

Figure 8.65: Electrically insulating the 7812 from the heatsink

When a heatsink is used, then the heat transfer path is:

1. from the device junction to the device case

2. from the device case to the heatsink (in general, the heat transfer system)

3. from the heatsink to the ambient temperature

To calculate the heat transfer and temperatures that occur at different points in the heat transfer path, an electrical equivalent circuit for the heat transfer circuit can be created. This is shown in Figure 8.66. The resistors model the thermal resistance of the parts in the heat transfer path. (Thermal resistance is the reciprocal of thermal conductance).

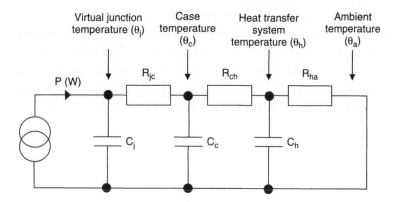

Figure 8.66: Heat transfer path equivalent model

In Figure 8.66, the following are defined:

- P is the internal power loss of the device (in watts, W).

- θ_j is the temperature of the device virtual junction (°C).

- θc is the temperature of the case (°C).

- θ_h is the temperature of the heat transfer system (°C).

- θ_a is the temperature of the ambient environment (°C).

- R_{jc} is the thermal resistance of the junction to the device case (°C/W).

- R_{ch} is the thermal resistance of the device case to the heat transfer system (°C/W).

- R_{ha} is the thermal resistance of the heat transfer system to the ambient (°C/W).

- C_j is the thermal capacitance of the junction (Ws/°C).

- C_c is the thermal capacitance of the device case (Ws/°C).

- C_h is the thermal capacitance of the heat transfer system (Ws/°C).

Under steady-state conditions, the thermal capacitances have no effect and are removed from the analysis of the model. When the thermal capacitance is ignored, the junction temperature is given by the series combination of the thermal resistances:

$$\theta_j = \theta_a + P(R_{jc} + R_{ch} + R_{ha})$$

The thermal resistance of each part of the path is given in the data sheet of the particular device or heatsink. For example, consider the 7812 IC shown in Figure 8.65. If the thermal resistance of the junction to the case of the 7812 is 3°C/W, the thermal resistance of the thermal pad is 0.2°C/W, the thermal resistance of the heatsink is 6.8°C/W, the ambient temperature is 25°C, and the power dissipated by the IC is 3W, then the junction temperature is:

$$\theta_j = 25 + 3 \cdot (3 + 0.2 + 6.8)$$
$$\theta_j = 55°C$$

8.6 Operational Amplifier Circuits

The op-amp is a form of differential amplifier. The op-amp is manufactured as a monolithic integrated circuit and is typically available as a packaged device that contains one, two, or four op-amps. The op-amp is a three-terminal device whose circuit symbol is shown in Figure 8.67.

Here:

- The + input is referred to as the noninverting input.
- The − terminal is referred to as the inverting input.
- A voltage (V+) is applied to the noninverting input.
- A voltage (V−) is applied to the inverting input.
- The output node produces a single-ended output.
- All voltages in the circuit must be referenced to a common (0 V) node.

The op-amp requires a DC power supply voltage (positive and negative). These connections are not shown in the figure; it is common to leave the power supply

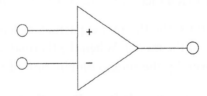

Figure 8.67: Op-amp symbol

connections out of the diagram. Nonetheless, the power supply connections are implicit to the design and cannot be left out of the circuit implementation.

The op-amp amplifies the difference in voltage between the two inputs and produces a single output voltage. A high-differential voltage gain (for many op-amps, in the order of 100,000) is required. However, the op-amp is typically used to incorporate feedback between the op-amp output and input terminals to produce linear signal processing circuits, nonlinear signal processing circuits, and signal generator circuits.

- *Linear signal processing circuit applications* include inverting amplifiers, noninverting amplifiers, unity gain buffers, summer amplifiers, difference amplifiers, low-pass active filters, high-pass active filters, band-pass active filters, and band-reject (notch) active filters.

- *Nonlinear signal processing circuit applications* include comparators, peak detectors, Schmitt triggers, precision half-wave rectifiers, precision full-wave rectifiers, and logarithmic amplifiers.

- *Signal generator circuit waveforms* include sine, cosine, square, triangular, and staircase.

Each op-amp circuit can be used alone or in combination to create a specific signal conditioning circuit arrangement.

References

[1] Grout, I. A., *Integrated Circuit Test Engineering Modern Techniques*, Springer, 2006, ISBN 1-84628-023-0.

[2] Bushnell, M. L., *Essentials of Electronic Testing for Digital, Memory and Mixed-Signal VLSI Circuits*, Kluwer Academic Publishers, 2000, ISBN 0-7923-7991-8.

[3] Burns, M., and Roberts, G. W., *An Introduction to Mixed-Signal IC Test and Measurement*, Oxford University Press, 2001, ISBN 0-19-514016-8.

[4] Jespers, P. G. A., *Integrated Converters D to A and A to D Architectures, Analysis and Simulation*, Oxford University Press, 2001, ISBN 0-19-856446-5.

[5] Jaegar, R. C., *Microelectronic Circuit Design*, McGraw-Hill, 1997, ISBN 0-07-114386-6.

[6] Tocci, R. J., Widmer, N. S., and Moss, G. L., *Digital Systems Principles and Applications*, Ninth Edition, Pearson Prentice Hall, 2001, ISBN 0-13-121931-6.

[7] Analog Devices Inc., *AD7524 CMOS 8-Bit Buffered Multiplying DAC* datasheet.

[8] Haskard, M. R., and May, I. C., *Analog VLSI Design nMOS and CMOS*, Prentice Hall, 1988, ISBN 0-7248-0027-1.

[9] Geerts, Y., Steyaert, M., and Sansen, W., *Design of Multi-Bit Delta-Sigma A/D Converters*, Kluwer Academic Publishers, 2002, ISBN 1-4020-7078-0.

[10] Meade, M. L., and Dillon, C. R., *Signals and Systems, Models and Behaviour*, Second Edition, Chapman & Hall, 1991, ISBN 0-412-40110-x.

[11] Schaumann, R., and Van Valkenburg, M., *Design and Analog Filters*, Oxford University Press, 2001, ISBN 0-19-511877-4.

[12] Bradley, D., *Power Electronics*, Van Nostrand Reinhold (UK), 1987, ISBN 0-442-31778-6.

[13] Horowitz, P., and Hill, W., *The Art of Electronics*, Second Edition, Cambridge University Press, 1989, ISBN 0-521-37095-7.

[14] Storey, N., *Electronics, A Systems Approach*, Second Edition, Addison-Wesley, 1998, ISBN 0-201-17796-X.

Student Exercises

8.1 Using the CPLD development board and the digital I/O board, develop a circuit that will flash a single 12 V bulb ON and OFF at a frequency of 1 Hz. Develop the interface circuitry to allow this by using a suitable opto-isolator.

8.2 Modify the design in Exercise 8.1 so that it now allows the user to set the light flash frequency to either OFF or 1 Hz for an array of eight bulbs (one bulb per digital output). The user will control the operation of the bulb by using the eight digital inputs (0 = bulb OFF, 1 = bulb flashing). Each digital input is to control one bulb.

8.3 Modify the design in Exercise 8.2 so that it now allows the user to control the lights from a PC via a UART within the CPLD.

8.4 How could the CPLD control a stepper motor? Choose a small stepper motor from a component supplier and, using the component data sheet, develop an interface circuit that will connect the stepper motor to the CPLD development board and the digital I/O board. Develop a VHDL code design that will allow the CPLD to control the rotation of the stepper motor.

8.5 Develop the VHDL code that will allow a user to control the stepper motor from a PC via a UART within the CPLD.

8.6 Identify and compare the key parameters of importance when choosing the following power electronic components:

- diode
- BJT
- MOSFET
- thyristor
- triac

8.7 In what circuit applications would a power MOSFET be preferred over a power BJT?

8.8 Identify the types of power electronic components used in the battery charging system for automobile applications. What are the differences between a dynamo-based and alternator-based battery charging system?

8.9 What is a Darlington pair? What types of circuit applications would use it?

8.10   A thyristor has a thermal resistance of 0.8°C/W between its virtual junction and the heat transfer system, and 1.8°C/W between the heat transfer system and the ambient. If the ambient temperature is 30°C, what is the power loss in the thyristor if the junction temperature is 125°C?

8.11   What types of power supply are used in a desktop PC? In a laptop PC?

CHAPTER 9

Testing the Electronic System

9.1 Introduction

If the world were perfect, there would be no need to test. A perfect design would not contain functional errors, and a manufacturing process would not include defects that lead to faults. However, the world is not perfect, and with ever-advancing design complexities, the use of smaller device geometries and new materials, there has never been a greater need for testing.

Testing an electronic system, both the hardware and software, is essential to ensuring that the design meets the requirements of the end-user, meets quality requirements, adheres to standards, and is actually complete [1–3]. The three basic engineering actions are design, manufacture, and testing (Figure 9.1); the testing process is integral to design and manufacture and cannot be seen as a stand-alone process [4]. The design process is primarily concerned with creating a design in hardware or software that meets the required design specifications, and doing so as quickly and

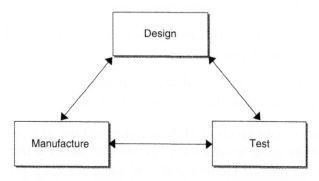

Figure 9.1: Design, manufacture, and testing

economically as possible. The manufacturing process is initially concerned with ensuring that the design can be manufactured, then with manufacturing it as quickly and economically as possible. The test process is concerned with ensuring initially that the design is error free (functionally correct) and secondly that it has been manufactured correctly, and again, as quickly and economically as possible. However, speed and cost in the design, manufacture, and test processes must not be at the expense of product quality.

Testing provides a means to ensure that the design:

- meets the requirements of the end-user (the customer)

- adheres to standards

- is technically correct

- is complete

- is reliable—the design works whenever called upon over its defined lifespan

- is durable—the design works over its defined lifespan without component failure

- is usable

- is capable of working in the intended environment

- is not overstressed by incorrect component specification so that it will fail early in the final application

- does not include software errors that will lead to a software failure (if applicable)

- does not include any hardware circuit faults

- provides feedback to the design process to identify any problems or potential problems with the design itself

- provides feedback to the manufacturing process to identify any problems or potential problems with the manufacturing process

Given the number of steps involved in the creation and use of a particular product, circuit or system testing will be undertaken at different points in time: during product development, during product manufacture, and while in service either as part of a normal routine service arrangement or to rectify faults as they occur.

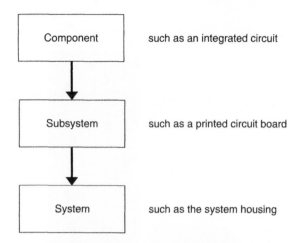

Figure 9.2: Component to system test

During product development, the testing is undertaken at different points in the fabrication process: at the component level, at the subsystem level, and at the system level.

Figure 9.2 identifies these points. Testing at the component level—integrated circuits (ICs), passive components (resistors, capacitors, inductors), discrete semiconductor devices (diodes, transistors, thyristors, and triacs), cables, etc.—is undertaken by the component manufacturer. These components are placed into a subsystem (e.g., soldered to a printed circuit board, PCB), and the subsystem will be placed within the overall system. An example is a mobile phone. The overall system is the phone, which is made up of a number of subsystems such as the main circuit board, power supply, display, keypad, and so on. Each subsystem consists of components; for example, the main circuit board consists of ICs, passive components, and connectors.

Testing is necessary to maintain and improve the overall product quality. Testing is an increasingly important activity at both the device level and the system level. The importance of testing in discovering faults in an electronic circuit or system hardware after fabrication and before use is summarized in the Rule of Ten, shown in Figure 9.3.

The cost multiplies by ten every time a faulty item is not detected before use in a large electronic circuit or system. Here, if the cost of detecting a faulty device when it is

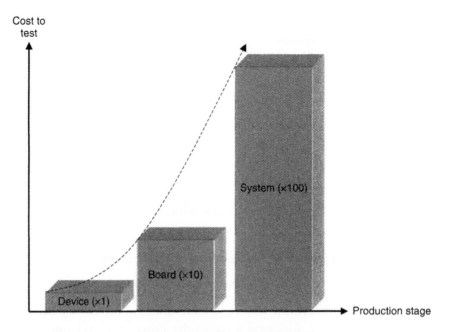

Figure 9.3: The Rule of Ten

produced is one unit, the cost of detecting that faulty device when used at the board level (PCB) is ten units, and the cost of detecting that faulty board after inserted in its system is 100 units, and so on. While this rule is a generalisation, is does provide an insight into the need for effective testing at the right point in the product manufacturing process. Preventing faulty components from being used (where they are faulty but the fault has not been detected and so are thought to be fault-free) is essential to developing cost effective and high quality products.

This chapter will consider integrated circuit testing, printed circuit board testing, boundary scan testing (both IC and PCB levels), and software testing.

Testing is primarily concerned with two aspects, controllability and observability:

- *Controllability* refers to controlling specific parts of a design to set specific values at specific points within the design. In hardware, this is a specific analogue voltage or current, or a digital logic level (0 or 1). In software, this is a specific variable value.

- *Observability* refers to observing a value within the design.

The controllability and observability are undertaken in the most economic way possible while providing the necessary test coverage. In hardware, the observation will be at the circuit nodes at the I/O and power supply pins of the PCB or IC, along with specific internal nodes within the design. This will be achieved with the use of suitable test equipment and increasingly by building testability aspects into the design itself.

Where testability is designed into the circuit, this is referred to as *design for testability* (DfT). This may be accomplished by adding specific test hardware or, in a processor-based system, by including software test routines. A part of DfT, but usually considered as a subject in its own right, is the built-in self-test (BIST). BIST is test circuitry included within a design to allow the design to test parts of itself. With the addition of test functionality, the design must be required to have two or more modes of operation: mission mode is the normal operating mode, and one or more test modes dedicated to testing. The end user might have access only to the normal operating mode and might not even be aware of the existence of test modes, whereas the developer would have access to all modes of operation.

The external test equipment [5] used to test the design is either manual test equipment or automatic test equipment (ATE). Manual test equipment is the type used in the laboratory and requires the user to control the instruments manually, either by the use of controls on the equipment itself or via a PC interface. Examples of such equipment include:

- basic instruments such as an AVO meter (which is used to measure voltage, current, and resistance) and digital multimeters (DMM)

- oscilloscopes, spectrum analyzers, and waveform generators

- logic analyzers

- special instruments such as signal distortion meters, high-voltage, and high-current meters.

Designs are also visually inspected to identify any problems or damage, such as broken components.

Automatic test equipment (ATE) is programmed for a specific operation and will automatically test a batch of designs with no or minimal human intervention. This allows short test times and is used in production testing of devices. Examples of such equipment include:

- *Vision systems* optically inspect the circuit by capturing an image of the circuit and analyzing the image for abnormalities. The types of vision system are:
  - automated optical inspection (AOI)
  - X-Ray inspection (used to inspect hidden features)
  - infra-red and laser scanning
- *In-circuit testers* (ICT) test the functionality of components within an assembled circuit. The ICT equipment accesses components one at a time, via a test fixture commonly referred to as a bed of nails fixture.
- *Flying probe testers*, also referred to as a fixtureless tester, access the circuit with probes that rapidly move between points on the circuit using an accurate, high-speed positioning system.
- *Functional testers* (FT) test the circuit via its primary connectors and additional spring-loaded probes.
- *Hot rigs* are special test facilities for testing the subassemblies (subsystems) in a rig.
- *Special testers* include other test equipment for the testing of specific system parameters.

The types of circuit and system tests identified above perform tests using either *electrical* or *optical* measurements. Another type of test which would be performed is a *mechanical* test, which tests the mechanical strength of the system for the required operating conditions (e.g., it tests the operation of the circuit or system when subjected to the mechanical vibrations that would occur in automotive applications).

Some of the tests to be undertaken are *destructive* in nature, so samples of a production run are taken with the intention of testing the circuit or system to destruction. When the tested system is required for use after testing, the tests need to be *non-destructive* and it is essential to ensure that the non-destructive tests do not cause damage to the circuit or system.

9.2 Integrated Circuit Testing

9.2.1 *Introduction*

An IC consists of a circuit on a die of semiconductor material such as silicon, which is secured to a protective package. IC testing must consider the circuit operation as well as the packaging.

Before considering what tests are to be undertaken on the IC, the physical structure of the IC should be considered, and the tests should ensure that the circuit design is correct, and the design has been fabricated without producing a circuit failure.

The ability to fabricate the circuit without producing a circuit failure is essential to ensuring that the end-user has received a working circuit. Ideally, the manufacturer can provide the customer with devices, knowing that 100 percent of all supplied devices are fully functional. However, the ability to test for every possible circuit fault is prohibitively expensive, and a test program to implement this may be impossible, so delivering 100 percent of fully functional devices is unlikely to happen. The goal for the manufacturer is to reduce the number of faulty devices to an acceptable level.

The defect level (DL) is used as a measure of the test quality and is expressed as the number of faulty ICs that exist within a group of ICs that has passed the test. This is normally expressed in terms of parts per million (ppm). Typically, a defect level of 0 ppm is the target, but 100 ppm would be considered a high-quality test. The defect level is determined from analysis of the returned faulty devices. These devices may have failed the customer incoming acceptance test (supplied devices are tested by the customer prior to use), may have failed an in-system test (the device has passed the customer incoming acceptance test but fails in the customer application), or may fail in the in-field test (either during normal operation or routine maintenance tests).

The IC packaging is of three main types:

- *Hermetic*, usually ceramic (hermetic-ceramic), is designed for high-performance applications and is usually the most costly. The die is environmentally decoupled from the external environment with a vacuum-tight enclosure.

- *Plastic* refers to a resin material (usually epoxy-based) in which the die is encapsulated. Over time the external environment penetrates the plastic.

- *Metal* is commonly used for discrete device (e.g., transistor) packaging.

Figure 9.4: IC package (84 CLCC) with the lid removed

An example packaged IC is shown in Figure 9.4. The silicon die is secured in a cavity in the center of the ceramic package and is electrically bonded to the package using bond wires. For use, the package is placed into a socket soldered to the application printed circuit board. The IC package has to:

- provide the necessary electrical connections for signal and power

- provide proper heat dissipation from the die

- provide a thermal expansion compatible with the die

- minimize signal delay and noise

- provide mechanical, environmental, and electromagnetic protection

In normal applications, a lid covers the top of the package and the die is not visible. Some packages, however, require a glass-covered hole to expose the die (e.g., for UV erasure of an EPROM).

The die is connected to the package using one of three methods:

- *Wire bond* is the most common bonding technique.

- *Flip-chip*, in which solder bumps on the die and substrates are joined to form the connections.

- *TAB* (tape-automated bonding).

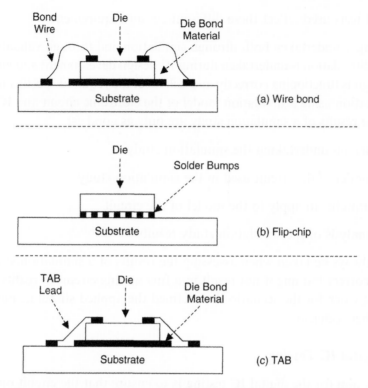

Figure 9.5: Bonding methods of circuit die to its package

These three bonding methods are shown in Figure 9.5. The circuit die (usually silicon) is secured to the package substrate, and electrical bonding connects the signal and power connections from the die to the package pins.

The circuit created on the die provides the required circuit functionality. The three basic circuit categories are:

- *Digital circuits* consider signals in the form of discrete values (usually binary, a logic 0 or 1 value) that change at discrete points in time. The logic levels and timing of their changes are of primary concern.

- *Analogue circuits* consider continuously varying signals (voltages and currents) over a range of values.

- *Mixed-signal circuits* combine the functionality of analogue and digital circuits, usually for interfacing analogue signals to a digital processor system.

The types of tests used reflect these different circuit requirements.

Design testing is undertaken both through simulation and through validation of the physical IC. Simulation is undertaken during IC design development and aims to ensure that the design is functioning correctly and will meet the required specifications prior to device fabrication using a simulation model of the electronic circuit and IC packaging. However, the results of a simulation study are only as good as:

- the person undertaking the simulation study

- the model of the circuit used in the simulation study

- the stimulus to apply to the model of the circuit

- the analysis of the simulation study results

Care must always be taken when assessing the results of a simulation study: they might look correct but might not result in a functioning circuit in reality in that the circuit might work for the scenario that defined the applied stimulus, but might not work for other scenarios.

9.2.2 Digital IC Testing

The primary aim for the digital IC testing is to ensure that the circuit operates correctly in terms of digital logic levels (0 and 1) and the timing of the changes to the logic levels. The basic arrangement for testing a packaged digital IC is shown in Figure 9.6. Here, an external ATE is used to apply the digital stimulus and capture the

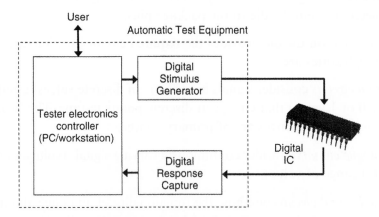

Figure 9.6: Digital IC test

digital response. Internally within the ATE, the response to the test stimulus is analyzed, and the test program determines whether the IC passes or fails the test for that particular stimulus.

The role of the test engineer is to:

- develop the test program requirements (the types of tests to undertake), commonly referred to as the *test specification*

- develop the software test program to run on the ATE (usually a C language based program)

- develop the hardware tester interface between the existing tester electronics and the particular digital IC

- analyze and deal with the results of the test program

Typically each of three types of test is performed to a greater or lesser level for a particular product:

1. A *functional test* exercises the operation of the design through the various functional operations that it is intended to undertake. For complex digital circuits and systems, this can be extremely time consuming and hence costly.

2. A *structural test* stimulates faults that may exist in the design resulting from fabrication defects. The idea is to apply suitable digital vectors that sensitize the particular fault considered so that the faulty circuit produces a different result than a fault-free circuit. This requires the creation of suitable fault models to model fabrication defects. These models are simulated in the design to identify the right set of digital vectors to apply to the actual fabricated circuit.

3. A *parametric test* addresses specific analogue parameters (voltage and current) that are not tested in the functional or structural tests, which address logic level and timing issues. Parametric testing focuses on the I/O cells within the periphery of the circuit die (Figure 9.7). The die consists of two main areas, the core and the periphery, which is a standard arrangement for ICs. The core contains the main circuitry; this is the circuitry targeted by functional and structural tests. The periphery contains the cells (the rectangles in the periphery) that contain circuitry and metal pads (black squares) to bond the die to the package pins. These connections are for signals and power. It is the circuitry in the signal periphery cells that the parametric tests target.

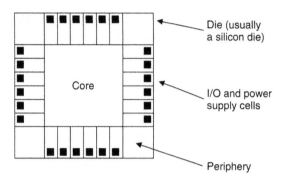

Figure 9.7: IC die structure

The example IC in Figure 9.8 shows the die within a package and the package lid removed. The silicon die is in the center of the image, and the bond wires for the signals and power radiate from all four sides.

Functional tests are required during device development (tests on physical prototypes prior to high-volume fabrication) to determine that the design operates according to the required design specification and that the results are consistent with the results from the simulation studies.

Structural tests are undertaken during production (high-volume) testing to stimulate faults if they exist (and hence allow for the observation of faulty circuit behavior). The resulting test program often requires less test time that an exhaustive functional test and hence costs less. However, specific functional tests to cover

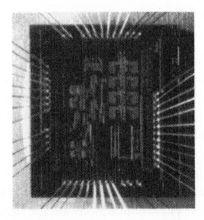

Figure 9.8: Example silicon die bonded to package

circuit operation that cannot be controlled with structural tests is also required. The structural tests assume that the design functionality is correct, and faults caused by manufacturing defects are to be detected. Structural tests are based on applying digital stimulus to the IC to stimulate specifically considered faults. These faults are based on fault models, and for digital circuits, the main fault models are stuck-at-fault, bridging fault, and delay fault.

The *stuck-at-fault* (SAF) considers a fault to create nodes within the design—either at the primary inputs/outputs (the circuit die I/O) or at the internal nodes—that get stuck at a logic level. This basic idea is shown in Figure 9.9. A node is either stuck-at-logic 0 (SA0) or stuck-at-logic 1 (SA1) and will always be at these fixed logic levels, even though the circuit might be attempting to set a different logic level. The premise is that a defect within the circuit causes this type of logical fault. The circuit is considered to operate in one of three scenarios:

- In *fault-free* operation: no fault exists in the circuit.

- In *single-stuck-at-fault* (SSAF) operation, the circuit contains a single stuck-at-fault (a single fault assumption).

- In *multiple-stuck-at-fault* (MSAF) operation, the circuit contains multiple stuck-at-faults.

It is common to consider only the SSAF scenario because it simplifies the test pattern generation to manageable levels. The stimulus (test pattern) generation process uses software tools and a procedure referred to as automatic test pattern generation (ATPG).

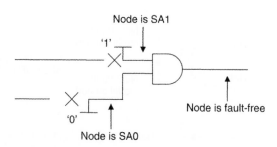

Figure 9.9: Stuck-at-fault

In the *bridging fault*, two (or more) nodes are considered to be unintentionally connected by a conductive material such as metal. An example of this is the resistive material in the metal interconnect layer connecting tracks, where tracks are placed close to each other on the physical circuit layout. A bridging fault is usually considered in terms of logic level behavior, although a resistive bridging fault can be considered in the analogue rather than digital domain. For the logical bridging fault, the two models are:

- **Wired-AND**, in which two nodes are considered, and the fault is modeled as logic 0 dominant. The nodes that are bridged are connected to the inputs of a two-input AND gate. When either or both bridged nets are logic 0, then the output is logic 0.

- **Wired-OR**, in which two nodes are considered, and the fault is modeled to be logic 1 dominant. The nodes that are bridged are connected to the inputs of a two-input OR gate. When either or both bridged nets are logic 1, then the output is logic 1.

A basic bridging fault is shown in Figure 9.10, in which a two-input NAND gate with an output nodes Net1 and Net2 that are close enough to result in a bridging fault.

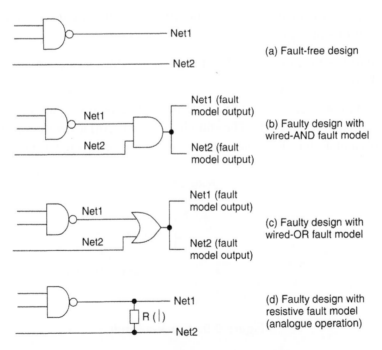

Figure 9.10: Bridging fault models

In a *delay fault*, the fault does not cause a logical error in the circuit output, but rather it causes an error in the timing. The output then reaches its final logical value at a later time than expected. Several delay fault models have been developed and are in use.

The *stuck-at-fault*, *bridging fault* and *delay fault* are examples of commonly considered faults that can be modelled using digital logic levels and timing. However, such faults operate by considering that the faults that could exist in the underlying analogue circuitry will exhibit such logical and timing faulty behaviour. An alternative look at the faults would look into the analogue circuit operation in more detail. Defects in the processing can create analogue faults such as open and short circuit faults that have both resistive and capacitive effects. For example, a short circuit between two nets can cause a *resistive* bridging fault that does not cause a pure logical bridging fault circuit behaviour (as previously described), but is more complex in nature. Similarly, an open circuit fault in a metal track (net) could be considered as either a high value resistance and/or a capacitance between the two parts of the broken track. Such a broken track could lead to a floating net situation where the voltage on the net will be set by the charge stored on the net metal track (the broken net then forming a capacitor between the net and the circuit common node). This can lead to a complex circuit behaviour that can be caused by a fault that might be difficult to detect. As the semiconductor processing technology evolves and finer process geometries are used (as in the high performance digital semiconductor processing technologies on the 90 nm, 65 nm, 45 nm, and lower technology nodes) in the fabrication of digital logic, so do the fabrication processing defects and the resulting circuit behaviour caused by these defects. This then leads on to the need to detect faults through suitable test programs that now exist in the finer semiconductor processing technologies that would not have existed in the coarser fabrication processing technologies.

9.2.3   Analogue IC Testing

The primary reason for testing analogue ICs is to ensure circuit operation is correct in terms of analogue voltage and current levels, and analogue parameters such as frequency response, impedance, and noise. The basic arrangement for testing a packaged analogue IC is shown in Figure 9.11. Here, an external ATE is used to apply the analogue stimulus and capture the analogue response. Internally within the ATE, the response to the test stimulus is analyzed, and the test program determines whether the IC passes or fails the test for the particular stimulus.

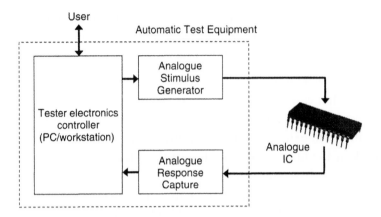

Figure 9.11: Analogue IC test

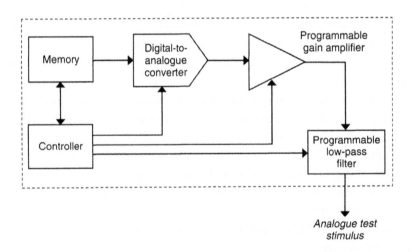

Figure 9.12: Analogue stimulus generator

With this set-up, and with the tester electronics controller operating in the digital domain, the tester requires a digital-to-analogue converter (DAC) to generate the analogue test stimulus from a digital representation of the analogue test stimulus, and an analogue-to-digital converter (ADC) to convert the captured analogue response into the digital domain. The basic building blocks of the analogue stimulus generator are shown in Figure 9.12, and those of the analogue response capture are shown in Figure 9.13.

In the analogue stimulus generator, the digital representation of the analogue test stimulus is held in memory within the tester. The contents of the memory are read at a

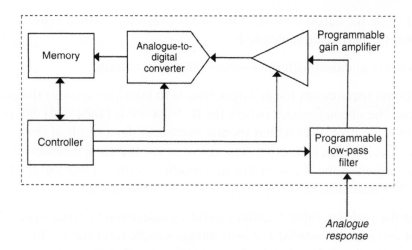

Figure 9.13: Analogue response capture

particular rate and applied to a DAC. The analogue output from the DAC is then applied to a programmable gain amplifier and a programmable low-pass filter. All of these blocks are controlled by a controller unit within the tester. It is also possible to select the amplifier and low-pass filter blocks to be used or to be bypassed. This programmability is typically achieved by using relays (the relays providing a low impedance switching of the analogue signals).

In the analogue response capture, the analogue input signal is applied to a programmable low-pass filter and programmable gain amplifier before becoming the input to an ADC. The digital output of the ADC is then stored in memory for results analysis. All of these blocks are controlled by a controller unit within the tester. It is also possible to select the amplifier and low-pass filter blocks to be used or to be bypassed. This programmability is typically achieved by using relays (the relays providing a low impedance switching of the analogue signals).

As in digital testing, the role of the test engineer is to:

- develop the test program requirements (the types of tests to undertake), commonly referred to as the test specification

- develop the software test program to run on the ATE (usually a C language based program)

- develop the hardware tester interface between the existing tester electronics and the particular analogue IC

- analyze and deal with the results of the test program

An additional requirement for analogue tests is to take into account the process variations. The silicon foundry (where the IC behavior is fabricated) will guarantee their fabrication process to within specific tolerances. As a result of these tolerances, no single IC is identical to another. The analogue test therefore considers the expected typical operation of the IC, as well as its operation at the extremes of the fabrication process. The tolerances of the ATE are also taken into account.

Sampling the analogue input requires careful consideration to avoid signal aliasing (if the input signal is not sampled at a high enough sample rate) and to obtain a coherent sampling of the signal. In general, when a signal is sampled, the sampling can be either coherent or noncoherent, and the results in the frequency domain (where the digital samples of the analogue signal are analysed using the fast fourier transform (FFT) and the spectral components of the sampled signal are identified) will be different. In coherent sampling, and considering a sine wave analogue input voltage, N samples are used to represent exactly one cycle of the signal (sine wave) and will repeat without disjunction from one cycle to the next. This idea is shown in Figure 9.14, where N = 4, and the sampling repeats at the same point on the sine wave.

If N samples are taken for one cycle of the signal frequency (f_{in}), then the sampling frequency (f_S), is given by:

$$f_{in} = (f_S/N)$$

In reality, a signal is not a single-frequency (single-tone) sine wave, so a certain amount of noncoherent sampling (where the sampling of a complete cycle of the analogue signal is not achieved) is expected. Postprocessing of the samples using windowing techniques will reduce the effects of this noncoherent sampling situation.

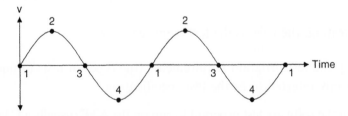

Figure 9.14: Coherent sampling of a sine wave

Unlike digital IC testing, in which structural test methods (particularly stuck-at-fault) are widely used, analogue IC testing is still predominantly functional. Attempts to develop structural test methods for analogue ICs has met with limited success.

9.2.4 Mixed-Signal IC Testing

Mixed-signal ICs combine both analogue and digital circuitry. Examples of mixed-signal ICs include:

- ADC

- DAC

- phase-locked loop (PLL)

- digitally programmable analogue amplifier

- comparator

- analogue switch

The primary reason for testing mixed-signal ICs is to ensure the circuit operation is correct in terms of the analogue voltage and current levels for the analogue parts, and the digital logic levels and timing for the digital parts. The testers used for mixed-signal testing must be suitable for both analogue and digital parts.

9.3 Printed Circuit Board Testing

Components such as ICs and passive components are soldered to a printed circuit board (PCB) to create the final electronic circuit or system. The PCB (discussed more fully in Chapter 3) provides the necessary electrical, thermal, and mechanical properties to ensure that the electronic circuit or system will function correctly. The technology drivers for the PCB to provide the right performance at the right cost include:

- greater reliability

- higher device density

- smaller device geometries

- higher interconnect density (finer track geometries and pitch between tracks)

- higher operating frequencies (e.g., clock frequency of modern microprocessors, RF signal frequencies for communications applications)

- more terminals and wiring at PCB interfaces

- preservation of signal integrity

- electromagnetic compatibility (EMC)

- multiple layers of interconnect

- move toward surface-mount packaging for ICs (from through-hole packages)

- move toward surface-mount replacement of discrete devices (mainly resistors and capacitors)

All these are required, but with lower costs and higher product quality!

The functions of PCB testing are to:

1. Ensure that the PCB itself (the substrate and interconnect) has been fabricated correctly prior to populating the board with components. Electrical tests on the interconnect detect any open or short circuits in the tracks, and a visual inspection of the board detects any faults such as missing protective coating and missing or lifted (away from the substrate) tracks and pads.

2. Ensure that the circuit operates properly after the PCB has been populated (as in the example shown in Figure 9.15). Electrical tests confirm that the circuit operates electrically as required, and a visual inspection identifies problems such as:

   - missing components

   - misplaced components

   - correct type of component but wrong value

   - short circuits resulting from too much solder

   - open circuits resulting from too little solder

   - badly formed solder joints (shape and color)

3. Ensure that the PCB has the correct thermal performance by operating it at elevated or reduced temperatures, and through thermal cycling where the PCB is cycled through a range of temperatures for a set time and under specific circuit operating conditions.

4. Ensure that the mechanical strength of the solder joints and the overall PCB is as required. Samples are taken from a production batch of PCBs and tested to destruction:

- vibration test, in which the board is shaken until destruction
- shear strength testing of solder joint strength
- visual inspection of microsectioned solder joints

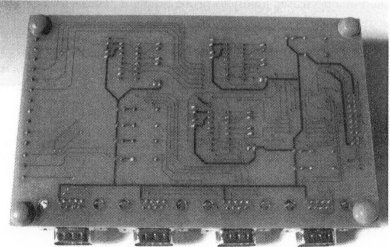

Figure 9.15: Example populated PCB top and bottom views

9.4 Boundary Scan Testing

Boundary scan tests [6–8] for digital ICs was introduced as IEEE Standard 1149.1-1990 in 1990 and now forms one of a family of standards covering boundary scanning for digital circuits, mixed-signal circuits, and system designs. The current standards are identified in Table 9.1.

Boundary scan testing is an extension of scan path testing that was developed for digital logic. Scan path testing provides test access to the core of the IC via the circuit bistables (e.g., D-type bistable), and to allow test data to be serially clocked into, and out of, the device under test (DUT) via the primary I/O (i.e., the device package pins) of the IC. Scan path testing is an easily adopted DfT approach for digital ICs that allows the controllability and observability of internal nodes within a digital IC to be maintained for the more complex ICs. Scan path testing is included within the core of the IC. However, the basic idea can be adopted for the peripheral cells of the IC and at the PCB level, as well. Boundary scan testing essentially allows a scan path to be set up in the peripheral cells of the IC and the PCB to provide test access to the individual ICs mounted onto the PCB.

The motivation for the creation and adoption of boundary scan testing lies in the increasing complexity of digital ICs, the reduction in IC package dimensions, and the increasing complexity of PCB designs. Two example PCBs are shown in Figure 9.16. Here, the size of the IC packages (number of pins and physical footprint) as well as the close placement on the PCB itself can be seen. These technological advances have resulted in a decreased ability to probe the PCB nodes with conventional test equipment such as the flying probe tester and the ICT.

To address the PCB test problem, the Joint Test Action Group (JTAG) was established in the 1980s. This led to the development of the IEEE Standard 1149.1-1990, also commonly referred to as JTAG boundary scan. The basic idea is to set up a

Table 9.1: IEEE Standard 1149

IEEE Std 1149.1-1990	IEEE standard test access port and boundary-scan architecture
IEEE Std 1149.1a-1993	IEEE standard test access port and boundary-scan architecture
IEEE Std 1149.1b-1994	Supplement to IEEE Std 1149.1-1990, IEEE standard test access port and boundary-scan architecture
IEEE Std 1149.1-2001	IEEE standard test access port and boundary-scan architecture
IEEE Std 1149.4-1999	IEEE standard for a mixed-signal test bus
IEEE Std 1149.5-1995	IEEE standard for module test and maintenance bus (MTM-Bus) protocol
IEEE Std 1149.6^{TM}-2003	IEEE standard for boundary-scan testing of advanced digital networks

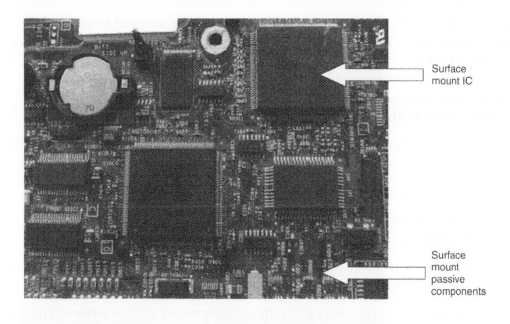

Surface
mount IC

Surface
mount
passive
components

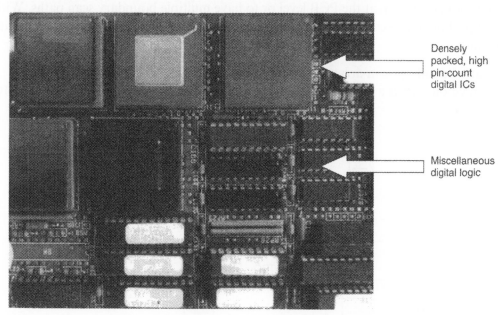

Densely
packed, high
pin-count
digital ICs

Miscellaneous
digital logic

Figure 9.16: Example complex digital PCBs

serial scan path in the periphery (I/O) of the device under test, and to use the IC interconnect on the PCB to set up the IC boundary scan test path for the application and retrieval of test data. In this approach, one or more digital ICs supporting boundary scan (boundary scan compliant ICs) are within the boundary scan path. This approach includes the interconnections (tracks) on the PCB and the ICs mounted on the PCB in the test.

Given that some ICs contain a boundary scan capability (boundary scan compliant) while others do not, one of the following scenarios exists at the PCB level:

1. No boundary scan capability is available.

2. A partial boundary scan capability is available where some of the ICs are boundary scan compliant, while others are not.

3. A full boundary scan capability is available where all of the ICs are boundary scan compliant.

Where a boundary scan capability exists, the designer must choose whether to create a single boundary scan at the PCB level or to have multiple boundary scan paths. With multiple boundary scan paths, the designer much choose between shared and separate signals.

A basic boundary scan arrangement is shown in Figure 9.17 where two ICs are mounted on a PCB and connected together in a single boundary scan path. Each IC

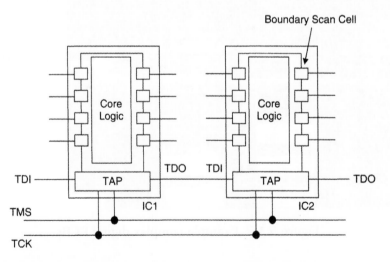

Figure 9.17: Basic boundary scan arrangement for two ICs

has two modes of operation: normal operating mode (the mission mode) and boundary scan mode.

In the *normal operating mode*, the boundary scan is bypassed and has no effect, However, in the *boundary scan mode*, test data can be serially clocked in and out of each IC in the scan path that is set up in the interconnect at the PCB level and in the I/O cells of the IC. Each I/O cell is a boundary scan cell and contains logic to control the operation of the cell between the normal operating mode and the boundary scan mode. Test access is via the TAP (test access port). A basic test access port will consist of the following:

1. TAP controller

2. Data registers (boundary scan register, bypass register, identification register and user defined registers)

3. Instruction register

In a basic arrangement, each IC requires four additional pins for the boundary scan:

- *test data input* (TDI) is used for providing the IC with both data and instructions

- *test clock* (TCK)

- *test data output* (TDO) is used for providing both data and instruction outputs from the IC

- *test mode select* (TMS) which, together with the TCK, is used to control the TAP controller, a 16-state finite state machine used as the controller of the boundary scan circuitry on the IC

An additional optional *test reset* (TRST) is sometimes included, but is not a requirement in the standard.

A boundary scan solution consists of a hardware part and a software part. The basic hardware part has been previously discussed. The software part is supported by the boundary scan description language (BSDL), which a subset of VHDL. BSDL provides the structured and machine-readable form needed to describe the parameters for an IEEE Standard 1149.1 compliant IC. However, VHDL is a large language, and not all of its available features are required for boundary scanning, hence BSDL is only required to be a subset of VHDL.

Although the boundary scan standard was created to solve test problems at the PCB level, it has the following additional uses:

- as part of a test solution at the PCB level where boundary scan circuitry is included at the digital IC level

- as part of a test solution for the digital IC itself in production test

- for configuring programmable logic devices, whereas the FPGAs and CPLDs available today are configured via a JTAG interface

For the configuration of programmable devices, IEEE Standard 1532-2002 (In-System Configuration of Programmable Devices) is available.

A more detailed look into a boundary scan compliant IC is provided in *figure 9.18*. Here the core logic (or system circuitry) provides the core of the IC that is surrounded by digital input and output obtained from the *boundary scan register*. This is formed using the *boundary scan cells*. Alongside this circuitry is the circuitry that forms the *test access port* (TAP). This is formed by the *TAP controller, instruction register* (and the *instruction decode logic*), *data registers* and *associated logic* that is required for the IC to be compliant with the standard. The two types of register are the *instruction register* (sets up the mode of operation of the IC) and the *data register*. Four data registers are shown here, with these registers being either mandatory (must be present) or optional (can be included if so required):

1. **Boundary scan register**: This is formed using the boundary scan cells in the I/O of the IC and allows for test data to be applied to the core inputs of the IC and to be retrieved from the core outputs.

2. **Bypass register**: This is a mandatory 1-bit register which allows the IC to be bypassed so that test data applied on the test data input (TDI) pin is passed to the test data output (TDO) pin and is not stored in either the instruction register or the other data registers.

3. **Device identification (ID) register**: This is the hardwired identification for the particular device which is formed using a 32-bit code. This is an optional register.

4. **User defined register**: This is an optional register (or registers) that are specified by the user for the particular IC.

Test data can be accessed through the TDO pin from any of the data registers or from the instruction register depending on the actions determined by the TAP controller.

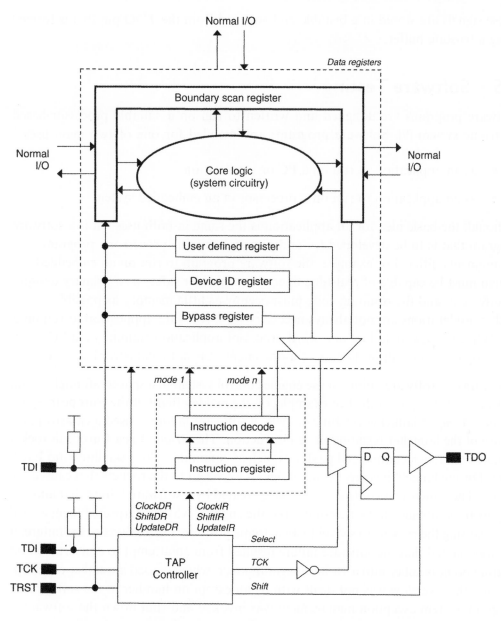

Figure 9.18: Example boundary scan complaint IC structure

These signals are stored in a bistable and accessed from the TDO pin that is formed using a tri-state buffer.

9.5   Software Testing

Software programs are designed and written to run on a suitable processor-based electronic system [9]. Software programs are intended for one of two main uses:

- as an application to run on a PC or workstation

- as an application to run on a processor in an embedded system

Although the basic idea for an application is the same in both uses—it is a software program that is to be developed for and run on a suitable processor—program requirements differ. For example, the software program to run on an embedded system must be capable of real-time operation, consider safety as a primary design requirement, and the resulting code must occupy as little memory as possible. Such considerations are not always necessary for a software application to run on a PC. While a software failure in a word processor application running on a PC might be annoying and result in the loss of a document, it is not a catastrophic failure.

By contrast, a software failure in the engine control system in a spacecraft might result in the loss of the spacecraft. For example, on the maiden flight of the European Space Agency *Ariane 5* launcher on June 4, 1996, a software failure caused a catastrophic failure of the launcher resulting in its destruction. The *Ariane 5* is a European rocket designed to launch commercial payloads such as communications satellites into Earth orbit. During launch, incorrect control signals were sent to the rockets that caused the failure. These incorrect control signals were created by a software failure that allowed diagnostic commands to be transmitted to the engines, which interpreted them as real data, causing the rockets to swivel to an extreme position. On analysis of the failure, it was ascertained that the software failure resulted from an attempt to convert a 64-bit floating point number into a 16-bit integer number, which caused a number overflow. Because this overflow was not dealt with by an exception handler in the software program, system exception management was invoked and shut down the software. This potential scenario was not considered during the software development, and so not handled within the software.

The development of a software program involves two main players, the developer and the customer. The developer develops the software program on behalf of the customer and to the customer's specifications. The software testing process must take into account the requirements of both players.

When writing the software program, both syntax and semantics are considered. Program syntax relates to the grammatical structure of the programming language used and the rules that determine whether the program can be compiled or interpreted. Program semantics relate to the meaning of the program as written. A program might be correct in its syntax but incorrect in its semantics, or vice versa. In programming languages such as C, syntax errors are removed when the program is compiled using the available program checking and debugging tools that support the code compiler.

As part of the design process, software engineers must consider that the developed program might contain faults, which may or may not cause a software failure. Hence, faults are distinguished from failures. A fault is an error in the semantics of the software program. A failure occurs when the program performs an action that it was not designed to perform. For example, a program might delete a required file on a PC. A program might operate correctly on the processor for which it was compiled, but might fail when compiled to a different processor. Thorough testing of the program should remove the possibility of a software failure.

Testing the software program is undertaken at different times within a program development process and again once the development is completed. The types of tests depend on what information is needed from the test process. When a software program is developed, the overall design problem is broken down into smaller, more manageable tasks. The code development is broken down into units, each of which is connected together to form the overall software program. The basic software tests consider this approach:

1. unit testing
2. unit interaction testing
3. integration testing
4. acceptance testing

Whenever a fault is located and corrected in the software program, then the program must be retested both to verify that the fault was corrected and to verify that correcting the error did not introduce new faults. It is essential to ensure that the test process is suitably planned and managed and that the right tests are undertaken at the right points in the process. A lack of testing could set up a situation in which faults and failures are not detected. An excess of testing will result in higher than necessary test costs. Planning and conducting software tests is a complex process, particularly when a team of software developers are involved in the overall design process.

Unit testing, also referred to as white box testing, considers the program code in detail. It is undertaken after an individual unit has been developed, and by the developer with knowledge of the internal code structure. Therefore all parts of the code (e.g., variable values) can be monitored. The tests are based on the program code detail, so if the code changes, then the tests probably have to be changed, as well. Once the tests for functional correctness have been completed, exception testing is conducted to test the unit under exceptional conditions. This ensures that the program behaves gracefully when the program experiences unexpected situations, such as when it receives a value outside the range for the declared variable type, and does not simply crash.

Unit interaction testing is undertaken after the units have been written and individually tested. This tests for the correct operation of the units working together, for example, correctly passing variable values between units.

Integration testing is then undertaken and checks that the whole system has been integrated successfully. Unit interaction testing should have already identified many of the integration problems and allowed for their correction at the unit level.

Acceptance testing is performed to ensure that the system is doing the right thing. These give the customer confidence that the program has the required features and behaves correctly. The program must pass all of the acceptance tests to be accepted by the customer. The tests that are specified and undertaken represent the interests and needs of the customer rather than the interests and needs of the developer. Acceptance tests are also sometimes referred to as functional tests. Two examples are:

- When the application is initially opened, the application log file named "application.log" opens in the directory "C:\bin\".

- When the application is closed, the application log file named "application.log" is closed.

The application is run and checked that these actions are correctly performed and, where multiple actions are undertaken, are performed in the correct order. Acceptance testing considers the software specification and is a form of black box testing. In black box testing, the tests consider an external perspective of the program and do not involve any knowledge of the internal structure of the program.

References

[1] Bushnell, M. L., *Essentials of Electronic Testing for Digital, Memory and Mixed-Signal VLSI Circuits*. Kluwer Academic Publishers, 2000, ISBN 0-7923-7991-8.

[2] Burns, M., and Roberts, G. W., *An Introduction to Mixed-Signal IC Test and Measurement*. Oxford University Press, 2001, ISBN 0-19-514016-8.

[3] Rajsuman, R., *System-on-a-Chip Design and Test*, Artech House Publishers, USA, 2000, ISBN 1-58053-107-5.

[4] Grout, I. A., *Integrated Circuit Test Engineering Modern Techniques, Springer, 2006*, ISBN 1-84628-023-0.

[5] O'Connor, P., *Test Engineering, A Concise Guide to Cost-effective Design, Development and Manufacture*, John Wiley & Sons Ltd., England, 2001, ISBN 0-471-49882-3.

[6] The Institute of Electrical and Electronics Engineers, *IEEE Standard Test Access Port and Boundary-Scan Architecture*, IEEE Standard 1149.1-2001, IEEE, USA.

[7] Van Treuren, B., and Miranda, J., "Embedded Boundary Scan," *IEEE Design and Test of Computers*, March–April 2003, pp. 20–25.

[8] Parker, K., *The Boundary-Scan Handbook, Analog and Digital*, Second Edition, Kluwer Academic Publishers, USA, 2000, ISBN 0-7923-8277-3.

[9] Ince, D. C., *Software Engineering*. Van Nostrand Reinhold International, 1989, ISBN 0-278-00079-7.

Student Exercises

9.1 What is a scan D-type bistable? How is it used in testing digital ICs?

9.2 What is the principle of I_{DDQ} testing? How is it used in testing digital ICs?

9.3 Built-in self-test (BIST) for digital circuits will normally be based on specific known circuit designs and operation in order to provide the necessary BIST functionality, but with a small circuit overhead (the amount of circuitry required to implement the BIST). One example of a commonly used circuit is the linear feedback shift register (LFSR). Using suitable texts identified in the reference list for this chapter, identify the operation of the LFSR and determine why it is used.

9.4 For the IEEE Standard 1149.1 boundary scan standard, identify the operation of the TAP controller. For this digital finite state machine, develop and simulate a VHDL code implementation for this design.

9.5 For each of the case study PCB designs identified in Appendix F, Case Study PCB Designs (see the last paragraph of the Preface for instructions regarding how to access this online content), develop a suitable test procedure that could be used to test the PCBs during design and fabrication.

CHAPTER 10

System-Level Design

10.1 Introduction

Increasingly, there is a need to develop more complex digital systems and more quickly to reduce development time and cost and to get a new product to market first. These requirements have highlighted the limitations that exist with traditional design approaches that were developed and suited to smaller designs where working at a more detailed level was part of the normal design process. In the more complex designs, working at the detailed design level is no longer viable, resulting in a need to work at a higher level of design abstraction. Here, the designer develops, validates, and verifies the operation of high-level design models, which are then automatically synthesized into the circuit design implementation. The designer can concentrate on getting the high-level operation right, then put the detailed implementation into the hands of software electronic design automation (EDA) tools [1] for automating the creation of the design details—essentially the synthesis of the system-level models into registered transfer logic (RTL) code (for hardware) and/or software source code (such as C/C++). Then the standard RTL logic synthesis design flow to utilize this RTL code is undertaken for the hardware, and source code compilation is undertaken for the software.

An electronic system-level design approach considers both hardware and software and the right mix of hardware and software. Much of the system-level and detailed design implementation work is undertaken using one or more EDA tools.

The implementation technologies available for use today—processor based, FPGA (field programmable gate array) or CPLD (complex programmable logic device) based, ASIC (application-specific integrated circuit) based—typically contain a number of libraries of predesigned blocks ranging in complexity (considering hardware) from basic logic gates to IP (intellectual property) blocks such as multipliers, processors, memory, and communications that must be brought together in the system design and must work together to create a working system.

The question does, however, need to be asked: Why use a system level design approach? There are many possible answers. For a designer working with smaller designs (What is small?), the traditional design flows and EDA tools appear to be adequate and well supported. But what if the needed design is several times larger in magnitude and is needed in less time? Would the traditional design flows and EDA tools be sufficient? Probably not. The designer is encountering a number of factors:

- increasing design complexity

- increased speed of operation of designs

- increasing range of possible implementation technologies

- use of third-party IP (intellectual property) blocks

- licensing and IP issues associated with the use of third-party IP blocks

- working with third-party suppliers and keeping up to date with IP supplier changes

- increased need to work in project teams and on a global scale in various time zones

- outsourcing of specific design, manufacturing, and test activities to third-party organizations

- need for cost-efficient and fast design prototyping

- reduced time to market (TTM)

- increased competition

- using best practices in design

- keeping up to date with the latest developments

- use of mature and new design flows

- use of mature and new EDA tools

- reduced costs

- ... and so on ...

Even with simpler designs, a number of issues must be considered before a design concept can be realized as a product. Working at the electronic system-level design, some of these identified issues become a more critical role in realizing a design concept:

- design implementation efficiency

- increasing distance between the design abstraction level and the design implementation

- increased potential for detail implementation issues (*the devil is in the detail*)

- test and testability issues

- DfX, design for X:

  ○ **DfA**, design for assembly

  ○ **DfD**, design for debugging

  ○ **DfM**, design for manufacturability

  ○ **DfR**, design for reliability

  ○ **DfT**, design for testability

  ○ **DfY**, design for yield

- correct, robust, and repeatable system operation

- automated documentation generation

- information at all stages of the process about actions taken automatically (by the EDA tools for example) so the designer can quickly and easily check specific design details

- target hardware design issues

- design team communications (on a local and global scale)

- robustness of the design to manufacturing process variations

- keeping up to date with existing and new legislation (e.g., the WEEE directive)

- ... and so on ...

The area of electronic system-level (ESL) design is still emerging, and various activities are undertaken in defining the direction for ESL design [2]. However, there

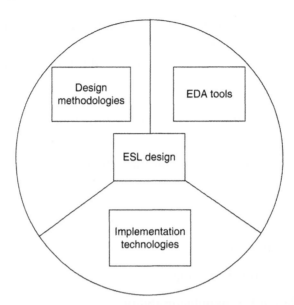

Figure 10.1: Combining design methodologies, EDA tools, and implementation technologies for ESL design

is a basic need to combine into a single and robust design methodology multiple design methods, EDA tools, and implementation technologies, as shown in Figure 10.1. How would these be combined? What changes are needed to enable the best integration into the ESL design? Or would it be better to start new and create new solutions for ESL design from scratch, without risking any limitations and compromises the old solutions might bring with them? These are only a few of the pertinent questions that are not easy questions to answer, but nevertheless must be asked.

The way in which an ESL design methodology will be developed and implemented has a major impact on how easily the methodology can be used and ultimately whether a product can be successfully developed. Two approaches are possible:

1. A *one method fits all* approach, in which a single, fixed methodology would exist, and any given design project would adapt to fit the model. If the methodology is not right for the particular project, this could ultimately limit the success of a project. However, this approach does facilitate monitoring and controlling a project.

2. A *general framework* that can adapt to particular project requirements. This simple framework would identify the key steps in design, and the details, the design modules, would be added for a particular project—to use a computer industry term, providing a plug-and-play capability. This allows the designer to create dynamic methodologies that can react to different project requirements and changes that inevitably occur.

Three points raised at the beginning of Chapter 2 bear repeating. Before considering any design:

1. **Always use common sense**. If something does not seem right, then it probably isn't.

2. **Never leave anything to chance**. What can go wrong will go wrong.

3. **There is almost always more than one way to solve a problem**. The choice for the designer is to determine the most appropriate solution. The first solution developed might not necessarily be the best.

A fourth point is also necessary:

4. **All systems are based on the same principles of physics**. The principles learned in the first year of any degree program in electronics and computer engineering will be always valid.

As design complexity increases, it is easy to lose sight of the requirement to implement the design, and rather for the designer to become so focused on the higher levels of design abstraction that common sense falls away.

Aside: Never underestimate the need for basic theorems such as Ohm's Law.

The designer therefore needs to remember that, although he or she may be working at a high level of design abstraction, the design must work in a manufactured system. This is not always easy to achieve, and it may be necessary for a team of designers working at different levels of abstraction to work closely together and use the same terminology, talk the same language. At the high level of design abstraction are the platform design engineers. At the detailed implementation,

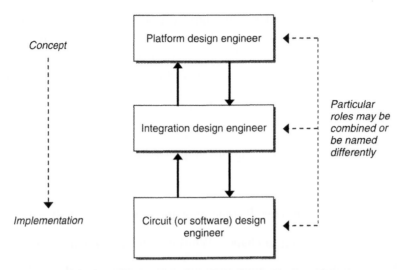

Figure 10.2: System level designer interaction

are the circuit (or software) design engineers. In between are the integration design engineers. Figure 10.2 identifies these three levels:

1. *Platform design engineers* work at the highest levels of abstraction and define the right architecture for the particular system, along with the interfacing between the components.

2. *Circuit (or software) design engineers* work on detailed implementation issues.

3. *Integration design engineers* work at a middle level of detail and provide a link interface between the specialties. The integration design engineer is not a specialist in one area, but rather has a general understanding of both specialties and the communications skills to achieve fluent and productive two-way communications.

A two-way flow of information among all three roles must be clear, concise, and efficient, with a common terminology and an appreciation of the needs of each role. The need is for cooperation rather than competition or conflict.

The separation of tasks must also be considered, given that any single individual cannot do everything at all stages of the design process. This parallels the modern approach taken in electronic circuit testing for the proactive involvement of the test engineer alongside the design engineer to assess test and design testability issues at an early stage in the design process—as early as the design specification stage. This has brought test

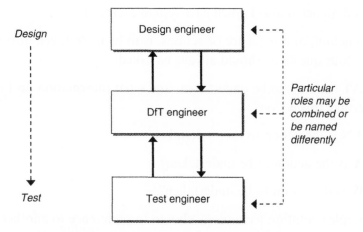

Figure 10.3: Design, test, and DfT

engineering closer to the design activities in that the test program development for an electronic or microelectronic circuit occurs at an early stage in the product development process and requires a basis in design. This overcomes the problems encountered when design and test activities were separate and distinct, an unnecessary barrier between two interrelated activities. In this design for testability (DfT), sometimes referred to as Design for Test approach, test activities can influence how a design is created by identifying testability issues and improving test access to specific circuitry within the design. Specialist engineers in both design and testing are supported by a generalist DfT engineer, shown in Figure 10.3, who bridges the gap between them. The need for specialists is based on the need for in-depth knowledge of specific design and test issues, roles which a single person could not realistically be expected to undertake.

The term *platform* is used in ESL design to refer to the architecture of the system, what different technologies are to be used for particular functions, and how these will be interconnected. It refers to both hardware and software components. For example, if an embedded system uses an 8051 microcontroller and associated chip-set, the platform that the embedded software will operate on will be the microcontroller and associated chip-set. A number of software EDA tools, both separate and integrated tools, will be used to design the platform, including:

- software compiler for the particular microcontroller
- PCB design software
- SPICE-based circuit simulator

The platform designer is also known as a system architect.

Whenever an action, or a sequence of actions, is to be undertaken during the design, the following four questions should always be asked:

1. **WHAT** action is to be undertaken, and what information do I need to undertake this action?

2. **WHEN** is the action to the undertaken?

3. **WHY** is the action to be undertaken?

4. **HOW** is the action to be undertaken?

How is one implementation technology chosen in preference to another?

To consider these questions and to identify some of the implementation issues, system-level design in this chapter is considered with reference to the development of VHDL (or VHSIC, very high-speed integrated circuit, hardware description language) code for implementing required digital algorithms from a system-level simulation model. In this case, a MATLAB®/Simulink® model is created and manually translated to VHDL code [3–6] by considering the required high-level functions and how they can then be implemented in RTL-level VHDL. Two case study designs are considered, one digital control algorithm for a motor control scenario and a digital filter design. They are worked through from design concept through to VHDL coding and simulation.

When looking at particular case studies, be aware that each particular problem considered has specific needs. Although a number of the issues identified and resolved for the particular case study may be generic, other issues may be specific to that particular problem. Attempting to generalize the problem for all possible applications is not trivial and requires careful consideration.

10.2   Electronic System-Level Design

Electronic system-level (ESL) design is emerging as an important design approach for more complex electronic system design problems [2]. Rather than focusing on design details, the ESL approach concentrates on the higher levels of design abstraction. It is not the same as behavioral modeling, rather it is advancement on what was attempted to be achieved by behavioral modeling. An important point is to note that ESL has a taxonomy. Taxonomy has a number of definitions, coming mainly from the natural sciences. One definition is the science of classification according to a predefined system. This produces a catalog with which to provide a conceptual framework that allows for discussion, analysis, and information retrieval by those who use the framework.

There are a number of reasons for considering an ESL approach. The work is undertaken at a high level of abstraction, allowing ideas to be rapidly developed and evaluated without time-consuming detail work. The design is created and the operation simulated using a high-level modeling and simulation language. The implementation (down to the basic digital logic hardware and software programs) is undertaken using a suitable EDA tool or set of tools

> *Aside*: A useful compilation of articles on various issues associated with ESL design is provided in an article entitled "Electronic System-Level Design" in the September–October 2006 issue of *Design & Test of Computers* (http://www.computer.org/dt).

A number of the companies and organizations currently involved in ESL design are identified in Table 10.1. As an emerging area, ESL is dynamically changing to the needs of the complex systems encountered today, as well as predicting the needs for the future. The need for the future is for insight rather than hindsight.

Table 10.1: Organizations currently involved in ESL

Company	URL
ARM (Advanced RISC Machines)	http://www.arm.com/
Altium™	http://www.altium.com
AutoESL Design Technologies	http://www.autoesl.com/
Cadence Design Systems	http://www.cadence.com/
Celoxica	http://www.celoxica.com/
Codetronix	http://www.codetronix.com
Coware	http://www.coware.com
Mentor Graphics	http://www.mentor.com/
The Mathworks Inc.	http://www.mathworks.com/
Synopsys	http://www.synopsys.com/
Esterel Technologies	http://www.esterel-technologies.com/
Bluespec	http://www.bluespec.com/
Cebatech	http://www.cebatech.com/
Impulse Accelerated Technologies	http://www.impulsec.com/
Forte Design Systems	http://www.forteds.com
Computer Engineering Group (GrecO) of the Informatics Center of Federal University of Pernambuco	http://www.pdesigner.org/
SystemCrafter Ltd.	http://www.systemcrafter.com

Within the semiconductor industry, which is driving much of global technological advances, is the International Technology Roadmap for Semiconductors (ITRS) [7, 8]. This roadmap is a set of public domain documentation that identifies industry trends, highlights technical obstacles, and provides companies with the information to align their product cycles with the developing technologies. The ITRS identifies the semiconductor industry technological challenges and needs over the next 15 years and is regularly updated. This cooperative effort of industry manufacturers and suppliers, government organizations, and universities is sponsored by the following organizations:

- European Semiconductor Industry Association (ESIA)

- Japan Electronics and Information Technology Industries Association (JEITA)

- Korean Semiconductor Industry Association (KSIA)

- Semiconductor Industry Association (SIA)

- Taiwan Semiconductor Industry Association (TSIA)

The ITRS considers design, fabrication, test, and EDA, but does not propose a fixed plan for the future. Rather it provides a means to identify the future challenges and where the organizations involved consider that effort is best expended.

ESL is a response to the emerging needs of the designers (both hardware and software) to support their need to develop more complex systems designs but in less time. This allows the designer to:

- raise the design entry point to a more abstract level to make the complex design problem manageable

- concentrate on the high-level design concept issues rather than the low-level design implementation issues

- reduce design time by automating time-consuming tasks that are suited to automation

- explore the design space at the abstraction level and explore trade-offs (in size, performance, power consumption) in the design decisions

ESL design is a response to designers working at a behavioral level, as has become increasingly common with behavioral-level modeling being developed for synthesis

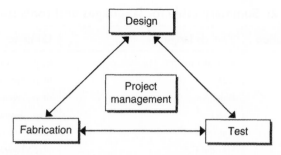

Figure 10.4: Design, fabrication, and testing

into logic. However, ESL design is required to overcome limitations of behavioral-level modeling by working at higher levels of abstraction and complexity.

The design aspects considered thus far constitute only one part of a three-part story: design, fabrication, and testing as shown in Figure 10.4. In the center of the triangle is project management, which provides the necessarily logistic support.

To work efficiently, ESL design methodology must integrate all three parts in a seamless and easy-to-use manner. For example, with the complexity of digital system-on-a-chip (SoC) designs being developed today, the test circuit is integrated into the design as part of the DfT strategy.

Table 10.2 provides a list of the main tools and languages currently available today, including source information. It is typical for the developers and suppliers to provide information such as product and language specifications, data sheets, and technical white papers that identify the principles of operation.

Everyone can see the usefulness of electronic systems in everyday life. It is unusual, for example, for anyone in the developed world not to either own, use, or at least have seen a mobile phone. Such devices have been in everyday use for the last twenty years, and now, every couple of months, new phones with additional features become available. Figure 10.5 shows one example, the NEC e228 phone. The left view shows the front of the phone (the user interface). As this is a portable device, the phone operates on a battery, shown on the right view of the side, which is rechargeable and lasts for a number of hours on a charge (the 1100 mAh rated lithium-ion battery provides 3.7 V DC). The main functions of the user interface are:

- keypad for hand input

- microphone for audio input

Table 10.2: Summary table for languages and tools used in ESL

Company or organization	Tool or language	Company URL
Agilent Technologies	Agilent Ptolemy	http://eesof.tm.agilent.com
Altera	C2H	http://www.altera.com
Bluespec Inc.	BlueSpec	http://www.bluespec.com/
Cadence Design Systems	Spectre®	http://www.cadence.com
Mentor Graphics	Catapult-C®	http://www.mentor.com
Celoxica	Handel-C	http://www.celoxica.com/
Forte Design Systems	Cynthesiszer	http://www.forteds.com
Accellera Verilog Analog Mixed-Signal Group	Verilog®-A	http://www.eda.org/verilog-ams/
	Verilog®-AMS	http://www.eda.org/verilog-ams/
IEEE	Verilog®-HDL	http://www.ieee.org
	VHDL	http://www.ieee.org
	VHDL-AMS	http://www.ieee.org
	Language SystemVerilog	http://www.ieee.org
Impulse Accelerated Technologies	Impulse-C®	http://www.impulsec.com/
Inria	Estererel	http://www-sop.inria.fr
Maplesoft	Maple	http://www.maplesoft.com
Mentor Graphics	SystsemVision	http://www.mentor.com
National Instruments	LabVIEW™	http://www.national.com
Scilab	Scilab and toolboxes such as Scicos	http://www.scilab.org
Sun Microsystems	JAVA™	http://www.sun.com/
Open SystemC Initiative	System-C®	http://www.systemc.org
The Mathworks Inc.	MATLAB® and toolboxes/blocksets (such as Simulink®)	http://www.themathworks.com
Wolfram Research	Mathematica	http://www.wolfram.com
International Organization for Standardization	ANSI-C	http://www.iso.org
	ANSI-C++	http://www.iso.org
UML Object Management Group	UML	http://www.uml.org/
University of California, Irvine	SpecC	http://www.ics.uci.edu
University of Kansas	Rosetta	http://wwwsldl.org
University of California at Berkeley	SPICE	http://bwrc.eecs.berkeley.edu/Classes/IcBook/SPICE/
Xilinx®	Platform Studio ISE and EDK (Embedded Development Kit)	http://www.xilinx.com http://www.xilinx.com

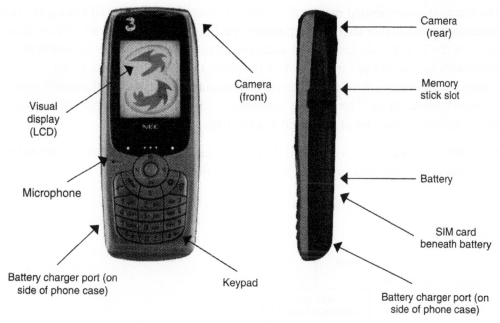

Figure 10.5: Mobile phone (portable electronics example) (Images courtesy of NEC, Copyright © NEC 2001–2004, no longer in stock)

- camera for visual input

- speaker for audio output

- LCD (liquid crystal display) for visual output

- battery for power supply

- port for battery recharge

- SIM card for personalizing the system

It is a useful exercise simply to take a look at the electronics in everyday use and to consider what functions they undertake, how they look, and then to consider what design decisions led to this.

As another example, consider a conceptual design that uses four separate processors. Each processor has associated memory (RAM and ROM). Part of the RAM used by the processor is available only to itself, whereas other parts of the RAM are shared.

The memory is distributed memory, so called because it is distributed among the main functional parts rather than being concentrated in one memory block. Given that some of the RAM is shared, RAM access control is needed to prevent multiple processors accessing the same RAM at the same time. Each processor can communicate with each of the other processors and with external electronic systems via wired connections. The wireless communications link is accessible only by one of the processors. This idea is shown in Figure 10.6.

The question arises, how would this design be realized and why? Asking the right questions is the first step toward a working solution.

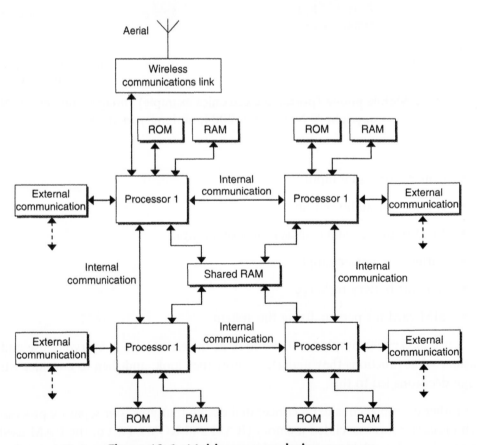

Figure 10.6: Multiprocessor design concept

10.3 Case Study 1: DC Motor Control

10.3.1 Introduction

In this case study, a digital controller is developed to control the speed of a DC electric motor. The overall control system model will be developed in MATLAB® [9] and its Simulink® [10] toolbox. The model of the control algorithm will then be manually converted to VHDL code using a set design translation flow for implementation as a digital controller using a CPLD. The design issues will be captured and presented in a way that allows the VHDL code to be generated automatically. The overall design flow is shown in Figure 10.7.

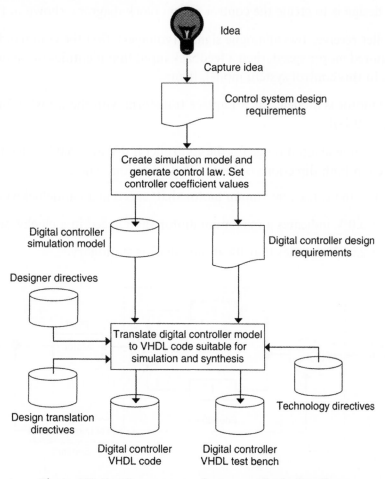

Figure 10.7: Motor control case study design flow

10.3.2   Motor Control System Overview

The control system is a closed-loop controller using PI (proportional plus integral) control [11, 12]. Other forms of control algorithm such as PID (proportional plus integral plus derivative) could be used, but the added complexity is unnecessary in this case. The particular control algorithm was chosen based on the requirements of the motor (the plant to control) and the required system response. As such, PI control provides zero steady-sate error in the motor speed (a motor speed steady-state error would exist if only proportional control was used) and a design simple to implement and easy to understand. The coefficients of each action within the PI control law are set to give a response that settles to a steady state in an adequately short time. The initial step in the design is to create the control system block diagram, shown in Figure 10.8.

The controller receives two analogue signals (voltages): first the command input that sets the required motor speed, then a feedback input that identifies the actual speed of the motor. In this control system model, then:

- The motor is modeled as a Laplace transform with the transfer function $[1/(1 + 0.1s)]$.

- The analogue input range for the controller is 0 V to $+5.0$ V, which indicates a speed in both directions of motor shaft rotation, where:

  ○ 0 V indicates a maximum motor shaft speed in an anticlockwise direction.

  ○ $+5.0$ V indicates a maximum motor shaft speed in a clockwise direction.

  ○ $+2.5$ V indicates that the motor shaft is stationary.

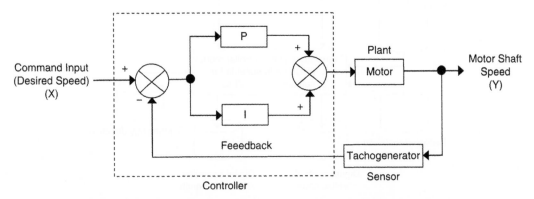

Figure 10.8: Motor control system example with PI control

- The proportional action (Kp) gain is $+2.0$, and the integral action gain (Ki) is $+8.0$ (not optimized).

- This is a high-level behavioral model (and a linear model of the system) that does not take into account nonlinear effects such as value limits, slew rate effects, and any existing motor dead zone.

- The motor model contains a tachogenerator (sensor) that produces an analogue voltage output in the range 0 V to $+5.0$ V.

- The command input (required speed) and actual motor speed outputs are considered to be voltages, and the motor shaft speed uses suitable units (e.g., rads/sec).

- The model uses the built-in Simulink® library continuous time blocks, and no design hierarchy has been developed.

- The digital controller is required to sample analogue signals and to undertake digital signal processing on the discrete time samples. The sampling frequency for this design is 100 Hz, a slow sampling frequency compared to many control systems, but adequate for this application.

- The model uses only continuous time blocks, so when the digital controller is created, the analogue model prototype must be converted to a digital approximation. Those parts of the controller to be mapped to a digital algorithm modeled in VHDL must therefore be identified.

The motor model used is a simple first-order Laplace transform that models the motor and tachogenerator as a single unit. This was created by monitoring the tachogenerator output voltage to a step change in motor speed command input voltage. This is reasonably representative of the motor reaction to larger step changes in command input, but does not model nonideal characteristics such as a motor dead zone around a null (zero) command input and the need to minimize the command input voltage required for the motor to react to a command input change.

A full analysis of the control system is undertaken to determine that the derived control algorithm is suitable for the application. This analysis is not, however, covered in this text.

At this level of design abstraction (i.e., a simplified model of the system), none of the implementation issues have been considered and only a mathematical model of the system exists. But of course, ultimately, the system must be built using electronic circuits. The basic arrangement created for such a control system is shown in Figure 10.9. Here, the CPLD implements the digital control algorithm and interfaces

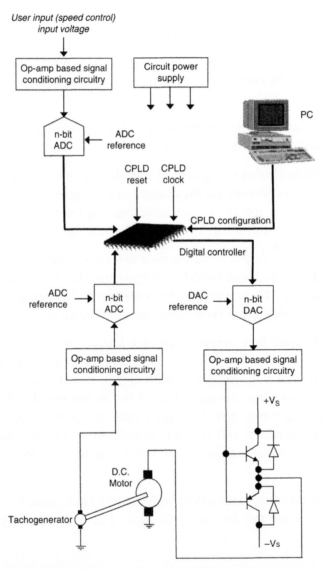

Figure 10.9: Motor speed control circuit arrangement

to two ADCs (analogue-to-digital converters, to sample the analogue input voltages for the command input and the feedback) and one DAC (digital-to-analogue converter, to output an analogue voltage to create the motor voltage). This DAC output voltage is applied to a transistor power amplifier (because the DAC would not be able to provide the necessary voltage and current levels required by the DC motor). Op-amp based analogue circuitry is used on the ADC inputs and DAC outputs as necessary to provide specific low-power analogue signal conditioning. A power supply unit provides the necessary voltage and current levels required by the overall circuit. Finally, a PC is used here to configure the CPLD.

10.3.3   MATLAB®/Simulink® Model Creation and Simulation

Before considering how controller is to be implemented, the control law (algorithm) must be developed and analyzed. An example Simulink® model for this system is shown in Figure 10.10.

The controller is placed within a single block (the controller block), and the motor (motor model) is modeled using as a first-order system a Laplace transform equation. The motor model also contains the tachogenerator output, so the output from the

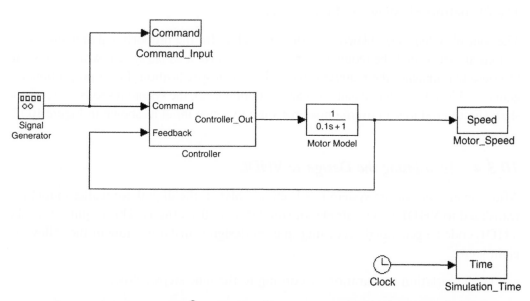

Figure 10.10: Simulink® model for the motor control system case study

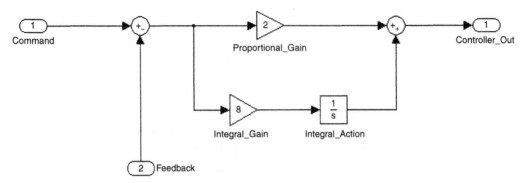

Figure 10.11: Simulink® model for the PI controller

system is modeled as the tachogenerator voltage (which represents the motor shaft speed). This equation was obtained from the motor itself by applying directly a step input voltage to the motor and observing the tachogenerator voltage. A signal generator (signal generator block) allows different signals to be applied to the system.

The design is analyzed using both hand calculations and the Simulink® simulator, with typical analogue input signals (step, sine wave, DC, triangle, and ramp) as part of the overall system analysis routine. A frequency response could be undertaken by generating a model for frequency analysis in MATLAB®.

The PI controller is shown in Figure 10.11.

The control system is simulated, and the gain values for the proportional and integral actions are set so that the required response is obtained: a stable system with a transient response that matches the requirements of the design specification. For a proportional gain of +2.0 and an integral gain of +8.0 (not optimized), the system response (i.e., motor shaft speed) produces an overdamped response to a step input as shown in Figure 10.12.

10.3.4   Translating the Design to VHDL

After the analysis of the system has been completed, the digital controller model is translated to VHDL code suitable for simulation and synthesis. This requires that the VHDL code be generated according to a set design translation flow in the following eight steps:

1. Translation preparation (according to the nine steps below).

2. Set the architecture details (according to the six steps below).

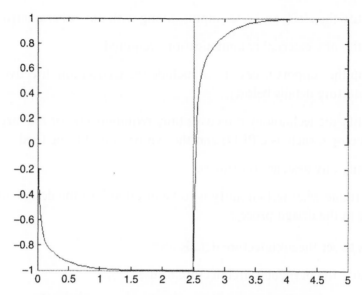

Figure 10.12: Simulation results for step change

3. Translation from Simulink® model to VHDL code by reading the Simulink® model, extracting the necessary design information, and generating the VHDL code.

4. Generate VHDL test bench.

5. Simulate the VHDL code and check for correct operation to validate the operation of the generated VHDL code.

6. Synthesize the VHDL code and resimulate the design to generate a structural design based on the particular target technology.

7. Configure the CPLD and validate the operation of the design.

8. Use the controller.

The nine steps of translation preparation are:

1. Identify the parts to be translated into digital (the controller).

2. Remove any unnecessary information, leaving only the controller model.

3. Identify the digital controller interfacing.

4. Identify the clock and reset inputs, along with any other control signals.

5. Identify any external communications required.

6. Set up the support necessary to include the translation directives (see architecture details below).

7. Identify the technology directives (any requirements for the target technology, such as CPLD) and the synthesis tool to be used.

8. Identify any designer directives.

9. Determine what test circuitry is to be inserted into the design and at what stage in the design process.

The six steps to set the architecture details are:

1. Identify the particular architecture to use.

2. Identify the internal wordlength within the digital signal processing part of the digital core.

3. Identify any specific circuits to avoid (e.g., specific VHDL code constructs).

4. Identify the control signals required by the I/O.

5. Identify the number system to use (e.g., 2s complement) in the arithmetic operations.

6. Identify any number scaling requirements to limit the required wordlength within the design.

The model translation must initially consider the architecture to use either a processor-based architecture running a software application (standard fixed architecture processor or a configurable processor) or a custom hardware architecture based directly on the model. This idea is shown in Figure 10.13.

If the translation is to be performed manually, this can be undertaken by visual reference to the graphical representation of the model (i.e., the block diagram). If the translation is to be performed automatically (by a software application), the translation can be performed using the underlying text based model (i.e., with the Simulink®.*mdl* file).

A fixed architecture processor is based on an existing CISC or RISC architecture, and its translation either will generate the hardware design (in HDL) and the processor

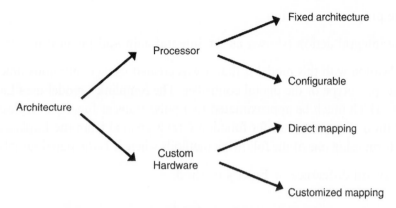

Figure 10.13: Controller architecture decisions

microcode together, or will generate only the processor microcode using an existing processor design. The configurable processor is a processor design that dynamically changes specific aspects of the architecture based on the particular application.

Direct mapping starts with the model as presented and directly translates its functions to a custom hardware HDL code equivalent. Customized mapping uses custom architecture based on the model, but then determines the most appropriate way to implement its functions (e.g., by using multiple multiplication blocks or a single multiplexed multiplier block) based on the application.

No matter what particular architecture is chosen, in addition to generating the required digital signal processing algorithm hardware (as identified in the system block diagram), then there would be the need to also generate the necessary interfacing signals for external circuitry such as ADCs and DACs, and the internal timing signals for the control of the signal processing operations, along with the storage and movement of data signals within the design. These interfacing and internal timing signals would need to be created by an additional circuit creating the functions of a *control unit* particular to the design.

In this case study, direct mapping of model functions will be considered, so the controller shown in Figure 10.11 will be translated. This requires the use of the following main functional blocks:

- one subtraction block
- one addition block

- one proportional action

- one integral action (shown as the integral gain and integrator action blocks)

One complication with this model is that it was created using continuous time blocks as an analogue prototype of the digital controller. The Simulink® model uses Laplace transforms, which much be approximated to a pulse transfer function for discrete time implementation. The pulse transfer function G(z) is created from the Laplace (s) transform form using one of the following methods where T is the signal sampling period:

1. Forward difference or Euler's method:

$$s = \frac{z-1}{T}$$

2. Backward difference method:

$$s = \frac{z-1}{zT}$$

3. Tustin's approximation (also referred to as the bilinear transform):

$$s = \frac{2}{T} \cdot \frac{z-1}{z+1}$$

These methods are readily applied by hand to transform from s to z.

In this case study, Tustin's approximation is used. It applies only to the integral action since the proportional action is simply a multiplication on the sampled data. The proportional action (using Z-transforms) is:

$$P(z) = Kp.X(z)$$

The integral action (using Z-transforms) is:

$$I(z) = \left(\left(\frac{KiT}{2} \right) \left(x(z) + x(z)z^{-1} \right) \right) + I(z)z^{-1}$$

The PI controller block diagram can be remodeled using Z-transforms, as shown in Figure 10.14. The two storage (z^{-1}) blocks have a common clock signal that controls when the inputs to the blocks are stored. This control signal must be created. The

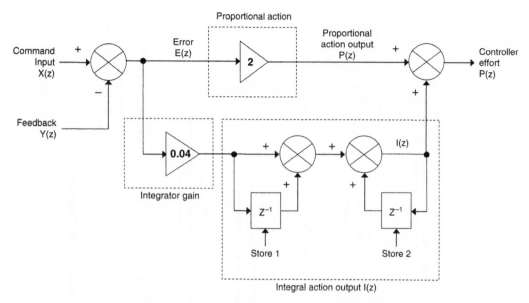

Figure 10.14: Discrete time PI controller

controller block shown in Figure 10.14 forms the digital signal processing core of the overall controller design. Figure 10.15 shows this core along with the necessary control unit that generates the internal control signals based on the timing requirements of the controller. The inputs to the controller are sampled at a sampling frequency of 100 Hz; this timing is generated from a master input clock. After the algorithm has been run on the current input signal (and previous inputs along with previous outputs), the current output is updated. Because these actions are performed in less time than the 100 Hz sampling frequency allows, the design must wait until the next sample is required. This idea is shown in Figure 10.16.

Signed arithmetic is used inside the control algorithm hardware (2s complement in this case study). To achieve this, and given that the input is straight binary, the sampled value must be stored (in a register) and converted to a 2s complement number, as shown in Table 10.3.

Finally, the interconnects between the main functional blocks must be considered. The inputs are analogue inputs sampled using two AD7575 eight-bit LC^{2}MOS (leadless chip carrier metal oxide semiconductor) successive approximation ADCs [13]. The output is an analogue signal created using a single AD7524 eight-bit buffered multiplying DAC [14]. The internal wordlength is 16 bits, so the eight-bit input and analogue output is transformed from 16-bit input and output. The eight-bit

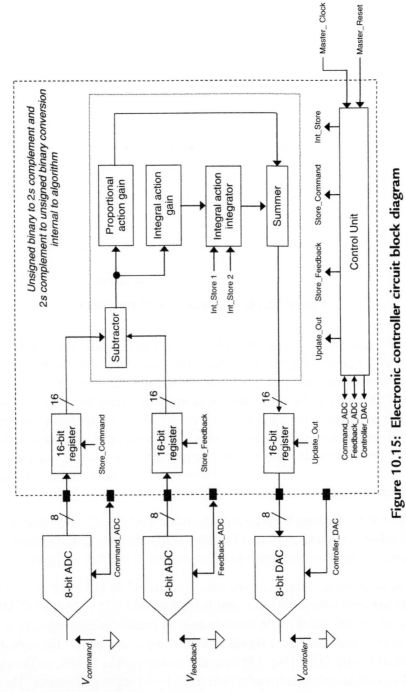

Figure 10.15: Electronic controller circuit block diagram

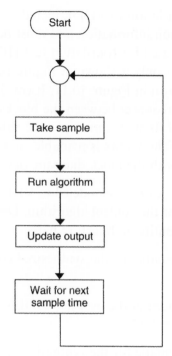

Figure 10.16: PI controller operation flowchart

output circuitry must also include value limiting because the 16-bit internal value exceeds the value limits set by the eight-bit output.

The Simulink® model for the overall control system must be reviewed and should contain:

- information for translation to VHDL

- information *not* for translation to VHDL

Table 10.3: Binary I/O to internal value mapping

Digital I/O code, decimal	Digital I/O code (8-bits), binary		Internal code, decimal	Internal code (8-bits), binary
0	00000000	→	−128	10000000
127	01111111	→	−1	11111111
128	10000000	→	0	00000000
255	11111111	→	+127	01111111

The information *not* for translation to VHDL includes information such as visual attributes and software version information and must be stripped from the representation of the model used for translation to VHDL. The Simulink® model code for the controller only is shown in Figure 10.17. This is the text description of the model shown in Figure 10.11. It consists of the blocks used, their attributes, and the interconnect between the blocks (lines). Interpreting this model requires knowledge of its syntax and how the values that can be modified by the user are represented. The syntax is readable, and the names used are identifiable by comparison with the block diagram view.

This model for the controller can be remodeled in VHDL, shown in Figure 10.18 as a structural description for the control algorithm. Detailed operation of each block is defined in separate entity-architecture pairs.

The Xilinx ISE™ RTL schematic for the synthesized controller design is shown in Figure 10.19.

This control algorithm is placed in the overall VHDL structural description of the controller, as shown in Figure 10.20.

The Xilinx ISE™ RTL schematic for the synthesized controller design is shown in Figure 10.21.

The final step is to generate and simulate a VHDL test bench for the controller. An example VHDL test bench is shown in Figure 10.22.

10.3.5   Concluding Remarks

This case study design was for a simple digital control algorithm, but it also shows the main operations required for typical digital control algorithms. The VHDL code to implement the design within a CPLD was created by mapping the original Simulink® block diagram to a VHDL code equivalent in which each of the main functional blocks was presented as a unique entity-architecture pair. The structural design of the controller top level and the control algorithm were presented, although the details of the individual operations are left for the reader to implement.

The block diagram was mapped directly to VHDL to implement a custom hardware design. In many cases, this would result in a large design, particularly when multiple multiplications are necessary. However here, the ease and rapid development of the

```
1    System {
2        Name                      "Controller"
3
4        Block {
5            BlockType             Inport
6            Name                  "Command"
7            Position              [20, 73, 50, 87]
8        }
9        Block {
10           BlockType             Inport
11           Name                  "Feedback"
12           Position              [140, 225, 170, 240]
13           Orientation           "up"
14           Port                  "2"
15       }
16       Block {
17           BlockType             Integrator
18           Name                  "Integral_Action"
19           Ports                 [1, 1]
20           Position              [355, 155, 385, 185]
21       }
22       Block {
23           BlockType             Gain
24           Name                  "Integral_Gain"
25           Position              [260, 155, 290, 185]
26           Gain                  "8"
27       }
28       Block {
29           BlockType             Gain
30           Name                  "Proportional_Gain"
31           Position              [285, 65, 315, 95]
32           Gain                  "2"
33       }
34       Block {
35           BlockType             Sum
36           Name                  "Sum"
37           Ports                 [2, 1]
38           Position              [145, 70, 165, 90]
39           ShowName              off
40           IconShape                     "round"
41           Inputs                "|+-"
42           InputSameDT           off
43           OutDataTypeMode       "Inherit via internal rule"
44       }
45       Block {
46           BlockType             Sum
47           Name                  "Sum1"
48           Ports                 [2, 1]
49           Position              [430, 70, 450, 90]
             ShowName              off
             IconShape                     "round"
             Inputs                "|++"
             InputSameDT           off
             OutDataTypeMode       "Inherit via internal rule"
         }
```

Figure 10.17: Simulink® model for the PI controller

```
50      Block {
51        BlockType              Outport
52        Name                   "Controller_Out"
53        Position               [525, 73, 555, 87]
54      }
55      Line {
56        SrcBlock               "Sum"
57        SrcPort                1
58        Points                 [0, 0; 45, 0]
59        Branch {
60          DstBlock               "Proportional_Gain"
61          DstPort                1
62        }
63        Branch {
64          Points                 [0, 90]
65          DstBlock               "Integral_Gain"
66          DstPort                1
67        }
68      }
69      Line {
70        SrcBlock               "Integral_Gain"
71        SrcPort                1
72        DstBlock               "Integral_Action"
73        DstPort                1
74      }
75      Line {
76        SrcBlock               "Proportional_Gain"
77        SrcPort                1
78        DstBlock               "Sum1"
79        DstPort                1
80      }
81      Line {
82        SrcBlock               "Integral_Action"
83        SrcPort                1
84        Points                 [50, 0]
85        DstBlock               "Sum1"
86        DstPort                2
87      }
88      Line {
89        SrcBlock               "Command"
90        SrcPort                1
91        DstBlock               "Sum"
92        DstPort                1
93      }
94      Line {
95        SrcBlock               "Feedback"
96        SrcPort                1
97        DstBlock               "Sum"
98        DstPort                2
99      }
100     Line {
101       SrcBlock               "Sum1"
102       SrcPort                1
103       DstBlock               "Controller_Out"
104       DstPort                1
105     }
        }
```

Figure 10.17: (Continued)

```
1   LIBRARY ieee;
2   USE ieee.std_logic_1164.all;
3   USE ieee.std_logic_arith.all;
4   USE ieee.std_logic_unsigned.all;
5
6
7   ENTITY Control_Algorithm IS
8    PORT ( Command            : IN    STD_LOGIC_VECTOR (15 downto 0);
9           Feedback           : IN    STD_LOGIC_VECTOR (15 downto 0);
10          Controller_Out     : OUT   STD_LOGIC_VECTOR (15 downto 0);
11          Integrator_Store_1 : IN    STD_LOGIC;
12          Integrator_Store_2 : IN    STD_LOGIC;
13          Reset              : IN    STD_LOGIC);
14   END ENTITY Control_Algorithm;
15
16
17   ARCHITECTURE Structural OF Control_Algorithm IS
18
19
20   SIGNAL    Error        :   STD_LOGIC_VECTOR (15 downto 0);
21   SIGNAL    Proportional :   STD_LOGIC_VECTOR (15 downto 0);
22   SIGNAL    Int_1        :   STD_LOGIC_VECTOR (15 downto 0);
23   SIGNAL    Int_2        :   STD_LOGIC_VECTOR (15 downto 0);
24   SIGNAL    Int_3        :   STD_LOGIC_VECTOR (15 downto 0);
25   SIGNAL    Int_4        :   STD_LOGIC_VECTOR (15 downto 0);
26   SIGNAL    Integral     :   STD_LOGIC_VECTOR (15 downto 0);
27
28
29   COMPONENT Adder IS
30    PORT ( Data_In_1 : IN    STD_LOGIC_VECTOR (15 downto 0);
31           Data_In_2 : IN    STD_LOGIC_VECTOR (15 downto 0);
32           Data_Out  : OUT   STD_LOGIC_VECTOR (15 downto 0));
33   END COMPONENT Adder;
34
35   COMPONENT Subtractor IS
36    PORT ( Data_In_1 : IN    STD_LOGIC_VECTOR (15 downto 0);
37           Data_In_2 : IN    STD_LOGIC_VECTOR (15 downto 0);
38           Data_Out  : OUT   STD_LOGIC_VECTOR (15 downto 0));
39   END COMPONENT Subtractor;
40
41   COMPONENT Integral_Gain IS
42    PORT ( Data_In  : IN    STD_LOGIC_VECTOR (15 downto 0);
43           Data_Out : OUT   STD_LOGIC_VECTOR (15 downto 0));
44   END COMPONENT Integral_Gain;
45
46   COMPONENT Proportional_Gain IS
47    PORT ( Data_In  : IN    STD_LOGIC_VECTOR (15 downto 0);
48           Data_Out : OUT   STD_LOGIC_VECTOR (15 downto 0));
49   END COMPONENT Proportional_Gain;
50
```

Figure 10.18: VHDL model for the control algorithm

```
51   COMPONENT Delay IS
52    PORT ( Data_In   :   IN    STD_LOGIC_VECTOR (15 downto 0);
53           Data_Out  :   OUT   STD_LOGIC_VECTOR (15 downto 0);
54           Store     :   IN    STD_LOGIC;
55           Reset     :   IN    STD_LOGIC);
56   END COMPONENT Delay;
57
58
59   BEGIN
60
61
62   I1 : Subtractor
63        PORT MAP  ( Data_In_1  =>   Command,
64                     Data_In_2  =>   Feedback,
65                     Data_Out   =>   Error);
66
67   I2 : Proportional_Gain
68        PORT MAP  ( Data_In    =>   Error,
69                     Data_Out   =>   Proportional);
70   I3 : Adder
71        PORT MAP  ( Data_In_1  =>   Proportional,
72                     Data_In_2  =>   Integral,
73                     Data_Out   =>   Controller_Out);
74
75   I4 : Integral_Gain
76        PORT MAP  ( Data_In    =>   Error,
77                     Data_Out   =>   Int_1);
78   I5 : Delay
79        PORT MAP  ( Data_In    =>   Int_1,
80                     Data_Out   =>   Int_2,
81                     Reset      =>   Reset,
82                     Store      =>   Integrator_Store_1);
83
84   I6 : Adder
85        PORT MAP  ( Data_In_1  =>   Int_1,
86                     Data_In_2  =>   Int_2,
87                     Data_Out   =>   Int_3);
88
89   I7 : Adder
90        PORT MAP  ( Data_In_1  =>   Int_3,
91                     Data_In_2  =>   Int_4,
92                     Data_Out   =>   Integral);
93
94   I8 : Delay
95        PORT MAP  ( Data_In    =>   Integral,
96                     Data_Out   =>   Int_4,
97                     Reset      =>   Reset,
98                     Store      =>   Integrator_Store_2);
99
100
101  END ARCHITECTURE Structural;
```

Figure 10.18: (Continued)

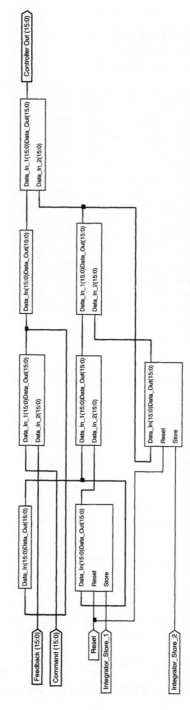

Figure 10.19: Digital control algorithm synthesis results (Coolrunner™-II CPLD)

```
1    LIBRARY ieee;
2    USE ieee.std_logic_1164.all;
3    USE ieee.std_logic_arith.all;
4    USE ieee.std_logic_unsigned.all;
5
6
7    ENTITY Controller IS
8      PORT ( Command_ADC_BUSY  : IN   STD_LOGIC;
9             Command_ADC_TP    : OUT  STD_LOGIC;
10            Command_ADC_RD    : OUT  STD_LOGIC;
11            Command_ADC_CS    : OUT  STD_LOGIC;
12            Command_ADC_Data  : IN   STD_LOGIC_VECTOR (7 downto 0);
13            Feedback_ADC_BUSY : IN   STD_LOGIC;
14            Feedback_ADC_TP   : OUT  STD_LOGIC;
15            Feedback_ADC_RD   : OUT  STD_LOGIC;
16            Feedback_ADC_CS   : OUT  STD_LOGIC;
17            Feedback_ADC_Data : IN   STD_LOGIC_VECTOR(7 downto 0);
18            Controller_DAC_WR : OUT  STD_LOGIC;
19            Controller_DAC_CS : OUT  STD_LOGIC;
20            Controller_Out    : OUT  STD_LOGIC_VECTOR(7 downto 0);
21            Master_Clock      : IN   STD_LOGIC;
22            Master_Reset      : IN   STD_LOGIC);
23   END ENTITY Controller;
24
25
26   ARCHITECTURE Structural OF Controller IS
27
28
29   SIGNAL Command_Int         :  STD_LOGIC_VECTOR (15 downto 0);
30   SIGNAL Feedback_Int        :  STD_LOGIC_VECTOR (15 downto 0);
31   SIGNAL Controller_Out_Int  :  STD_LOGIC_VECTOR (15 downto 0);
32   SIGNAL Store_Command       :  STD_LOGIC;
33   SIGNAL Store_Feedback      :  STD_LOGIC;
34   SIGNAL Update_Out          :  STD_LOGIC;
35   SIGNAL Integrator_Store_1  :  STD_LOGIC;
36   SIGNAL Integrator_Store_2  :  STD_LOGIC;
37
38
39   COMPONENT Control_Algorithm IS
40     PORT ( Command           : IN  STD_LOGIC_VECTOR(15 downto 0);
41            Feedback          : IN  STD_LOGIC_VECTOR(15 downto 0);
42            Controller_Out    : OUT STD_LOGIC_VECTOR(15 downto 0);
43            Integrator_Store_1 : IN  STD_LOGIC;
44            Integrator_Store_2 : IN  STD_LOGIC;
45            Reset             : IN  STD_LOGIC);
46   END COMPONENT Control_Algorithm;
```

Figure 10.20: VHDL model for the controller

```
47
48   COMPONENT Control_Unit IS
49      PORT ( Master_Clock            : IN    STD_LOGIC;
50             Master_Reset            : IN    STD_LOGIC;
51             Command_ADC_BUSY        : IN    STD_LOGIC;
52             Command_ADC_TP          : OUT   STD_LOGIC;
53             Command_ADC_RD          : OUT   STD_LOGIC;
54             Command_ADC_CS          : OUT   STD_LOGIC;
55             Feedback_ADC_BUSY       : IN    STD_LOGIC;
56             Feedback_ADC_TP         : OUT   STD_LOGIC;
57             Feedback_ADC_RD         : OUT   STD_LOGIC;
58             Feedback_ADC_CS         : OUT   STD_LOGIC;
59             Controller_DAC_WR       : OUT   STD_LOGIC;
60             Controller_DAC_CS       : OUT   STD_LOGIC;
61             Store_Command           : OUT   STD_LOGIC;
62             Store_Feedback          : OUT   STD_LOGIC;
63             Update_Out              : OUT   STD_LOGIC;
64             Integrator_Store_1      : OUT   STD_LOGIC;
65             Integrator_Store_2      : OUT   STD_LOGIC);
66   END COMPONENT Control_Unit;
67
68
69   COMPONENT Input_Register IS
70      PORT ( Data_In  : IN    STD_LOGIC_VECTOR (7 downto 0);
71             Data_Out : OUT   STD_LOGIC_VECTOR (15 downto 0);
72             Store    : IN    STD_LOGIC;
73             Reset    : IN    STD_LOGIC);
74   END COMPONENT Input_Register;
75
76
77   COMPONENT Output_Register IS
78      PORT ( Data_In  : IN    STD_LOGIC_VECTOR (15 downto 0);
79             Data_Out : OUT   STD_LOGIC_VECTOR (7 downto 0);
80             Store    : IN    STD_LOGIC;
81             Reset    : IN    STD_LOGIC);
82   END COMPONENT Output_Register;
83
84
85   BEGIN
86
87   I1 : Control_Algorithm
88      PORT MAP(  Command              =>   Command_Int,
89                 Feedback             =>   Feedback_Int,
90                 Controller_Out       =>   Controller_Out_Int,
91                 Integrator_Store_1   =>   Integrator_Store_1,
92                 Integrator_Store_2   =>   Integrator_Store_2,
```

Figure 10.20: (Continued)

```
93                      Reset               =>   Master_Reset);
94
95    I2 : Control_Unit
96        PORT MAP(  Master_Clock        =>   Master_Clock,
97                   Master_Reset        =>   Master_Reset,
98                   Command_ADC_BUSY    =>   Command_ADC_BUSY,
99                   Command_ADC_TP      =>   Command_ADC_TP,
100                  Command_ADC_RD      =>   Command_ADC_RD,
101                  Command_ADC_CS      =>   Command_ADC_CS,
102                  Feedback_ADC_BUSY   =>   Feedback_ADC_BUSY,
103                  Feedback_ADC_TP     =>   Feedback_ADC_TP,
104                  Feedback_ADC_RD     =>   Feedback_ADC_RD,
105                  Feedback_ADC_CS     =>   Feedback_ADC_CS,
106                  Controller_DAC_WR   =>   Controller_DAC_WR,
107                  Controller_DAC_CS   =>   Controller_DAC_CS,
108                  Store_Command       =>   Store_Command,
109                  Store_Feedback      =>   Store_Feedback,
110                  Update_Out          =>   Update_Out,
111                  Integrator_Store_1  =>   Integrator_Store_1,
112                  Integrator_Store_2  =>   Integrator_Store_2);
113
114   I3 : Input_Register
115       PORT MAP ( Data_In             =>   Command_ADC_Data,
116                  Data_Out            =>   Command_Int,
117                  Store               =>   Store_Command,
118                  Reset               =>   Master_Reset);
119
120
121   I4 : Input_Register
122       PORT MAP ( Data_In             =>   Feedback_ADC_Data,
123                  Data_Out            =>   Feedback_Int,
124                  Store               =>   Store_Feedback,
125                  Reset               =>   Master_Reset);
126
127   I5 : Output_Register
128       PORT MAP ( Data_In             =>   Controller_Out_Int,
129                  Data_Out            =>   Controller_Out,
130                  Store               =>   Update_Out,
131                  Reset               =>   Master_Reset);
132
133
134   END ARCHITECTURE Structural;
```

Figure 10.20: (Continued)

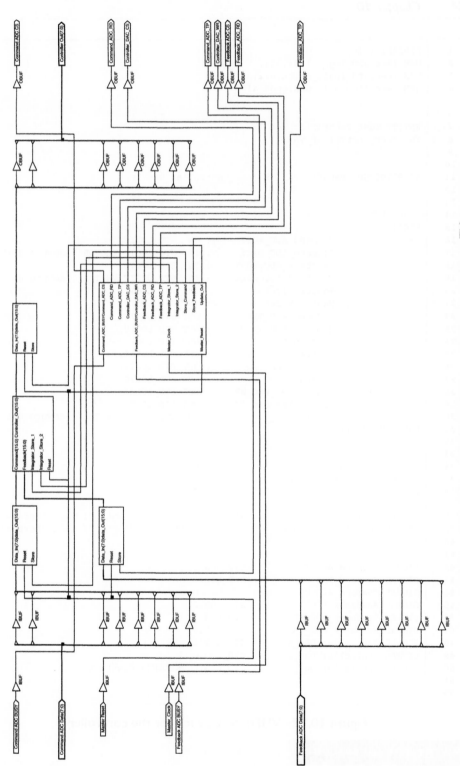

Figure 10.21: Digital controller synthesis results (Coolrunner™-II CPLD)

```
1    LIBRARY ieee;
2    USE ieee.std_logic_1164.all;
3    USE ieee.std_logic_arith.all;
4    USE ieee.std_logic_unsigned.all;
5
6
7    ENTITY Test_Controller_vhd IS
8    END Test_Controller_vhd;
9
10
11   ARCHITECTURE Behavioural OF Test_Controller_vhd IS
12
13
14   COMPONENT Controller
15   PORT(
16                   Command_ADC_BUSY      : IN  STD_LOGIC;
17                   Command_ADC_Data      : IN  STD_LOGIC_VECTOR(7 downto 0);
18                   Feedback_ADC_BUSY     : IN  STD_LOGIC;
19                   Feedback_ADC_Data     : IN  STD_LOGIC_VECTOR(7 downto 0);
20                   Master_Clock          : IN  STD_LOGIC;
21                   Master_Reset          : IN  STD_LOGIC;
22                   Command_ADC_TP        : OUT STD_LOGIC;
23                   Command_ADC_RD        : OUT STD_LOGIC;
24                   Command_ADC_CS        : OUT STD_LOGIC;
25                   Feedback_ADC_TP       : OUT STD_LOGIC;
26                   Feedback_ADC_RD       : OUT STD_LOGIC;
27                   Feedback_ADC_CS       : OUT STD_LOGIC;
28                   Controller_DAC_WR     : OUT STD_LOGIC;
29                   Controller_DAC_CS     : OUT STD_LOGIC;
30                   Controller_Out        : OUT STD_LOGIC_VECTOR(7 downto 0));
31   END COMPONENT;
32
33
34   SIGNAL Command_ADC_BUSY    :  STD_LOGIC := '0';
35   SIGNAL Feedback_ADC_BUSY   :  STD_LOGIC := '0';
36   SIGNAL Master_Clock        :  STD_LOGIC := '0';
37   SIGNAL Master_Reset        :  STD_LOGIC := '0';
38   SIGNAL Command_ADC_Data    :  STD_LOGIC_VECTOR(7 downto 0)  := (others=>'0');
39   SIGNAL Feedback_ADC_Data   :  STD_LOGIC_VECTOR(7 downto 0)  := (others=>'0');
40
41
42   SIGNAL Command_ADC_TP      :  STD_LOGIC;
43   SIGNAL Command_ADC_RD      :  STD_LOGIC;
44   SIGNAL Command_ADC_CS      :  STD_LOGIC;
45   SIGNAL Feedback_ADC_TP     :  STD_LOGIC;
46   SIGNAL Feedback_ADC_RD     :  STD_LOGIC;
47   SIGNAL Feedback_ADC_CS     :  STD_LOGIC;
48   SIGNAL Controller_DAC_WR   :  STD_LOGIC;
49   SIGNAL Controller_DAC_CS   :  STD_LOGIC;
50   SIGNAL Controller_Out      :  STD_LOGIC_VECTOR(7 downto 0);
51
52
```

Figure 10.22: VHDL test bench for the controller

```
53      BEGIN
54
55
56      uut: Controller PORT MAP(
57                      Command_ADC_BUSY    =>   Command_ADC_BUSY,
58                      Command_ADC_TP      =>   Command_ADC_TP,
59                      Command_ADC_RD      =>   Command_ADC_RD,
60                      Command_ADC_CS      =>   Command_ADC_CS,
61                      Command_ADC_Data    =>   Command_ADC_Data,
62                      Feedback_ADC_BUSY   =>   Feedback_ADC_BUSY,
63                      Feedback_ADC_TP     =>   Feedback_ADC_TP,
64                      Feedback_ADC_RD     =>   Feedback_ADC_RD,
65                      Feedback_ADC_CS     =>   Feedback_ADC_CS,
66                      Feedback_ADC_Data   =>   Feedback_ADC_Data,
67                      Controller_DAC_WR   =>   Controller_DAC_WR,
68                      Controller_DAC_CS   =>   Controller_DAC_CS,
69                      Controller_Out      =>   Controller_Out,
70                      Master_Clock        =>   Master_Clock,
71                      Master_Reset        =>   Master_Reset);
72
73
74      Reset_Process : PROCESS
75          BEGIN
76
77          Wait for 0 ns;  Master_Reset <= '0';
78          Wait for 5 ns;  Master_Reset <= '1';
79          Wait;
80
81      END PROCESS;
82
83
84      Clock_Process : PROCESS
85          BEGIN
86
87          Wait for 0 ns;  Master_Clock <= '0';
88          Wait for 10 ns; Master_Clock <= '1';
89          Wait for 10 ns; Master_Clock <= '0';
90
91      END PROCESS;
92
93
94      ADC_Data_Process : PROCESS
95          BEGIN
96
97          Wait for 0 ns;    Command_ADC_Data <= "00000000";
98                            Feedback_ADC_Data <= "00000000";
99          Wait;
100
101     END PROCESS;
102
103
104     ADC_Busy_Process : PROCESS
```

Figure 10.22: (Continued)

```
105          BEGIN
106
107          Wait for 0 ns;     Command_ADC_BUSY <= '0';
108                             Feedback_ADC_BUSY <= '0';
109          Wait;
110
111    END PROCESS;
112
113
114    END ARCHITECTURE Behavioural;
```

Figure 10.22: (Continued)

VHDL code by a direct mapping for this small design reduced the design time. The multiplications were undertaken within the *Proportional_Gain* and *Integrator_Gain* blocks. The design can use either a full 16×16 multiplier design or a shift-and-add approach. Given that the multiplications are fixed and relatively simple, a full multiplier design can be expected to produce a larger hardware design than necessary.

An internal wordlength of 16 bits is used in this case study and must be considered in the calculations performed. Where the potential for number overflow existed, this was prevented either by ensuring that the internal values are never large enough to create an overflow, or if an overflow situation does occur, by saturating the output from a computation to the limits set by the wordlength. The internal multiplication within the integrator gain also produces a number with integer and fractional parts. Therefore, for a fixed-point calculation, the lower part of the 16-bit wordlength must be used to represent the fractional part, and the upper part must be used to represent the integer part. Placing the decimal point in the number is a design decision. If the finite wordlength creates errors in calculations, that information is fed back to the original simulation model for the control system and used to modify the controller.

10.4   Case Study 2: Digital Filter Design

10.4.1   *Introduction*

Digital filters perform the operations of addition, subtraction, multiplication, and division on sampled data. Among the types of digital filter are the infinite impulse response (IIR) filter, the finite impulse response (FIR) filter [15], and the computationally efficient cascaded integrator comb (CIC) filter [16]. The CIC filter is widely used in decimation and interpolation in communications systems:

- *Decimation* is the process of sample rate reduction. Where a signal is sampled at a particular sampling rate, a decimator reduces the original sample rate (f_s) to a lower rate (f_s/M). For example, if a signal is sampled at 10 kHz and M = 5, a decimator outputs a value once every M samples and discards the other (M – 1) samples. When M = 5, the sample rate is reduced from 10 kHz to 2 kHz.

- *Interpolation* is the process of sample rate increase. Where a signal is sampled at a particular sampling rate, the interpolation process increases the sample rate (f_s) to a higher rate (Lf_s). For example, if a signal is sampled at 10 kHz and L = 5, an interpolator outputs a value at an increased rate of 50 kHz. The input sample is output once every L output values, and the interpolator will fill the remaining (L – 1) output values with a zero value.

This idea is shown in Figure 10.23.

Decimation and interpolation functions are used in communications systems and in circuits such as the digital signal conditioning circuitry within sigma-delta modulator architecture ADCs and DACs. For example, CIC filters are suited for digital anti-aliasing filtering prior to decimation; a typical arrangement is shown in Figure 10.24. Here, the input is applied to the CIC filter, and the output from the CIC filter is applied to an FIR filter.

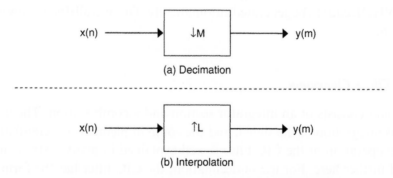

Figure 10.23: Decimation and interpolation

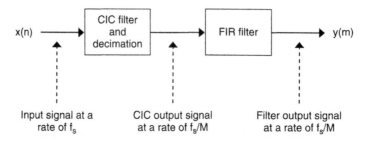

Figure 10.24: CIC filter in decimation

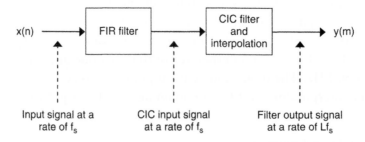

Figure 10.25: CIC filter in interpolation

A CIC filter used in interpolation is shown in Figure 10.25. Here, the input is applied to the FIR filter, and the output from the FIR filter is applied to an CIC filter.

In this case study, a third-order digital CIC filter will be developed to filter a single-bit bitstream pattern. The overall filter model will be developed in MATLAB® [9] and its Simulink® toolbox [10]. The model of the filter algorithm will then be manually converted to VHDL code using a set design translation flow for implementation as a digital filter using a CPLD. The design issues will be captured and presented in a way that allows the VHDL code to be generated automatically. The overall design flow is shown in Figure 10.26.

10.4.2 Filter Overview

The CIC filter consists of an integrator section and a comb section. The integrator implements integration of the signal, and the comb implements differentiation on the signal. The operation of the CIC filter is well explained in many texts, so it is not considered further here. For use in decimation, the CIC filter has the form shown in Figure 10.27. This design is for a third-order CIC filter with three integrator and three

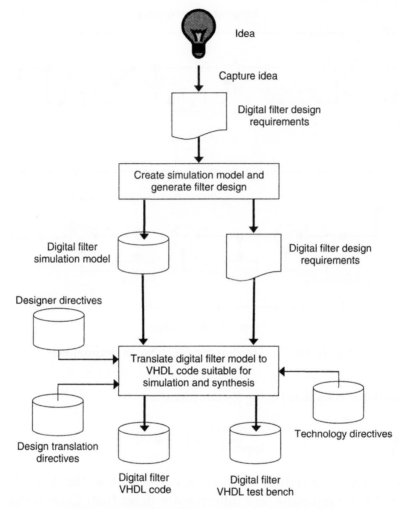

Figure 10.26: CIC filter case study design flow

comb circuits. (A fourth-order CIC filter would use four integrators and four comb circuits, etc.) Variations on this basic structure are possible. Note that the decimator is placed between the integrator and comb parts of the design.

The integrator is modeled using Z-transforms as:

$$\frac{\text{Output}(z)}{\text{Input}(z)} = \left(\frac{1}{1 - z^{-1}} \right)$$

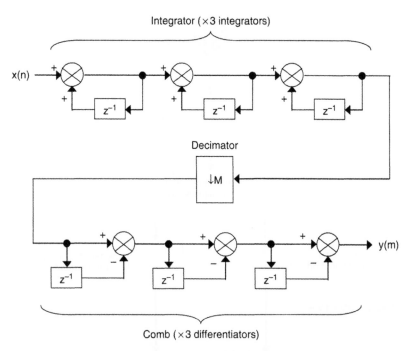

Figure 10.27: Third-order CIC filter in decimation

The differentiator is modeled using Z-transforms as:

$$\frac{\text{Output(z)}}{\text{Input(z)}} = \left(1 - z^{-1}\right)$$

Each delay block has a control signal to store the input to the delay. Note that these forms for the integrator and differentiator differ from those presented in Chapter 7.

10.4.3   MATLAB®/Simulink® Model Creation and Simulation

Before considering how the controller is to be implemented, the algorithm must be developed and analyzed. An example Simulink® model for this system is shown in Figure 10.28.

Here, the CIC filter is separated into the integrator and differentiator parts. Within the integrators, the inputs are sampled at a sampling rate of f_s. Within

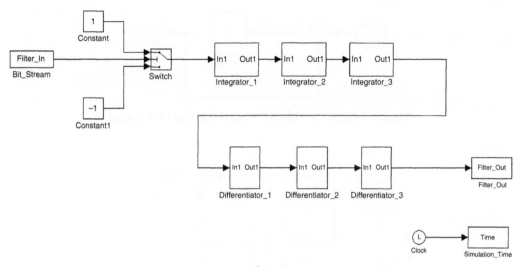

Figure 10.28: Simulink® model for the CIC filter

the differentiators, the inputs are sampled at a sampling rate of (f_s/M). In this model, the CIC filter is intended for use in the digital signal conditioning circuitry within a sigma-delta ADC design. A single-bit bitstream pattern (Filter_In) is applied to the filter input, and a 16-bit output from the CIC filter (Filter_Out) is created. This is achieved by a switch block at the input of the filter such that:

- When the input is a logic 0, then using 2s complement arithmetic, a value of -1_{10} is applied to the filter input.

- When the input is a logic 1, then using 2s complement arithmetic, a value of $+1_{10}$ is applied to the filter input.

The integrator design is shown in Figure 10.29, and the differentiator design is shown in Figure 10.30.

The design is analyzed using both hand calculations and the Simulink® simulator, with typical bitstream patterns representing different signal frequencies as part of the overall system analysis routine. A frequency response could be undertaken by generating a model for frequency analysis in MATLAB®.

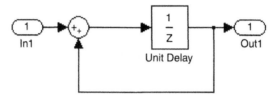

Figure 10.29: Simulink® model for the integrator

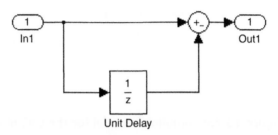

Figure 10.30: Simulink® model for the differentiator

10.4.4 Translating the Design to VHDL

After system analysis of the system has been completed, the digital filter model is translated to VHDL code suitable for simulation and synthesis. This requires that the VHDL code be generated according to a set design translation in the following eight steps:

1. Translation preparation (according to the nine steps below).

2. Set the architecture details (according to the six steps below).

3. Translation from Simulink® model to VHDL code by reading the Simulink® model, extracting the necessary design information, and generating the VHDL code.

4. Generate VHDL test bench.

5. Simulate the VHDL code and check for correct operation to validate the operation of the generated VHDL code.

6. Synthesize the VHDL code and resimulate the design to generate a structural design based on the particular target technology.

7. Configure the CPLD and validate the operation of the design.

8. Use the filter.

The nine steps of translation preparation are:

1. Identify the parts to be translated into digital (the filter).

2. Remove any unnecessary information, leaving only the filter model.

3. Identify the digital filter interfacing.

4. Identify the clock and reset inputs, along with any other filter signals.

5. Identify any external communications required.

6. Set up the support necessary to include the translation directives (see architecture details below).

7. Identify the technology directives (any requirements for the target technology, such as CPLD) and the synthesis tool to be used.

8. Identify any designer directives.

9. Determine what test circuitry is to be inserted into the design and at what stage in the design process.

The six steps to set the architecture details are:

1. Identify the particular architecture to use.

2. Identify the internal wordlength within the digital signal processing part of the digital core.

3. Identify any specific circuits to avoid (e.g., specific VHDL code constructs).

4. Identify the control signals required by the I/O.

5. Identify the number system to use (e.g., 2s complement) in the arithmetic operations.

6. Identify any number scaling requirements to limit the required wordlength within the design.

The model translation must initially consider which architecture to use, either a processor-based architecture running a software application (standard fixed

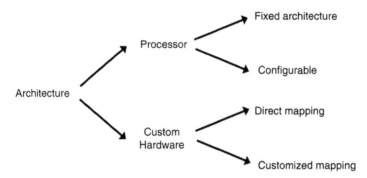

Figure 10.31: Filter architecture decisions

architecture processor or a configurable processor) or a custom hardware architecture based directly on the model. This idea is shown in Figure 10.31.

If the translation were performed manually, this could be accomplished by visual reference to the graphical representation of the model (i.e., the block diagram). If the translation were performed automatically (by a software application), it could be accomplished using the underlying text based model (i.e., with the Simulink®*.mdl* file).

A fixed architecture processor is based on an existing CISC or RISC architecture, and the translation either will generate the hardware design (in HDL) and the processor microcode together, or will use an existing processor design and only generate the processor microcode. The configurable processor is a processor design that dynamically changes specific aspects of the architecture based on the particular application.

Direct mapping starts with the model as presented and directly translates its functions to a custom hardware HDL code equivalent. Customized mapping uses custom architecture based on the model, but then determines the most appropriate way to implement its functions (e.g., by using multiple multiplication blocks or a single multiplexed multiplier block) based on the application.

No matter what particular architecture is chosen, in addition to generating the required digital signal processing algorithm hardware (as identified in the system block diagram), then there would be the need to also generate the necessary interfacing signals for external circuitry such as ADCs and DACs, and the internal timing signals for the control of the signal processing operations, along with the storage and movement of data signals within the design. These interfacing and internal timing signals would need to be created by an additional circuit creating the functions of a *control unit* particular to the design.

In this case study, direct mapping of model functions will be considered, so the filter shown in Figure 10.28 will be translated. This requires the use of the following main functional blocks:

- three integrator blocks

- three differentiator blocks

- one switch block

- two constant values

The input is a single-bit bitstream pattern, and the output is a 16-bit pattern. The Simulink® model for the overall control system must be reviewed and should contain:

- information for translation to VHDL

- information *not* for translation to VHDL

The information *not* for translation to VHDL includes information such as visual attributes and software version information, which must be stripped from the representation of the model used for translation to VHDL. The Simulink® model code for the filter only is shown in Figure 10.32. This is the text description of the model shown in Figure 10.28. It consists of the blocks used, their attributes, and the interconnect between the blocks (lines). Interpreting this model requires knowledge of its model syntax and how the values that can be modified by the user are represented in the model. The syntax is readable, and the names used can be identified by comparison with the block diagram view.

To create a digital design to implement the filter, a control unit is needed within the design to generate the necessary timing signals to control the operation of the filter parts from master clock and reset inputs. The basic structure for this is shown in Figure 10.33.

```
1     System {
2        Name                    "comb_filter_1"
3        Block {
4           BlockType                    FromWorkspace
5           Name                    "Bit_Stream"
6           Position                [25, 78, 90, 102]
7           VariableName            "Filter_In"
8           SampleTime              "0"
9        }
10       Block {
11          BlockType                    Constant
12          Name                    "Constant"
13          Position                [135, 20, 165, 50]
14       }
15       Block {
16          BlockType                    Constant
17          Name                    "Constant1"
18          Position                [135, 145, 165, 175]
19          Value                   "-1"
20       }
21       Block {
22          BlockType                    SubSystem
23          Name                    "Differentiator_1"
24          Ports                   [1, 1]
25          Position                [355, 215, 395, 275]
26          TreatAsAtomicUnit            off
27       }
28       Block {
29          BlockType                    SubSystem
30          Name                    "Differentiator_2"
31          Ports                   [1, 1]
32          Position                [455, 215, 495, 275]
33          TreatAsAtomicUnit            off
34       }
35       Block {
36          BlockType                    SubSystem
37          Name                    "Differentiator_3"
38          Ports                   [1, 1]
39          Position                [550, 215, 590, 275]
40          TreatAsAtomicUnit            off
41       }
42       Block {
43          BlockType                    ToWorkspace
44          Name                    "Filter_Out"
45          Position                [715, 230, 775, 260]
46          VariableName            "Filter_Out"
47          MaxDataPoints           "inf"
48          SampleTime              "-1"
49          SaveFormat              "Structure"
50       }
51       Block {
52          BlockType                    SubSystem
53          Name                    "Integrator_1"
54          Ports                   [1, 1]
55          Position                [330, 66, 395, 114]
56          TreatAsAtomicUnit            off
57       }
```

Figure 10.32: Simulink® model for the CIC filter

```
58      Block {
59        BlockType                 SubSystem
60        Name                      "Integrator_2"
61        Ports                     [1, 1]
62        Position                  [430, 66, 495, 114]
63        TreatAsAtomicUnit         off
64      }
65      Block {
66        BlockType                 SubSystem
67        Name                      "Integrator_3"
68        Ports                     [1, 1]
69        Position                  [530, 66, 595, 114]
70        TreatAsAtomicUnit         off
71      }
72      Block {
73        BlockType                 Switch
74        Name                      "Switch"
75        Position                  [235, 75, 265, 105]
76        InputSameDT               off
77      }
78      Line {
79        SrcBlock                  "Bit_Stream"
80        SrcPort                   1
81        DstBlock                  "Switch"
82        DstPort                   2
83      }
84      Line {
85        SrcBlock                  "Constant"
86        SrcPort                   1
87        Points                    [25, 0; 0, 45]
88        DstBlock                  "Switch"
89        DstPort                   1
90      }
91      Line {
92        SrcBlock                  "Constant1"
93        SrcPort                   1
94        Points                    [25, 0; 0, -60]
95        DstBlock                  "Switch"
96        DstPort                   3
97      }
98      Line {
99        SrcBlock                  "Switch"
100       SrcPort                   1
101       DstBlock                  "Integrator_1"
102       DstPort                   1
103     }
104     Line {
105       SrcBlock                  "Integrator_1"
106       SrcPort                   1
107       DstBlock                  "Integrator_2"
108       DstPort                   1
109     }
110     Line {
111       SrcBlock                  "Integrator_2"
112       SrcPort                   1
113       DstBlock                  "Integrator_3"
114       DstPort                   1
115     }
```

Figure 10.32: (Continued)

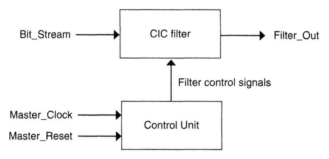

Figure 10.33: Digital filter control

The CIC filter can be remodeled in VHDL, shown in Figure 10.34 as a structural description for the filter. Detailed operation of each of the blocks is defined in separate entity-architecture pairs.

10.4.5 Concluding Remarks

In this case study, a third-order CIC digital filter was developed as a Simulink® block diagram and translated to a VHDL model for implementation within a CPLD. The structural VHDL description for the CIC filter section of a digital core was developed. The following VHDL code is also needed to configure the CPLD:

- top-level design containing the CIC filter and the control unit

- switch block details

- integrator details

- differentiator details

The block diagram was mapped directly to VHDL to implement a custom hardware design. In many cases, this would result in a large design, particularly where multiple repeated operations are needed. However, the ease and rapid development of the VHDL code by direct mapping for this small design reduced design time. This design included no multiplications, so the multiplier implementation required in other digital filter designs was not needed.

An internal wordlength of 16 bits was required for this design, which has to be accommodated in the calculations. When number overflow was possible in the

```
1     library IEEE;
2    use IEEE.STD_LOGIC_1164.ALL;
3    use IEEE.STD_LOGIC_ARITH.ALL;
4    use IEEE.STD_LOGIC_UNSIGNED.ALL;
5
6
7    ENTITY CIC_Filter IS
8        Port ( Bit_Stream    : IN    STD_LOGIC;
9               Master_Clock  : IN    STD_LOGIC;
10              Master_Reset  : IN    STD_LOGIC;
11              Filter_Out    : OUT   STD_LOGIC_VECTOR (15 downto 0));
12   END ENTITY CIC_Filter;
13
14
15
16   ARCHITECTURE Structural OF CIC_Filter IS
17
18
19   SIGNAL Internal_1              :   STD_LOGIC_VECTOR(15 downto 0);
20   SIGNAL Internal_2              :   STD_LOGIC_VECTOR(15 downto 0);
21   SIGNAL Internal_3              :   STD_LOGIC_VECTOR(15 downto 0);
22   SIGNAL Internal_4              :   STD_LOGIC_VECTOR(15 downto 0);
23   SIGNAL Internal_5              :   STD_LOGIC_VECTOR(15 downto 0);
24   SIGNAL Internal_6              :   STD_LOGIC_VECTOR(15 downto 0);
25   SIGNAL Internal_7              :   STD_LOGIC_VECTOR(15 downto 0);
26   SIGNAL Internal_8              :   STD_LOGIC_VECTOR(15 downto 0);
27   SIGNAL Integrator_1_Store      :   STD_LOGIC;
28   SIGNAL Integrator_2_Store      :   STD_LOGIC;
29   SIGNAL Integrator_3_Store      :   STD_LOGIC;
30   SIGNAL Differentiator_1_Store  :   STD_LOGIC;
31   SIGNAL Differentiator_2_Store  :   STD_LOGIC;
32   SIGNAL Differentiator_3_Store  :   STD_LOGIC;
33
34
35   COMPONENT Switch IS
36       PORT ( Bit_Stream    : IN    STD_LOGIC;
37              Data_In_1     : IN    STD_LOGIC_VECTOR (15 downto 0);
38              Data_In_2     : IN    STD_LOGIC_VECTOR (15 downto 0);
39              Data_Out      : OUT   STD_LOGIC_VECTOR (15 downto 0));
40   END COMPONENT Switch;
41
42
43   COMPONENT Plus_One IS
44       PORT ( Data_Out  : OUT   STD_LOGIC_VECTOR (15 downto 0));
45   END COMPONENT Plus_One;
46
47
```

Figure 10.34: VHDL model for the CIC filter

```
48   COMPONENT Minus_One IS
49       PORT ( Data_Out  : OUT  STD_LOGIC_VECTOR (15 downto 0));
50   END COMPONENT Minus_One;
51
52
53   COMPONENT Integrator IS
54       PORT ( Data_In   : IN   STD_LOGIC_VECTOR (15 downto 0);
55              Data_Out  : OUT  STD_LOGIC_VECTOR (15 downto 0);
56              Store     : IN   STD_LOGIC;
57              Reset     : IN   STD_LOGIC);
58   END COMPONENT Integrator;
59
60
61   COMPONENT Differentiator IS
62       PORT ( Data_In   : IN   STD_LOGIC_VECTOR (15 downto 0);
63              Data_Out  : OUT  STD_LOGIC_VECTOR (15 downto 0);
64              Store     : IN   STD_LOGIC;
65              Reset     : IN   STD_LOGIC);
66   END COMPONENT Differentiator;
67
68
69   COMPONENT Control_Unit is
70       PORT ( Master_Clock           : IN   STD_LOGIC;
71              Master_Reset           : IN   STD_LOGIC;
72              Integrator_1_Store     : OUT  STD_LOGIC;
73              Integrator_2_Store     : OUT  STD_LOGIC;
74              Integrator_3_Store     : OUT  STD_LOGIC;
75              Differentiator_1_Store : OUT  STD_LOGIC;
76              Differentiator_2_Store : OUT  STD_LOGIC;
77              Differentiator_3_Store : OUT  STD_LOGIC);
78   END COMPONENT Control_Unit;
79
80
81   BEGIN
82
83   I1: Switch
84       PORT MAP ( Bit_Stream          =>  Bit_Stream,
85                  Data_In_1           =>  Internal_1,
86                  Data_In_2           =>  Internal_2,
87                  Data_Out            =>  Internal_3);
88
89   I2 : Plus_One
90       PORT MAP ( Data_Out            =>  Internal_1);
91
92   I3 : Minus_One
93       PORT MAP ( Data_Out            =>  Internal_2);
94
```

Figure 10.34: (Continued)

```
95    I4 : Integrator
96         PORT MAP ( Data_In              =>   Internal_3,
97                    Data_Out             =>   Internal_4,
98                    Store                =>   Integrator_1_Store,
99                    Reset                =>   Master_Reset);
100
101   I5 : Integrator
102        PORT MAP ( Data_In              =>   Internal_4,
103                   Data_Out             =>   Internal_5,
104                   Store                =>   Integrator_2_Store,
105                   Reset                =>   Master_Reset);
106
107   I6 : Integrator
108        PORT MAP ( Data_In              =>   Internal_5,
109                   Data_Out             =>   Internal_6,
110                   Store                =>   Integrator_3_Store,
111                   Reset                =>   Master_Reset);
112
113   I7 : Differentiator
114        PORT MAP ( Data_In              =>   Internal_6,
115                   Data_Out             =>   Internal_7,
116                   Store                =>   Differentiator_1_Store,
117                   Reset                =>   Master_Reset);
118
119   I8 : Differentiator
120        PORT MAP ( Data_In              =>   Internal_7,
121                   Data_Out             =>   Internal_8,
122                   Store                =>   Differentiator_2_Store,
123                   Reset                =>   Master_Reset);
124
125   I9 : Differentiator
126        PORT MAP ( Data_In              =>   Internal_8,
127                   Data_Out              => Filter_Out,
128                   Store                 => Differentiator_3_Store,
129                   Reset                 => Master_Reset);
130
131   I10 : Control_Unit
132         PORT MAP ( Master_Clock             => Master_Clock,
133                    Master_Reset             => Master_Reset,
134                    Integrator_1_Store       => Integrator_1_Store,
135                    Integrator_2_Store       => Integrator_2_Store,
136                    Integrator_3_Store       => Integrator_3_Store,
137                    Differentiator_1_Store => Differentiator_1_Store,
138                    Differentiator_2_Store => Differentiator_2_Store,
139                    Differentiator_3_Store => Differentiator_3_Store);
140
141   END ARCHITECTURE Structural;
```

Figure 10.34: (Continued)

integrators, it was prevented either by ensuring that the internal values encountered are never large enough to create an overflow situation, or if an overflow situation could occur, saturating the output from a computation to the limits set by the wordlength.

10.5   Automating the Translation

The two case studies presented provide a snapshot of two possible target applications for the automatic generation of VHDL code from a system-level simulation model. A number of design implementation issues were raised and solved for these two scenarios. However, for automating the translation process into VHDL, the translation steps must be adaptable to a more generic application. Any possible approach to automating model translation, however, must:

1. be capable of being manually undertaken (i.e., by hand) if required

2. allow the designer to enter specific requirements for the particular application

3. be presented to the designer in a way that is familiar to his or her particular engineering domain and technical language

4. not intentionally restrict the designer to such an extent that the translation tool cannot be used

5. be aware that different versions of the software can vary the syntax of the underlying text file containing the model description, so a translation tool written for one version of the simulation software must be validated for a different version

6. select a software programming language appropriate to the end use of the application

7. select an architecture appropriate to the required operation and coding styles in VHDL

8. be developed in a modular manner so that the translation tool can be readily modified and enhanced

9. consider timing issues in the underlying digital logic

10. consider testability issues for the designs to be implemented

11. Effectively and efficiently deal with design hierarchy

12. Consider the circuit functions that are required to support the algorithm modelled at the high level of description, but which are not modelled at this level. For example, in digital designs there would be the need to include some form of circuit *control*. This would be required to perform synchronisation of signals around the circuit and ensure that the correct data flow is provided. In addition, specific control signals for signal sampling (e.g. through and ADC) and output updating (e.g. through a DAC) would be required

13. Include the capability for automatic documentation creation as part of the translation process. This can be in the form of document formats such as plain text, postscript, portable document format (PDF) and hypertext markup language (HTML)

10.6 Future Directions

The area of ESL design is still emerging, and various activities are undertaken in defining the direction for ESL design. However, there is a basic need to combine into a single and robust design methodology multiple design methods, EDA tools, and implementation technologies. With the area of ESL design dynamically changing, designers must be aware of the technologies, ESL design methodologies, and EDA tools that are becoming available to provide the right approach for the types of complex electronic systems being developed. This will come from the collaboration between the developers and the design community. Initially, a number of different approaches will be adopted; those showing the most promise will ultimately become industry standards, adopted and formally developed by one or more of the professional bodies.

Alongside the systems-level design methods and EDA tools being developed to solve the complex problems encountered today and expected in the future, there is still the need for electronic circuit and computer software designers who work at the most detailed level of design. Advances at this detailed level allow more complex systems to be developed of smaller size, in less time, and at lower cost. No matter how complex a system becomes, the devil will always remain in the details, so the need for effective communication and collaboration among the designers working at all levels of abstraction will always exist.

References

[1] MacMillen, D., et al., "An Industrial View of Electronic Design Automation," *IEEE Transactions on Computer Aided Design of Integrated Circuits and Systems,* Vol. 19, No. 12, December 2000, pp. 1428–1448.

[2] Bailey, B., Martin, G., and Piziali, A., *ESL Design and Verification,* Morgan Kaufmann Publishers, 2007, ISBN 0-12-373551-3.

[3] Grout, I. A., and Keane, K., "A Matlab to VHDL Conversion Toolbox for Digital Control," IFAC Symposium on Computer Aided Control System Design (CACSD2000), Salford, UK, September 2000.

[4] Grout, I. A., and O'Shea, T., "MATLAB/VHDL-AMS Modelling and Simulation Support for Microelectronic Circuit Design and Test," *Proceedings of the 10th International Mixed-Signal Testing Workshop,* 2004, pp. 178–183.

[5] Simulink® HDL Coder, The Mathworks Inc., http://www.themathworks.com

[6] Karnofsky, K., "Simulink Brings Model-Based Design to Embedded Signal Processing," *Xcell Journal,* Xilinx Inc., Winter 2004.

[7] International Technology Roadmap for Semiconductors, 2003 Edition, "Executive Summary."

[8] Edenfeld, D., et al., "2003 Technology Roadmap for Semiconductors," *Computer,* IEEE Computer Society, January 2004, pp. 47–56.

[9] MATLAB®, The Mathworks Inc., http://www.themathworks.com

[10] Simulink®, The Mathworks Inc., http://www.themathworks.com

[11] Golden, J., and Verwer, A., *Control System Design and Simulation,* McGraw-Hill, 1991, ISBN 0-07-707412-2.

[12] Astrom, K. J., and Wittenmark, B., *Computer-Controlled Systems Theory and Design,* Second Edition, Prentice Hall International, 1990, ISBN 0-13-172784-2.

[13] Analog Devices Inc., AD7575 LC^2MOS Successive Approximation ADC datasheet.

[14] Analog Devices Inc., AD7524 CMOS 8-Bit Buffered Multiplying DAC datasheet.

[15] Ifeachor, E. C., and Jervis, B. W., *Digital Signal Processing: A Practical Approach,* Pearson Education Ltd., 2002, ISBN 0-201-59619-9.

[16] Crochiere, R. E., "Interpolation and Decimation of Digital Signals—A Tutorial Review," *Proceedings of the IEEE,* Vol. 69, No. 3, March 1981, pp. 300–331.

Student Exercises

10.1   For the PI controller design case study, develop the VHDL code to implement the detailed operation of the different functional blocks that constitute the system.

10.2   For the PI controller design case study, develop a new design that uses a standard processor architecture. What differences are there in the design flow and the design effort involved?

10.3   For the CIC filter design case study, develop the VHDL code to implement the detailed operation of the different functional blocks that constitute the system.

10.4   For the CIC filter design case study, develop the control unit model and the overall digital core for the filter design.

10.5   For the CIC filter design case study, develop a new design that uses a standard processor architecture. What differences are there in the design flow and the design effort involved?

Additional References

Books

Astrom, K. J., and Wittenmark, B., *Computer-Controlled Systems Theory and Design, Second Edition*, Prentice Hall International, 1990, ISBN 0-13-172784-2.

Balarin, F., et al., *Hardware-Software Co-Design of Embedded Systems: The Polis Approach*, Kluwer Academic Publishers, 1997, ISBN 079239936.

Barron, D., *The World of Scripting Languages*, Wiley, 2000, ISBN 0-471-99886-9.

Bennett, S., Skelton, J., and Lunn, K., *UML*, McGraw-Hill, 2001, ISBN 0-07-709673-8.

Bolton, W., *Mechatronics: Electronic Control Systems in Mechanical Engineering, Second Edition*, Longman, 1999, ISBN 0582357055.

Bradley, D., *Power Electronics*, Van Nostrand Reinhold (UK), 1987, ISBN 0-442-31778-6.

Bradley, D., Seward, D., Dawson, D., and Burge, S., *Mechatronics and the Design of Intelligent Machines and Systems*, Stanley Thornes, 2000, ISBN 0-7487-5443-1.

Burns, M., and Roberts, G. W., *An Introduction to Mixed-Signal IC Test and Measurement*, Oxford University Press, 2001, ISBN 0-19-514016-8.

Bushnell, M., and Agrawal, V., *Essentials of Electronic Testing for Digital, Memory and Mixed-Signal VLSI Circuits*, Kluwer Academic Publishers, 2000, ISBN 0-7923-7991-8.

Cadenhead, R., and Lemay, L., *SAMS Teach Yourself Java^{TM} 2 in 21 days*, SAMS, 2004, ISBN 0-672-32628-0.

Deitel, H. M., and Deitel, P. J., *C, How to Program*, Fourth Edition, Prentice Hall, 2004, ISBN 0-13-122543-X.

Doane, D. A., and Franzon, P. D., *Multichip Module Technologies and Alternatives, the Basics*, Van Nostrand Reinhold, New York, 1993, ISBN 0-442-01236-5.

Floyd, T., *Electronics Fundamentals, Circuits, Devices and Applications,* Fifth Edition, 2001, Prentice Hall, ISBN 0-13-085236-8.

Geerts, Y., Steyaert, M., and Sansen, W., *Design of Multi-Bit Delta-Sigma A/D Converters,* Kluwer Academic Publishers, 2002, ISBN 1-4020-7078-0.

Golten, J., and Verwer, A., *Control System Design and Simulation,* McGraw-Hill, 1991, ISBN 0-07-707412-2.

Grant, M., Bailey, B., and Piziali, A., *ESL Design and Verification: A Prescription for Electronic System Level Methodology,* Morgan Kaufmann Publishers Inc., 2007, ISBN 0123735513.

Grotker, T. et al., *System Design with SystemC,* Kluwer Academic Publishers, 2004, ISBN 1-4020-7072-1.

Grout, I. A., *Integrated Circuit Test Engineering Modern Techniques,* Springer, 2006, ISBN 1-84628-023-0.

Hanselman, D., and Littlefield, B., *Mastering MATLAB® 6: A Comprehensive Tutorial and Reference,* Prentice Hall Inc., 2001, ISBN 0-13-019468-9.

Haskard, M. R., and May, I. C., *Analog VLSI Design nMOS and CMOS,* Prentice Hall, 1988, ISBN 0-7248-0027-1.

Horowitz, P., and Hill, W., *The Art of Electronics, Second Edition,* Cambridge University Press, 1989, ISBN 0-521-37095-7.

Hughes, E., *Electrical and Electronic Technology,* Ninth Edition, Pearson Education, 2005, ISBN 0-13-114397-2.

Ifeachor, E. C., and Jervis, B. W., *Digital Signal Processing: A Practical Approach,* Pearson Education Ltd., 2002, ISBN 0-201-59619-9.

Ince, D. C., *Software Engineering,* Van Nostrand Reinhold (International), 1989, ISBN 0-278-00079-7.

Jaegar, R. C., *Microelectronic Circuit Design,* McGraw-Hill, 1997, ISBN 0-07-114386-6.

Jespers, P., *Integrated Converters D to A and A to D Architectures, Analysis and Simulation,* Oxford University Press, 2001, ISBN 0-19-856446-5.

Kamen, E. W., and Heck, B. S., *Fundamentals of Signals and Systems Using the Web and MATLAB®,* Pearson Education Ltd., 2007, ISBN 0-13-168737-9.

Kang, S., and Leblebici, Y., *CMOS Digital Integrated Circuits Analysis and Design,* McGraw-Hill International Editions, Singapore, 1996, ISBN 0-07-114423-4.

Kropf, T., *Introduction to Formal Hardware Verification*, Springer 1999, ISBN 3-540-65445-3.

Lutz, M., *Programming Python, Second Edition*, O'Reilly, 2006, ISBN 0-596-00925-9.

Maxfield, C., *The Design Warrior's Guide to FPGAs*, Elsevier, 2004, ISBN 0-7506-7604-3.

Meade, M. L., and Dillon, C. R., *Signals and Systems Models and Behaviour*, Chapman & Hall, 1991, ISBN 0-412-40110-X.

Meloni, J. C., *SAMS Teach Yourself PHP, MySQL^{TM} and Apache in 24 Hours*, 2003, ISBN 0-672-32489-X.

Mintz, M. and Ekendahl, R., *Hardware Verification with System Verilog: An Object-Oriented Framework*, Springer-Verlag New York, May 2007, ISBN 9780387717388.

Mueller, S., *Upgrading and Repairing PCs*, Sixteenth Edition, Que Publishing, 2005, ISBN 0-7897-3210-6.

Navabi, Z., *VHDL Analysis and Modeling of Digital Systems*, McGraw-Hill International Editions, 1993, ISBN 0-07-112732-1.

O'Connor, P., *Test Engineering, A Concise Guide to Cost-Effective Design, Development and Manufacture*, John Wiley & Sons, Ltd, 2001, ISBN 0-471-49882-3.

Parhi, K., *VLSI Digital Signal Processing Systems, Design and Implementation*, John Wiley & Sons, Inc., 1999, ISBN 0-471-24186-5.

Parker, K., *The Boundary-Scan Handbook, Analog and Digital*, Second Edition, Kluwer Academic Publishers, USA, 2000, ISBN 0-7923-8277-3.

Rajsuman, R., *System-on-a-Chip Design and Test*, Artech House Publishers, USA, 2000, ISBN 1-58053-107-5.

Salcic, Z., and Smailagic, A., *Digital Systems Design and Prototyping Using Field Programmable Logic*, Kluwer Academic Publishers, 1998, ISBN 0-7923-9935-8.

Sastry, V., and Sastry, L., *SAMS Teach Yourself Tcl/Tk in 24 Hours*, SAMS, 2000, ISBN 0-672-31749-4.

Schaumann, R., and Van Valkenburg, M., *Design and Analog Filters*, Oxford University Press, 2001, ISBN 0-19-511877-4.

Sears, F., Zemansky, M., and Young, H., *University Physics*, Seventh Edition, Addison-Wesley Publishing Company, 1987, ISBN 0-201-06694-7.

Skahill, K., *VHDL for Programmable Logic*, Addison-Wesley, 1996, ISBN 0-201-89573-0.

Smith, M., *Application Specific Integrated Circuits*, Addison-Wesley, 1999, ISBN 0-201-50022-1.

Soanes, C., and Stevenson, A. (Eds.), *Oxford Dictionary of English*, Second Edition, Revised, Oxford University Press, 2005, ISBN 0-19-861057-2.

Stonham, T. J., *Digital Logic Techniques Principles and Practice*, Second Edition, Van Nostrand Reinhold (UK), 1988, ISBN 0-278-00011-8.

Storey, N., *Electronics, a Systems Approach*, Second Edition, Addison-Wesley, 1998, ISBN 0-201-17796-X.

Sutherland, S., Davidmann, S. and Flake, P., *SystemVerilog for Design: A Guide to Using SystemVerilog for Hardware Design and Modeling*, Second Edition, Springer, 2006, ISBN 0-387-3399-1.

Terrell, T. J., *Introduction to Digital Filters*, The MacMillan Press Ltd, 1980, ISBN 0-333-24671-3.

Tocci, R. J., Widmer, N. S., and Moss, G. L. K., *Digital Systems,* Ninth Edition, Pearson Education International, USA, 2004, ISBN 0-13-121931-6.

Tuinenga, P., *SPICE, A Guide to Circuit Simulation and Analysis Using Pspice*, Third Edition, Prentice Hall, 1995, ISBN 0-13-158775-7.

Wilson, C., *Intellectual Property Law*, Second Edition, Sweet & Maxwell, 2005, ISBN 0-421-89150-5.

Zwolinski, M., *Digital System Design with VHDL*, Pearson Education Limited, 2000, England, ISBN 0-201-36063-2.

Journals, Conferences, and Symposium Papers

Chapin, N., "Flowcharting With the ANSI Standard: A Tutorial," ACM Computing Surveys (CSUR), Vol. 2, Issue 2, June 1970, pp. 119–146.

Cooley, J. W., and Tukey, J. W., "An Algorithm for the Machine Computation of the Complex Fourier Series," *Mathematics of Computation*, Vol. 19, April 1965, pp. 297–301.

Densmore, D., et al., "A Platform-Based Taxonomy for ESL Design," *IEEE Design & Test of Computers*, September–October 2006, pp. 359–374.

Deubzer, O., Hamano, H., Suga, T., and Griese, H., "Lead-free soldering-toxicity, energy and resource consumption," *Proceedings of the 2001 IEEE International Symposium on Electronics and the Environment*, 7–9 May 2001, pp. 290–295.

Edwards, S. A., "The Challenges of Synthesizing Hardware from C-Like Languages," *IEEE Design & Test of Computers*, September–October 2006, pp. 375–386.

Gajski, D. D., and Ramachandran, L., "Introduction to high-level synthesis," *IEEE Design & Test of Computers*, Vol. 11, Issue 4, Winter 1994, pp. 44–54.

Gajski, D. D., and Kuhn, R. H., "New VLSI Tools," *Computer*, Vol. 16, Issue 12, Dec. 1983, pp. 11–14.

Gajski, D. D., and Vahid, F., "Specification and design of embedded hardware-software systems," *IEEE Design & Test of Computers*, Vol. 12, Issue 1, Spring 1995, pp. 53–67.

Ghose, A. K., Mandal, S. K., and Deb, G. K., "PCB Design with low EMI," *Proceedings of the International Conference on Electromagnetic Interference and Compatibility*, 6–8 December 1995, pp. 69–76.

Hemani, A., "Charting the EDA Roadmap," *IEEE Circuits and Devices Magazine*, Vol. 20, Issue 6, November–December 2004, pp. 5–10.

John, W., "Remarks to the solution of EMC-problems on printed-circuit-boards," *Proceedings of the Seventh International Conference on Electromagnetic Compatibility*, 28–31 August 1990, pp. 68–72.

MacMillen, D., et al., "An Industrial View of Electronic Design Automation," *IEEE Transactions on Computer-Aided Design of Integrated Circuits and Systems*, Vol. 19, No. 12, December 2000, pp. 1428–1448.

Mancini, R., "How to read a semiconductor datasheet," *EDN*, 14 April 2005, pp. 85–90, http://www.edn.com.

Marculescu, R., and Eles, P., "Guest Editors' Introduction: Designing Real-Time Embedded Multimedia Systems," *IEEE Design & Test of Computers*, September–October 2004, pp. 354–356.

Pecheux, F., Lallement, C., and Vachoux, A., "VHDL-AMS and Verilog-AMS as Alternative Hardware Description Languages for Efficient Modeling of Multidiscipline Systems," *IEEE Transactions on Computer-Aided Design of Integrated Circuits and Systems*, Vol. 24, No. 2, February 2005.

Ran, Y., and Marek-Sadowska, M., "Designing Via Configurable Logic Blocks for Regular Fabric," *IEEE Transactions on Very Large Scale Integration (VLSI) Systems*, Jan. 2006, pp. 1–14.

Reeser, S., "Design for in-circuit testability," *11th International IEEE/CHMT Electronics Manufacturing Technology Symposium*, 16–18 September 1991, pp. 325–328.

Rickett, P., "Cell Phone Integration: SiP, SoC and PoP," *IEEE Design & Test of Computers*, May–June 2006, pp. 188–195.

Sharawi, M. S., "Practical issues in high speed PCB design," *IEEE Potentials*, Vol. 23, Issue 2, April–May 2004, pp. 24–27.

Smithson, G., "Practical RF printed circuit board design," IEE Training Course on "How to Design RF Circuits" (Ref. No. 2000/027), IEE, 5 April 2000, pp. 11/1–11/6.

Van Treuren, B., and Miranda, J., "Embedded Boundary Scan," *IEEE Design and Test of Computers*, March–April 2003, pp. 20–25.

Verma, A., "Optimizing test strategies during PCB design for boards with limited ICT access," *27th International IEEE/SEMI Annual Electronics Manufacturing Technology Symposium* (IEMT 2002), 17–18 July 2002, pp. 364–371.

Walters, R. M., Bradley, D. A., and Dorey, A. P., "The High-Level Design of Electronic Systems for Mechatronic Applications," *IEE Colloquium on Structured Methods for Hardware Systems Design*, 1994, pp. 1/1–1/4.

Wolf, W. H., "Hardware-software co-design of embedded systems," *Proceedings of the IEEE*, Vol. 82, Issue 7, July 1994, pp. 967–989.

XiaoKun Zhu, Bo Qi, Xin Qu, JiaJi Wang, Taekoo Lee, and Hui Wang, "Mechanical test and analysis on reliability of lead-free BGA assembly," *Proceedings of the 6th International Conference on Electronic Packaging Technology*, 20 August–2 September 2005, pp. 498–502.

Zahiri, B., "Structured ASICs: Opportunities and Challenges," *Proceedings of the 21st International Conference on Computer Design*, Oct. 2003, pp. 404–409.

Internet Resources

Accelera Verilog Analog Mixed-Signal Group, http://www.verilog.org/verilog-ams/

Accellera, http://www.accellera.org

American National Standards Institute, INCITS/ISO/IEC 9899–1999 (R2005), Programming languages—C (formerly ANSI/ISO/IEC 9899–1999), http://www.ansi.org

Bell Laboratories (Bell Labs), http:www.bell-labs.com/

Cadence Design Systems Inc., USA, http::://www.cadence.com

Department for Trade and Industry (United Kingdom), http://www.dti.gov.uk/innovation/strd/cemark/page11646.html

European Commission, Guide to the Implementation of Directives Based on New Approach and Global Approach, http://ec.europa.eu/enterprise/newapproach/legislation/guide/

Federal Communications Commission (USA), http://www.fcc.gov/

The Institute of Electrical and Electronics Engineers, *IEEE Std 1076-2002, IEEE Standard Verilog Hardware Description Language*, http://www.ieee.org

The Institute of Electrical and Electronics Engineers, *IEEE Std 1364-2001, IEEE Standard VHDL Language Reference Manual*, http://www.ieee.org

The Institute of Electrical and Electronics Engineers, *IEEE Std 1076.1-1999. IEEE Standard VHDL Analog and Mixed-Signal Extensions*, http://www.ieee.org

IPC, http://www.ipc.org

Joint Electronic Device Engineering Council (JEDEC), http://www.jedec.org/

Maplesoft, http://www.maplesoft.com

The Mathworks Inc., http://www.themathworks.com

Microsoft® Corporation, Microsoft® Visual C++®, http://www.microsoft.com

Microsoft® Corporation, Microsoft® Visual Basic™, http://www.microsoft.com

Modelica Association, http://www.modelica.org/

Open Verilog International, *Verilog-A Language Reference Manual Analog Extensions to Verilog HDL*, Version 1.0, August 1996, http://www.verilog.org/

Scilab, http://www.scilab.org

Sun Microsystems, *Java Platform, Standard Edition* (J2SE) http://java.sun.com/j2se/

SystemC, http://www.systemc.org

SystemVerilog, http://www.systemverilog.org

Wolfram Research, http://www.wolfram.com/

Xilinx Inc., USA, http://www.xilinx.com

Datasheets

Analog Devices Inc., *AD7524 CMOS 8-Bit Buffered Multiplying DAC* datasheet.

Analog Devices Inc., *AD7575 LC²MOS 5 µs 8-Bit ADC with Track/Hold.*

Harris Semiconductor, "CDP6402, CDP6402C CMOS Universal Asynchronous Receiver/Transmitter (UART)," product datasheet, March 1997.

Maxim Integrated Products, "MAX232-CPE RS-232 Transceiver," product datasheet, 2000.

Standards

American National Standards Institute, INCITS/ISO/IEC 14882-2003, Programming languages – C++, http://www.ansi.org

IEEE, IEEE Std 91-1984, *Graphics Symbols for Logic Functions*, IEEE, USA.

IEEE Std 1076™-2002, *IEEE Standard VHDL Language Reference Manual*, IEEE, USA.

IEEE Std 1666™-2005, IEEE Standard SystemC® Language Reference Manual, IEEE, http://www.ieee.org

IEEE Std 1800™-2005, IEEE Standard for SystemVerilog – Unified Hardware Design, Specification, and Verification Language, IEEE, http://www.ieee.org

IEEE Std 1364™-2005, IEEE Standard for Verilog® Hardware Description Language, IEEE, http://www.ieee.org

IEEE Std 1149.1-2001, *IEEE standard test access port and boundary-scan architecture*, IEEE, USA.

Overview of IEEE Standard 91-1984, *Explanation of Logic Symbols, 1996, Texas Instruments*, USA.

Other Documents

European Union, Directive 2002/96/EC on waste electrical and electronic equipment (WEEE).

European Union, Directive 2002/95/EC on the restriction of use of certain hazardous substances.

International Technology Roadmap for Semiconductors, 2006 Edition.

International Technology Roadmap for Semiconductors (ITRS), 2003 Edition, "Assembly and Packaging."

Maxim Integrated Products, "Power-On Reset and Related Supervisory Functions," application note 3227, 11 May 2004.

SPICE: Simulation Program with Integrated Circuit Emphasis, Version 3f5, University of California, Berkeley, USA.

Index

Printed and bound by CPI Group (UK) Ltd, Croydon, CR0 4YY

03/10/2024

01040335-0005